Electrical

Level Three

Trainee Guide
2008 *NEC*® Revision

PEARSON

Prentice
Hall

Upper Saddle River, New Jersey
Columbus, Ohio

National Center for Construction Education and Research
President: Don Whyte
Director of Product Development: Daniele Stacey
Electrical Project Manager: Daniele Stacey
Production Manager: Tim Davis
Quality Assurance Coordinator: Debie Ness
Desktop Publishing Coordinator: James McKay
Editors: Rob Richardson, Matt Tischler, Brendan Coote

Writing and development services provided by Topaz Publications, Liverpool, NY
Lead Writer/Project Manager: Veronica Westfall
Desktop Publisher: Joanne Hart
Art Director: Megan Paye
Permissions Editors: Andrea LaBarge, Jackie Vidler
Writers: Tom Burke, Gerald Shannon, Nancy Brown, Charles Rogers

Pearson Education, Inc.
Product Manager: Lori Cowen
Product Development Editor: Janet Ryerson
Project Managers: Stephen C. Robb, Christina M. Taylor
AV Project Manager: Janet Portisch
Senior Operations Supervisor: Pat Tonneman
Art Director: Diane Y. Ernsberger
Text Designer: Kristina D. Holmes
Cover Designer: Kristina D. Holmes
Cover Photo: Tim Davis
Director of Marketing: David Gesell
Executive Marketing Manager: Derril Trakalo
Senior Marketing Coordinator: Alicia Dysert
Copyeditor: Sheryl Rose
Proofreader: Bret Workman

This book was set in Palatino and Helvetica by S4Carlisle Publishing Services and was printed and bound by Courier Kendallville, Inc. The cover was printed by Phoenix Color Corp.

Pearson Education Ltd., London
Pearson Education Singapore Pte. Ltd.
Pearson Education Canada, Inc.
Pearson Education—Japan

Pearson Education Australia Pty. Limited
Pearson Education North Asia Ltd., Hong Kong
Pearson Educación de Mexico, S.A. de C.V.
Pearson Education Malaysia Pte. Ltd.

10 9 8 7 6 5 4

Perfect bound   ISBN-13: 978-0-13-604471-0
                 ISBN-10:   0-13-604471-9
Loose leaf   ISBN-13: 978-0-13-604472-7
                 ISBN-10:   0-13-604472-7

Preface

Electricity powers the applications that make our daily lives more productive and efficient. Thirst for electricity has led to vast job opportunities in the electrical field. Electricians constitute one of the largest construction occupations in the United States, and they are among the highest-paid workers in the construction industry. According to the U.S. Bureau of Labor Statistics, job opportunities for electricians are expected to be very good as the demand for skilled craftspeople is projected to outpace the supply of trained electricians.

Electricians install electrical systems in structures. They install wiring and other electrical components, such as circuit breaker panels, switches, and light fixtures. Electricians follow blueprints, the *National Electrical Code®*, and state and local codes. They use specialized tools and testing equipment, such as ammeters, ohmmeters, and voltmeters. Electricians learn their trade through craft and apprenticeship programs. These programs provide classroom instruction and on-the-job training with experienced electricians.

We wish you success as you embark on your second year of training in the electrical craft and hope that you will continue your training beyond this textbook. There are more than 700,000 people employed in electrical work in the United States, and, as most of them can tell you, there are many opportunities awaiting those with the skills and desire to move forward in the construction industry.

NEW WITH *ELECTRICAL LEVEL THREE*

NCCER and Pearson Education, Inc. are pleased to present the sixth edition of *Electrical Level Three*. This edition includes an enhanced "Practical Applications of Lighting" module that incorporates induction and LED lighting. A new "Voice, Data, and Video" module covers the role of the electrician in the installation of various types of communications cable. Also, a code-driven worksheet complements the "Motor Calculations" module. This edition has been fully updated to meet the 2008 *National Electrical Code®*.

The opening page of each of the eleven modules in this textbook features photos of interesting and inspiring electrical works, ranging from green projects such as the Statue of Liberty being lighted by 100% green power to award-winning construction projects from Associated Builders and Contractors, Inc. and The Associated General Contractors of America.

We invite you to visit the NCCER website at **www.nccer.org** for the latest releases, training information, newsletter, and much more. You can also reference the Contren® product catalog online at that site. Your feedback is welcome. Email your comments to **curriculum@nccer.org** or send general comments and inquiries to **info@nccer.org**.

CONTREN® LEARNING SERIES

The National Center for Construction Education and Research (NCCER) is a not-for-profit 501(c)(3) education foundation established in 1996 by the world's largest and most progressive construction companies and national construction associations. It was founded to address the severe workforce shortage facing the industry and to develop a standardized training process and curricula. Today, NCCER is supported by hundreds of leading construction and maintenance companies, manufacturers, and national associations. The Contren® Learning Series was developed by NCCER in partnership with Pearson Education, Inc., the world's largest educational publisher.

Some features of NCCER's Contren® Learning Series are as follows:

- An industry-proven record of success
- Curricula developed by the industry for the industry
- National standardization providing portability of learned job skills and educational credits
- Compliance with the Office of Apprenticeship requirements for related classroom training (*CFR 29:29*)
- Well-illustrated, up-to-date, and practical information

NCCER also maintains a National Registry that provides transcripts, certificates, and wallet cards to individuals who have successfully completed modules of NCCER's Contren® Learning Series. *Training programs must be delivered by an NCCER Accredited Training Sponsor in order to receive these credentials.*

Special Features of This Book

In an effort to provide a comprehensive, user-friendly training resource, we have incorporated many different features for your use. Whether you are a visual or hands-on learner, this book will provide you with the proper tools to get started in the electrical industry.

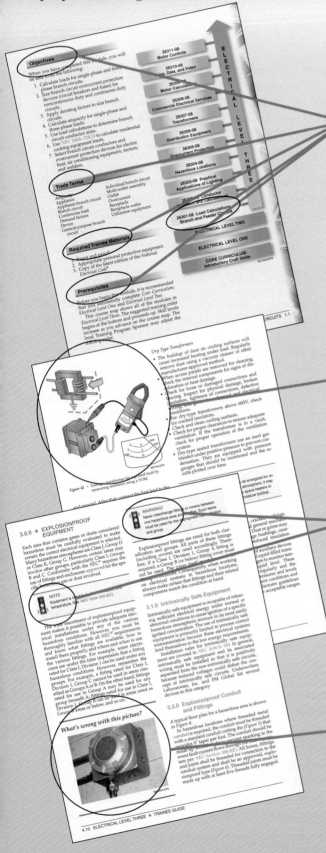

Introduction Page

This page is found at the beginning of each module and lists the Objectives, Trade Terms, Required Trainee Materials, Prerequisites, and Course Map for that module. The Objectives list the skills and knowledge you will need in order to complete the module successfully. The list of Trade Terms identifies important terms you will need to know by the end of the module. Required Trainee Materials list the materials and supplies needed for the module. The Prerequisites for the module are listed and illustrated in the Course Map. The Course Map also gives a visual overview of the entire course and a suggested learning sequence for you to follow.

Color Illustrations and Photographs

Full-color illustrations and photographs are used throughout each module to provide vivid detail. These figures highlight important concepts from the text and provide clarity for complex instructions. Each figure reference is denoted in the text in *italic type* for easy reference.

Notes, Cautions, and Warnings

Safety features are set off from the main text in highlighted boxes and are organized into three categories based on the potential danger of the issue being addressed. Notes simply provide additional information on the topic area. Cautions alert you of a danger that does not present potential injury but may cause damage to equipment. Warnings stress a potentially dangerous situation that may cause injury to you or a co-worker.

What's wrong with this picture?

What's wrong with this picture? features include photos of actual code violations for identification and encourage you to approach each installation with a critical eye.

Inside Track

Inside Track features provide a head start for those entering the electrical field by presenting technical tips and professional practices from master electricians in a variety of disciplines. Inside Tracks often include real-life scenarios similar to those you might encounter on the job site.

Think About It

Think About It features use "What if?" questions to help you apply theory to real-world experiences and put your ideas into action.

Trade Terms

Each module presents a list of Trade Terms that are discussed within the text and defined in the Glossary at the end of the module. These terms are denoted in the text with **blue bold type** upon their first occurrence. To make searches for key information easier, a comprehensive Glossary of Trade Terms from all modules is located at the back of this book.

Going Green

Going Green looks at ways to preserve the environment, save energy, and make good choices regarding the health of the planet. Through the introduction of new construction practices and products, you will see how the "greening of America" has already taken root.

Case History

Case History features emphasize the importance of safety by citing examples of the costly (and often devastating) consequences of ignoring *National Electrical Code®* or OSHA regulations.

Review Questions

Review Questions are provided to reinforce the knowledge you have gained. This makes them a useful tool for measuring what you have learned.

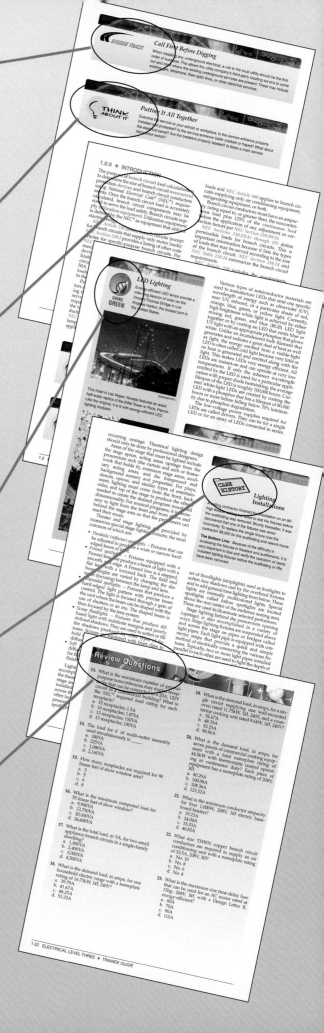

Contren® Curricula

NCCER's training programs comprise over 50 construction, maintenance, and pipeline areas and include skills assessments, safety training, and management education.

Boilermaking
Cabinetmaking
Carpentry
Concrete Finishing
Construction Craft Laborer
Construction Technology
Core Curriculum: Introductory Craft Skills
Drywall
Electrical
Electronic Systems Technician
Heating, Ventilating, and Air Conditioning
Heavy Equipment Operations
Highway/Heavy Construction
Hydroblasting
Industrial Maintenance Electrical and
    Instrumentation Technician
Industrial Maintenance Mechanic
Instrumentation
Insulating
Ironworking
Masonry
Millwright
Mobile Crane Operations
Painting
Painting, Industrial
Pipefitting
Pipelayer
Plumbing
Reinforcing Ironwork
Rigging
Scaffolding
Sheet Metal
Site Layout
Sprinkler Fitting
Welding

Pipeline

Control Center Operations, Liquid
Corrosion Control
Electrical and Instrumentation
Field Operations, Liquid
Field Operations, Gas
Maintenance
Mechanical

Safety

Field Safety
Safety Orientation
Safety Technology

Management

Introductory Skills for the Crew Leader
Project Management
Project Supervision

Spanish Translations

Andamios
Currículo Básico: Habilidades Introductorias del
    Oficio
Instalación de Rociadores Nivel Uno
Introducción a la Carpinteria
Orientación de Seguridad
Seguridad de Campo

Supplemental Titles

Applied Construction Math
Careers in Construction
Your Role in the Green Environment

Acknowledgments

This curriculum was revised as a result of the farsightedness and leadership of the following sponsors:

ABC of Iowa
ABC of New Mexico
ABC Pelican Chapter Southwest, Westlake, LA
Baker Electric
Beacon Electric Company
Cuyahoga Valley Career Center
M.C. Dean Inc.
Duck Creek Engineering
Hamilton Electric Construction Company
IMTI of New York and Connecticut

Lamphear Electric
Madison Comprehensive High School/Central Ohio ABC
Pumba Electric LLC
Putnam Career & Technical Center
Rust Constructors Inc.
TIC Industrial
Tri-City Electrical Contractors, Inc.
Trident Technical College
Vector Electric and Controls Inc.

This curriculum would not exist were it not for the dedication and unselfish energy of those volunteers who served on the Authoring Team. A sincere thanks is extended to the following:

John S. Autrey
Clarence "Ed" Cockrell
Scott Davis
Tim Dean
Gary Edgington
Tim Ely
Al Hamilton
William "Billy" Hussey
E. L. Jarrell
Dan Lamphear

Leonard R. "Skip" Layne
L. J. LeBlanc
David Lewis
Neil Matthes
Jim Mitchem
Christine Porter
Michael J. Powers
Wayne Stratton
Marcel Veronneau
Irene Ward

A final note: This book is the result of a collaborative effort involving the production, editorial, and development staff at Pearson Education, Inc., and the National Center for Construction Education and Research. Thanks to all of the dedicated people involved in the many stages of this project.

PARTNERING ASSOCIATIONS

ABC Texas Gulf Coast Chapter
American Fire Sprinkler Association
Associated Builders & Contractors, Inc.
Associated General Contractors of America
Association for Career and Technical Education
Association for Skilled & Technical Sciences
Carolinas AGC, Inc.
Carolinas Electrical Contractors Association
Center for the Improvement of Construction Management and Processes
Construction Industry Institute
Construction Users Roundtable
Design Build Institute of America
Electronic Systems Industry Consortium
Merit Contractors Association of Canada
Metal Building Manufacturers Association
NACE International

National Association of Minority Contractors
National Association of Women in Construction
National Insulation Association
National Ready Mixed Concrete Association
National Systems Contractors Association
National Technical Honor Society
National Utility Contractors Association
NAWIC Education Foundation
North American Crane Bureau
North American Technician Excellence
Painting & Decorating Contractors of America
Portland Cement Association
SkillsUSA
Steel Erectors Association of America
U.S. Army Corps of Engineers
University of Florida
Women Construction Owners & Executives, USA

Contents

Load Calculations – Branch and Feeder Circuits

Power usage world-wide is growing at a staggering rate

Worldwide power use nearly doubled from 1985 to 2005, according to the U.S. Energy Information Administration. Consumption jumped from 8.6 trillion to more than 15.7 trillion kilowatt-hours. An extreme example is power use in Hong Kong (shown here), which nearly tripled in that decade.

26301-08

26301-08
Load Calculations – Branch and Feeder Circuits

Topics to be presented in this module include:

Overview

Before electrical components can be installed, they must be selected based on voltage and current ratings. The operating voltage is determined by the commercial power available in the area of service. The branch circuit loads connected to each feeder circuit determine the feeder circuit rating, since each feeder circuit must be able to safely handle the cumulative current draw of the branch circuits connected to it. The service-entrance equipment must be rated to supply the total loads connected to all feeder circuits.

The branch and feeder loads are used to calculate the service size. Electricians who regularly perform load calculations must be familiar with the minimum load ratings regulated by the *NEC*®. Some of the *NEC*®-regulated loads include general lighting loads based on square footage, commercial receptacle and show window loads, residential small appliance and laundry circuits, and cooking appliance loads. Calculating high-demand or continuous loads such as electric heat, motors, and electric welders requires review of the code sections specifically assigned to these loads.

Objectives

When you have completed this module, you will be able to do the following:

1. Calculate loads for single-phase and three-phase branch circuits.
2. Size branch circuit overcurrent protection devices (circuit breakers and fuses) for noncontinuous duty and continuous duty circuits.
3. Apply derating factors to size branch circuits.
4. Calculate ampacity for single-phase and three-phase loads.
5. Use load calculations to determine branch circuit conductor sizes.
6. Use *NEC Table 220.55* to calculate residential cooking equipment loads.
7. Select branch circuit conductors and overcurrent protection devices for electric heat, air conditioning equipment, motors, and welders.

Trade Terms

Ampacity
Appliance
Appliance branch circuit
Branch circuit
Continuous load
Demand factors
Device
General-purpose branch circuit

Individual branch circuit
Multi-outlet assembly
Outlet
Overcurrent
Receptacle
Receptacle outlet
Utilization equipment

Required Trainee Materials

1. Paper and pencil
2. Appropriate personal protective equipment
3. Copy of the latest edition of the *National Electrical Code®*

Prerequisites

Before you begin this module, it is recommended that you successfully complete *Core Curriculum; Electrical Level One;* and *Electrical Level Two*.

This course map shows all of the modules in *Electrical Level Three*. The suggested training order begins at the bottom and proceeds up. Skill levels increase as you advance on the course map. The local Training Program Sponsor may adjust the training order.

E L E C T R I C A L   L E V E L   T H R E E

26311-08
Motor Controls

26310-08
Voice, Data, and Video

26309-08
Motor Calculations

26308-08
Commercial Electrical Services

26307-08
Transformers

26306-08
Distribution Equipment

26305-08
Overcurrent Protection

26304-08
Hazardous Locations

26303-08 Practical Applications of Lighting

26302-08 Conductor Selection and Calculations

26301-08 Load Calculations – Branch and Feeder Circuits

ELECTRICAL LEVEL TWO

ELECTRICAL LEVEL ONE

CORE CURRICULUM: Introductory Craft Skills

301CMAP.EPS

1.0.0 ◆ INTRODUCTION

The purpose of **branch circuit** load calculations is to determine the size of branch circuit **overcurrent** protection **devices** and branch circuit conductors using *National Electrical Code®* (*NEC®*) requirements. Once the branch circuit load is accurately calculated, branch circuit components may be sized to serve the load safely. Branch circuits supply **utilization equipment**. Utilization equipment is defined by the *NEC®* as equipment that utilizes electric energy.

NEC Article 210 covers branch circuits (except for branch circuits that supply only motor loads). *NEC Section 210.2* provides a listing of other code articles for specific-purpose branch circuits. Per *NEC Section 210.3,* branch circuits are rated by the maximum rating or setting of the overcurrent device. Except for circuits serving individual utilization equipment (dedicated circuits), branch circuits shall be rated 15A, 20A, 30A, 40A, and 50A. Branch circuits designed to serve individual loads can supply any size load with no restrictions to the ampere rating of the circuit.

Per *NEC Section 210.19,* branch circuit conductors are required to be sized with an **ampacity** rating that is no less than the maximum load to be served. Branch circuit overcurrent protection is required to have a rating or setting not exceeding the rating specified in *NEC Section 240.4* for conductors, *NEC Section 240.3* for equipment, and *NEC Section 210.21* for **outlet** devices including lamp holders and **receptacles**. *NEC Article 430* applies to branch circuits supplying only motor loads and *NEC Article 440* applies to branch circuits supplying only air conditioning equipment, refrigerating equipment, or both.

Branch circuit conductors must have an ampacity rating equal to, or greater than, the noncontinuous load plus 125% of the **continuous load** before the application of any adjustment or correction factors per *NEC Section 210.19(A).*

NEC Sections 210.23(A) through (D) define permissible loads for branch circuits. This is important information because it lists the types of loads that may be served according to the size of the branch circuit. *NEC Section 210.24* and *NEC Table 210.24* summarize the branch circuit requirements.

NEC Article 220 includes the requirements used to determine the number of branch circuits required and the requirements used to compute branch circuit, feeder, and service loads. *NEC Section 210.20(A)* states that the rating of a branch circuit overcurrent protection device shall not be less than the noncontinuous load plus 125% of the continuous load. *NEC Table 220.12* gives general lighting loads listed by types of occupancies. These general lighting loads are expressed as a unit load per square foot in volt-amperes (VA).

For example, the unit lighting load for a barber shop is 3VA/sq ft, a store is also 3VA/sq ft, and a storage warehouse is ¼VA/sq ft. *NEC Sections 220.14(A) through 220.14(L)* list minimum loads for outlets used in all occupancies—these outlets include general-use receptacles and outlets not used for general illumination.

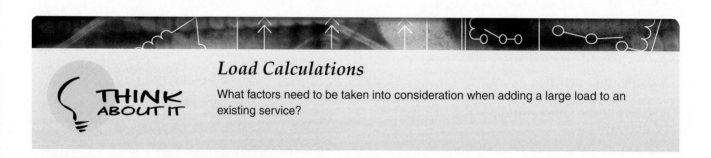

Load Calculations

What factors need to be taken into consideration when adding a large load to an existing service?

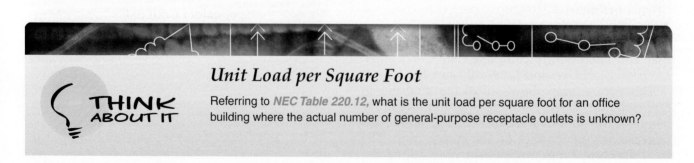

Unit Load per Square Foot

Referring to *NEC Table 220.12,* what is the unit load per square foot for an office building where the actual number of general-purpose receptacle outlets is unknown?

NOTE

Local codes may require different values than the minimum *NEC*® values. For instance, local codes may limit the number of outlets on a branch circuit to less than the calculated value, or they may require dedicated circuits other than those that are listed in the *NEC*®. Always check local codes before beginning any installation.

1.1.0 Branch Circuit Ratings

The maximum load that a single-phase branch circuit may serve is determined by multiplying the rating or setting of the overcurrent protection device (circuit breaker or fuse) by the circuit voltage. For example, the maximum load that may be supplied by a 20A, two-wire, 120V circuit is calculated by multiplying 20A by 120V.

$$20A \times 120V = 2,400VA$$

The maximum load that can be supplied by a 20A, three-wire, 120/240V circuit is:

$$20A \times 240V = 4,800VA$$

The maximum load that can be supplied by a 20A, 208V, three-phase circuit is determined by multiplying 20A by 208V by $\sqrt{3}$.

NOTE

The square root of 3 (written as $\sqrt{3}$) is approximately 1.732. You may wish to make a note of this for use in future calculations.

$$20A \times 208V \times 1.732 = 7,205.12VA$$

Branch circuits may supply noncontinuous loads, continuous loads, or a combination of the two. Per *NEC Section 210.20(A)*, the branch circuit rating shall not be less than the noncontinuous load plus 125% of the continuous load. A continuous load is defined by *NEC Article 100* as a load whose maximum current is expected to continue for three hours or more. Continuous loads are calculated at 125% of the maximum current rating of the load. Both *NEC Sections 210.19(A) Exception and 210.20(A) Exception* state that, except where the assembly is listed for operation at 100% of its rating (exception), overcurrent devices located in panelboards shall not exceed 80% of their rating where the overcurrent device supplies a continuous duty load.

For 15A and 20A branch circuits, *NEC Section 210.23(A)(2)* allows fastened-in-place utilization equipment to be connected in the same circuit with lighting units, cord- and plug-connected utilization equipment not fastened in place, or both. Under this condition, the fastened-in-place utilization equipment shall not exceed 50% of the branch circuit ampere rating.

Example 1:

A store has fluorescent lighting fixtures consisting of nine fluorescent ballasts rated at 1.5A at 120V. The fixtures will operate continuously during normal business hours from 9:00 A.M. until 9:00 P.M. daily. What is the minimum size circuit breaker required for a branch circuit to serve this load?

Determine the branch circuit load:

$$9 \times 1.5A = 13.5A$$

Determine the continuous duty load (this is a continuous duty load because the lighting fixtures will stay on for more than three hours):

$$13.5A \times 125\% = 16.88A$$

The minimum size circuit breaker required is 20A.

Example 2:

What is the maximum continuous load that may be connected to a 30A, 120V fuse?

Per *NEC Section 210.20(A)*:

$$30A \times 80\% = 24A$$

The maximum continuous load cannot exceed 24A.

Example 3:

How many **receptacle outlets** in other-than-residential occupancies can be connected to a 20A, two-wire, 120V circuit? (The receptacle outlets serve noncontinuous duty loads.)

Determine branch circuit capacity:

$$20A \times 120V = 2,400VA$$

Per *NEC Section 220.14(I)*, each outlet is assigned a load of 180VA.

$$2,400VA \div 180VA = 13.33$$

Thirteen receptacle outlets can be connected to this circuit.

Example 4:

An office manager has purchased a new state-of-the-art copy machine. The nameplate rating on the copy machine is 17A, 120V.

What is the minimum size branch circuit required to serve this equipment?

This is not a continuous load; however, per *NEC Sections 210.23(A)(1) and (B),* the rating of any one cord-connected utilization equipment shall not exceed 80% of the branch circuit ampere rating for 15A, 20A, and 30A circuits. Since the load is greater than 15A, determine the current-carrying capacity of a 20A (the smallest logical size) circuit to serve this equipment:

$$20A \times 80\% = 16A$$

A 20A circuit is not sufficient; determine the current-carrying capacity of a 30A circuit:

$$30A \times 80\% = 24A$$

The minimum size branch circuit required is 30A.

Note that this solution assumes that only multireceptacle 20A and 30A branch circuits exist in the office. An alternate solution would be to install a 20A **individual branch circuit** exclusively for the copy machine.

Example 5:

What is the maximum lighting load that may be connected to a 20A branch circuit supplying a piece of fixed equipment that has a rating of 8.5A, 120V?

The equipment rating is smaller than 50% of the rating of the 20A branch circuit. Therefore, 11.5A of noncontinuous lighting may be added.

$$20A - 8.5A = 11.5A$$

Example 6:

A restaurant dishwasher has a nameplate rating of 14.7A, 208V, 3Ø. During busy times in the restaurant, it is anticipated that the dishwasher will be turned on and operated for more than three hours at a time. What is the minimum size branch circuit required to supply this equipment?

This equipment is considered a continuous load (operated for more than three hours). Therefore, the load is to be multiplied by 125% to determine the branch circuit size.

$$14.7A \times 125\% = 18.38A$$

The minimum size branch circuit required is 20A.

1.2.0 Derating

The current-carrying capacity (ampacity) of branch circuit conductors must be derated when any of the following circumstances apply:

- There are more than three current-carrying conductors in a raceway per *NEC Section 310.15(B)(2)(a).*
- The ambient temperature that the conductors will pass through exceeds the temperature ratings for conductors listed in *NEC Table 310.16.*
- A voltage drop exists that exceeds the recommended 3% for branch circuits or 5% for the combination of the feeder and branch circuits that is caused when the distance of branch circuit conductors becomes excessive, per *NEC Section 210.19(A), FPN No. 4.*
- You bundle any cable assemblies more than 24" in length per *NEC Section 310.15(B)(2)(a).*

Conductors must be derated when installed in conduit on or above rooftops.

NEC Table 310.16 lists allowable ampacities for insulated conductors rated 0 to 2,000V. The ampacities listed apply when no more than three current-carrying conductors are installed in a raceway, are part of a cable assembly, or are for direct-buried conductors. The ampacities are based upon an ambient temperature of 30°C (86°F). If the number of current-carrying conductors in the conduit or cable exceeds three, the ampacity of the conductors must be derated using *NEC Section 310.15(B)(2)(a).* If the ambient temperature exceeds 30°C (86°F), the ampacity of the conductors must be derated using the temperature correction factors listed at the bottom of *NEC Table 310.16.* Note that the multipliers listed at the bottom of *NEC Table 310.16* are called correction factors. *NEC*® references to adjustment and correction factors apply to ambient temperatures other than 30°C and more than three current-carrying conductors in a raceway or cable. Where conduit is exposed to sunlight on or above rooftops, the conductors must be derated additionally by applying the factors in *NEC Table 310.15(B)(2)(c)* to the ambient temperature.

1.2.1 Temperature Derating

NEC Section 110.14(C) states that the lowest temperature rating of any component in a branch circuit (circuit breaker, fuse, receptacle, conductors, etc.) must be used to determine the ampacity rating of the branch circuit conductors. Conductors that have a higher temperature rating per *NEC Table 310.16* can be used for ampacity adjustment, correction, or both. For example, if the termination rating of a circuit breaker is 60°C, the ampacity rating of the branch circuit conductor rating cannot exceed the value given in the 60°C column of *NEC Table 310.16* for overcurrent protection. However, if the branch circuit conductor has a higher temperature rating (i.e., 90°C THHN), the higher ampacity rating can be used for ampacity adjustment, correction, or both.

Example 1:

It is determined that by combining branch circuits we can eliminate two runs of conduit. The branch circuits are to be pulled in a single conduit. All of the conductors in the conduit will be current-carrying conductors of the same size and insulation type.

What is the derated ampacity for each of eight No. 12 THHN copper conductors when pulled in a single conduit?

Per *NEC Table 310.16*, the ampacity of No. 12 THHN is 30A. Per *NEC Table 310.15(B)(2)(a)*, the ampacity shown in *NEC Table 310.16* must be adjusted to 70% of its value (seven to nine current-carrying conductors).

$$30A \times 70\% = 21A$$

Example 2:

What is the maximum load that may be connected on each of six No. 2 current-carrying THWN copper conductors in a single conduit?

Per *NEC Table 310.16*, the ampacity of No. 2 THWN is 115A. Per *NEC Table 310.15(B)(2)(a)*, the ampacity shown in *NEC Table 310.16* must be adjusted to 80% of its value (four to six current-carrying conductors).

$$115A \times 80\% = 92A$$

The maximum load cannot exceed 92A.

Example 3:

A branch circuit is required to supply a noncontinuous equipment load with a nameplate rating of 55A. The equipment is located in a room with plastic extrusion equipment. The ambient temperature in the room is 92°F.

What is the minimum size THHN copper conductors required to supply the equipment?

Per *NEC Table 310.16*, No. 8 THHN copper conductors have an ampacity rating of 55A. Per correction factors to *NEC Table 310.16*, the ampacity of the conductors must be multiplied by a factor of 0.96.

$$55A \times 0.96 = 52.8A$$

The ampacity of No. 8 THHN copper conductors in an ambient temperature of 92°F will not meet the requirements of *NEC Table 310.16* for a load of 55A. A No. 6 THHN copper conductor has a rating of 75A.

$$75A \times 0.96 = 72A$$

No. 6 THHN copper conductors will be required.

Example 4:

A branch circuit supplies 16A of continuous loads. The conductors to be used are Type THHN copper. The conductors will be installed in a raceway with a total of 12 current-carrying conductors and run in an ambient temperature of 105°F. What is the minimum wire size and overcurrent protective device required for this circuit?

Per *NEC Section 210.19(A)(1)* for conductors and *NEC Section 210.20(A)* for overcurrent, the minimum current rating for the conductor and the overcurrent protective device is calculated as follows:

$$16A \times 125\% = 20A \text{ (for a continuous load)}$$

Per *NEC Table 310.16* using the 60° column, a No. 12 THHN conductor is required since *NEC Section 240.4(D)(3)* prohibits a No. 14 from having an overcurrent protective device rated higher than 15A.

The correction factor for 12 current-carrying conductors is 50%. The correction factor for 105°F is 0.87. The allowable ampacity for a No. 12 THHN would be:

$$30 \times 50\% \times 0.87 = 13.05A$$

The allowable ampacity for a No. 10 THHN would be:

$$40 \times 50\% \times 0.87 = 17.4A$$

The No. 10 would be required as the minimum safe conductor and must be protected at 20A.

1.2.2 Voltage Drop Derating for Single-Phase Circuits

NEC Section 210.19(A), FPN No. 4 states that reasonable efficiency will be provided by branch circuits in which the voltage drop for branch circuit conductors does not exceed 3% to the farthest outlet and 5% for the combination of feeder and branch circuit distance to the farthest outlet. There are two formulas used to calculate voltage drop for single-phase circuits.

Formula 1:

$$VD = \frac{2 \times L \times R \times I}{1,000}$$

Where:

L = load center or total length in feet

R = conductor resistance per *NEC Chapter 9, Table 8*

I = current

The number 1,000 represents 1,000' or less of conductor.

Formula 2:

$$VD = \frac{2 \times L \times K \times I}{CM}$$

Where:

L = load center or total length in feet

K = constant (12.9 for copper conductors; 21.2 for aluminum conductors)

I = current

CM = area of the conductor in circular mils (from *NEC Chapter 9, Table 8*)

The result of these formulas is in volts. The result is divided by circuit voltage, and the second result is multiplied by 100 to determine the percent voltage drop. If the branch circuit voltage drop exceeds 3%, the conductor size should be increased to compensate. For a 120V circuit, the maximum voltage drop (in volts) is 120V × 3% = 3.6V. For a 240V single-phase circuit, the maximum voltage drop (in volts) is 240V × 3% = 7.2V.

In commercial or industrial fixed load applications, if the branch circuit load is not concentrated at the end of the branch circuit but is spread out along the circuit due to multiple outlets, the load center length of the circuit should be calculated and used in the formulas just listed. This is because the total current does not flow the complete length of the circuit. If the full length is used in computing the voltage drop, the drop determined would be greater than what would actually occur. The load center length of a circuit is that point in the circuit where, if the load were concentrated at that point, the voltage drop would be the same as the voltage drop to the farthest load in the actual circuit. To determine the load center length of a branch circuit with multiple outlets, as shown in *Figure 1*, multiply each outlet load by its actual physical routing distance from the supply end of the circuit. Add these products for all loads fed from the circuit and divide this sum by the sum of the individual loads. The resulting distance is the load center length (L) for the total load (I) of the branch circuit.

Example 1:

The length of a 120V, two-wire branch circuit is 95'. The noncontinuous load is 14.5A.

If No. 12 THHN solid copper conductors are used, will the voltage drop for this branch circuit exceed 3%?

Use the first voltage drop formula to determine voltage drop for this branch circuit. Look up the resistance for No. 12 solid copper in *NEC Chapter 9, Table 8*. It is 1.93Ω.

$$VD = \frac{2 \times L \times R \times I}{1,000}$$

$$VD = \frac{2 \times 95' \times 1.93\Omega \times 14.5A}{1,000}$$

$$VD = 5.32V$$

$$5.32V \div 120V = 0.0443 \times 100 = 4.43\%$$

Since this percentage exceeds the 3% allowable voltage drop, the answer to the question is yes,

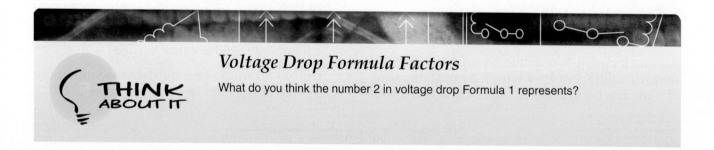

Voltage Drop Formula Factors

THINK ABOUT IT

What do you think the number 2 in voltage drop Formula 1 represents?

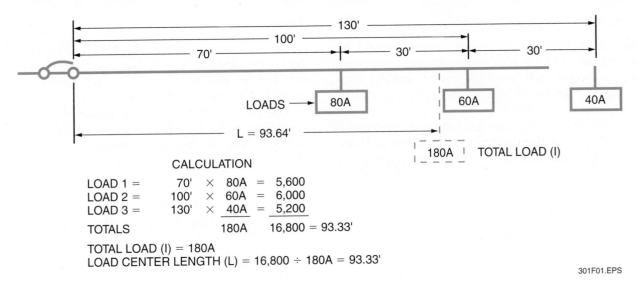

CALCULATION

LOAD 1 =	70'	× 80A =	5,600
LOAD 2 =	100'	× 60A =	6,000
LOAD 3 =	130'	× 40A =	5,200
TOTALS		180A	16,800 = 93.33'

TOTAL LOAD (I) = 180A
LOAD CENTER LENGTH (L) = 16,800 ÷ 180A = 93.33'

301F01.EPS

Figure 1 ◆ Calculating a load center length and total load for multiple fixed loads on a circuit.

and larger conductors would be required for this circuit. Note that this solution uses the first formula for single-phase (1Ø) voltage drop.

Example 2:

What size THHN solid copper branch circuit conductors would be recommended for a noncontinuous branch circuit load of 23A, 240V, 1Ø? The length of the circuit is 130'.

Use the second voltage drop formula to determine the voltage drop for this branch circuit. Since No. 10 THHN copper would be the smallest size permitted for a branch circuit load of 23A, this size will be evaluated first. Look up the area in

CM for No. 10 solid conductor in *NEC Chapter 9, Table 8.*

$$VD = \frac{2 \times L \times K \times I}{CM}$$

$$VD = \frac{2 \times 130' \times 12.9 \times 23A}{10,380}$$

$$VD = 7.43$$

$$7.43V \div 240V = 0.031 \times 100 = 3.10\%$$

This percentage exceeds 3%; therefore, No. 10 THHN copper conductors would not meet *NEC*® recommendations, and the next larger size would be required, which in this case is No. 8 THHN.

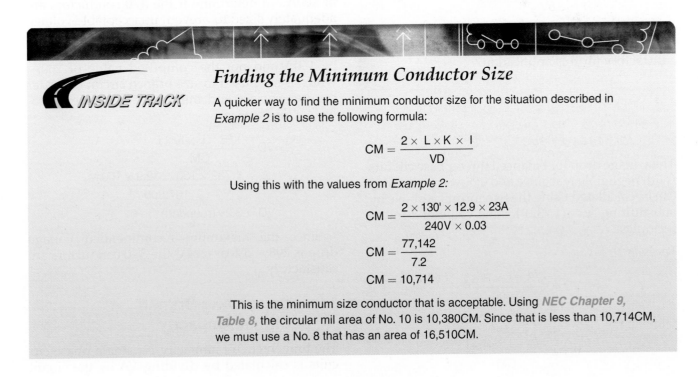

Finding the Minimum Conductor Size

INSIDE TRACK

A quicker way to find the minimum conductor size for the situation described in *Example 2* is to use the following formula:

$$CM = \frac{2 \times L \times K \times I}{VD}$$

Using this with the values from *Example 2*:

$$CM = \frac{2 \times 130' \times 12.9 \times 23A}{240V \times 0.03}$$

$$CM = \frac{77,142}{7.2}$$

$$CM = 10,714$$

This is the minimum size conductor that is acceptable. Using *NEC Chapter 9, Table 8,* the circular mil area of No. 10 is 10,380CM. Since that is less than 10,714CM, we must use a No. 8 that has an area of 16,510CM.

Example 3:

What size THHN copper conductors would be recommended for a 240V, single-phase, three-wire branch circuit with multiple power outlets consisting of three fixed 240V loads of 30A at 60', 30A at 80', and 20A at 100'?

Determine the load center length of the circuit by multiplying the outlet loads by their distance from the circuit source and then by dividing the sum of the three products by the sum of the three loads as follows:

$$Outlet\ 1 = 60' \times 30A = 1,800$$
$$Outlet\ 2 = 80' \times 30A = 2,400$$
$$Outlet\ 3 = 100' \times 20A = 2,000$$

Sum the products:

$$1,800 + 2,400 + 2,000 = 6,200$$

Divide by the sum of the loads:

$$30A + 30A + 20A = 80A$$
$$6,200 \div 80A = 77.5'\ for\ the\ load\ center$$

For a load of 80A (per *NEC Table 310.16*), No. 4 copper THWN/THHN conductors at 75°C will be selected. Use the second formula and substitute values to check voltage drop. In this example, all terminals are rated for 75°C.

$$VD = \frac{2 \times L \times K \times I}{CM}$$
$$VD = \frac{2 \times 77.5' \times 12.9 \times 80A}{41,740}$$
$$VD = 3.8V$$

Since the permissible voltage drop is 240V × 3% or 7.2V, the No. 4 THWN/THHN conductors are satisfactory for this application.

1.2.3 Voltage Drop Derating for Three-Phase Circuits

The voltage drop for balanced three-phase circuits (with negligible reactance and a power factor of 1) can be calculated using the two formulas above by substituting $\sqrt{3}$ (1.732) for the value of 2 in the formulas:

Formula 1:

$$VD = \frac{\sqrt{3} \times L \times R \times I}{1,000}$$

Formula 2:

$$VD = \frac{\sqrt{3} \times L \times K \times I}{CM}$$

Example 1:

What is the voltage drop of a 208V, 3Ø branch circuit with a load of 32A, a distance from the circuit breaker to the load of 115', and using No. 8 stranded copper conductors?

Use the first three-phase voltage drop formula to determine voltage drop for this 3Ø circuit. Look up the resistance for No. 8 stranded copper conductor in *NEC Chapter 9, Table 8.* It is 0.778Ω.

$$VD = \frac{\sqrt{3} \times L \times R \times I}{1,000}$$
$$VD = \frac{1.732 \times 115' \times 0.778\Omega \times 32A}{1,000}$$
$$VD = 4.96V$$

The maximum voltage drop permitted for a 208V circuit is:

$$208V \times 3\% = 6.24V$$

Since, in this problem, 4.96V is less than the maximum allowable voltage drop, this circuit will not have a voltage drop problem.

Example 2:

What size THHN copper conductors would be recommended for a 208V, 3Ø, four-wire feeder with a length of 150' from the source to a fixed continuous load of 160A?

Determine the minimum ampacity required by multiplying 160 by 125% to obtain 200A. Using *NEC Table 310.16,* determine that 3/0 THHN copper conductors at 75°C are satisfactory for the load of 200A. To determine if the 3/0 conductors are adequately sized to prevent unacceptable voltage drop, use the second 3Ø voltage drop formula and substitute values. Note that the current used will be 160A and not the computed value of 200A, which is only used to size branch circuit or feeder conductor sizes and minimum overcurrent protective device sizes.

$$VD = \frac{\sqrt{3} \times L \times K \times I}{CM}$$
$$VD = \frac{1.732 \times 150' \times 12.9 \times 160A}{167,800}$$
$$VD = 3.2V\ (rounded)$$

Since the maximum recommended voltage drop is 208 × 3% or 6.24V, the 3/0 conductors are satisfactory.

1.3.0 Calculating Branch Circuit Ampacity

The branch circuit ampacity for single-phase circuits is calculated by dividing VA by the circuit voltage. For example, the ampacity of a 120V, 1Ø load rated at 1,600VA is determined by dividing 1,600VA by 120V.

$$1{,}600VA \div 120V = 13.33A$$

The ampacity of a 3,450VA load at 277V, 1Ø is 12.45A.

$$3{,}450VA \div 277V = 12.45A$$

The branch circuit ampacity for three-phase circuits is calculated by dividing VA by the circuit voltage times $\sqrt{3}$ (1.732). For example, the ampacity (I) of a 208V, three-phase load rated at 5,200VA is determined as follows:

$$I = \frac{VA}{V \times \sqrt{3}}$$

$$I = \frac{5{,}200VA}{208V \times 1.732}$$

$$I = 14.43A$$

Using the same equation, the ampacity of a 14,000VA three-phase load at 480V is:

$$\frac{14{,}000VA}{480V \times 1.732} = 16.84A$$

Example 1:

What is the ampacity of a single-phase load with a nameplate rating of 5.5kW, 240V?

Multiply 5.5kW by 1,000 to determine VA (watts = volt-amperes) and then divide the result by 240V.

$$5.5kW \times 1{,}000 = 5{,}500VA \div 240V = 22.92A$$

The ampacity of this load is 22.92A.

Example 2:

What is the ampacity of a three-phase electric water heater with a nameplate rating of 20kW, 208V?

Multiply 20kW by 1,000 and divide the result by (208V × 1.732).

$$20kW \times 1{,}000 = 20{,}000W$$

$$20{,}000W \div (208V \times 1.732) = 55.52A$$

The ampacity of this load is 55.52A.

2.0.0 ◆ LIGHTING LOADS

Lighting load branch circuit calculations are based upon the type of lighting (incandescent or electric discharge), the branch circuit voltage, and whether or not the lighting is to be used for more than three hours without an off period (continuous duty).

Branch circuit loads for incandescent lighting are determined by adding the total incandescent load (for a purely resistive load such as incandescent lighting, watts = VA) and dividing the total by the circuit voltage. For example, three 500W quartz lamps connected on a 120V circuit would equal 12.5A, as shown here.

$$3 \times 500VA = 1{,}500VA \div 120V = 12.5A$$

Branch circuit loads for electric discharge lighting (lighting units having ballasts, transformers, or autotransformers) are determined by multiplying the number of fixtures times the ampacity rating of the ballast, assuming that all of the ballasts are identical. Per *NEC Section 220.18(B),* the calculated load shall be based upon the total ampere rating of the fixture (ballast or ballasts) and not the total watts of the lamps. For example, if fifteen 150W, high-pressure sodium fixtures connected on a 277V circuit each have a ballast amperage of 0.79A, the total load is 11.85A, as shown in the following equation.

$$15 \times 0.79A = 11.85A$$

Example 1:

Incandescent lighting utilizing 150W medium base lamps is required for temporary lighting on a construction site. This lighting will remain on all day and all night.

How many 150W lamps can be connected on a 20A, 120V circuit?

Since the lighting will remain on for longer than three hours, this is a continuous load. For continuous duty, multiply 150VA (watts) × 125% = 187.5VA. Determine the total capacity (in VA) for the 20A circuit:

$$20A \times 120V = 2{,}400VA$$

Divide the circuit ampacity by the lamp demand load:

$$2{,}400VA \div 187.5VA = 12.8 \text{ lamps}$$

Twelve 150W lamps may be safely connected to a 20A, 120V circuit.

Example 2:

Two 400W metal halide high bay fixtures are required to provide additional lighting for a production machine. There is an existing 20A, 277V lighting circuit available near the equipment and, after investigation, it is determined that the circuit has six of the same type of fixtures connected. The nameplate ampere rating for these fixtures is 2.0A at 277V. During normal operation, the fixtures are turned on for 12 hours every day.

Can two 400W fixtures be safely added to the existing circuit?

This is a continuous lighting load. The maximum continuous load that may be connected to the 20A circuit is 16A.

$$20A \times 80\% = 16A$$

The load for eight fixtures (six existing and two new fixtures) is 16A.

$$8 \text{ fixtures} \times 2.0A = 16A$$

Per *NEC*® requirements, the two new fixtures may safely be added to this circuit.

2.1.0 Recessed Lighting

The load for recessed lighting fixtures (excluding the residential general lighting load that is calculated at 3VA per square foot) is calculated per *NEC Section 220.12* by using the maximum VA rating of the equipment (fixture) and lamps. For example, a load of 250VA would be used to calculate the load for a recessed incandescent fixture rated at 250W. To calculate the load for a recessed fixture that uses compact fluorescent lamps, we would need to know the voltage and ampacity rating of the fluorescent ballast.

Example 1:

What is the load in amps for seven incandescent recessed cans that have a nameplate indicating a maximum lamp size of 150W, and the fixtures are to be connected to a 120V circuit?

The number of fixtures is multiplied by the VA rating of each fixture to obtain the total VA load.

$$7 \times 150VA \text{ (watts)} = 1,050VA$$

To determine ampacity, the total VA rating is divided by the circuit voltage.

$$1,050VA \div 120V = 8.75A$$

Example 2:

What is the total load in amps for seven fluorescent recessed fixtures operating continuously? Each fixture has a ballast ampacity of 0.20A at 277V, and each fixture takes two compact fluorescent lamps rated at 26 watts each.

Because these fixtures are of the electric discharge type, we need to know the ampacity of the fixture and the number of fixtures in order to calculate the total ampacity.

$$7 \times 0.20A = 1.4A \times 125\%$$

(for a continuous load) = 1.75A

2.2.0 Heavy-Duty Lamp Holder Outlets

Per *NEC Section 220.14(E),* outlets for heavy-duty lamp holders are calculated at a minimum of 600VA each. For example, a noncontinuous load consisting of four 120V outlets for heavy-duty lamp holders could be connected to a 20A branch circuit:

$$2,400VA \div 600VA = 4 \text{ outlets}$$

3.0.0 ◆ RECEPTACLE LOADS

When the exact VA rating of a load that is to be cord- and plug-connected to a receptacle outlet is not known, a VA rating of not less than 180VA per outlet is used per *NEC Section 220.14(I).* For noncontinuous receptacle loads, the total noncontinuous load is calculated at 100%. For continuous loads, the total continuous load is calculated at 125%. This *NEC*® reference does not apply to residential receptacles. Residential receptacles are included in the general illumination load (general-purpose branch circuits) or in specific residential loads such as receptacles required for small appliance and laundry loads.

For example, a 15A, 120V circuit is to be added to supply general-purpose receptacles. The receptacles are to be located above workbenches that are to be positioned against a wall. The owner states that the workers using the benches will plug small tools into the outlets, but the tools will only be occasionally used for short periods of time.

How many general-purpose duplex receptacles can be connected to a 15A, 120V circuit when the outlets are to be rated for noncontinuous duty?

Per *NEC Section 220.14(I),* each receptacle is assigned a value of 180VA. The capacity of a 15A, 120V circuit is 1,800VA.

$$15A \times 120V = 1,800VA$$

The total circuit capacity is then divided by the rating per receptacle to determine the total number that may be connected.

$$1,800VA \div 180VA = 10 \text{ receptacles}$$

4.0.0 ◆ MULTI-OUTLET ASSEMBLIES

A multi-outlet assembly is defined by *NEC Article 100* as a type of surface or flush raceway designed to hold conductors and receptacles, assembled in the field or at the factory. Multi-outlet assemblies may consist of single outlets wired to one or more circuits and typically spaced equally apart at distances of 6", 12", 18", etc. The *NEC*® rules for calculating loads for multi-outlet assemblies do not apply to dwelling units.

Per *NEC Section 220.14(H),* each 5' of multi-outlet assembly is considered as one outlet of at least 180VA capacity. In locations where many appliances are likely to be used at one time, each

1' of multi-outlet assembly is considered as one outlet of at least 180VA capacity.

Example 1:

A total of 40' of Plugmold® includes one single receptacle per foot of the assembly. The multi-outlet assembly is to be connected to a single 120V circuit.

What is the minimum size 120V circuit that would safely supply this light-duty multi-outlet assembly?

Since this is a light-duty application, the load is to be calculated at 180VA per 5' of multi-outlet assembly. Divide 40' of multi-outlet assembly by 5' and multiply the result by 180VA to obtain the total VA load. Divide the total VA load by 120V to determine the circuit ampacity and minimum circuit size.

$$40' \div 5' = 8$$
$$8 \times 180VA = 1,440VA$$
$$1,440VA \div 120V = 12A$$

The minimum 120V circuit size would be a 15A circuit.

Example 2:

The conditions listed for the previous example have changed, and it is necessary to feed this Plugmold® with three circuits. The reason for the change is that now more workers are using equipment plugged into the assembly. This change means that the assembly will now be rated to supply several loads simultaneously.

How many 20A, 120V circuits would be required to safely supply this heavy-duty multi-outlet assembly?

Since this is a heavy-duty application, the load is to be calculated at 180VA per foot of multi-outlet assembly. First, determine the VA capacity for a 20A, 120V circuit.

$$20A \times 120V = 2,400VA$$

Multiply 40' of multi-outlet assembly by 180VA to obtain the total VA load.

$$40' \times 180VA = 7,200VA \text{ (total VA)}$$

Divide the total VA by 2,400VA (20A circuit capacity) to determine the number of circuits required.

$$7,200VA \div 2,400VA = 3$$

Three 20A, 120V circuits would be required to supply this assembly.

5.0.0 ◆ SHOW WINDOW LOADS

The *NEC*® addresses three areas of requirements with regard to show window lighting: receptacle requirements, branch circuit load calculations, and feeder or service load calculations.

NEC Section 210.62 requires that at least one receptacle outlet be installed directly above a show window for each 12 linear feet of window, or major fraction thereof.

NEC Section 220.14(G) provides two options for calculating branch circuit loads for show

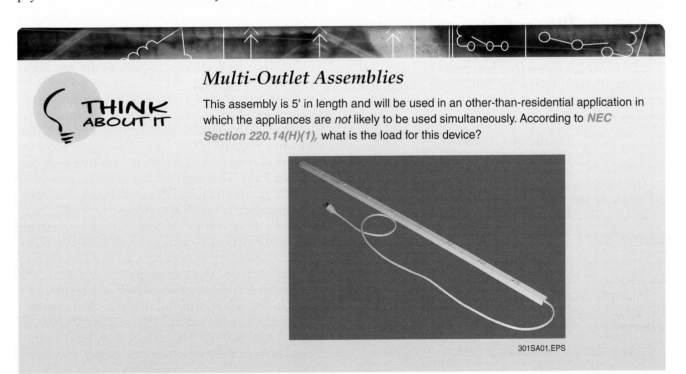

Multi-Outlet Assemblies

This assembly is 5' in length and will be used in an other-than-residential application in which the appliances are *not* likely to be used simultaneously. According to *NEC Section 220.14(H)(1)*, what is the load for this device?

301SA01.EPS

window lighting: either using the computed receptacle load of 180VA per receptacle, or computing the load at 200VA per linear foot of show window.

NOTE

Since show window lighting is likely to be on for three hours or more, it must be considered as a continuous load when computing the branch circuit load.

NEC Section 220.43 addresses the computed load of show windows when calculating feeder or service loads and requires that the total load be calculated at not less than 200VA for each linear foot of show window.

Example 1:

According to *NEC Section 210.62,* how many receptacles would be required for a show window area that measures 67' in length?

Divide the length of show window by 12 to determine the number of receptacles required.

$$67 \div 12 = 5.58$$

Since *NEC Section 210.62* requires one receptacle for each 12 linear feet or major fraction thereof, six receptacles would be required.

Example 2:

Using *NEC Section 220.14(G)(1),* what is the total load for the receptacles required in the previous example?

NEC Section 220.14(G)(1) refers to *NEC Section 220.14(I),* which states that you must multiply the number of receptacles by 180VA per receptacle and multiply the result by 125% to determine the load.

$$6 \times 180VA = 1,080VA$$

$$1,080VA \times 125\% = 1,350VA$$

The load for these show window receptacles is 1,350VA. If the voltage is 120V, the load will be 11.25A (1,350VA ÷ 120V).

Example 3:

Applying *NEC Section 220.14(G)(2),* what is the load for a 30'-long show window?

Multiply the total length of show window area by 200VA and multiply the result by 125% to determine the load.

$$30' \times 200VA = 6,000VA \times 125\% = 7,500VA$$

THINK ABOUT IT

Show Window Load Calculations

These individual show windows are each 5' in length. Assuming that we are using the requirements given in *NEC Section 220.14(G)(2),* what is the branch circuit load for each window?

301SA02.EPS

If the voltage is 120V, the load will be 62.5A (7,500VA ÷ 120V).

6.0.0 ◆ SIGN LOAD

NEC Article 600 covers requirements for signs and outline lighting. *NEC Sections 600.5(B)(1) and (2)* specify that sign circuits that supply incandescent and fluorescent lighting shall not exceed 20A, and sign circuits that supply neon tubing shall not exceed 30A. *NEC Section 600.5(A)* requires at least one 20A sign circuit that supplies no other load. This sign circuit must be provided for each commercial building and each commercial occupancy accessible to pedestrians. *NEC Section 220.14(F)* requires the load for the sign circuit to be computed at a minimum of 1,200VA. Since signs for commercial occupancies are expected to operate for more than three hours at a time, the sign circuit is typically considered as a continuous load. Under this condition, the branch circuit rating must be calculated at 125% of 1,200VA, which is equal to 1,500VA. The actual sign load cannot exceed 80% of the branch circuit rating.

For example, what is the maximum continuous load in VA that may be connected to a 20A, 120V sign circuit?

Multiply 20A by 120V to determine the total VA load for a 20A circuit.

$$20A \times 120V = 2,400VA$$

Multiply the result by 80% to determine the maximum continuous VA load.

$$2,400VA \times 80\% = 1,920VA$$

7.0.0 ◆ RESIDENTIAL BRANCH CIRCUITS

There are a number of branch circuits that are required for dwelling units. The *NEC®* lists the requirements for these circuits including the requirements for calculating branch circuit loads for specific dwelling unit (residential) loads. These loads, which are unique to dwelling units, will be covered here.

7.1.0 Small Appliance Load

NEC Sections 210.11(C)(1) and 210.52(B) require at least two 20A small **appliance branch circuits** to be installed to supply receptacles installed in the kitchen, pantry, breakfast room, and dining room. *NEC Section 220.52(A)* requires that the feeder load be computed at 1,500VA for each required small appliance branch circuit. No small appliance branch circuit may serve more than one kitchen.

NOTE

Most jurisdictions interpret this to include only the two circuits required. Others require a unit load of 1,500VA for each small appliance branch circuit required.

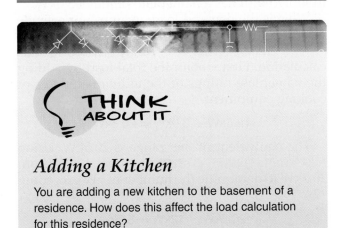

THINK ABOUT IT

Adding a Kitchen

You are adding a new kitchen to the basement of a residence. How does this affect the load calculation for this residence?

For example, what is the total feeder load for two small appliance branch circuits rated at 20A, 120V?

Multiply the number of small appliance branch circuits by 1,500VA to determine the total load.

$$2 \text{ circuits} \times 1,500VA = 3,000VA$$

7.2.0 Laundry Circuit

NEC Sections 210.11(C)(2) and 210.52(F) require the installation of at least one receptacle to supply laundry equipment. *NEC Section 220.52(B)* requires that the feeder load be computed at 1,500VA for each laundry circuit.

7.3.0 Dryers

As specified in *NEC Section 220.54,* the load for household electric dryers is calculated at 5,000VA or the nameplate rating of the dryer, whichever is larger. *NEC Table 220.54* lists **demand factors** for dryers. Note that this table applies to residential dryers used in single-family dwelling units or in multi-family dwelling units. The demand factors listed do not apply to commercial dryers used in commercial facilities.

THINK ABOUT IT

Residential Laundry Room Circuit

Does the *NEC®* specifically require a dedicated branch circuit for a washing machine?

301SA03.EPS

Example 1:

What is the demand load for one household electric dryer with a nameplate rating of 5,500 watts?

Per *NEC Table 220.54,* the demand factor for one dryer is 100%. Therefore, the load would be calculated as 5,500VA.

Example 2:

A new apartment building is planned. It consists of 10 dwelling units. Each unit has an electric dryer with a nameplate rating of 5,500 watts. There will be one utility transformer and one service entrance serving the building.

What is the demand load, in VA, for these 10 dryers?

Per *NEC Table 220.54,* the demand factor for 10 dryers is 50%. Add the total VA for the 10 dryers and multiply the result by 50%.

$$10 \text{ dryers} \times 5,500\text{W} = 55,000\text{VA}$$

$$55,000\text{VA} \times 50\% = 27,500\text{VA}$$

7.4.0 Cooking Appliances

Loads for ranges, wall-mounted ovens, counter-mounted cooking units, and other household cooking appliances are calculated using the demand factors listed in *NEC Table 220.55* and the notes to *NEC Table 220.55.* The most important single piece of information required to size the circuit for residential cooking equipment is the nameplate rating of the equipment. Although *NEC Table 220.55* lists loads in kW, *NEC Section 220.55* states that kVA shall be considered equivalent to kW for loads calculated using *NEC Table 220.55.*

Demand factors are calculated using *NEC Table 220.55* based upon the number of appliances, the maximum demand listed in Column C, and a demand factor percentage that is found using Columns A and B. Column C is used when the nameplate rating is over 8¾kW but not over 12kW. Column C is also used to calculate larger range loads per Notes 1 and 2. Column A is used when the nameplate rating is less than 3½kW. Column B is used when the nameplate rating is from 3½kW to 8¾kW. Note 3 provides a method to calculate the demand for multiple ranges, each of which has a nameplate rating of more than 1¾kW but less than 8¾kW. Note 4 provides a method to determine the size for a single branch circuit supplying one counter-mounted unit and no more than two wall-mounted ovens, provided the equipment is located in the same room.

Example 1:

What is the demand load for one range with a nameplate rating of 11.3kW?

Since the nameplate rating is greater than 8¾kW and not over 12kW, Column C is used. The demand rating for one range, not over 12kW, is listed as 8kW.

Example 2:

What is the demand load, in amps, for one range with a nameplate rating of 15.7kW, 1Ø, 240V?

Since the nameplate rating exceeds 12kW, *NEC Table 220.55, Note 1* is used. Subtract 12kW from the nameplate rating of the range to determine the number of kW exceeding 12.

$$15.7\text{kW} - 12\text{kW} = 3.7\text{kW}$$

Since 0.7 is a major fraction (≥0.5), the result is rounded off to 4kW. Multiply 5% by 4 to obtain the demand increase percentage.

$$4\text{kW} \times 5\% = 20\%$$

The maximum demand listed in Column C for one range (8kW) is then multiplied by 120% to determine the demand load for this range.

$$8\text{kW} \times 120\% = 9.6\text{kW}$$

To determine the ampacity of this load, first convert to VA (kW × 1,000), then divide VA by the circuit voltage:

$$9.6\text{kW} \times 1,000 = 9,600\text{VA}$$

$$9,600\text{VA} \div 240\text{V} = 40\text{A}$$

Example 3:

One wall-mounted oven and one counter-mounted cooking unit are to be installed in the kitchen. To reduce the cost of the electrical installation, both pieces of cooking equipment are to be connected to the same 240V, 1Ø circuit. The oven has a nameplate rating of 12.5kW and the counter-mounted unit has a nameplate rating of 8.0kW.

What is the demand load, in amps, for a single circuit to supply this cooking equipment?

NEC Table 220.55, Note 4 may be used for this calculation. First, obtain the total load by adding the nameplate ratings of the individual pieces of cooking equipment.

$$12.5\text{kW} + 8.0\text{kW} = 20.5\text{kW}$$

The equivalent of one range is 20.5kW. Using *NEC Table 220.55, Note 1,* subtract 12kW from 20.5kW to determine the number of kW exceeding 12.

$$20.5\text{kW} - 12\text{kW} = 8.5\text{kW}$$

One Cooktop and One Oven

Which note under *NEC Table 220.19* refers to this installation?

301SA04.EPS

Since 0.5 is a major fraction, the result is rounded off to 9kW. Multiply 5% by 9:

$$9 \times 5\% = 45\%$$

The maximum demand in Column C for one range (8kW) is then multiplied by 145% to determine the demand load.

$$8kW \times 145\% = 11.6kW$$

Divide total VA by circuit voltage to determine branch circuit ampacity.

$$11,600VA \div 240V = 48.33A$$

Example 4:

Using the resulting ampacity for the branch circuit in the previous example, what is the minimum size conductor that may be used for this branch circuit, and what size circuit breaker is required?

According to *NEC Section 110.14(C)(1)*, branch circuits of 100A or less must be terminated using the 60°C column of *NEC Table 310.16*. Therefore, we will select a No. 6 copper conductor rated at 55A. A 50A circuit breaker would protect this circuit.

Calculating Range Loads

NEC Table 220.55, Note 1 states that when calculating demand loads for individual ranges over 12kW but not more than 27kW, the maximum demand in Column C of the table must be increased by 5% for each additional kilowatt of rating (or major fraction thereof) over 12kW.

Demand Loads

You are adding a hot tub with a nameplate rating of 4,000VA and have three other fixed appliances with a total rating of 13,000VA. Assuming that this feeder does not also serve an electric range, clothes dryer, space heating units, or air conditioning equipment, how will this addition affect the total demand load?

8.0.0 ◆ COMMERCIAL KITCHEN EQUIPMENT

Loads for commercial kitchen equipment are calculated based upon the nameplate rating of the equipment. *NEC Table 220.56* lists demand factors for commercial kitchen equipment. *NEC Section 220.56* applies to commercial kitchen equipment, including cooking equipment, dishwasher booster heaters, water heaters, and other kitchen equipment.

Example 1:

What is the load, in amps, for one dishwasher booster heater? The booster heater has a nameplate rating of 15kW, 208V, 3Ø.

Under this condition, kW = kVA. Divide total kVA by voltage × $\sqrt{3}$ (1.732).

$$15kVA \times 1,000 = 15,000VA$$

$$\frac{15,000VA}{208V \times 1.732} = 41.64A$$

Example 2:

What is the demand load for six pieces of commercial kitchen equipment with a total VA rating of 47,000VA?

Per *NEC Table 220.56,* the demand factor for six units is 65%. Multiply the total VA rating by 65%.

$$47,000VA \times 65\% = 30,550VA$$

The demand load is 30,550VA.

9.0.0 ◆ WATER HEATERS

As specified in *NEC Section 422.13,* fixed storage water heaters having a storage capacity of 120 gallons or less shall have the branch circuit rated at no less than 125% of the nameplate rating of the water heater because it is considered a continuous load.

For example, what are the minimum size circuit breaker and branch circuit conductors using Type NM cable for a water heater with a nameplate rating of 9,000VA at 240V?

Divide total VA by voltage to determine ampacity; then multiply the result by 125%.

$$9,000VA \div 240V = 37.5A$$

$$37.5A \times 125\% = 46.88A$$

Per *NEC Table 310.16,* 60°C column, No. 6 copper conductor is rated (for NM cable) at 55A. The circuit breaker size would be 50A.

10.0.0 ◆ ELECTRIC HEATING LOADS

NEC Article 424 covers fixed electric space heating equipment, including heating cable, unit heaters, boilers, central systems, or other approved heating equipment. This article does not apply to process heating or room air conditioning. Per *NEC Section 424.3(A),* branch circuits are permitted to supply any size fixed electrical space heating equipment. If two or more outlets for heating equipment are supplied, the branch circuit size is limited to 15A, 20A, or 30A. When

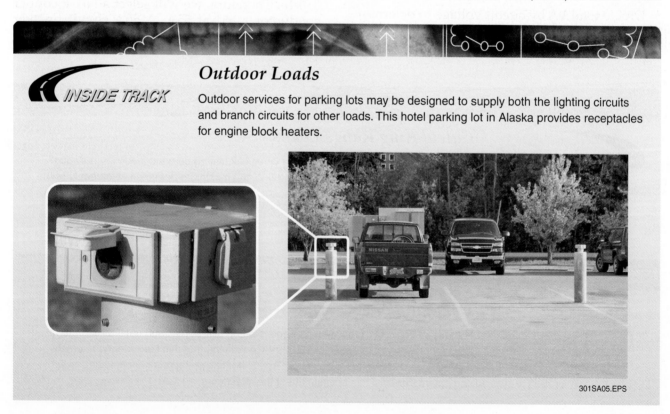

INSIDE TRACK

Outdoor Loads

Outdoor services for parking lots may be designed to supply both the lighting circuits and branch circuits for other loads. This hotel parking lot in Alaska provides receptacles for engine block heaters.

301SA05.EPS

space heating equipment utilizes electric resistance heating elements, protection for the resistance elements shall not exceed 60A (*NEC Section 424.22 [B]*). If the equipment is rated at more than 48A, the heating elements must be subdivided and each subdivided load must not exceed 48A.

For space heating equipment consisting of resistance heating elements with or without a blower motor, the ampacity of the branch circuit shall not be less than 125% of the total load per *NEC Section 424.3(B)*.

Example 1:

What is the minimum conductor ampacity for the following baseboard electric heaters rated 240V, 1Ø: one 1,500W unit, one 1,000W unit, and two 500W units?

First, determine the connected load by adding the load of each baseboard unit and then divide the total load in VA by the circuit voltage.

$$1,500W + 1,000W + 500W + 500W = 3,500W (VA)$$

$$3,500VA \div 240V = 14.58A$$

Next, multiply the load by 125% to determine the minimum conductor ampacity.

$$14.58A \times 125\% = 18.23A$$

Example 2:

What size NM conductors and circuit breaker would be required to supply the load for the previous question?

Per *NEC Table 310.16,* 60°C column, No. 12 copper conductor (for Type NM) is rated at 25A and may be connected to a 20A circuit breaker. The load, 18.23A, would require a 20A circuit breaker.

Example 3:

What is the minimum conductor ampacity for an electric forced-air furnace with heating elements totaling 25kW and a blower motor with a nameplate full-load current of 4.9A? The resistance heat and the motor are rated at 240V, 1Ø.

Divide the total of the resistance heat, in VA, by the circuit voltage to determine the ampacity for the resistance heat. Add the motor ampacity to the result and multiply the total by 125% to determine total circuit ampacity.

$$25,000VA \div 240V = 104.17A$$

$$104.17A + 4.9A = 109.07A$$

The total circuit ampacity is 109.07A.

$$109.07A \times 125\% = 136.34A$$

Note that since this equipment is rated at more than 48A, the heating elements must be subdivided, and each subdivided load cannot exceed 48A.

11.0.0 ◆ AIR CONDITIONING LOADS

Branch circuit conductors supplying a single motor compressor are rated at least 125% of the air conditioning compressor full-load current or the branch circuit selection current, whichever is greater, per *NEC Section 440.32*. Branch circuit selection current is determined by the equipment manufacturer, and this current is required to be greater than or equal to the compressor full-load current. If the branch circuit selection current is higher, its rating shall be used instead of the compressor full-load current. Per *NEC Section 440.22(A),* overcurrent protection devices shall not exceed 175% of the compressor full-load current or the branch circuit selection current. If this breaker is not sufficient to hold the starting current of the motor, the rating is permitted to be increased but cannot exceed 225% of the motor-rated load current or the branch circuit selection current, whichever is greater.

Example 1:

What size THHN copper conductors are required for a branch circuit supplying an air conditioner with a nameplate rating of 45A, 480V, 3Ø?

Multiply the nameplate ampacity by 125% and, using the result, select the proper THHN conductor size from *NEC Table 310.16*.

$$45A \times 125\% = 56.25A$$

NEC Table 310.16 gives an ampacity of 70A for No. 4 THHN copper. Use the 60°C column—see *NEC Section 110.14(C)*. Therefore, the conductor size would be No. 4 THHN copper.

Example 2:

What size Type NM cable and circuit breaker would be required to supply an air conditioner with a nameplate rating of 16.5A, 240V, 1Ø?

Multiply the nameplate ampacity by 125% and, using the result, select the proper NM conductor size from the 60°C column of *NEC Table 310.16*.

$$16.5A \times 125\% = 20.63A$$

The 60°C column gives a rating of 25A for No. 12 copper conductors (for Type NM). Multiply the nameplate ampacity by 175% to determine the circuit breaker size.

$$16.5A \times 175\% = 28.88A$$

Since *NEC Section 440.22(A)* states that the rating of the overcurrent device cannot exceed 175%, a 25A circuit breaker would be required.

Air Conditioning Nameplates

The nameplate provides information about the compressor's operating frequency, voltage, and current. An important piece of information given on the nameplate is the RLA or FLA value. RLA stands for rated-load amps and FLA stands for full-load amps. The two terms are used interchangeably. Nameplates for hermetic compressors built after 1972 are marked with the RLA, as shown here.

RUUD AIR CONDITIONER		
MODEL NO. UPMC–042JAZ	·	MFD 04/99
SERIAL NO. 6288 M1799 22780		OUTDOOR USE
VOLTS 208–230	PHASE 1	HERTZ 60
COMPRESSOR R.L.A. 18.6/18.6		L.R.A. 109
OUTDOOR FAN MOTOR F.L.A. 2.0	HP(WATTS)1/3 ( )	
MIN. SUPPLY CIRCUIT AMPACITY		26/26 AMP
MAX. FUSE OR CKT. BRK. SIZE*		40/40 AMP
MIN. FUSE OR CKT. BRK. SIZE*		30/30 AMP
DESIGN PRESSURE HIGH		300 PSIG
DESIGN PRESSURE LOW		150 PSIG
OUTDOOR UNITS FACTORY CHARGE		214 OZ. R22
TOTAL SYSTEM CHARGE		OZ. R22
SEE INSTRUCTIONS INSIDE ACCESS PANEL		
RUUD AIR CONDITIONING DIVISION FORT SMITH, ARKANSAS		MADE IN THE USA

*HACR TYPE BREAKER FOR U.S.A.

301SA06.EPS

12.0.0 ◆ MOTOR LOADS

Branch circuit conductors and overcurrent protection devices for motors are determined by using *NEC Table 430.248* for single-phase motors and *NEC Table 430.250* for three-phase motors. To use the *NEC®* motor tables, it is necessary to know the motor horsepower, voltage, phase, and design letter. Motor information is obtained from the motor nameplate. *NEC Section 430.6(A)(1)* requires that the values given in *NEC Tables 430.247 through 430.250* be used to determine the ampacity of branch circuit conductors and the ampere rating of switches, branch circuit short circuit and ground fault protection, etc. The motor nameplate current data is not permitted to be used in sizing these components unless otherwise specified.

To determine the branch circuit conductor size for motor circuits, the motor full-load current taken from the appropriate table is multiplied by 125% and the resulting ampacity is used to select branch circuit conductors. To determine motor

Buck-and-Boost Transformers

This 240V commercial kitchen mixer is connected to a standard 120V/208V service using a buck-and-boost transformer. Will this affect the branch circuit load calculation? What would happen if a buck-and-boost transformer was *not* used?

301SA07.EPS

full-load current using the tables, it is necessary to know the motor size in horsepower, motor voltage, and motor phase. If the motor nameplate includes a value for amps but not horsepower, *NEC Section 430.6(A)* states that the horsepower rating shall be assumed to be the value given in the tables. To assume a horsepower value, the horsepower value that most closely corresponds to the ampere value in the table is selected.

Example 1:

What is the motor full-load current for a 2hp, 230V, 1Ø motor?

NEC Table 430.248 is used to determine full-load current for single-phase AC motors. The left column in the table is used to find the correct motor hp and the ampacity is then determined based upon the motor voltage. A 2hp, 230V motor has a full-load current of 12A.

Example 2:

What is the motor full-load current for a 10hp, 460V, 3Ø motor?

NEC Table 430.250 is used to determine full-load current for three-phase AC motors. The left column is used to locate the motor hp and the ampacity is then obtained under the voltage column. A 10hp, 460V, 3Ø motor has a full-load current of 14A.

Example 3:

What size copper THWN branch circuit conductors would be required to supply a 25hp, 3Ø, 460V motor?

Using *NEC Table 430.250,* the full-load current for this motor is 34A. Multiply the motor full-load current by 125% to determine the minimum ampacity for branch circuit conductors.

$$34A \times 125\% = 42.5A$$

Branch circuit conductor size is then determined using *NEC Table 310.16.* No. 6 THWN copper is rated at 55A using the 60°C column. The branch circuit conductors required to supply this 25hp motor would be No. 6 THWN copper.

Motor short circuit and ground fault protection is sized using *NEC Table 430.52.* To use this table, it is helpful to know the type of motor, horsepower, phase, motor ampacity (full-load current), and the design letter, if any.

Example 4:

What size dual-element time-delay fuse would be required for a squirrel cage motor with the following rating: 50hp, 3Ø, 460V, Design Letter D?

The motor full-load current is first determined using *NEC Table 430.250.* Then, using *NEC Table 430.52,* the full-load current is multiplied by the proper percentage listed for the characteristics of the motor. From *NEC Table 430.250,* the full-load current for a 50hp, 3Ø, 460V motor is 65A. From *NEC Table 430.52,* the full-load current is multiplied by 175% to select a time-delay fuse for a polyphase motor (3Ø) with a design letter other than B, energy-efficient.

$$65A \times 175\% = 113.75A$$

Per *NEC Section 430.52(C)(1), Exception 1,* the next higher standard fuse size is 125A. A fuse with a 125A rating would be used. If the 125A fuse proves to be inadequate to carry the load, *NEC Section 430.52(C)(1), Exception 2* allows the next higher size fuse to be used. In no case, however, can the rating exceed 225% of the full-load current.

Example 5:

What size inverse-time circuit breaker would be required for a squirrel cage motor with the following rating: 10hp, 3Ø, 208V, Design Letter B, energy-efficient?

The motor full-load current is 30.8A per *NEC Table 430.250.* The full-load current is multiplied by 250% per *NEC Table 430.52* to size an inverse-time circuit breaker for a 3Ø motor with Design Letter B, energy-efficient.

$$30.8A \times 250\% = 77A$$

The next higher size circuit breaker is 80A, and *NEC Section 430.52(C)(1), Exception 2* would permit a 110A circuit breaker if the 80A breaker is inadequate for the load.

Power factor can be a significant issue with larger motor installations. In facilities with multiple large motors, the running current of a specific motor will depend on the power factor of the facility. *NEC Table 430.250* provides adjustment factors for 80% and 90% power factors. For an 80% power factor, the running current must be increased by a factor of 1.25 to achieve 100% of the full-load current. With a 90% power factor, it is 110% or 1.1.

13.0.0 ◆ WELDERS

Branch circuit conductors and overcurrent protection devices for welders are sized using the primary current and duty cycle for the welder. This information is obtained from the nameplate on the welder. Using the relevant *NEC*® section, the multiplier (demand factor) is determined by using the nameplate duty cycle of the welder. The primary current is multiplied by the *NEC*® multiplier to determine the ampacity that is then used to size branch circuit conductors and overcurrent protection. Typical welders include transformer arc welders, motor-generator welders, and resistance welders.

NEC Article 630 covers electric welders. Part B covers arc welders with or without motor-generators (*Figure 2*); Part C covers resistance welders. Each *NEC®* section lists duty cycle multipliers for individual welders based upon the type of welder. *NEC Section 630.11(A)* provides multipliers for arc welders and *NEC Section 630.31(A)(2)* provides multipliers for resistance welders.

Per *NEC Section 630.12(A)*, the overcurrent protection device for an individual welder branch circuit shall have a rating or setting not exceeding 200% of the primary current rating of the welder. This *NEC®* section applies to arc welders with or without motor-generators. For resistance welders, *NEC Section 630.32(A)* provides that the rating or setting for the overcurrent protection device shall not exceed 300% of the primary current rating of the welder.

NOTE

It is advisable to check with the local inspection authority or project specifications regarding the setting or rating for welder branch circuit overcurrent protection devices. Many authorities will not permit the overcurrent device rating to exceed the rating of the branch circuit conductors.

Example 1:

What size THWN copper branch circuit conductors would be required to supply an individual arc welder with a nonmotor-generator having a nameplate primary current of 70A and a duty cycle of 80%?

The nameplate primary current is multiplied by 0.89. This multiplier is obtained from *NEC Section 630.11(A)* for an arc welder with a nonmotor-generator and a duty cycle of 80%.

$$70A \times 0.89 = 62.3A$$

The copper THWN conductor size is selected per *NEC Table 310.16.* No. 4 THWN copper conductors, with a rating of 70A, would be required using the 60°C column.

Example 2:

What size THWN copper branch circuit conductors and what size circuit breaker would be required to supply a resistance welder with a nameplate primary current of 125A and a duty cycle of 40%?

The nameplate primary current is multiplied by 0.63. This multiplier is obtained from *NEC Section 630.31(A)(2)* for an arc welder with a duty cycle of 40%.

$$125A \times 0.63 = 78.75A$$

Per *NEC Table 310.16,* No. 3 THWN conductors (85A using the 60°C column) would be used. Per *NEC Section 630.32(A),* the primary current is multiplied by 300% to size the overcurrent protection device. The nameplate rating for this resistance welder is 125A.

$$125A \times 300\% = 375A$$

Since this value exceeds the rating of a standard 300A overcurrent protection device, *NEC Section 630.32* permits the next higher standard size listed in *NEC Section 240.6(A)* (400A) to be used.

(A) TRANSFORMER WELDING MACHINE

(B) ENGINE-DRIVEN WELDING MACHINE

301F02.EPS

Figure 2 ◆ Welding machines.

1. The general lighting load for a store is _____ VA/sq ft.
   a. 3
   b. 18
   c. 100
   d. 180

2. The general lighting load (expressed as a unit load per square foot) for a warehouse is _____ VA/sq ft.
   a. ⅛
   b. ¼
   c. 1
   d. 1½

3. A continuous load is defined by the *NEC*® as a load whose maximum current is expected to continue for at least _____.
   a. 30 minutes
   b. 1 hour
   c. 2 hours
   d. 3 hours

4. What is the capacity, in amperes, for a 20A circuit breaker supplying a continuous load?
   a. 10A
   b. 12A
   c. 16A
   d. 20A

5. Branch circuit conductors supplying continuous duty loads are calculated at _____ % of the rated load.
   a. 100
   b. 125
   c. 150
   d. 175

6. What is the current-carrying ampacity for each of 10 No. 10 THHN copper conductors installed in a single conduit?
   a. 20A
   b. 28A
   c. 32A
   d. 40A

7. The voltage drop should not exceed _____ % to the farthest outlet in a branch circuit.
   a. 2
   b. 3
   c. 5
   d. 10

8. The voltage drop for a 208V, 1Ø circuit with a load of 26.5A, using No. 8 stranded copper conductors at a circuit length of 145' is _____.
   a. 5.87V
   b. 5.98V
   c. 6.08V
   d. 6.18V

9. The voltage drop for a 480V, 3Ø circuit with a load of 26.5A, using No. 8 stranded copper conductors at a circuit length of 145' is _____.
   a. 4.87V
   b. 4.98V
   c. 5.08V
   d. 5.18V

10. Using the resistance formula, the voltage drop for an 18A, 208V, 3Ø load with a total circuit length of 105' using No. 10 solid copper conductors is _____.
    a. 3.96V
    b. 4.57V
    c. 5.06V
    d. 5.69V

11. Load calculations for circuits supplying lighting units with ballasts are based upon _____.
    a. the total wattage of all lamps in the fixtures
    b. the circuit voltage times lamp wattage
    c. the ampere ratings of the ballasts
    d. 180VA per ballast

12. What is the calculated load, in amps, for a 120V circuit supplying four 150W recessed incandescent fixtures and six recessed fluorescent fixtures operating continuously? (Each fluorescent ballast is rated at 0.65A.)
    a. 3.9A
    b. 6.075A
    c. 10.5A
    d. 11.125A

13. What is the maximum number of general-purpose noncontinuous duty duplex receptacles that can be connected to a 20A, 120V circuit in a commercial building? What is the *NEC*® required load rating for each receptacle?
    a. 10 receptacles; 1.5A
    b. 10 receptacles; 1.875A
    c. 13 receptacles; 150VA
    d. 13 receptacles; 180VA

14. The load for 6' of multi-outlet assembly used simultaneously is _____.
    a. 180VA
    b. 225VA
    c. 1,080VA
    d. 2,160VA

15. How many receptacles are required for 90 linear feet of show window area?
    a. 3
    b. 5
    c. 6
    d. 8

16. What is the maximum computed load for 55 linear feet of show window?
    a. 9,900VA
    b. 13,750VA
    c. 20,000VA
    d. 26,600VA

17. What is the total load, in VA, for two small appliance branch circuits in a single-family dwelling?
    a. 1,800VA
    b. 2,400VA
    c. 3,000VA
    d. 4,500VA

18. What is the demand load, in amps, for one household electric range with a nameplate rating of 16.75kW, 1Ø, 240V?
    a. 39.79A
    b. 41.67A
    c. 48.25A
    d. 53.33A

19. What is the demand load, in amps, for a single circuit supplying one wall-mounted oven rated 11.75kW, 1Ø, 240V, and a countertop cooking unit rated 9.6kW, 1Ø, 240V?
    a. 36.67A
    b. 48.33A
    c. 53.25A
    d. 88.96A

20. What is the demand load, in amps, for seven pieces of commercial cooking equipment with a total nameplate rating of 44.5kW with thermostatic control operating at continuous duty? Each piece of equipment has a nameplate rating of 208V, 3Ø.
    a. 80.29A
    b. 100.08A
    c. 108.36A
    d. 123.52A

21. What is the minimum conductor ampacity for four 1,000W, 208V, 1Ø electric baseboard heaters?
    a. 19.23A
    b. 24.04A
    c. 33.33A
    d. 40.83A

22. What size THWN copper branch circuit conductors are required to supply an air conditioning unit with a nameplate rating of 33.5A, 208V, 3Ø?
    a. No. 10
    b. No. 8
    c. No. 6
    d. No. 4

23. What is the maximum size time-delay fuse that can be used for an AC motor rated at 15hp, 208V, 3Ø, with a Design Letter B, energy-efficient?
    a. 60A
    b. 70A
    c. 90A
    d. 110A

24. What is the motor full-load current for a single-phase motor rated at 5hp, 230V, and what is the minimum ampacity for branch circuit conductors?

   a. 15.2A; 28A

   b. 16.7A; 16.7A

   c. 28A; 35A

   d. 30.8A ; 28A

25. An AC nonmotor-generator arc welder on an individual circuit has a nameplate primary current of 100A and a duty cycle of 70%. The branch circuit conductors must be a minimum size _____ AWG THHN copper.

   a. No. 3

   b. No. 4

   c. No. 6

   d. No. 8

Summary

In order to determine the size of branch circuit overcurrent protection devices and branch circuit conductors, it is important to accurately calculate branch circuit loads using *NEC®* requirements. When the branch circuit load is calculated, branch circuit components can be sized to safely serve the load.

NEC® articles cover branch circuits ranging from lighting loads to welders. As a trainee in the electrical field, you must understand branch circuit requirements.

Notes

Trade Terms
Introduced in This Module

Ampacity: The current in amperes that a conductor can carry continuously under the conditions of use without exceeding its temperature rating.

Appliance: Utilization equipment, generally other than industrial, normally built in standardized sizes or types, that is installed or connected as a unit to perform one or more functions such as clothes washing, air conditioning, food mixing, deep frying, etc.

Appliance branch circuit: A branch circuit supplying energy to one or more outlets to which appliances are to be connected. Such circuits are to have no permanently connected lighting fixtures that are not part of an appliance.

Branch circuit: The circuit conductors between the final overcurrent device protecting the circuit and the outlet(s).

Continuous load: A load in which the maximum current is expected to continue for three hours or more.

Demand factors: The ratio of the maximum demands of a system, or part of a system, to the total connected load of a system or the part of the system under consideration.

Device: A unit of an electrical system that is intended to carry but not utilize electric energy.

General-purpose branch circuit: A branch circuit that supplies a number of outlets for lighting and appliances.

Individual branch circuit: A branch circuit that supplies only one piece of utilization equipment.

Multi-outlet assembly: A type of surface or flush raceway designed to hold conductors and receptacles, assembled in the field or at the factory.

Outlet: A point on the wiring system at which current is taken to supply utilization equipment.

Overcurrent: Any current in excess of the rated current of equipment or the ampacity of a conductor. It may result from overload, short circuit, or ground fault.

Receptacle: A contact device installed at an outlet for connection as a single contact device. A single receptacle is a single contact device with no other contact device on the same yoke. A multiple receptacle is a single device containing two (duplex) or more receptacles.

Receptacle outlet: An outlet where one or more receptacles are installed.

Utilization equipment: Equipment that utilizes electric energy for electronic, chemical, heating, lighting, electromechanical, or similar purposes.

This module is intended to present thorough resources for task training. The following reference work is suggested for further study. This is optional material for continued education rather than for task training.

National Electrical Code® Handbook, Latest Edition. Quincy, MA: National Fire Protection Association.

NCCER makes every effort to keep these textbooks up-to-date and free of technical errors. We appreciate your help in this process. If you have an idea for improving this textbook, or if you find an error, a typographical mistake, or an inaccuracy in NCCER's Contren® textbooks, please write us, using this form or a photocopy. Be sure to include the exact module number, page number, a detailed description, and the correction, if applicable. Your input will be brought to the attention of the Technical Review Committee. Thank you for your assistance.

Instructors – If you found that additional materials were necessary in order to teach this module effectively, please let us know so that we may include them in the Equipment/Materials list in the Annotated Instructor's Guide.

Write: Product Development and Revision
            National Center for Construction Education and Research
            3600 NW 43rd St., Bldg. G, Gainesville, FL 32606

Fax: 352-334-0932

E-mail: curriculum@nccer.org

Craft _____ Module Name _____

Copyright Date _____ Module Number _____ Page Number(s) _____

Description _____

_____

_____

_____

(Optional) Correction _____

_____

_____

(Optional) Your Name and Address _____

_____

_____

Conductor Selection
and Calculations

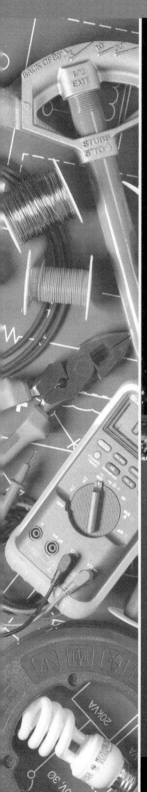

Seattle – A city focused on saving power

Seattle has incorporated LEED Certification into its City Plan. All city-funded projects and renovations with over 5,000 square feet of occupied space must achieve LEED Silver rating compliance. Builders can obtain city staff guidance in creating and building green projects.

26302-08

26302-08
Conductor Selection and Calculations

Overview

Conductors cannot be selected based on size alone. Many other factors must be considered when determining the right conductor for the job. For example, since conductors must be installed in all types of environments, there must be a variety of insulations available that are able to withstand these environments while still protecting the conductor inside.

Loads determine conductor size and conductor size determines overcurrent protection size. Overcurrent protective devices are distributed throughout electrical systems. The *NEC®* regulates the location of overcurrent protective devices.

Voltage drop in conductors is caused by excessive lengths of current-carrying conductors. The smaller the diameter of the conductor, the greater its resistance to current flow. The *NEC®* does not regulate voltage drop, but does recommend that the total voltage drop in feeder and branch circuits combined should not exceed 5% of the supply voltage. In order to calculate the potential voltage drop in a planned circuit, you must apply voltage drop formulas that are based on conductor length. Conductor resistance values and other conductor properties can be found in *NEC Chapter 9*.

Objectives

When you have completed this module, you will be able to do the following:

1. Select electrical conductors for specific applications.
2. Calculate voltage drop in both single-phase and three-phase applications.
3. Apply *National Electrical Code*® (*NEC*®) regulations governing conductors to a specific application.
4. Calculate and apply *NEC*® tap rules to a specific application.
5. Size conductors for the load.
6. Derate conductors for fill, temperature, and voltage drop.
7. Select conductors for various temperature ranges and atmospheres.

Trade Terms

American Wire Gauge (AWG)
Cable

Required Trainee Materials

1. Pencil and paper
2. Appropriate personal protective equipment
3. Copy of the latest edition of the *National Electrical Code*®

Prerequisites

Before you begin this module, it is recommended that you successfully complete *Core Curriculum; Electrical Level One; Electrical Level Two; Electrical Level Three*, Module 26301-08.

This course map shows all of the modules in *Electrical Level Three*. The suggested training order begins at the bottom and proceeds up. Skill levels increase as you advance on the course map. The local Training Program Sponsor may adjust the training order.

26311-08
Motor Controls

26310-08
Voice, Data, and Video

26309-08
Motor Calculations

26308-08
Commercial Electrical Services

26307-08
Transformers

26306-08
Distribution Equipment

26305-08
Overcurrent Protection

26304-08
Hazardous Locations

26303-08 Practical Applications of Lighting

26302-08 Conductor Selection and Calculations

26301-08 Load Calculations – Branch and Feeder Circuits

ELECTRICAL LEVEL TWO

ELECTRICAL LEVEL ONE

CORE CURRICULUM:
Introductory Craft Skills

ELECTRICAL LEVEL THREE

302CMAP.EPS

1.0.0 ◆ INTRODUCTION

A variety of materials may be used to transmit electrical energy, but copper, due to its excellent cost-to-conductivity ratio, still remains the most ideal conductor. Electrolytic copper, the type used in most electrical conductors, has the following general characteristics:

- *Method of stranding* – Stranding refers to the relative flexibility of the conductor and may consist of only one strand or many strands, depending on the rigidity or flexibility required for a specific need. *NEC Section 310.3* requires all conductors size No. 8 and larger to be stranded (more than one strand) when installed in a raceway. For example, a small-gauge wire that is to be used in a fixed installation is normally solid (one strand), whereas a wire that will be constantly flexed requires a high degree of flexibility and would contain many strands.
  - Solid wire is the least flexible form of a conductor and is merely one strand of copper.
  - Stranded refers to more than one strand in a given conductor and may vary widely, depending on size. See *Figure 1* and *NEC Chapter 9, Table 8*.
- *Degree of hardness (temper)* – Temper refers to the relative hardness of the conductor and is noted as soft drawn (SD), medium hard drawn (MHD), and hard drawn (HD). Again, the specific need of

an installation will determine the required temper. Where greater tensile strength is indicated, MHD would be used over SD, and so on.

- *Bare or coated* – Bare copper is plain copper that is available in either solid or stranded types and in the various tempers just described. In this form, it is often referred to as red copper. Bare copper is also available with a coating of tin, silver, or nickel to facilitate soldering, impede corrosion, and prevent adhesion of the copper conductor to rubber or other types of conductor insulation. The various coatings will also affect the electrical characteristics of copper.

The **American Wire Gauge (AWG)** is used in the United States to identify the sizes of wire and **cable** up to and including No. 4/0 (0000), which is commonly pronounced in the electrical trade as four-aught or four-naught. These numbers run in reverse order as to size; that is, No. 14 AWG is smaller than No. 12 AWG and so on up to size No. 1 AWG. Up to this size (No. 1 AWG), the larger the gauge number, the smaller the size of the conductor. However, the next larger size after No. 1 AWG is No. 1/0 AWG, then 2/0 AWG, 3/0 AWG, and 4/0 AWG. At this point, the AWG designations end and the larger sizes of conductors are identified by circular mils (CM or cmil). From this point, the larger the size of wire, the larger the number of circular mils. For example, 300,000 cmil is larger than 250,000 cmil. In writing these sizes in circular mils, the thousand decimal is replaced by the letter *k*, and instead of writing, say, 500,000 cmil, it is usually written 500 kcmil—pronounced five-hundred kay-cee-mil. See *Figure 2* for a comparison of the different wire sizes.

1.1.0 Compact Conductors

Compact conductors are those conductors that have been compressed so as to reduce the air space between the strands. *Figure 3* shows a cross section of a 37-strand compact aluminum conductor.

INSIDE TRACK

Solid Conductors

According to *NEC Chapter 9, Table 8,* a solid conductor is considered a one-stranded conductor.

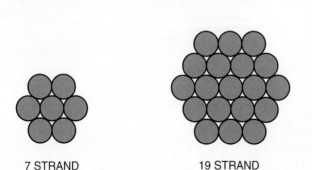

7 STRAND   19 STRAND

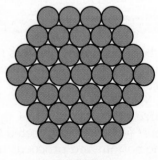

37 STRAND

302F01.EPS

Figure 1 ◆ Common strand configurations.

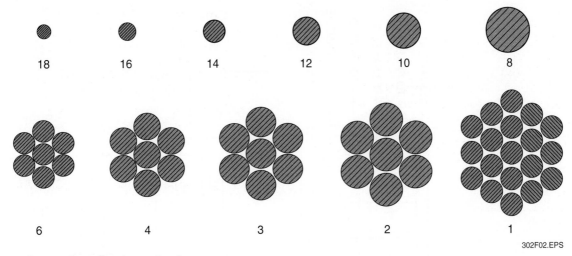

| 18 | 16 | 14 | 12 | 10 | 8 |

| 6 | 4 | 3 | 2 | 1 |

302F02.EPS

Figure 2 ◆ Comparison of various wire sizes.

The purpose of compact conductors is to reduce the overall diameter of the cable so that it may be installed in a conduit that is smaller than that required for standard stranded conductors of the same wire size. Compact conductors are especially useful when increasing the ampacity of an existing service or feeder circuits.

For example, an existing service is rated at 250A and is fed with four 350 kcmil THW conductors in 3″ conduit. Should it become necessary to increase the ampacity of the service to 300A, 500 kcmil THW compact conductors may replace the 350 kcmil conductors without increasing the size of the conduit. Compact conductors are listed in *NEC Chapter 9, Table 5A.*

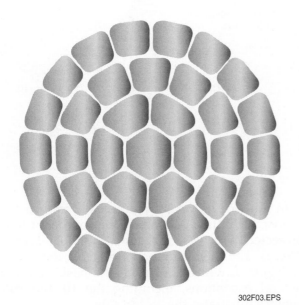

302F03.EPS

Figure 3 ◆ Cross section of a 37-strand compact conductor.

Both standard and compressed conductors are covered in this module, including various types of conductor insulation and the practical applications of each type.

2.0.0 ◆ CONDUCTOR APPLICATIONS

NEC Article 100 defines feeders as the circuit conductors between the service equipment (or other power supply source) and the final branch circuit overcurrent device. A branch circuit is defined as the circuit conductors between the final overcurrent device protecting the circuit and the outlet(s). The power riser diagram in *Figure 4* shows examples of both feeders and branch circuits.

When current-carrying conductors are used in an electrical system, the *NEC*® requires that each ungrounded conductor be protected from damage by an overcurrent protective device such as a fuse or circuit breaker. The conductors must also be identified so that the ungrounded conductors may be distinguished from the grounded and grounding conductors. Minimum sizes or required ampacity for any of these conductors used in a circuit are selected based on *NEC*® rules and tables. The *NEC*® also provides tables that list the physical and electrical properties of conductors to allow selection of the proper conductor for any application.

NEC Section 240.4 covers the requirements for protecting conductors from excess current caused by overloads, short circuits, or ground faults. The setting or sizes of the protective device are based on the ampacity of the conductors as listed in *NEC Tables 310.16 through 310.19.* Under certain conditions, the overcurrent device setting may be larger than the ampacity rating of the conductors as listed in the exceptions to the basic rules. For

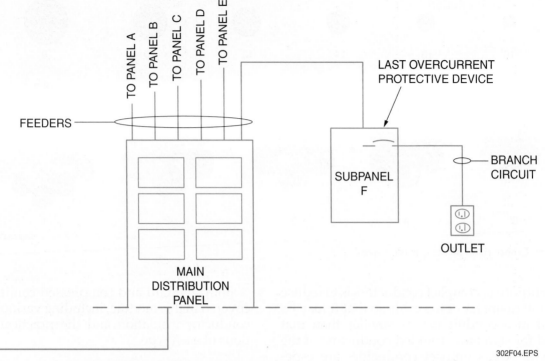

Figure 4 ◆ Power riser diagram showing feeders and branch circuits.

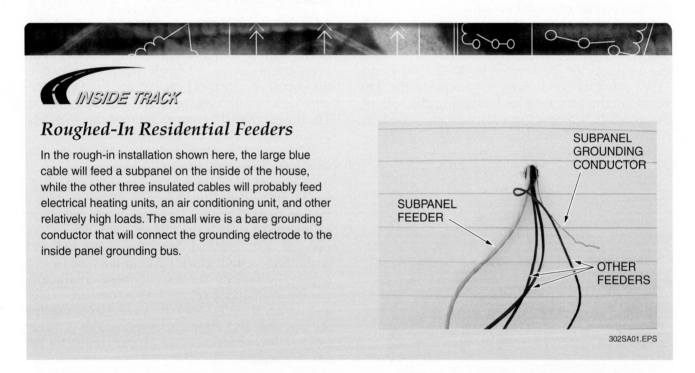

Roughed-In Residential Feeders

In the rough-in installation shown here, the large blue cable will feed a subpanel on the inside of the house, while the other three insulated cables will probably feed electrical heating units, an air conditioning unit, and other relatively high loads. The small wire is a bare grounding conductor that will connect the grounding electrode to the inside panel grounding bus.

302SA01.EPS

convenience, a standard rating of a fuse or circuit breaker may be used even if this rating exceeds the ampacity of the conductor as long as the next higher standard rating does not exceed 800A. For example, a branch circuit with a load of 56A may be protected with a 60A overcurrent device, because 56A is not a standard overcurrent device size.

In most cases, an overcurrent device must be connected at the point where the conductor to be protected receives its supply *(NEC Section 240.21)*. The most common situations are shown in *Figure 5*, which illustrates the basic rule and several exceptions, including the 10' tap rule. *Figure 6* illustrates the 25' tap rule.

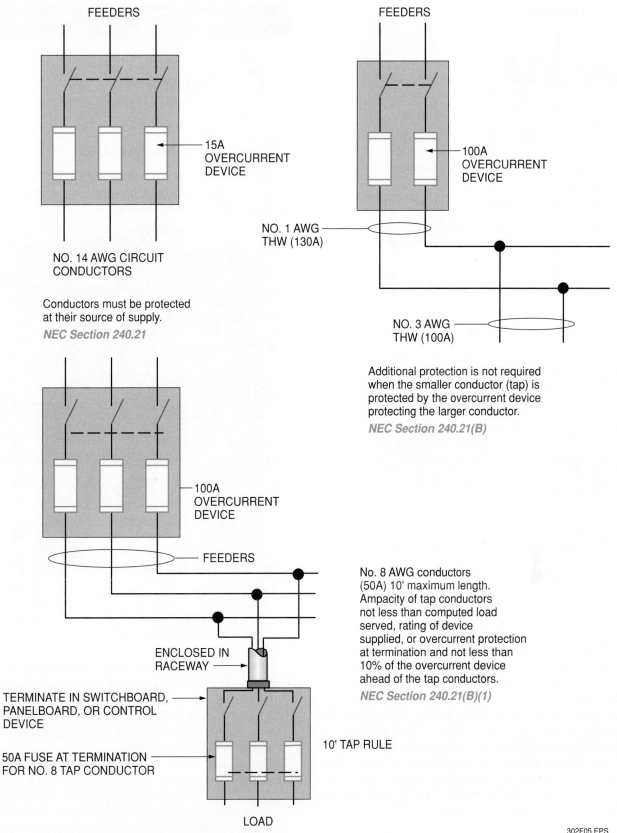

FEEDERS

15A OVERCURRENT DEVICE

NO. 14 AWG CIRCUIT CONDUCTORS

Conductors must be protected at their source of supply.
NEC Section 240.21

FEEDERS

100A OVERCURRENT DEVICE

NO. 1 AWG THW (130A)

NO. 3 AWG THW (100A)

Additional protection is not required when the smaller conductor (tap) is protected by the overcurrent device protecting the larger conductor.
NEC Section 240.21(B)

100A OVERCURRENT DEVICE

FEEDERS

ENCLOSED IN RACEWAY

TERMINATE IN SWITCHBOARD, PANELBOARD, OR CONTROL DEVICE

50A FUSE AT TERMINATION FOR NO. 8 TAP CONDUCTOR

No. 8 AWG conductors (50A) 10' maximum length. Ampacity of tap conductors not less than computed load served, rating of device supplied, or overcurrent protection at termination and not less than 10% of the overcurrent device ahead of the tap conductors.
NEC Section 240.21(B)(1)

10' TAP RULE

LOAD

302F05.EPS

Figure 5 ◆ Location of overcurrent protection in circuits.

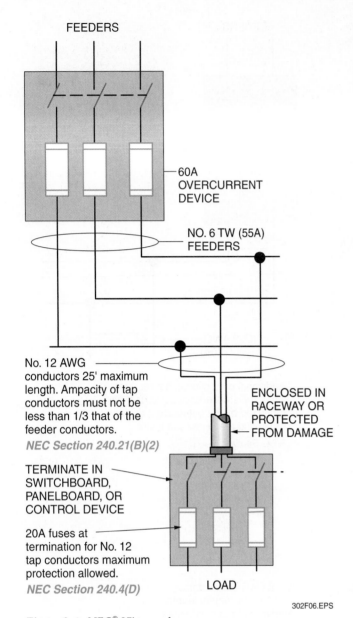

FEEDERS

60A OVERCURRENT DEVICE

NO. 6 TW (55A) FEEDERS

No. 12 AWG conductors 25' maximum length. Ampacity of tap conductors must not be less than 1/3 that of the feeder conductors.
NEC Section 240.21(B)(2)

ENCLOSED IN RACEWAY OR PROTECTED FROM DAMAGE

TERMINATE IN SWITCHBOARD, PANELBOARD, OR CONTROL DEVICE

20A fuses at termination for No. 12 tap conductors maximum protection allowed.
NEC Section 240.4(D)

LOAD

302F06.EPS

Figure 6 ◆ *NEC®* 25' tap rule.

Conductor Identification

Can you identify the ungrounded feeder conductors, the grounded feeder conductor, and the grounding feeder conductor in this picture?

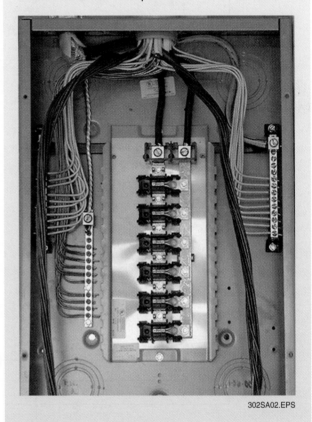

302SA02.EPS

Wiring systems require a grounded conductor in most installations. A grounded conductor, such as a neutral, or a grounding conductor must be identified either by the color of its insulation, by markings at the terminals, or by other suitable means *(NEC Section 200.6)*. In general, a grounded conductor must have a white or gray finish. When this is not practical for conductors larger than No. 6 AWG, marking the terminations white is an acceptable method of identifying the conductor. Tagging is also acceptable.

2.1.0 Branch Circuits

Now that an overview of *NEC®* overcurrent protection has been presented, we will examine the *NEC®* installation requirements for conductors.

In general, the ampacity (current-carrying capacity) of a conductor must not be less than the maximum load served. However, there are exceptions to this rule, namely, a branch circuit supplying a motor. Motors and motor circuits are covered in *NEC Article 430*.

The rating of the branch circuit overcurrent device determines the rating of the branch circuit *(NEC Section 210.3)*. For example, if a No. 10 AWG, 30A conductor is protected by a 20A circuit breaker, then the circuit is considered a 20A branch circuit.

Furthermore, the current-carrying capacity of branch circuit conductors must not be less than the maximum load to be served. Where the branch circuit supplies receptacle outlets for use with cord- and plug-connected appliances and other utilization equipment, the conductor's ampacity must not be less than the rating of the branch circuit overcurrent device.

As mentioned previously, when the ampacity of the conductor does not match up with a standard rating of fuses or circuit breakers, the next higher standard-size overcurrent device may be used, provided that the overcurrent device does not exceed 800A *[NEC Section 240.4(B)]*. This exception is not permitted, however, when the branch circuit supplies receptacles where cord-and plug-connected appliances and similar electrical equipment could be used, because too many loads plugged into the circuit could result in an overload condition. The next standard size fuse or circuit breaker may be used only when the circuit supplies a fixed load.

The allowable ampacities of conductors used on most electrical systems are found in *NEC Tables 310.16 through 310.19*. However, the ampacities are subject to correction factors that must be applied where the ambient temperature for the conductor location exceeds 30°C (86°F). This reduction is required even if the reduction for more than three conductors in a raceway is also applied.

> **NOTE**
> *NEC Section 310.15(B)(2)(c)* provides temperature adjustments that must be added to the average ambient temperature for the ampacity derating of conductors installed in conduit exposed to direct sunlight on or above rooftops. A nationwide map showing average ambient temperatures is located in the *Appendix*.

For example, if six No. 10 AWG, TW current-carrying conductors are installed in a single raceway where the ambient temperature is 40°C, the ampacity of 30A must be derated or reduced to 80% of its value because of the number of current-carrying conductors in the raceway per *NEC Table 310.15(B)(2)(a)* and then reduced again by a correction factor of 0.82 *(NEC Table 310.16)* because of the ambient temperature. Therefore, when more than three current-carrying conductors are installed in a single raceway or cable or bundled cables in an ambient temperature of 40°C, the allowable ampacity for this condition is calculated as follows:

$$30A \times 0.80 \times 0.82 = 19.68A$$

In this situation, the listed ampacities must be reduced because of the heating effect of more than three current-carrying conductors. Grounding conductors are not counted as current-carrying conductors.

INSIDE TRACK

Minimum Branch Circuit Conductor Size

The smallest size conductor permitted by *NEC Table 310.5* for branch circuits in residential, commercial, and industrial locations is No. 14 copper. However, some local codes may require No. 12 or larger.

The rating of the branch circuit overcurrent device serving continuous loads must be not less than the noncontinuous load plus 125% of the continuous load per *NEC Section 210.20(A)*.

> **NOTE**
> *NEC Article 100* defines a continuous load as a load where the maximum current is expected to continue for three hours or more.

2.2.0 Conductor Protection

According to the *NEC®*, conductors must be installed and protected from damage (both physically and electrically). Additional requirements specify the use of boxes or fittings for certain connections, specify how connections are made to terminals, and restrict the use of parallel conductors. When conductors are installed in enclosures or raceways, additional rules apply. Finally, if conductors are installed underground, the burial depth and other installation requirements are specified by the *NEC®*. All conductors must be protected against overcurrent in accordance with their ampacities as set forth in the *NEC®*. They must also be protected against overloads and ground fault/short circuit current damage.

According to *NEC Section 240.6,* standard overcurrent device sizes are 15A, 20A, 25A, 30A, 35A, 40A, 45A, 50A, 60A, 70A, 80A, 90A, 100A, 110A, 125A, 150A, 175A, 200A, 225A, 250A, 300A, 350A, 400A, 450A, 500A, 600A, 700A, 800A, 1,000A, 1,200A, 1,600A, 2,000A, 2,500A, 3,000A, 4,000A, 5,000A, and 6,000A. Additional standard ratings for fuses are 1A, 3A, 6A, 10A, and 601A.

> **NOTE**
> The small fuse ratings of 1A, 3A, 6A, and 10A were added to the *NEC®* to provide more effective overcurrent protection for small loads.

Protection of conductors under short circuit conditions is accomplished by obtaining the maximum short-circuit current available at the supply end of the conductor, the short circuit withstand rating of the conductor, and the short circuit let-through characteristics of the overcurrent device.

When a noncurrent-limiting device is used for short circuit protection, the conductor's short circuit withstand rating must be properly selected based on the overcurrent protective device's ability to protect the circuit. See *Figure 7*.

It is necessary to check the energy let-through of the overcurrent device under short circuit conditions. Select a wire size of sufficient short circuit withstand ability.

In contrast, the use of a current-limiting device permits a device to be selected that limits short-circuit current to a level substantially less than obtainable in the same circuit if the overcurrent protection device were replaced with a solid conductor having comparable impedance (resistance)—doing away with the need for oversized ampacity conductors. See *Figure 8*.

In many applications, it is desirable to use the convenience of a circuit breaker for a disconnecting means and general overcurrent protection, supplemented by current-limiting devices at strategic points in the circuits.

Per *NEC Section 240.5(A)*, flexible cords, including tinsel cords and extension cords, must be protected against overcurrent in accordance with their ampacities as listed in *NEC Table 400.5(A)*. Supplementary overcurrent protection is acceptable. For example, with No. 18 AWG fixture wire that is 50' in length or more, a 6A overcurrent device would provide the necessary protection. For No. 16 AWG fixture wire of 100' or more, an 8A overcurrent device would provide

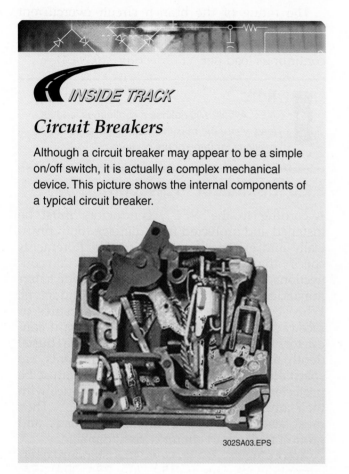

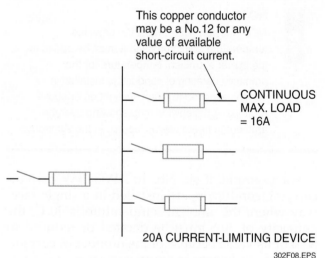

This copper conductor may be a No.12 for any value of available short-circuit current.

CONTINUOUS MAX. LOAD = 16A

20A CURRENT-LIMITING DEVICE

302F08.EPS

Figure 8 ◆ Current-limiting device.

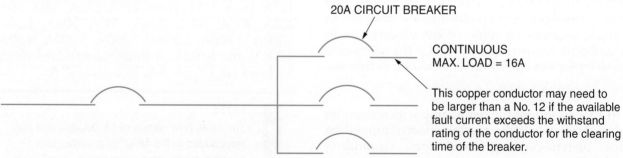

20A CIRCUIT BREAKER

CONTINUOUS MAX. LOAD = 16A

This copper conductor may need to be larger than a No. 12 if the available fault current exceeds the withstand rating of the conductor for the clearing time of the breaker.

302F07.EPS

Figure 7 ◆ Noncurrent-limiting device.

the necessary protection. For No. 18 AWG extension cords, a 10A overcurrent device would provide the necessary protection for a cord where only two conductors are carrying current; a 7A overcurrent device would provide the necessary protection for a cord where three conductors are carrying current.

2.2.1 Location of Overcurrent Protection in Circuits

In general, overcurrent protection must be installed at points where the conductors receive their supply, that is, at the beginning or line side of a branch circuit or feeder. See *NEC Section 240.21*. The exceptions to this rule are known as tap rules. They are as follows:

- Per *NEC Section 240.21(B)(1)*, overcurrent protection is not required at the conductor supply if a feed tap conductor is not over 10' long; is enclosed in raceway; does not extend beyond the switchboard, panelboard, disconnecting means, or control device that it supplies; and has an ampacity not less than the combined computed loads supplied and not less than the rating of the device supplied by the tap conductors or not less than the rating of the overcurrent device at the tap conductor termination. For field-installed taps, the ampacity of the overcurrent device on the line side of the tap conductor cannot exceed 10 times the ampacity of the tap conductor.
- Per *NEC Section 240.21(B)(2)*, overcurrent protection is not required at the conductor supply if a feeder tap conductor is not over 25' long; is protected from physical damage; has an ampacity not less than one-third that of the overcurrent device protecting the feeder conductors; and terminates in a single overcurrent device.
- Per *NEC Section 240.21(B)(3)*, overcurrent protection is not required at the conductor supply if a transformer feeder tap has primary conductors at least ⅓ ampacity and/or secondary conductors at least ⅓ ampacity when multiplied by the approximate transformer turns ratio of the overcurrent device protecting the feeder conductors; the total length of one primary plus one secondary conductor (excluding any portion of the primary conductor that is protected at its ampacity) is not over 25' in length; the secondary conductors terminate in a single overcurrent device rated at the ampacity of the tap conductors; and if the primary and secondary conductors are suitably protected from physical damage.
- Per *NEC Section 240.21(B)(4)*, overcurrent protection is not required at the conductor supply in high bay manufacturing buildings over 35' high at walls when only qualified persons will service such a system; if the tap conductors are not over 25' long horizontally and not over 100' long total length; the ampacity of the tap conductors is not less than ⅓ of the rating of the overcurrent device protecting the feeder conductors; terminate in a single overcurrent device; are suitably protected from physical damage or are enclosed in a raceway; are at least No. 6 AWG copper or No. 4 AWG aluminum; are continuous from end-to-end and contain no splices; do not penetrate walls, floors, or ceilings; and are made no less than 30' from the floor.

 WARNING!
Smaller conductors tapped to larger conductors can be a serious hazard. If not protected against short circuit conditions, these unprotected conductors can vaporize or incur severe insulation damage.

- Per *NEC Section 240.21(B)(5)*, overcurrent protection is not required at the conductor supply when the conductors are protected from physical damage by approved means, and they terminate in a disconnecting means that is either immediately outside or inside the building with an overcurrent device that limits the load to the ampacity of the secondary conductors. In this instance, the length of the conductors outside of the buildings is not limited.
- Per *NEC Section 240.21(C)(2)*, overcurrent protection is not required at a transformer secondary if the secondary conductor does not exceed 10', and the ampacity of the conductor is not less than the combined computed loads on the circuits supplied by the secondary conductors and not less than the rating of the device supplied by the secondary conductors or not less than the rating of the overcurrent protective device at the termination of the secondary conductors and not less than one-tenth of the rating of the overcurrent device of the primary, multiplied by the primary-to-secondary transformer voltage ratio. In addition, the secondary conductors cannot extend beyond the switchboard, panelboard, disconnecting means, or control devices they supply. The secondary conductors must also be enclosed in a raceway that extends from the transformer to the enclosure of an enclosed switchboard, panelboard, or control devices or to the back of an open switchboard.

- Per *NEC Section 240.21(C)(6),* overcurrent protection is not required at a transformer secondary if the secondary conductor does not exceed 25', and the secondary conductors have an ampacity that, when multiplied by the ratio of the primary-to-secondary voltage, multiplies by one-third the rating of the overcurrent device protecting the primary of the transformer. In addition, the secondary conductors must be suitably protected from physical damage by approved means and must terminate in a single circuit breaker or set of fuses that limit the load current to not more than the conductor ampacity permitted by *NEC Section 310.15.*
- Other tap provisions for single-phase, two-wire, and three-phase delta transformer secondaries are found in *NEC Section 240.21(C)(1),* while *NEC Section 240.21(C)(3)* covers secondary conductors in industrial installations.

> **NOTE**
>
> Switchboard and panelboard protection, along with transformer protection, must still be observed. See *NEC Sections 408.36 and 450.3.*

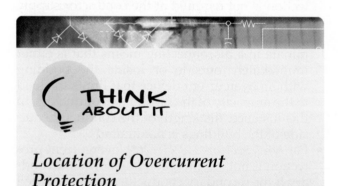

THINK ABOUT IT

Location of Overcurrent Protection

Is overcurrent protection required for the secondary conductors at the transformer terminals shown here?

302SA04.EPS

3.0.0 ◆ PROPERTIES OF CONDUCTORS

Various *NEC®* tables define the physical and electrical properties of conductors. Electricians use these tables to select the type of conductor and the size of conduit, or other raceway, to enclose the conductors in specific applications. *NEC®* tables list the properties of conductors as follows:

- Name
- Operating temperature
- Application
- Insulation
- Physical properties
- Electrical resistance
- AC resistance and reactance

NEC Table 310.13(A) lists the name, maximum operating temperature, applications, and insulation of various types of conductors. *NEC Chapter 9, Table 8* gives the physical properties and electrical resistance.

To gain an understanding of these tables and how they are used in practical applications, we will take a 4/0 THHN copper conductor and see what properties may be determined from the *NEC®* tables.

Step 1 Turn to *NEC Table 310.13(A)* and scan down the second column from the left (Type Letter) to find THHN. Scan to the left in this row to see that the trade name of this conductor is heat-resistant thermoplastic.

Step 2 Scanning to the right in this row, note that the maximum operating temperature for this wire type is 90°C (194°F). Continuing to the right in this row, under the column headed Application Provisions, we find that this wire type is suitable for use in dry and damp locations. The next column reveals that the insulation is flame-retardant, heat-resistant thermoplastic.

Step 3 Continuing to the right in this row, the next column lists insulation thickness for various AWG or kcmil wire sizes. The insulation thicknesses for Type THHN wire, from No. 14 AWG to 1,000 kcmil, are as follows:
- 14–12   15 mils
- 10   20 mils
- 8–6   30 mils
- 4–2   40 mils
- 1–4/0   50 mils
- 250–500   60 mils
- 501–1,000 70 mils

Consequently, the insulation thickness for 4/0 THHN is 50 mils.

Conductor Derating

THINK ABOUT IT

If six current-carrying conductors are bundled tightly together with no spacing, but are installed in an open tray instead of an enclosed conduit, do they have to be derated?

Step 4 Looking in the rightmost column (Outer Covering), we see that this wire type has a nylon jacket or equivalent.

If you want to find the maximum current-carrying capacity of this conductor when used in a raceway, turn to *NEC Table 310.16* and proceed as follows:

Step 1 Scan down the left-hand column until the wire size is found.

Step 2 Scan to the right in this row until the 90°C column is found. This column covers the insulation types that are rated for 90°C maximum operating temperature and includes Type THHN conductor insulation.

Step 3 Note that the current-carrying rating for this size and type of conductor is 260A. This rating, however, is for not more than three conductors in a raceway. If we have more than three conductors, such as four conductors in a three-phase, four-wire feeder to a subpanel, this figure (260A) must be derated.

Step 4 Refer to *NEC Section 310.15(B)(2)(a),* which states that where the number of current-carrying conductors in a raceway or cable exceeds three, the allowable ampacities shall be reduced as shown in *Table 1.*

Since we want to know the allowable maximum current-carrying capacity of four 4/0 THHN conductors in one raceway, it is necessary to multiply the previous amperage (260A) by 80% or 0.80.

$$260 \times 0.80 = 208A$$

These amperage tables are also based on the conductors being installed in areas where the ambient air temperature is 30°C (86°F). If the conductors are installed in areas with different ambient temperatures, a further deduction is required. For example, if this same set of four 4/0 THHN conductors were installed in an industrial area where the ambient temperature averaged 35°C, we would look in the correction factor tables at the bottom of *NEC Tables 310.16 through 310.19.* In doing so, we would find that the correction or derating factor for our situation is 0.96. Consequently, our present current-carrying capacity of 208A must be multiplied by 0.96 to obtain the actual current-carrying capacity of the four conductors:

$$208 \times 0.96 = 199.68A$$

Other sizes and types of conductors are handled in a similar manner; that is, find the appropriate table, determine the listed ampacity, and then multiply this ampacity by the appropriate factors in the correction factor tables.

It sometimes becomes necessary to know additional properties of conductors for some conductor calculations, especially for voltage drop calculations, which will appear in a later section in this module. There are many useful tables in *NEC Chapter 9.* Examples of their practical use are presented later in this module.

3.1.0 Identifying Conductors

The *NEC®* specifies certain methods of identifying conductors used in wiring systems of all types.

Table 1 Adjustment Factors for More Than Three Current-Carrying Conductors in a Raceway or Cable [Data from *NEC Table 310.15(B)(2)(a)*]

Number of Current-Carrying Conductors	Percent of Values in Tables as Adjusted for Ambient Temperature if Necessary
4 through 6	80
7 through 9	70
10 through 20	50
21 through 30	45
31 through 40	40
41 and above	35

High Leg Connection

NEC Section 408.3(E) requires that panelboards supplied by a three-phase, four-wire delta service have the high leg connected to center phase B.

For example, the high leg of a 120/240V, grounded three-phase, four-wire delta system must be marked with an orange color for identification; a grounded conductor must be identified either by the color of its insulation, by markings at the terminals, or by other suitable means. Unless allowed by *NEC®* exceptions, a grounded conductor must have a white or gray finish. When this is not practical for conductors larger than No. 6 AWG, marking the terminals white is an acceptable method of identifying the conductors.

3.1.1 Color Coding

Conductors contained in cables are color-coded for easy identification at each access point. When conductors are installed in raceway systems, any color insulation is permitted for the ungrounded phase conductors except the following:

* White or gray, which is reserved for use as the grounded circuit conductor
* Green, which is reserved for use as a grounding conductor only
* Orange, which is reserved for identifying the high leg of a three-phase, four-wire delta system

3.1.2 Changing Colors

Should it become necessary to change the actual color of a conductor to meet *NEC®* requirements or to facilitate maintenance of circuits and equipment, the conductors may be re-identified with nonconductive colored tape or paint.

For example, assume that a two-wire cable containing a black and white conductor is used to feed a 240V, two-wire, single-phase motor. Since the white-colored conductor is supposed to be reserved for the grounded conductor, and none is required in this circuit, the white conductor may be marked with a piece of black tape at each end of the circuit so that everyone will know that this wire is not a grounded conductor.

4.0.0 ◆ VOLTAGE DROP

In all electrical systems, the conductors should be sized so that the voltage drop does not exceed 3% for power, heating, and lighting loads, or combinations of these. Furthermore, the maximum total voltage drop for conductors, feeders, and branch circuits combined should not exceed 5%. These percentages are recommended by industry standards and *NEC Section 210.19(A)(1) FPN No. 4* but are not requirements. However, it is considered to be good practice to incorporate these percentages into every electrical installation.

In some applications, such as for circuits feeding hospital X-ray equipment, the voltage drop is even more critical—requiring a maximum of 2% voltage drop throughout. With the higher ratings on the newer types of insulation, it is extremely important to keep volt loss in mind; otherwise, some very unsatisfactory problems are likely to be encountered.

For example, the resistance and voltage drop on long conductor runs may be great enough to seriously interfere with the efficient operation of the connected equipment. Resistance elements, such as those used in incandescent lamps and electric heating units, are particularly critical in this respect; a drop of just a few volts greatly reduces their efficiency.

Electric motors are not affected by small voltage variations to the same degree as pure resistance loads, but motors will not operate at their rated horsepower if the voltage is below that at which they are rated. When loaded motors are operated at reduced voltage, the current flow actually increases, as it requires more amperes to produce a given wattage and horsepower at low voltage than at normal voltage. This current increase is also caused by the fact that the opposition of the motor windings to current flow decreases as the motor speed decreases.

From the foregoing, we can see that it is very important to have all conductors of the proper size to avoid excessive heating and voltage drop, and that, in the case of long runs, it is necessary to determine the wire size by consideration of resistance and voltage drop, rather than by the heating effect or *NEC Tables 310.16 through 310.19* alone.

To solve the ordinary problems of voltage drop requires only a knowledge of a few simple facts about the areas and resistances of conductors and the application of a few mathematical equations.

4.1.0 Wire Sizes Based on Resistance

Earlier modules covered wire sizes and how conductors are normally specified in kcmil or AWG sizes. This numbering system was originated by the Brown & Sharpe Company and was originally called the B & S gauge. However, the B & S gauge quickly evolved into the American Wire Gauge (AWG) and is now standard in the United States for indicating sizes of round wires and conductors.

AWG numbers are arranged according to the resistance of the wires, with the larger numbers representing the wires of greatest resistance and smallest area. A handy rule to remember is that decreasing the gauge by three numbers gives a wire of approximately twice the area and half the resistance. Conversely, increasing the gauge by three numbers gives a wire of approximately half the area and twice the resistance.

For example, if we increase the wire gauge from No. 3 AWG, which has a resistance of 0.245 ohm per 1,000', to a No. 6 AWG, we find it has a resistance of 0.491 ohm per 1,000', which is about double. See *NEC Chapter 9, Table 8.*

Although *NEC Chapter 9, Table 8* only lists sizes down to No. 18 AWG, the American Wire Gauge numbers range from 0000 (4/0) down in size to No. 40. The 4/0 conductor is more than ½" in overall diameter and the No. 40 is as fine as a thin hair.

The most common sizes used for light and power installations range from 4/0 to No. 14 AWG. Lighting fixture wires are frequently size No. 16 or No. 18 AWG, and low-voltage control wiring sometimes drops down to size No. 22 AWG.

4.1.1 Circular Mil—Unit of Conductor Area

In addition to AWG gauge numbers, a unit called the mil is also used for measuring the diameter and area of conductors. The mil is equal to $\frac{1}{1000}$ of an inch, so it is small enough to measure and express these sizes very accurately. For example, instead of saying a wire has a diameter of 0.055", or fifty-five thousandths of an inch, we can simply call it 55 mils. Consequently, a wire of 250 mils in diameter is also 0.250", or ¼" in diameter.

Since the resistance and current-carrying capacity of conductors both depend on the conductor's cross-sectional area, a unit is necessary to express this area. Electrical conductors are made in several different shapes (*Figure 9*). For square conductors, such as busbars, the square mil is used, which is a square that is $\frac{1}{1000}$" on each side. For round conductors, the circular mil unit is used, which is the area of a circle with a diameter of $\frac{1}{1000}$".

These units greatly simplify conductor calculations. For example, to determine the area of a square conductor, as shown in *Figure 9(B)*, multiply one side by the other, measuring the sides in either mils or thousandths of an inch.

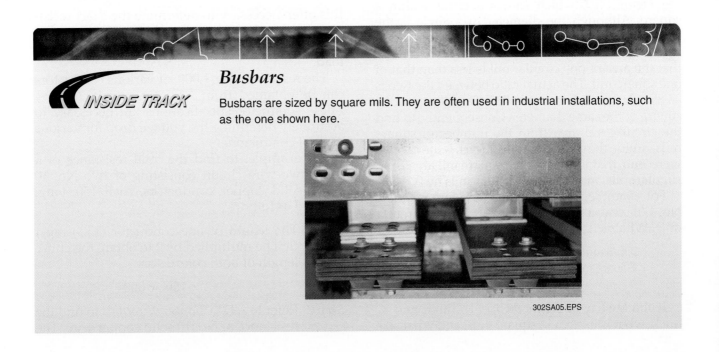

INSIDE TRACK

Busbars

Busbars are sized by square mils. They are often used in industrial installations, such as the one shown here.

302SA05.EPS

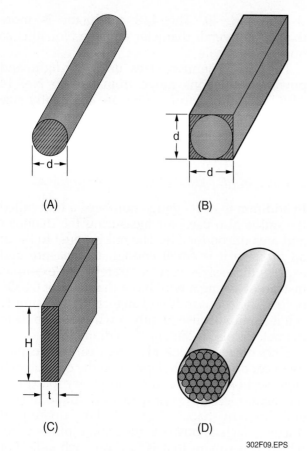

Figure 9 ◆ Electrical conductors in various shapes.

302F09.EPS

4.1.2 Conversion of Square Mils to Circular Mils

When comparing round and square conductors, remember that the square mil and the circular mil are not quite the same units of area. For example, see *Figure 9(B)*, which shows a circle within a square. Although the diameter of a circle is the same as the width of a square, the corners of the square make it larger in area. From this, we can say that the area of one circular mil is less than that of one square mil. The actual ratio between the two is 0.7854, or the circle has only 78.54% of the area of a square of the same diameter. Consequently, to find the circular mil area from the number of square mils, divide the square mils by 0.7854. If the reverse were true, that is, finding the square mil area from circular mils, multiply the circular mils by 0.7854.

For example, the conductor in *Figure 9(A)* is a No. 4/0 conductor with an area of 211,600 circular mils. What is its area in square mils?

> Circular mil area = 211,600 circular mils
> Square mil area = 211,600 × 0.7854
> = 166,191 square mils

If the busbar in *Figure 9(C)* is 1½" high and ¼" thick, what is its area in square mils, and what size

of round conductor would be necessary to carry the same current as this busbar? First, the dimensions of a ¼" × 1½" busbar, stated in mils, are 250 mils × 1,500 mils. Therefore, the area in square mils may be determined by the following equation:

$$250 \times 1,500 = 375,000 \text{ square mils}$$

To find this area in circular mils, divide the square mil area (375,000) by 0.7854; the results are as follows:

$$375,000 \div 0.7854 = 477,464 \text{ circular mils}$$

The nearest standard size to this is a round conductor 500,000 circular mils in size (500 kcmil).

Busbars of the shape shown in *Figure 9(C)* are commonly used in panelboards and switchgear. These bars normally range in thickness from 0.250" to 0.375" or more, and in heights from 1" to 12".

Stranded conductors such as the one shown in *Figure 9(D)* are used on all conductor sizes No. 6 AWG and larger. Since these conductors are not solid throughout, their area cannot be determined accurately. This diameter also varies somewhat with the twist or lay of the strands.

To determine the cross-sectional area of such conductors, first determine the area of each strand, either from a wire table or by calculation from its diameter, and then multiply this by the number of strands to get the total area of the cable in circular mils. *NEC Chapter 9, Table 8* can also be used to find the area of most conductors in circular mils. It also provides information necessary for other wiring calculations.

4.2.0 Resistance of Conductors

It is often necessary to determine the exact resistance of a conductor of a certain length in order to calculate the voltage drop under a certain current load.

The resistance per 1,000' of various conductors can be obtained from *NEC Chapter 9, Table 8.* These conductor specifications are necessary to accurately determine the voltage drop for various sizes of conductors.

For example, to find the total resistance of a 120V, two-wire circuit consisting of two No. 10 AWG solid copper conductors, each 150' long, proceed as follows:

Step 1 The length of one conductor (150') must first be multiplied by 2 to obtain the entire length of both conductors:

$$2 \times 150' = 300'$$

Step 2 Refer to *NEC Chapter 9, Table 8* to find the resistance of No. 10 solid copper wire. The table gives a resistance of 1.21 ohms per 1,000' for uncoated copper wire at 75°C.

Step 3 Since our circuit is less than 1,000', we must determine the resistance of 300'. This is accomplished by dividing the actual footage (300' in this case) by 1,000'.

$$\frac{300'}{1,000'} = 0.30$$

Step 4 Multiply this result (0.30) times the resistance of 1.21Ω:

$$0.30 \times 1.21\Omega = 0.363\Omega$$

In another situation, we want to install an outside 120V, two-wire line between two buildings a distance of 1,650' using No. 1 AWG copper wire. What would be the total resistance of this circuit?

Step 1 Determine the total length of both conductors by multiplying the length (one way) by 2:

$$1,650' \times 2 = 3,300'$$

Step 2 Referring again to *NEC Chapter 9, Table 8*, we see that No. 1 AWG uncoated copper wire has a resistance of 0.154 ohm per 1,000'.

Step 3 The total length of 3,300' must be divided by 1,000'.

$$\frac{3,300'}{1,000'} = 3.3$$

Step 4 Multiply this result (3.3) times the resistance found in *NEC Chapter 9, Table 8* (0.154Ω).

$$3.3 \times 0.154\Omega = 0.5082\Omega$$

Now we will see what happens if we apply the full current to these conductors allowed by *NEC Table 310.16*. Assuming that Type THHN conductors are used, *NEC Table 310.16* allows a maximum load on these conductors of 150A at 90°C. (Always use the 90°C column in this table when derating conductors.) If this much current flowed through this circuit for a distance of 3,300' (both ways), the voltage drop in the circuit would be:

$$\begin{aligned} \text{Voltage drop} &= I \times R \\ &= 150A \times 0.5082\Omega \\ &= 76.2V \end{aligned}$$

If the initial voltage is only 120V, this means that the voltage at the far end of the circuit will be only 120V − 76.2V or 43.8V and few, if any, 120V loads will operate at this low voltage. Consequently, the load will have to be reduced to make this circuit of any use.

Our goal is to keep the voltage drop within 3% of the original voltage. Therefore, since the original voltage is 120V, the allowable voltage drop may be found by using the following equation:

$$120V \times 0.03 = 3.6V$$

Now, what size load may be applied to this circuit to stay within this 3% range? The equation for finding the maximum current on this circuit to keep the voltage drop within 3% or 3.6V is as follows:

Maximum current (I) × total resistance (Ω)
= original voltage (V) × allowable voltage drop%

Substituting our known values in this equation, we have:

$$I \times 0.5082\Omega = 3.6V$$

To solve for I, we divide both sides of the equation by 0.5082, which results in the following:

$$I = \frac{3.6V}{0.5082\Omega} = 7.084A$$

Therefore, to keep the voltage drop within 3% for this length of circuit with only 120V applied, the amperage must be held to 7.084A or below.

If a larger load is connected to this circuit, and we need to hold the voltage drop to within 3%, then either a larger conductor will have to be used, or else the voltage will have to be increased. For example, assume that this circuit feeds a 120/240V dual-voltage pump motor. If the motor connections were rewired to accept a 240V branch circuit, then the allowable voltage drop (at 3%) would be:

$$0.03 \times 240V = 7.2V$$

Continuing as before:

$$I \times \frac{7.2V}{0.5082\Omega} = 14.17A$$

If 480V single phase were applied to the circuit (even though it is not feasible with this two-wire arrangement), the amperage would again double. These examples should show why long electric transmission lines utilize extremely high voltages—up to 250,000V or more—to keep the current, resulting voltage drop, and conductor size to the bare minimum.

4.3.0 Resistance of Copper per Mil Foot

In many cases, it may be necessary to calculate the resistance of a certain length of wire or a busbar of a given size.

This can be done very easily if the unit resistance of copper is known. A unit called the mil foot may be used. A mil foot represents a piece of round wire that is 1 mil in diameter and 1' in length and is a small enough unit to be very accurate for all practical calculations. A round wire of 1 mil in diameter has an area of 1 circular mil, as the diameter multiplied by itself, or squared, is:

$$1 \times 1 = 1 \text{ circular mil}$$

The resistance of ordinary copper is 12.9Ω per mil foot at 75°C. This constant is important and should be remembered. It is normally represented by the letter K.

Suppose we want to determine the resistance of a piece of No. 12 copper wire that is 50' long. We know that the resistance of any conductor increases as its length increases and decreases as its area increases. So, for a wire that is 50' long, we first multiply and get $50' \times 12.9\Omega = 645\Omega/\text{mil ft}$, which would be the resistance of a wire that is 1 circular mil in area and 50' long. Then we find in *NEC Chapter 9, Table 8* that the area of a No. 12 wire is 6,530 circular mils, which will reduce the resistance in proportion. So we now divide:

$$\frac{645}{6,530} = 0.0988\Omega/\text{mil ft}$$

In another case, we wish to find the resistance of a coil containing 3,000' of No. 18 stranded wire. We would first multiply and get $3,000' \times 12.9\Omega = 38,700\Omega/\text{mil ft}$. Since the area of No. 18 wire is 1,620 circular mils, the total resistance may be found using the following equation:

$$\frac{38,700}{1,620} = 23.89\Omega$$

Checking this with *NEC Chapter 9, Table 8,* we find a resistance of 7.95Ω per 1,000' for No. 18 stranded AWG wire. For 3,000', we use the following equation:

$$3 \times 7.95\Omega = 23.85\Omega$$

The small difference between this answer and the one obtained by the first calculation is caused by using approximate figures instead of lengthy decimal places.

The mil foot unit and its resistance of 12.9Ω for copper may also be used to calculate the resistance of square busbars by simply using the figure 0.7854 to convert from square mils to circular mils.

Suppose we wish to find the resistance of a square busbar that is $\frac{1}{4}" \times 2"$ and 100' long. The dimensions in mils will be $250 \times 2,000 = 500,000$ square mils. To find the circular mil area, we divide 500,000 by 0.7854 and get 636,618.3 circular mils. Then we multiply and get $100' \times 12.9\Omega = $

$1,290\Omega$, or the resistance of 100' of copper that is 1 mil in area. Because the area of this bar is 636,618.3 circular mils, we divide:

$$\frac{1,290\Omega}{636,618.3} = 0.002026\Omega \text{ total resistance}$$

NEC Tables 310.16 through 310.19 list the allowable current-carrying capacities of conductors with various types of insulation. These tables, however, do not take into consideration the length of the conductors or the voltage drop. Consequently, it is often necessary to use a larger size conductor than is indicated in the tables.

4.4.0 Equations for Voltage Drop Using Conductor Area or Conductor Resistance

The size of the conductor required to connect an electrical load to the source of supply is determined by several factors, including:

- Load current in amperes
- Permissible voltage drop between source and load
- Total length of conductor
- Type of wire (i.e., copper, aluminum, copper-clad aluminum, etc., and its permissible load-carrying capability based on *NEC Tables 310.16 through 310.19)*

To perform voltage drop calculations, it is also necessary to recall the following:

- The resistance of a wire varies directly with its length:

  $R = \text{resistance per foot (K)} \times \text{length in feet (L)}$

- The resistance varies inversely with its cross-sectional area:

$$R = \frac{1}{A}$$

Combining both statements, we obtain the following equation:

$$R = \frac{L \times K}{A}$$

Where:

A = wire size (first from *NEC Tables 310.16 through 310.19* for a given load, and then, if necessary, from *NEC Chapter 9, Table 8* for circular mils)

L = length of wire in feet

R = total resistance of wire

K = resistance per mil foot

K is a constant whose value depends upon the units chosen and the type of wire (12.9 for copper and 21.2 for aluminum). Using the foot as the unit of length and the circular mil as the unit of area, the values for K represent the resistance in ohms per mil foot at 75°C.

The length (L) in the equation just given is the total length of a single current-carrying conductor. For a two-conductor, 120V circuit or a 240V, balanced three-wire circuit (neutral current equals zero), the length in the equation must be multiplied by 2. The equation now becomes:

$$R = \frac{2 \times L \times K}{A} \text{ for single-phase circuits}$$

For a three-phase, four-wire balanced circuit with a power factor near the value of 1, the equation must be multiplied by $\sqrt{3} \div 2$. (The value of $\sqrt{3}$ is approximately 1.732; you may wish to make a note of this for use in future calculations.) The equation now becomes:

$$R = \frac{\sqrt{3}}{2} \times \frac{2 \times L \times K}{A}$$

The 2s cancel each other out and the equation reduces to:

$$R = \frac{\sqrt{3} \times L \times K}{A} \text{ for three-phase balanced circuits}$$

Since R = E ÷ I (where E = voltage drop [VD]), substitute VD ÷ I for R in the previous equations:

$$\frac{VD}{I} = \frac{2 \times L \times K}{A} \text{ for single-phase circuits}$$

and

$$\frac{VD}{I} = \frac{\sqrt{3} \times L \times K}{A} \text{ for three-phase balanced circuits}$$

Multiply each side of both equations by I:

$$VD = \frac{2 \times L \times K \times I}{A} \text{ for single-phase circuits}$$

and

$$VD = \frac{\sqrt{3} \times L \times K \times I}{A}$$

for three-phase balanced circuits

Substitute circular mils (CM) for A where CM is initially determined from the wire size ampacities listed in *NEC Tables 310.16 through 310.19* for the desired load and where *NEC Chapter 9, Table 8* is used to convert the wire size so determined to circular mils, if required. The equations become:

$$VD = \frac{2 \times L \times K \times I}{CM} \text{ for single-phase circuits}$$

and

$$VD = \frac{\sqrt{3} \times L \times K \times I}{CM}$$

for three-phase balanced circuits

Voltage Drops

This parking lot fixture has three 1,000W lamps, is 300' from the panel, uses No. 12 wire, and is supplied by a 20A, 208V circuit. The fixture manufacturer specifies a maximum voltage drop of 3%. Does this circuit adequately support this load?

302SA06.EPS

You should recognize these equations as the same ones used to calculate voltage drop for branch circuits in the previous module. Using an exercise similar to the one just listed, the following equations (also used in the previous module) can be derived for a given wire size resistance and load:

$$VD = \frac{2 \times L \times R \times I}{1,000} \text{ for single-phase circuits}$$

and

$$VD = \frac{\sqrt{3} \times L \times R \times I}{1,000}$$

for three-phase balanced circuits

4.5.0 Use of Voltage Drop Equations

If we want to determine the voltage drop on either an existing 120V, two-wire installation or a proposed project, we can find the voltage drop of the circuit by using one of the single-phase formulas previously given:

$$VD = \frac{2 \times L \times K \times I}{CM}$$

For example, assume we have a 120V, two-wire circuit feeding a total load of 30A and the branch circuit length is 120'.

NEC Table 310.16 allows a No. 10 AWG conductor at 60°C to carry this load. (This does not consider voltage drop or derating factors.) Referring to *NEC Chapter 9, Table 8,* we find that the area (A) of a No. 10 AWG conductor is 10,380 circular mils. Then, substituting these values and a value of 12.9 (copper) for K in the equation, we have:

$$VD = \frac{2 \times 120' \times 12.9 \times 30A}{10,380}$$

$$VD = 8.95V$$

The voltage at the load is 120V − 8.95V = 111.05V, which is usually not acceptable on most installations since it exceeds the 3% recommended maximum voltage drop; that is, 3% of 120 = 3.6V. In fact, it is more than double. There are three ways to correct this situation:

- Reduce the load
- Increase the conductor size
- Increase the voltage

4.5.1 Miscellaneous Voltage Drop Equations

The final voltage drop (VD) equations discussed earlier can be solved for length (L), yielding the following equations:

$$L = \frac{VD \times CM}{2 \times K \times I} \text{ or } L = \frac{1,000 \times VD}{2 \times R \times I}$$

for single-phase circuits

and

$$L = \frac{VD \times CM}{\sqrt{3} \times K \times I} \text{ or } L = \frac{1,000 \times VD}{\sqrt{3} \times R \times I}$$

for three-phase balanced circuits

These equations can be used to determine the maximum circuit length for a specified voltage drop at a given load using a wire size selected from *NEC Tables 310.16 through 310.19* for the ampacity of the load. *NEC Chapter 9, Table 8* may be used to convert the selected wire size to circular mils, if required.

For example, determine the maximum circuit length for a feeder serving a 208V balanced three-phase load of 400A at a voltage drop not to exceed 3%. From *NEC Table 310.16,* determine that a 750 kcmil conductor at 60°C is rated to carry 400A. Using the three-phase length equation above containing the CM term, substitute values and calculate length:

$$L = \frac{VD \times CM}{\sqrt{3} \times K \times I}$$

$$L = \frac{208V \times 3\% \times 750,000 \text{ CM}}{1.732 \times 12.9 \times 400A}$$

$$L = 523.6'$$

Therefore, in this example the 400A load may be positioned up to 523.6' from the source and the voltage drop will be 3% or less using 750 kcmil conductors at 60°C.

The two final voltage drop equations given earlier may also be solved for wire size in circular mils (CM) to directly determine, under certain conditions, the wire size for a given voltage drop, wire length, and load. Solving for CM yields:

$$CM = \frac{2 \times L \times K \times I}{VD} \text{ for single-phase circuits}$$

and

$$CM = \frac{\sqrt{3} \times L \times K \times I}{VD} \text{ for three-phase circuits}$$

The results from these two CM equations *must* be used in accordance with the following two rules:

- *Rule 1* – If the wire size calculated using the CM equation results in a wire size with less ampacity than the given load as determined from *NEC Tables 310.16 through 310.19,* discard the calculated result and use the wire size with the rated ampacity for the given load.

- *Rule 2* – If the wire size calculated using the CM equation results in a wire size equal to or greater than the wire size determined from *NEC Tables 310.16 through 310.19* for the given load, use the calculated wire size or the next larger standard wire size.

The reason for these rules is that the equation results are for non-derated bare wire. The wire sizes determined from the *NEC®* tables are derated for the number of wires, temperature, and other factors. Calculated sizes that are smaller than those sizes rated at the desired load ampacity cannot be used. However, calculated wire sizes that are larger may be used when excessive voltage drop occurs that is a function of only the size of wire (that has already been derated) and that is caused by additional length beyond the maximum length for a specified voltage drop and load ([L] equations previously shown).

To illustrate, we will solve for CM using the terms of the example given for determining length (L) (i.e., a 400A, three-phase load at 523.6'):

$$CM = \frac{\sqrt{3} \times L \times K \times I}{VD}$$

Substituting values, we have:

$$CM = \frac{1.732 \times 523.6' \times 12.9 \times 400A}{208V \times 3\%}$$

$$CM = 749,916 \text{ or } 750 \text{ kcmil}$$

In this instance, of course, the length and load correlated exactly with the 3% voltage drop and a 750 kcmil conductor was obtained from the calculation. In accordance with *Rule 2*, the size obtained equaled the size of a 60°C conductor with a rated ampacity of 400A. Note that longer lengths for this same load will result in larger conductor sizes, each of which will be proven correct if cross-checked using the VD equations.

Now take the same problem but change the 523.6' to 100' and solve for CM:

$$CM = \frac{\sqrt{3} \times L \times K \times I}{VD}$$

$$CM = \frac{1.732 \times 100' \times 12.9 \times 400A}{208V \times 3\%}$$

$$CM = 143,223$$

From *NEC Chapter 9, Table 8,* the next largest standard conductor is 167,800 CM or a No. 3/0 AWG conductor. However, when *NEC Table 310.16* is checked, the maximum permissible ampacity of a No. 3/0 AWG conductor is found to be only 165A—far below the load of 400A.

Therefore, in accordance with *Rule 1*, the 3/0 AWG conductor solution would be discarded and 750 kcmil 60°C conductors, determined from *NEC Table 310.16* for the 400A load, would be used for the 100' run.

Even when sizing wire for low-voltage (24V) control circuits, the voltage drop should be limited to 3% because excessive voltage drop causes:

- Failure of control coil to activate
- Control contact chatter
- Erratic operation of controls
- Control coil burnout
- Contact burnout

The voltage drop calculations described previously may also be used for low-voltage wiring, but tables are quite common and can save much calculation time. One example is shown in *Table 2*.

To use *Table 2*, assume a load of 35VA with a 50' run for a 24V control circuit. Referring to the table, scan the 50' column. Note that No. 18 AWG wire will carry 29VA and No. 16 wire will carry 43VA while still maintaining a maximum of 3% voltage drop. In this case, No. 16 wire would be the size to use.

The 3% voltage drop limitation is imposed to assure proper operation when the power supply is below the rated voltage. For example, if the

Table 2 Table for Calculating Voltage Drop in Low-Voltage (24V) Wiring

AWG Wire Size	Length of Circuit (One Way in Feet)											
	25	50	75	100	125	150	175	200	225	250	275	300
20	29	14	10	7.2	5.8	4.8	4.1	3.8	3.2	2.9	2.8	2.4
18	58	29	19	14	11	9.6	8.2	7.2	6.4	5.8	5.2	4.8
16	86	43	29	22	17	14	12	11	9.6	8.7	7.8	7.2
14	133	67	44	33	27	22	19	17	15	13	12	11

Voltage Drops in Low-Voltage Devices

THINK ABOUT IT

You are adding a 12V/12W head 50' away from an emergency pack. Use the voltage drop calculations to determine what size wire you should run.

rated 240V primary supply is 10% low (216V), the transformer secondary side does not produce 24V but rather 21.6V. When normal voltage drop is taken from this 21.6V, it approaches the lower operating limit of most controls. If it is assured that the primary voltage to the transformer will always be at rated value or above, the control circuit will operate satisfactorily with more than 3% voltage drop.

In most installations, several lines connect the transformer to the control circuit. One line usually carries the full load of the control circuit from the secondary side of the transformer to one control, with the return perhaps coming through several lines of the various other controls. Therefore, the line from the secondary side of the transformer is the most critical regarding voltage drop and VA capacity and must be properly sized.

When low-voltage lines are installed, it is suggested that one extra line be run for emergency purposes. This can be substituted for any one of the existing lines that may be defective.

Hollow Transmission Line

INSIDE TRACK

This copper cable was part of the original 287.5kV transmission line that supplied electricity to Los Angeles from the generators at the Hoover Dam. Its spiral-wound, interlocked design provided additional strength and flexibility while reducing the effects of thermal expansion. The original cable had a tube within a tube (not shown). This reduced the weight of the conductors and aided in heat dissipation. The hollow design enhanced the conductive quality of the cable due to a phenomenon known as skin effect. Because it travels in sine waves, alternating current tends to create internal interference within a conductor (eddy currents), causing most of the current to travel on the external surface or skin of the conductor rather than through the core. The skin effect of modern high-voltage transmission lines is often compensated for by the use of specially braided cable with a steel reinforcing core.

302SA07.EPS

1. All things considered, the most ideal conductor is made of _____.
   a. bronze
   b. brass
   c. aluminum
   d. copper

2. Each of the following is a standard configuration for stranded wire *except* _____.
   a. 5-strand
   b. 7-strand
   c. 19-strand
   d. 37-strand

3. Of the following, the largest standard AWG wire size is _____.
   a. No. 3 AWG
   b. No. 1 AWG
   c. No. 1/0 AWG
   d. No. 4/0 AWG

4. A branch circuit is best described as _____.
   a. the circuit conductors between the service equipment and the final branch circuit overcurrent device
   b. the circuit conductors between the final overcurrent device protecting the circuit and the outlet(s)
   c. a conductor tapped onto another conductor
   d. two conductors run in parallel

5. The rating of a branch circuit is determined by the _____.
   a. type and size of the branch circuit conductors
   b. rating of the branch circuit overcurrent device
   c. total connected load
   d. characteristics of the supply circuit

6. A continuous load is best described as one in which the maximum current is expected to continue for _____ hour(s) or more.
   a. one
   b. two
   c. three
   d. four

7. All of the following are standard ratings for fuses *except* _____.
   a. 600A
   b. 601A
   c. 650A
   d. 700A

8. You would find information on conductor applications and descriptions of insulation types in _____.
   a. *NEC Table 310.15(B)(2)(a)*
   b. *NEC Table 310.13(A)*
   c. *NEC Table 310.16*
   d. *NEC Chapter 9, Table 8*

9. Use *NEC Table 310.16* to find the maximum ampacity for a No. 2 AWG copper wire at 90°C (assume THHN insulation).
   a. 95A
   b. 110A
   c. 130A
   d. 150A

10. The constant (K) for the resistance of ordinary copper is _____ per mil foot.
    a. 12.9Ω
    b. 21.2Ω
    c. 1.732Ω
    d. 3.141Ω

Summary

A great deal of valuable information about conductors, such as dimensions, resistance, and current-carrying capacity, can be obtained from convenient *NEC*® tables. Use these tables whenever possible; they are great time-saving devices.

Sometimes appropriate tables are not available, or do not give the exact information needed for a certain project or situation. This is when a knowledge of simple conductor calculations pays off.

For example, *NEC Tables 310.16 through 310.19* give the allowable current-carrying capacities of conductors with various types of insulation based on the heating of the conductors, but these tables do not account for voltage drop due to resistance of long circuit runs. This is very important and should always be kept in mind when planning or installing any electrical installation.

Notes

Trade Terms Introduced in This Module

American Wire Gauge (AWG): The United States standard for measuring wires.

Cable: An assembly of two or more insulated or bare wires.

Average Ambient Temperatures

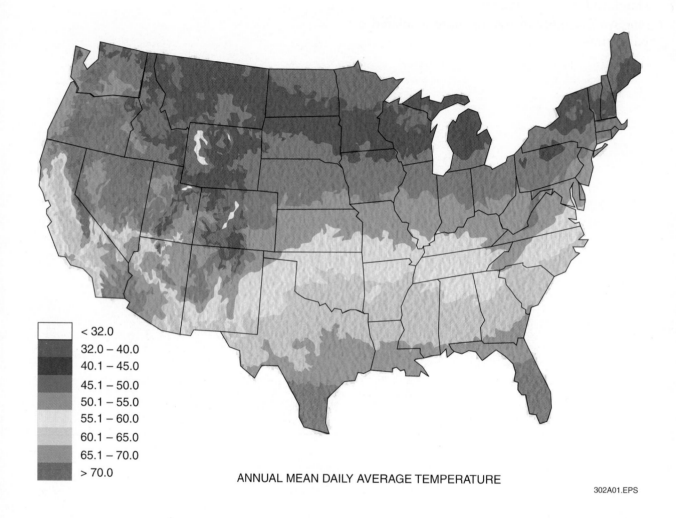

< 32.0
32.0 – 40.0
40.1 – 45.0
45.1 – 50.0
50.1 – 55.0
55.1 – 60.0
60.1 – 65.0
65.1 – 70.0
> 70.0

ANNUAL MEAN DAILY AVERAGE TEMPERATURE

302A01.EPS

This module is intended to present thorough resources for task training. The following reference work is suggested for further study. This is optional material for continued education rather than for task training.

National Electrical Code® Handbook, Latest Edition. Quincy, MA: National Fire Protection Association.

CONTREN® LEARNING SERIES – USER UPDATE

NCCER makes every effort to keep these textbooks up-to-date and free of technical errors. We appreciate your help in this process. If you have an idea for improving this textbook, or if you find an error, a typographical mistake, or an inaccuracy in NCCER's Contren® textbooks, please write us, using this form or a photocopy. Be sure to include the exact module number, page number, a detailed description, and the correction, if applicable. Your input will be brought to the attention of the Technical Review Committee. Thank you for your assistance.

Instructors – If you found that additional materials were necessary in order to teach this module effectively, please let us know so that we may include them in the Equipment/Materials list in the Annotated Instructor's Guide.

Write: Product Development and Revision
National Center for Construction Education and Research
3600 NW 43rd St., Bldg. G, Gainesville, FL 32606

Fax: 352-334-0932

E-mail: curriculum@nccer.org

Craft

Module Name

Copyright Date

Module Number

Page Number(s)

Description

(Optional) Correction

(Optional) Your Name and Address

Practical Applications
of Lighting

Brooklyn Bridge

The Brooklyn Bridge will soon go green with LED lights as part of New York City's plan to save energy and cut greenhouse gases. Each of the 160 LED fixtures consumes only 24 watts of power instead of the 100 watts per each mercury bulb in place now. Atmospheric release of carbon dioxide will drop by about 24 tons each year.

26303-08

26303-08
Practical Applications of Lighting

Topics to be presented in this module include:

Overview

The lighting system is a major component of any electrical installation. Selecting the appropriate lighting for a particular application is based on more than the aesthetics of the fixture itself, although luminaires should be appealing to the eye.

Luminaires should be selected with several factors in mind: their primary purpose, the environment in which they will be installed, their availability, their cost, and their operating characteristics. Some examples of primary lighting purposes include task illumination, accent, emergency lighting, and security. Specialty lighting is available for any situation where illumination, indication, or accent is needed. Lighting control system complexity can range from a single switch to a complex remote system operated by sensing and timing devices.

Objectives

When you have completed this module, you will be able to do the following:

1. Explain how the lighting terms lumen, candlepower, and footcandle relate to one another.
2. Classify lighting fixtures by type and application.
3. Identify the general lighting pattern produced by each type of fixture.
4. Identify the lighting requirements associated with lighting systems used in selected applications such as office buildings, schools, theaters, hazardous areas, etc.
5. Identify various dimming systems and their components.
6. Use manufacturers' lighting fixture catalogs to select the appropriate lighting fixtures for specific lighting applications.

Trade Terms

Candela (cd)
Candlepower (cp)
Diffused lighting
Diffuser
Footcandle (fc)
Work plane

Required Trainee Materials

1. Pencil and paper
2. Appropriate personal protective equipment
3. Copy of the latest edition of the *National Electrical Code*®

Prerequisites

Before you begin this module, it is recommended that you successfully complete *Core Curriculum; Electrical Level One; Electrical Level Two; Electrical Level Three*, Modules 26301-08 and 26302-08.

This course map shows all of the modules in *Electrical Level Three*. The suggested training order begins at the bottom and proceeds up. Skill levels increase as you advance on the course map. The local Training Program Sponsor may adjust the training order.

ELECTRICAL LEVEL THREE

26311-08
Motor Controls

26310-08
Voice, Data, and Video

26309-08
Motor Calculations

26308-08
Commercial Electrical Services

26307-08
Transformers

26306-08
Distribution Equipment

26305-08
Overcurrent Protection

26304-08
Hazardous Locations

26303-08 Practical Applications of Lighting

26302-08 Conductor Selection and Calculations

26301-08 Load Calculations – Branch and Feeder Circuits

ELECTRICAL LEVEL TWO

ELECTRICAL LEVEL ONE

CORE CURRICULUM:
Introductory Craft Skills

302CMAP.EPS

1.0.0 ◆ INTRODUCTION

This module is the second of two modules that cover the subject of electrical lighting. It builds on the information and lighting principles previously covered in the *Electrical Level Two* module, *Electric Lighting*.

This module describes specific applications for the different designs of incandescent, fluorescent, LED, induction, and HID lighting fixtures. This is followed by an overview of the major applications and requirements for lighting systems such as those used in office buildings, theaters, sports arenas, landscaping, etc. Special wiring systems and dimming systems commonly incorporated into lighting systems are also introduced.

2.0.0 ◆ LUMENS, CANDLEPOWER, AND FOOTCANDLES

Before addressing the specific content of this module, the terms lumen, candlepower (cp), and footcandle (fc) should be reviewed. In lighting, the basic measure of the amount of light produced by a light source is called a lumen. As with any unit of measure, one must start with a standard. The starting point for lumens is historically a candle. If it is of a specific composition and size, it emits one candela (cd).

If this candle is placed in a sphere with a one-foot radius, by definition, one lumen will fall on one square foot of the sphere's surface (*Figure 1*). Since the total surface area of a sphere with a one-foot radius is 12.57 square feet, there are 12.57 lumens given off by that candle. One lumen falling on one square foot of a surface produces an illumination of one footcandle (*Figure 2*).

The terms footcandle (fc) and candlepower (cp) are often confused. The footcandle is the unit used to measure the illuminance or amount of total light per unit area that falls on a surface, such as a wall or table. Candlepower measures the intensity of a light source in a given direction.

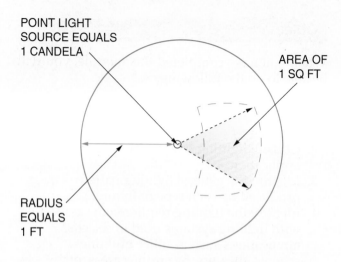

One lumen equals the amount of light cast upon one square foot of the inner surface of a hollow sphere of one-foot radius when a light source of one candela is placed at the center of the sphere.

303F01.EPS

Figure 1 ◆ Definition of one lumen.

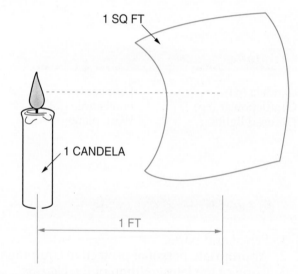

One lumen falling on one square foot of surface illuminates the surface to an intensity of one footcandle.

303F02.EPS

Figure 2 ◆ Definition of one footcandle.

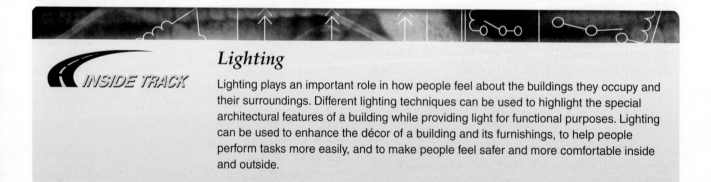

Lighting

Lighting plays an important role in how people feel about the buildings they occupy and their surroundings. Different lighting techniques can be used to highlight the special architectural features of a building while providing light for functional purposes. Lighting can be used to enhance the décor of a building and its furnishings, to help people perform tasks more easily, and to make people feel safer and more comfortable inside and outside.

Designers determine light needs by using suggested footcandle levels for different activities and areas. Candlepower measurements are independent of any object or surface that is being lit.

3.0.0 ◆ CLASSIFICATION OF LIGHTING FIXTURES

In the previous modules covering electric lighting, lighting fixtures (luminaires) were classified by their methods of installation, such as surface-mounted, recessed, or track-mounted. When describing lighting schemes and the application of lighting fixtures, other ways for classifying lighting fixtures are also used. These include classification by layout and location, luminaire type, and type of service. Definitions of these classifications are given here so that you will understand them when they are used in this module or in lighting-related product literature.

3.1.0 Layout and Location

There are several ways to classify lighting schemes by layout and location. These include:

- *General lighting* – Lighting designed to provide a uniform level of illumination for large rooms, entrances, corridors, lobbies, etc. See *Figure 3(A)*.

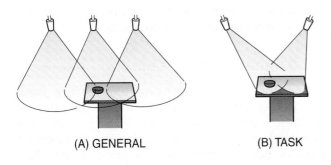

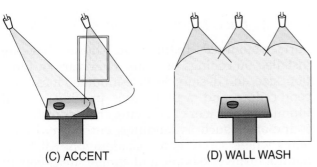

Figure 3 ◆ Classification of lighting by layout and location.

303F03.EPS

- *Localized general lighting* – Lighting that uses fixtures above the visual task area that also contribute to the lighting in the local area.
- *Local lighting* – Lighting that provides illuminance over a small area or confined space without providing any significant general lighting in the local area.
- *Task lighting* – Lighting that directs well-focused illumination to a specific surface or area such as a work surface, desk, or counter. See *Figure 3(B)*.
- *Accent (directional) lighting* – Lighting used to point out the best selling features of merchandise, pinpoint high-key decor, emphasize artwork, or make a room look larger by adding contrast and shadows. See *Figure 3(C)*.
- *Wall wash* – Lighting that can illuminate vertical surfaces from the ceiling to the floor. It can be used to expand space visually, define a space, light a wall of pictures, or emphasize color or texture. See *Figure 3(D)*.
- *Egress lighting* – Lighting that provides the minimum light levels needed to assist people when exiting a building.

3.2.0 Luminaire Types

Lighting fixtures for general lighting are classified by the Commission International de l'Eclairage (CIE) by the percentages of total light output emitted above and below the horizontal plane. The illumination patterns may take many forms within the limits of upward and downward distribution, depending on the type of light source and the design of the lighting fixture. The different classifications include:

- *Direct lighting* – Lighting by fixtures that distribute 90% to 100% of the emitted light below the horizontal plane. See *Figure 4(A)*.
- *Semi-direct lighting* – Lighting by fixtures that distribute 60% to 90% of the emitted light downward, but also distribute a small component (10% to 40%) upward to illuminate the ceiling and upper walls. See *Figure 4(B)*.
- *General diffuse lighting* – Lighting by fixtures that distribute 40% to 60% of the emitted light downward and the balance upward, sometimes with a strong component at the horizontal plane. See *Figure 4(C)*.
- *Direct-indirect lighting* – Lighting by fixtures that distribute 40% to 60% of the emitted light downward and the balance upward. Unlike the general diffuse type, very little light is emitted at horizontal angles. See *Figure 4(D)*.

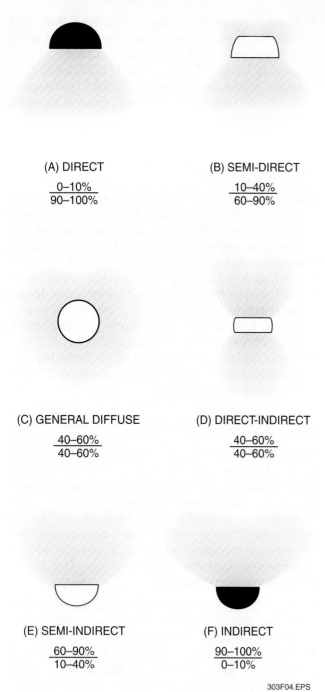

(A) DIRECT

$\dfrac{0–10\%}{90–100\%}$

(B) SEMI-DIRECT

$\dfrac{10–40\%}{60–90\%}$

(C) GENERAL DIFFUSE

$\dfrac{40–60\%}{40–60\%}$

(D) DIRECT-INDIRECT

$\dfrac{40–60\%}{40–60\%}$

(E) SEMI-INDIRECT

$\dfrac{60–90\%}{10–40\%}$

(F) INDIRECT

$\dfrac{90–100\%}{0–10\%}$

303F04.EPS

Figure 4 ◆ CIE classifications of lighting fixtures.

- *Semi-indirect lighting* – Lighting by fixtures that distribute 60% to 90% of the emitted light upward. See *Figure 4(E)*. The downward component usually produces a luminance that closely matches that of the ceiling.
- *Indirect lighting* – Lighting by fixtures that distribute 90% to 100% of the emitted light upward. See *Figure 4(F)*.

3.3.0 Type of Service

There are five categories used to classify lighting fixtures by service. These categories are as follows:

- *Industrial service* – Used in factories and manufacturing facilities
- *Commercial service* – Used in offices, stores, and public buildings
- *Residential service* – Used in houses and apartments
- *Street lighting* – Lighting of streets, roads, and highways
- *Security/floodlighting* – Lighting of outdoor areas such as recreational grounds, parking lots and other spaces, and building exteriors

4.0.0 ◆ PRACTICAL APPLICATIONS OF LIGHTING FIXTURES

Lighting fixtures are designed to meet a specific application. The type of construction, aesthetics, finish, lighting distribution, glare, heat control, size, and cost are only a few of the characteristics that yield a wide variety of fixtures.

Lighting fixture manufacturers describe each of their lighting products in their catalogs and give recommended applications for their use. When selecting a lighting fixture for a specific application, consult these catalogs and follow the manufacturer's recommendations. This section provides an overview of common types of fixtures and their typical applications.

4.1.0 Incandescent Lighting Fixtures

Today, incandescent lighting fixtures are used mainly for indoor residential and commercial lighting applications. They can be surface-mounted, recessed, suspended, or track-mounted fixtures. Some fixtures are designed so that they can be used in either a surface-mounted or suspended application.

4.1.1 Surface-Mounted Lighting Fixtures

Surface-mounted fixtures are those that are installed on ceilings or walls. They vary in function, size, and design. Surface-mounted incandescent ceiling and wall fixtures provide general illumination in work and living areas as well as in traffic areas such as landings, entries, and hallways where safety is a consideration. Kitchens, bathrooms, workshops, and similar areas benefit from the added light provided by ceiling fixtures used in conjunction with task lighting on work surfaces.

Portable Lamps

The focus of this module is on the various types of permanently installed lighting fixtures used to provide indoor and outdoor lighting. However, lighting designers and interior decorators also make extensive use of decorative portable lamps (plug-in lamps) as part of a building's lighting scheme in order to meet both lighting and decorating requirements. This is especially true in residential and professional office applications. Plug-in lamps include floor lamps, table lamps, desk lamps, task lamps, spotlights, picture lamps, and curio and kitchen cabinet lamps.

Ceiling-mounted fixtures equipped with **diffusers** (*Figure 5*) direct their light in a wide pattern and are used for general **diffused lighting**. Normally, the lamp is not visible through the fixture. The position of these fixtures allows them to produce more light than similar fixtures hung lower into the line of sight. Ceiling-mounted directional fixtures, called downlights, are typically used to provide accent lighting or task illumination.

Incandescent wall fixtures (*Figure 6*) are used to provide either diffused or directional lighting. Some have exposed clear lamps for decorative purposes. Wall fixtures designed as downlights are typically used to provide accent and/or display lighting. Those designed as uplights are used to provide indirect light. Incandescent wall sconces are commonly used in hallways and stairways or as accent lights in a room to match larger fixtures. By design, they cast their light toward the wall, diffuse light into a room, or direct light toward the ceiling.

4.1.2 Recessed Ceiling Lighting Fixtures

Recessed incandescent fixtures (downlights) offer light without the intrusion of a visible fixture

DIFFUSED FIXTURES

DOWNLIGHTS

303F05.EPS

Figure 5 ◆ Incandescent ceiling-mounted fixtures.

303F06.EPS

Figure 6 ◆ Incandescent wall-mounted fixtures.

Recessed Lighting

Recessed fixtures are widely used for general lighting as well as for task and accent lighting. When used for general lighting, downlights are typically spaced evenly in groups of four or more at intervals of 6' to 8' apart. In kitchens or other work areas, placing downlights above the front edge of the countertop will effectively light the workspace.

(*Figure 7*). For this reason, they are widely used in rooms with low ceilings. They can provide a variety of lighting effects including general lighting, accent or task lighting, and wall washing.

A recessed fixture is basically a domed reflector with a lamp set in the top (*Figure 8*). Depending on design, the shape of the reflector can be straight (can), cone-shaped, ellipsoidal, or parabolic. Some are adjustable to allow for precise positioning of the light beam. The most obvious adjustable fixtures are the eyeball types with a partially recessed sphere that can be rotated and tilted to direct light for wall washing and accent lighting purposes.

Recessed downlight fixtures can accept several different trims to achieve various lighting effects. This allows the trims to be changed with others produced by the same manufacturer should the lighting needs change. A flange trim can be a self-trim attached to or adjacent to the reflector, or it can be a plastic or metal trim ring that covers and seals the edges of the hole in the ceiling for round fixtures.

When used over sinks, countertops, desks, workstations, etc., an open downlight spreads a strong task light over the work surfaces. Open downlight fixtures are also used for lighting stairways and entries. Equipped as a wall washer fixture, a recessed downlight throws light onto a nearby wall. A series of such fixtures can be used to produce even, balanced lighting on a wall of artwork or bookcases. Low-voltage (12V) downlights, especially those that use MR-16 tungsten halogen lamps and black baffles, are widely used for accent lighting.

Because glare can be a problem with downlighting, reflective interior finishes, baffles, lenses or diffusers, and/or louvers are often incorporated into downlights to direct the light away from sight lines. In general, fixtures with deeply inset lamps will produce less glare. Baffles are a series of light-absorbing ridges that ring the lower portion of the fixture interior, shielding the lamp's brightness when viewed at an angle. Lenses or diffusers are glass or plastic coverings that are

SQUARE LENS   REFLECTOR CONE

BAFFLE   EYEBALL

PULL-DOWN ADJUSTABLE

(A) STANDARD DOWNLIGHTS

ARCHITECTURAL   ADJUSTABLE

PINHOLE APERTURE

(B) LOW-VOLTAGE DOWNLIGHTS

303F07.EPS

Figure 7 ◆ Recessed incandescent downlights (below ceiling view).

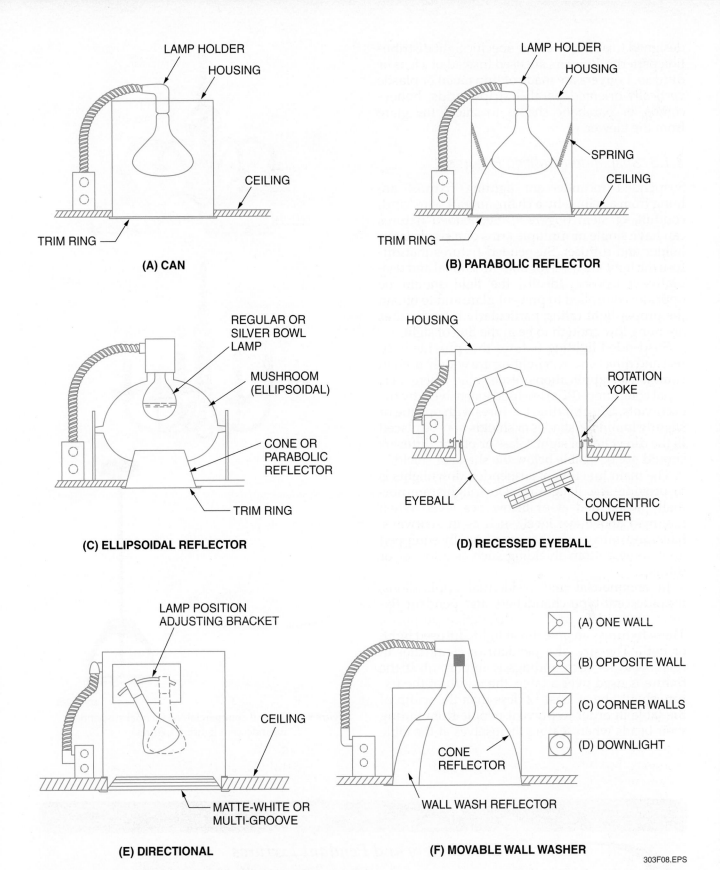

Figure 8 ◆ Recessed incandescent downlights (outline views).

303F08.EPS

designed to optically create specific light distribution patterns. Louvers are used instead of a lens or diffuser. Louvers are made of thin metal or plastic vertically-oriented parallel strips, grids, honeycombs, or parabolic shapes to shield the glare from the viewer.

4.1.3 Suspended Lighting Fixtures

Suspended incandescent lighting fixtures are hung from a ceiling by a chain, aircraft wire, cord, conduit, or stem (*Figure 9*). Suspended fixtures can have single or multiple arms or a single lamp holder and diffuser. Suspended fixtures that diffuse the light are used for both functional and decorative purposes. Ideally, the light should be optically controlled to prevent glare and to obtain the proper light ratios, particularly if the fixtures are hung low enough to be in the line of sight.

Suspended lighting fixtures that provide indirect and direct-indirect lighting are widely used in commercial applications, especially where personal computers (PCs) are used extensively by the occupants. Direct-indirect fixtures with opaque or slightly luminous shades or shields may be placed in the direct line of sight. Glass or plastic diffusers should not be visible below the shade or shield.

The main function of suspended downlights is to provide accent lighting. Downlights are normally suspended at or below eye level. When mounted above eye level, such as in stairways, halls, and entryways, they are normally equipped with some bottom shielding such as a louver or diffuser.

In commercial and residential applications, incandescent-type chandeliers and pendant fixtures are widely used for decorative purposes. These fixtures can give direct light, diffused light, or both. The size of a pendant or chandelier in relation to its surroundings is important. If the fixture is used over a table, the width of the fixture should be at least 12" less than the width of the table in order to prevent people from hitting their heads when seating themselves at the table

303F09.EPS

Figure 9 ◆ Typical commercial/residential suspended incandescent lighting fixtures.

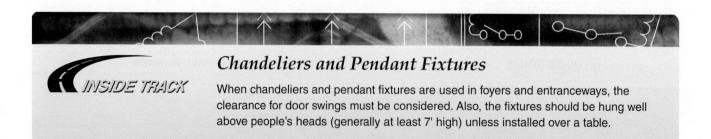

Chandeliers and Pendant Fixtures

INSIDE TRACK

When chandeliers and pendant fixtures are used in foyers and entranceways, the clearance for door swings must be considered. Also, the fixtures should be hung well above people's heads (generally at least 7' high) unless installed over a table.

and to prevent collisions with the fixture by those who pass by. For rooms with eight-foot ceilings, the fixtures are normally hung from 30" to 32" above the table to help prevent glare. For ceilings higher than 8', the fixture may look best when hung higher than 32". Suspended incandescent fixtures equipped with enamel steel reflectors (*Figure 10*) are made in several shapes for commercial applications.

4.1.4 Track-Mounted Incandescent Lighting Fixtures

Track lighting allows light to be directed where needed to achieve a variety of effects: general lighting, accent lighting, task lighting, and wall washing. A variety of track-mounted lighting fixtures are made. By design, most have a modern look (*Figure 11*). Track fixtures rotate 358° and adjust horizontally 90° to provide precise, effective lighting control. Low-voltage (12V) fixtures designed for use with MR-16 and PAR-36 tungsten halogen lamps are also made. Some of these fixtures snap into a low-voltage transformer, which locks into the track section. Others have a low-voltage transformer built into the fixture.

FLAT CONE   SHALLOW BOWL

RLM DOME   DEEP BOWL

(A) ENAMEL STEEL REFLECTORS

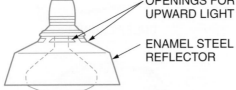

OPENINGS FOR UPWARD LIGHT

ENAMEL STEEL REFLECTOR

(B) GLASS AND STEEL DIFFUSER FIXTURE

303F10.EPS

Figure 10 ◆ Incandescent suspended lighting fixtures.

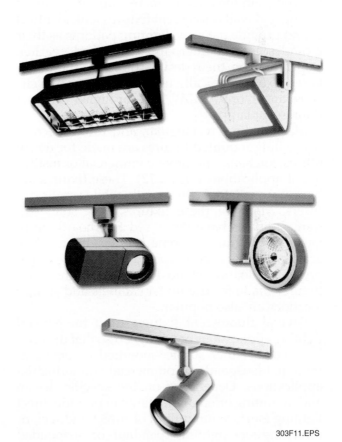

303F11.EPS

Figure 11 ◆ Typical incandescent track lighting fixtures.

Track Lighting

Track lighting is used for accent or task lighting intended to light a specific area or object with a pool of light. To light a specific object, the track fixture should be positioned so its light is at a 60° angle from the horizontal surface of the object. Track lighting should be installed 2' to 3' from the wall. Installing track lighting in the center of a room will cause glare and shadows on the task area.

4.2.0 Fluorescent Lighting Fixtures

Fluorescent lighting fixtures are generally used in indoor surface-mounted, suspended, or recessed applications. They provide for general lighting, task lighting, or accent lighting.

4.2.1 Surface-Mounted and Suspended Lighting Fixtures

Viewed from the outside, many surface-mounted ceiling and wall fluorescent fixtures look identical to and provide the same lighting patterns as their incandescent counterparts described earlier. Depending on their design and size, these ballasted fixtures typically use either one or two circline lamps, or 9W, 13W, 26W, or higher compact fluorescent lamps.

Many square, rectangular, and round decorative ceiling-mounted fixtures are made for use in offices, kitchens, and other commercial or institutional applications (*Figure 12*). These fixtures are typically available with smooth, white acrylic diffusers. Many of these fixtures have solid oak frames; others have processed fiberboard simulated wood frames. Some models have white acrylic diffusers with tapered round edges that present a floating cloud look. Surface-mounted fixtures made for use under cabinets and similar locations are also popular.

Several fluorescent fixtures used for general lighting applications are designed so that they can be used in either surface-mounted or pendant-mounted (suspended) commercial and industrial applications. Depending on the specific design and mounting orientation, they can provide direct lighting (surface mounting) or direct, indirect, or direct-indirect lighting (pendant or suspended mounting). These lighting fixtures include:

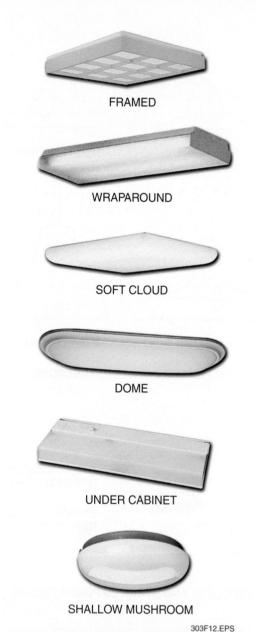

FRAMED

WRAPAROUND

SOFT CLOUD

DOME

UNDER CABINET

SHALLOW MUSHROOM

303F12.EPS

Figure 12 ◆ Decorative/functional fluorescent surface-mounted lighting fixtures.

- *Strip or channel fixtures* – Strip or channel fixtures, shown in *Figure 13(A)*, are used in areas in which a high amount of light is needed and aesthetics are not important, such as in storage rooms, warehouses, etc. They are made in single- and multi-lamp assemblies of various lengths. When used in continuous-row installations, the fixtures are connected together using channel connectors. These fixtures can also be used to produce aesthetic effects like indirect lighting in coves, cornices, and above draperies, where they are wall-mounted behind a decorative, opaque plate. The result is a glare-free, indirect wall light throughout its length, except for shadows that are produced at the areas in which the ends meet.
- *Staggered strips* – Staggered strips are similar to the strip or channel fixtures described above. However, the lamps are staggered so that the light from at least one lamp is always available at each joint to help eliminate shadows.
- *Wraparound fixtures* – Wraparound fixtures, shown in *Figure 13(B)*, are low-profile, two-, three-, or four-lamp fixtures commonly used when horizontal and vertical light from aesthetically pleasing, moderately bright, area-type sources is required. The narrow profile is well suited to areas such as corridors, stairwells, hallways, and between aisles.
- *Task area fluorescent fixtures* – Task area fluorescent fixtures, shown in *Figure 13(C)*, are generally used where low-mounted task lighting is needed in work stations, storage areas, and utility rooms. Typically, they are two-, three-, or four-lamp fixtures made in 4' and 8' lengths. Fixtures are connected together using channel connectors when used in continuous-row installations. All fixtures have reflectors that reflect light for more efficient illumination as well as to provide protection for the lamps. The reflectors can be of the non-apertured or apertured type. Non-apertured reflectors are solid reflectors designed to reflect 100% of the light directly downward toward the work surface. Apertured reflectors have slots cut into their tops, allowing about 10% to 20% of the light to be directed upward toward the ceiling. This uplighting helps improve the overall lighting in the work area, especially when used with a highly-reflective ceiling. The slots also allow air to circulate through the reflector, keeping the fixture cooler and cleaner than with a non-apertured type.

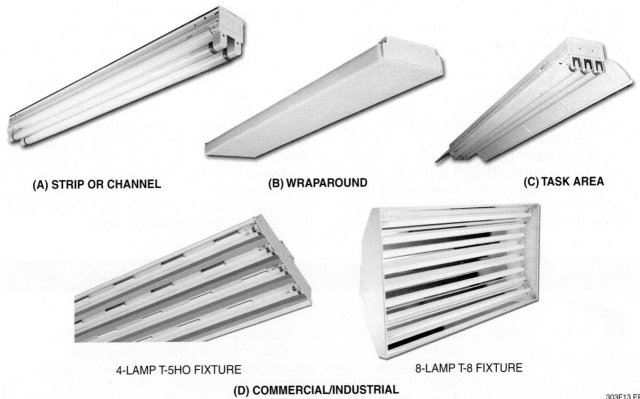

(A) STRIP OR CHANNEL   (B) WRAPAROUND   (C) TASK AREA

4-LAMP T-5HO FIXTURE   8-LAMP T-8 FIXTURE

(D) COMMERCIAL/INDUSTRIAL

303F13.EPS

Figure 13 ◆ Surface/suspension-mounted fluorescent lighting fixtures.

- *Commercial/industrial fluorescent fixtures* – Commercial/industrial linear fluorescent fixtures, shown in *Figure 13(D)*, are typically used as general downlighting in low-bay and high-bay applications that include factories, schools, warehouses, and retail stores. Because of advances in high light output and a variety of color temperatures for fluorescent tubes, fluorescent fixtures are currently being used in place of HID lighting to reduce energy consumption by about 50% and to improve color rendition. Light output from an HID source also degrades more rapidly than from a fluorescent source. Low-bay applications are usually defined as fixture heights of 20' or less and high-bay applications as fixture heights of over 20'. As a general rule, low-bay applications use standard linear T-8 tubes because they produce less glare and a better quality of light. High-bay applications generally use smaller diameter linear T-5HO (high output) tubes that allow more focused downlighting. Linear tubes are normally used because they provide a longer useful life than other tube configurations. For fixture heights between 18' and 25', either high-lumen extended-performance T-8 tubes or T-5HO tubes can be used. Commercial/industrial T-8 tube fixtures are normally available in two- to eight-tube configurations as 4' suspended, modular, or recessed units. The eight-tube 4' fixture with standard T-8 tubes shown in *Figure 13(D)* is equipped with a high-frequency electronic ballast to eliminate 60Hz flicker and can be obtained wired as a switch-controlled two-level lighting fixture (four tubes or eight tubes on). A six-tube unit with special high-lumen, extended-performance T-8s will replace a 400W HID lamp for high-bay applications. The high-bay four-tube 4' T-5HO fixture, also shown in *Figure 13(D)*, has nearly the same light output as a 400W HID lamp while consuming only 216W of power. These T-5HO fixtures typically use instant-strike tubes along with programmed-start electronic ballasts that reduce inrush current, extend tube life, and allow numerous on-off switches to conserve energy. Some units are equipped with motion sensors for switching. Dimmers can also be used with T-5HO tubes to control lighting levels. In contrast, HID lighting is usually left on because of the restrike time required. Even though linear T-5HO tubes are only available in 2' or 4' lengths, the T-5HO fixtures are available in 4' or 8' lengths with two- to twelve-tube configurations. They are mounted like HID high-bay fixtures described later, or as bracket-mounted units.

- *Vaportight* – A vaportight fixture is an enclosed and gasketed fixture used in areas where dust and/or moisture are present or where sanitation is important, such as in kitchens, food processing plants, and industrial and manufacturing environments.

NOTE

Fluorescent fixtures in commercial/industrial locations that utilize double-ended lamps and contain ballasts that can be serviced in place require an accessible disconnect. See *NEC Section 410.130(G)* for details and exceptions.

4.2.2 Recessed Fluorescent Lighting Fixtures

Recessed fluorescent troffers (*Figure 14*) are widely used to provide uniform general lighting and good, even lighting over specific work/task areas in office buildings and other commercial applications. These square (2' × 2') or rectangular (1' × 4' and 2' × 4') fixtures can be installed in the recesses above plaster or drywall ceilings, or they may be installed in suspended grid-type ceilings. Models using up to four lamps are available (T-5, T-8, T-10, or T-12). Some fluorescent troffers are equipped with large-cell parabolic louvers that provide sharper-cutoff, lower-brightness lighting. These units usually produce the same or higher light output than lens bottom units.

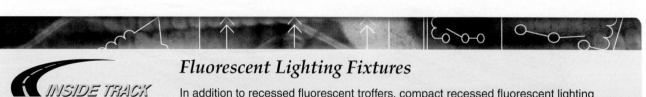

Fluorescent Lighting Fixtures

In addition to recessed fluorescent troffers, compact recessed fluorescent lighting fixtures are also available. They are similar in construction to the incandescent recessed fixtures described earlier and are widely used in downlighting applications, such as in corridors, lobbies, stores, and other public places. Depending on their design, these fixtures typically use lamps of 9W and above.

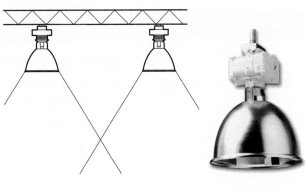

HIGH-BAY

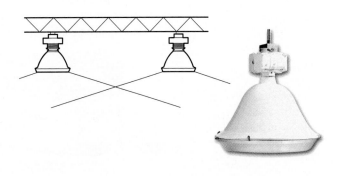

LOW-BAY

303F15.EPS

Figure 15 ◆ High-bay and low-bay HID fixtures.

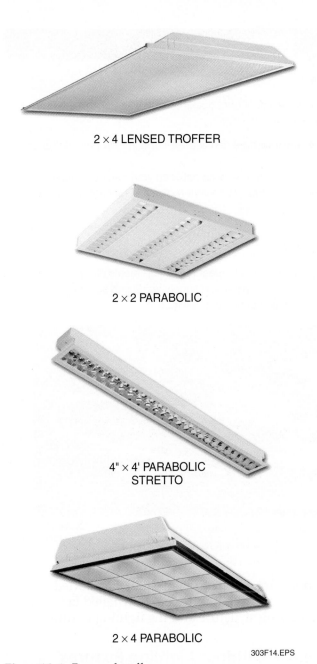

2 × 4 LENSED TROFFER

2 × 2 PARABOLIC

4" × 4' PARABOLIC
STRETTO

2 × 4 PARABOLIC

303F14.EPS

Figure 14 ◆ Recessed troffers.

4.3.0 High-Intensity Discharge (HID) Lighting Fixtures

Decorative HID indoor lighting fixtures that use metal halide lamps are made in many of the same styles and designs as described earlier for similar incandescent and compact fluorescent-type fixtures. These ballasted fixtures, like the other types of fixtures, are used in surface-mounted, suspended, or recessed commercial or residential applications.

High-bay and low-bay HID lighting fixtures (*Figure 15*) are used extensively in industrial applications. These fixtures generally use metal

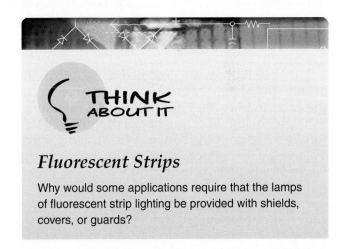

THINK ABOUT IT

Fluorescent Strips

Why would some applications require that the lamps of fluorescent strip lighting be provided with shields, covers, or guards?

halide or high-pressure sodium lamps. High-bay fixtures are designed to deliver the maximum amount of light onto the work plane below from mounting heights above 20'. For use at these higher mounting heights, high-bay fixtures have reflectors that produce a concentrated light beam in order to produce effective horizontal illumination. If mounted below 20', high-bay fixtures can cause glare, eye fatigue, harsh shadows, and dark spots on the work plane.

Color Temperatures

The color temperature is one way manufacturers designate the color tone of a lamp. Color temperatures fall into three main categories:

- Color temperatures of 3,000k and lower are considered warm in tone (they enhance reds and yellows).
- A color temperature of 3,500k provides a balance between warm and cool tones.
- Color temperatures of 4,100k and higher are considered cool in tone (they enhance blues and greens).

Fluorescent Disconnects

The disconnect shown here can be installed to meet the requirements of *NEC Section 410.130(G)*. It is connected between the ballast and supply leads and provides a safe and convenient method of disconnecting power at the fixture.

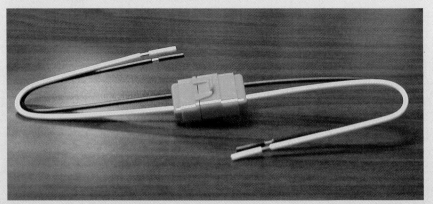

303SA01.EPS

Low-bay fixtures are designed to deliver more evenly spread light onto the work plane below. As the name implies, they are designed for mounting at heights below 20'. The reflector of a low-bay fixture has a lens on the bottom to diffuse the light. This helps to reduce glare, while providing a wider beam pattern to accommodate the lower mounting height.

HID high-bay and low-bay fixtures are typically surface-mounted or pendant-mounted. Some are suspended from and connected to the building's lighting distribution system by power hook assemblies. A power hook assembly consists of a hook with an attached power cord and plug that is connected between the lighting fixture and related power receptacle in the lighting power distribution system. The hook, via its attachment to a loop on the receptacle outlet box cover, supports the suspended fixture. For added protection, safety chains are often installed between the building structure and the lighting fixture.

4.4.0 Outdoor Lighting Fixtures

Because of their efficiency and long life, HID fixtures are used extensively for outdoor lighting applications. Today, most of these fixtures use either high-pressure sodium lamps or metal halide lamps. A few fixtures equipped with mercury vapor or low-pressure sodium lamps are still used for parking lots and street lighting. Some residential and small commercial installations may use incandescent or compact fluorescent lamps. Viewed from the outside, these fixtures look similar to their HID counterparts.

HID outdoor ceiling-mounted and wall-mounted fixtures (*Figure 16*) are widely used to illuminate building entranceways, parking areas, and lots next to buildings and parking garages. A wall-mounted fixture, commonly called a wall pack, has an uneven (asymmetrical) distribution to light nearby areas. Fixtures installed on the walls or ceilings inside parking garages or similar structures are designed to produce high levels of inter-reflected light in the structure.

Pole-mounted lighting fixtures (*Figure 17*) are used for roadway and parking lot lighting. These fixtures have wide light distribution patterns in order to permit greater spacing between poles. Fixtures with oval-shaped reflectors mounted on long arms are commonly used for roadway applications. Because of their appearance, they are often referred to as cobra head fixtures. Parking lot lighting is often accomplished using fixtures with flat-bottomed lenses that are mounted on short arms. These can be mounted in single, twin, triple, or quad arrangements. Small fixtures mounted on short poles can also be used to provide walkway and grounds lighting.

Walkway and grounds lighting is frequently done using localized lighting supplied by bollard fixtures. As shown in *Figure 17*, this type of fixture is shaped like a short, thick post with a lamp installed at the top.

In addition to the usual applications for sporting event lighting, floodlight or spotlight fixtures are commonly used to light the outsides (facades) of buildings. Note that a spotlight is a form of floodlight, usually equipped with lenses and reflectors to give a fixed or adjustable narrow beam. Building facade lighting uses fixtures with either narrow or wide distribution patterns. The pattern used depends on the portion of the building to be lighted and how far away the lighting fixture is located from the building. Lighting large areas with a fixture mounted near the building requires the use of a very wide distribution pattern. On the other hand, column lighting or accent lighting using a fixture located at a distance from the building requires the use of fixtures with narrow distribution patterns.

POLE-MOUNTED

BOLLARD

FLOODLIGHT

STEP

LANDSCAPE

303F17.EPS

Figure 17 ◆ Pole-mounted and other outdoor fixtures.

WALL MOUNT

PARKING GARAGE

303F16.EPS

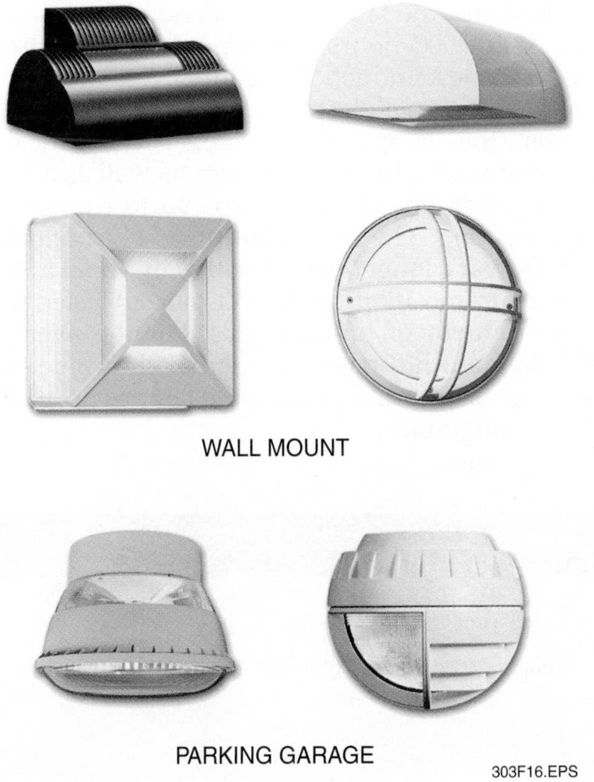

Figure 16 ◆ Typical HID outdoor surface-mounted fixtures.

Another type of outside fixture frequently used is a step light. These are used to provide safety lighting and to accent obstacles around steps, corridors, walkways, and entrances. Step fixtures are normally mounted in brick, concrete, or other commonly used building materials.

4.5.0 Emergency and Exit Lighting Fixtures

Emergency lights are designed to provide egress (departure) lighting in critical areas during power failures. Although local codes may be more specific, emergency lights should provide illumination in halls, stairways, and other public areas. Some installations may require a remote lighting head at each exit. Similarly, exit signs are designed to provide egress marking to help direct occupants from public areas of buildings. Local codes may vary, but exit signs should provide directional information in halls, stairways, and corridors.

Emergency and exit lighting fixtures (*Figure 18*) can either be separate fixtures or they can be combined in one assembly. Normally, this type of fixture consists of a self-contained rechargeable battery, a solid-state battery charger, a battery status indicator, a transfer device, a test switch and pilot lamp, and provisions for either integral or remote lamps or both. This type of fixture charges a battery when normal AC power is applied and transfers battery power to the emergency light source during a power outage. Exit fixture circuitry operates so that the lamps remain on during both normal and emergency situations. Emergency lighting fixtures are designed to turn on only during power failures or when being tested. Depending on the design of the emergency lighting fixture, halogen or incandescent sealed beam-type lamps are frequently used. A variety of low-voltage lamp and reflector combinations are also commonly used. Exit signs typically use incandescent or compact fluorescent lamps or light-emitting diode (LED) lamp modules.

303F18.EPS

Figure 18 ◆ Typical emergency and exit lighting fixtures.

4.6.0 Induction Lighting Systems

Induction lighting systems are basically a form of fluorescent lighting that does not use electrodes. The electrodes used inside standard fluorescent and HID lamps are subject to disintegration and breakage over time. This normally results in lamp failure at an average of 10,000 to 20,000 hours, depending on the number of on-off cycles. Because electrodes are not used in induction lighting systems, the lamp life is extended to an average of 100,000 hours when used with minimal on-off cycles. Induction lighting systems use a separate external electronic system to excite and power the lamp. This results in drastically reduced relamping maintenance as well as a 30% reduction in energy use compared to equivalent HID lighting. The sealed lamp vessel is actually usable over a longer period with only about a 30% loss of luminosity provided it is not subjected to a large number of on-off cycles. For the average 100,000-hour lifespan, the external electronic system that excites and powers the lamp is normally the item that fails.

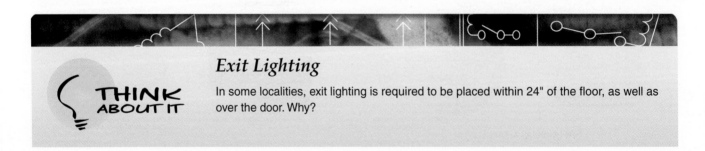

Exit Lighting

THINK ABOUT IT

In some localities, exit lighting is required to be placed within 24" of the floor, as well as over the door. Why?

Mounting Plates

This special induction fixture mounting plate eliminates the need to open the fixture during installation. The fixture is shipped complete with the lamp installed. To install the fixture, just mount the plate to the splice box, pull the wires through the access hole, and hang the fixture on the hook provided.

303SA02.EPS

Regardless of the shape or wattage of induction lighting systems, they all operate on the same principle. *Figures 19* and *20* show two popular types—the Philips QL system and the Sylvania ICETRON® system. The systems are made up of three parts: a sealed glass vessel containing a low-pressure mercury-vapor discharge gas and a phosphor coating, ferrite-core induction coil(s), and a high-frequency (HF) generator that powers the lamp. The HF energy sent to the ferrite-core induction coil(s) by the HF generator through a coaxial cable creates a rapidly alternating magnetic field that extends into the sealed vessel and causes current flow through the discharge gas. Like a standard fluorescent lamp, this excites the mercury atoms in the discharge gas that, in turn, radiate UV rays into a phosphor coating on the inside surface of the vessel, causing it to glow. A mercury amalgam within the sealed vessels stabilizes the mercury vapor pressure in the vessel over a wide operating temperature range (−30°F to 130°F) with starting temperatures as low as −40°F. Both systems produce a crisp white light.

The QL system has a choice of either a 3,000k or 4,000k color temperature and the ICETRON® system has a choice of a 3,500k, 4,100k, or 5,000k color temperature. In the QL system, any one of the three parts can be replaced individually. In the ICETRON® system, the sealed vessel with the induction coils or the HF generator can be replaced. Proper heatsinking of the HF generator in both systems is critical, as is heatsinking of the induction coil base in the QL system. This requires fixtures that are specially designed for these systems.

Typically, these systems are used for the following purposes to achieve operation in low temperatures with very low maintenance costs:

- Hazardous areas
- Freezer rooms
- Hard-to-reach applications
- High-bay lighting applications
- Street and parking lot lighting
- Bridges, tunnels, and parking garages

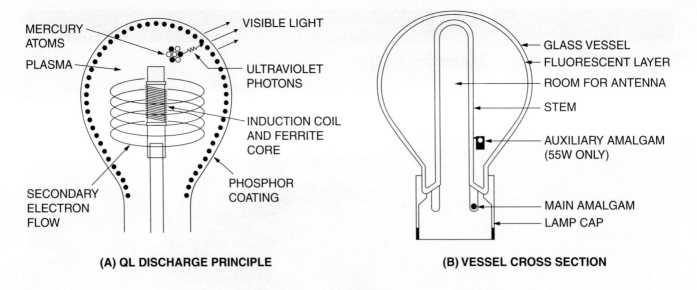

(A) QL DISCHARGE PRINCIPLE

(B) VESSEL CROSS SECTION

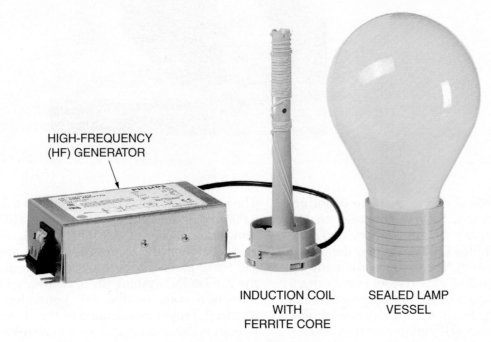

(C) QL SYSTEM COMPONENTS

303F19.EPS

Figure 19 ◆ QL induction lighting system.

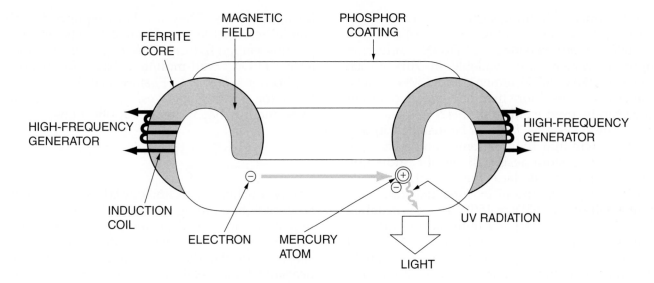

(A) ICETRON® DISCHARGE PRINCIPLE

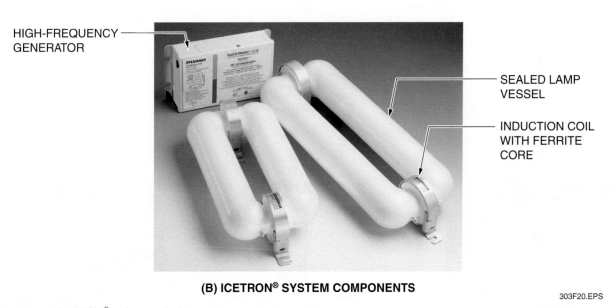

(B) ICETRON® SYSTEM COMPONENTS

303F20.EPS

Figure 20 ◆ ICETRON® induction lighting system.

QL induction lighting systems are available in 55W (3,500 lumen), 85W (6,000 lumen) or 165W (12,000 lumen) versions. ICETRON® induction lighting systems are available in 70W (6,500 lumen), 100W (8,500 lumen), and 150W (11,000 lumen) versions. More than one system of either type can be used inside a properly designed fixture. These systems can also be installed in any position within a fixture. Some companies specialize in custom-designed retrofit induction lighting kits to replace fluorescent, HID, or sodium vapor lamps in existing fixtures. Note that a pair of the 165W or 150W systems installed in a fixture can provide a light output approximately equal to a 400W HID lamp. *Figure 21* shows some of the applications listed previously. As indicated, the fixtures in *Figure 21(A)* and *21(B)* show ICETRON® suspended and post-top fixtures suitable for street, walkway, and parking lot applications. They are both equipped with vandal-resistant lenses. *Figure 21(C)* shows a QL parking garage fixture, while *Figure 21(D)* shows an ICETRON® wall-mounted fixture for exterior applications. *Figure 21(E)* shows QL low-bay fixtures for parking garages, certain commercial uses, and low-bay industrial applications.

4.7.0 Light-Emitting Diode (LED) Technology

This type of lighting technology is based on various types of light-emitting diodes (LEDs) that are continually being refined or developed to provide a high-intensity (brightness) output. When the substrate of the diode circuit chip is properly heatsinked, some of the latest ultra high-brightness LEDs can output light with an illumination efficiency of 100 lumens (lm) per watt at 350mA.

(A) ICETRON® STREET, WALKWAY, OR PARKING LOT FIXTURE

(B) ICETRON® POST-TOP FIXTURE

(C) QL PARKING GARAGE FIXTURE

(D) ICETRON® WALL-MOUNTED FIXTURE

(E) QL LOW-BAY FIXTURES

303F21.EPS

Figure 21 ◆ QL and ICETRON® induction lighting system applications.

LED Lighting

GOING GREEN

Energy-efficient LED lamps provide a mile-long beacon of color on the Vincent Thomas Bridge in Los Angeles Harbor, the busiest port in the United States.

303SA03A.EPS

This hotel in Las Vegas, Nevada features an exact half-scale replica of the Eiffel Tower in Paris, France. Like the original, it is lit with energy-efficient LED lighting modules.

303SA03B.EPS

Various types of semiconductor materials are used to manufacture LEDs that emit one specific wavelength of energy such as ultraviolet (UV), near UV, infrared, or a particular shade of red, orange, blue, green, or yellow light. Currently, high-brightness white light is achieved by either mixing red, green, and blue (RGB) LED light together or by coating an LED that emits blue or UV light with an appropriate phosphor that glows white. Unlike an incandescent bulb filament that generates and radiates a great deal of heat as well as light, the energy emitted from a visible-light LED is often called cold light because very little or no heat is generated and emitted along with the light. This makes LEDs very energy efficient. All LEDs are instant-on and can operate at very low temperatures. If only the inherent wavelength emitted by the LED is used for a particular application with proper diode heatsinking, the average lifespan of the LED is roughly 100,000 hours. Current white-light LEDs are created by coating the LED with a phosphor that has a lifespan of 80,000 hours or more before falling below 70% luminosity due to phosphor degradation.

The low-voltage power supplies required for LEDs are called drivers. They can be for a single LED or for an array of LEDs connected in series.

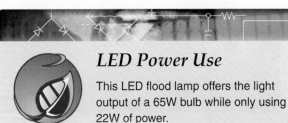

LED Power Use

GOING GREEN

This LED flood lamp offers the light output of a 65W bulb while only using 22W of power.

303SA04.EPS

They function much like a ballast for a fluorescent or HID lamp. The driver supplies the proper starting voltage and regulates the current flowing through LEDs after startup. Drivers are available with a dimming capability that allows the LEDs to be dimmed from 0% to 100% with only a minimal color shift in the light output. This is usually accomplished by pulse width modulation (PWM) of the low-voltage DC at high frequencies and does not affect efficiency. Unlike dimmable fluorescent lamps, dimming can increase LED life by reducing the operating temperature of the LED substrate.

High-brightness LEDs are now being used for traffic lights, exit and emergency lighting, step lighting, and very large direct-emission RGB video displays (*Figure* 22). They are also used for accent lighting, sign lighting, light bars on emergency vehicles, flashlights, vehicle taillights, Christmas tree lights, rope lights, backlighting for LCD television sets, infrared remote controls, optical-fiber communications, and a host of other applications.

4.7.1 Architectural LED Lighting

Color changing RGB LEDs and fixtures can be used for architectural or special effects lighting such as wall-wash or accent lighting in homes, restaurants, bars, and other retail establishments. This is accomplished by using dimming drivers for color changing or sequencing. One method is to just dim a mix of RGB LEDs in an array. Another method is to use the driver in conjunction with a color sequencer, which receives the driver output and converts it into a controlled three-channel output that can be applied in the

(A) LARGE RGB VIDEO DISPLAY

(B) TRAFFIC SIGNAL

(C) EXIT AND EMERGENCY
LIGHTING

(D) PATH AND STAIR
STEP LIGHTING

303F22.EPS

Figure 22 ◆ Typical high-brightness LED applications.

correct ratio to the RGB LEDs to obtain a wide, dynamically changing range of colors. The sequencer is programmed to control color changes at a speed determined by the user. A third method involves the use of a digital controller using DMX and digital addressable lighting interface (DALI) protocols to dynamically dim up or down thousands of LEDs individually to create a large spectrum of colors.

Other types of self-contained color-changing RGB lamps are available for individual use in existing fixtures and do not require a digital controller. *Figure 23* shows some of these lamps. The RGB lamps shown in *Figure 23(A)* are 12V or 120VAC indoor/outdoor units that can also be used in pool lights. The smaller one is a 72-LED unit and the larger one is a 202-LED unit. Both have seven preprogrammed colors and two color change modes. A standard wall switch and switch on/off codes are used to select a color or one of the two color change modes. All such lamps connected to the same wall

switch will be programmed simultaneously. The 120V RGB lamps shown in *Figure 23(B)* have the same color and mode selections as the lamps used for digitally controlled systems. The only difference is that the color and color modes for each lamp on the same light circuit can be selected with a standard light switch or individually with a handheld remote control. The same type of lamp is available as a 12V version with an MR16 base. Both types are also available as fixed color lamps, including white in three temperatures (3,400k, 5,500k, and 6,500k).

Various other fixed-color LED lamps are available for architectural accent or downlighting, including those shown in *Figure 24*. They are designed for use in existing fixtures and are available in white as well as colors. Spotlights are also available with diffusers that create circular, oval, and diamond-shape beams.

LED FLAME TIP BULB

(A) RGB 30° SELF-CONTAINED LAMPS

LED G25 GLOBE BULB

(B) RGB 30° AND 60° SELF-CONTAINED LAMPS

303F23.EPS

Figure 23 ◆ Individual color-changing LED lamps.

G11 GLOBE BULBS

303F24.EPS

Figure 24 ◆ Individual fixed-color LED lamps.

4.7.2 Solid-State Lighting (SSL)

Because of rapidly increasing light output, the newest application of high-brightness and ultra high-brightness LED technology is in fixtures and products for general downlighting. This latest application is called solid-state lighting (SSL). SSL is a rapidly developing field that is being pursued by a large number of new and existing manufacturers. Some of the latest products available or in development are shown in *Figures 25* through *28*.

Figure 25 shows edge-lit LED cove lighting. *Figure 26* shows special radio frequency interference (RFI)-free ceiling lighting used for magnetic resonance imaging (MRI) suites. It is especially suited for high magnetic field 3T scanners and is directly backlit by high-brightness white LEDs with a color temperature of 6,000k in standard size troffer units. In this case, a sky-type translucent panel is used for patient relaxation. The units can also be used with any color or white translucent panel for other applications. The power consumption for a 2' × 4' fixture of this type is 50W.

Figure 27 shows a low-bay LED lighting fixture suitable for a variety of commercial applications. Each fixture provides 2,500 lumens with a power consumption of 75W. Larger pole-mounted versions of these fixtures are also being used for parking lot illumination.

Figure 28 shows a streetlight design that was recently developed to meet all Illumination Engineering Society of North America (IESNA) streetlight pattern standards. A high-brightness LED array of this type provides over 8,000 lumens with a power consumption of about 75W and a lifespan of 100,000 hours.

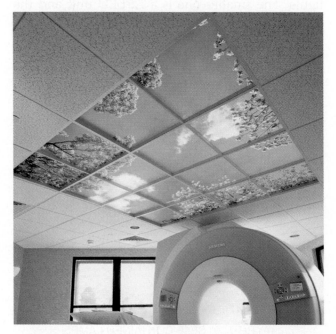

LUMINOUS SKYCEILING WITH RFI-FREE LED LIGHTING OVER MRI MACHINE

LED TROFFER

303F26.EPS

Figure 26 ◆ RFI-free LED ceiling lighting fixtures.

303F25.EPS

Figure 25 ◆ Edge-lit LED cove lighting.

Figure 27 ◆ Low-bay LED lighting fixture.

303F27.EPS

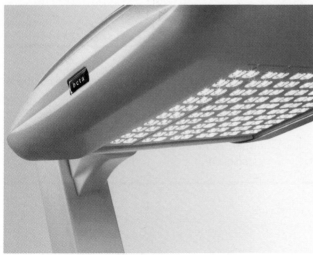

303F28.EPS

Figure 28 ◆ LED streetlights.

4.8.0 Hazardous and Adverse Location Lighting Fixtures

Numerous types of special lighting fixtures are made for use in hazardous locations and in locations with harsh (adverse) environments. An explosionproof fixture is one example of a fixture used in hazardous areas. It is a completely enclosed fixture capable of preventing the ignition of a gas or vapor surrounding the enclosure by sparks, flashes, or explosion of the gas or vapor within. Also, it must operate at an external temperature that will not cause a surrounding flammable atmosphere to be ignited. Obviously, to maintain the explosionproof integrity of such a fixture, all machine screw threads must be in good shape and fully engaged, all gaskets and seals must be intact, and the globe must be fully threaded into the fixture housing. *Figure 29* shows examples of enclosed and gasketed hazardous location lighting fixtures.

NEC Article 500 defines, categorizes, and provides the basic rules for the application and installation of lighting fixtures in hazardous locations. Hazardous locations are defined in terms of class, division, and group per the *NEC®*. The definitions for each class are as follows:

* *Class I locations* – Locations in which flammable gases or vapors are or may be present in the air in quantities sufficient to produce explosive or ignitable mixtures.
* *Class II locations* – Locations that are hazardous because of the presence of combustible dust.
* *Class III locations* – Locations that are hazardous because of the presence of easily ignitable fibers or other flying materials, but in which such materials are not likely to be in suspension in the air in quantities sufficient to produce ignitable mixtures.

WARNING LIGHTS

LIGHTING FIXTURE

303F29.EPS

Figure 29 ◆ Typical hazardous/adverse location lighting fixtures.

LED Fixtures for Commercial Use

GOING
GREEN

These photos show the conversion of a parking garage from sodium vapor low-bay lighting to long-life white LED lighting using fixtures with multiple LED arrays.

OLD SODIUM VAPOR LIGHTING

NEW LED LIGHTING

303SA05.EPS

Each class is further defined as being either Division 1 or Division 2. Division 1 is an environment that is normally hazardous; Division 2 is an environment that is not normally hazardous, but could become hazardous. Each of these divisions can be further classified according to the particular gas, vapor, or dust by defining the areas by groups.

NOTE

The classification of a hazardous area is not the responsibility of the electrician. The classification of a given area as to class, division, and group is the responsibility of the owner, insurance company, or authority having jurisdiction.

In many industries, lighting fixtures must operate under conditions that are considered adverse, such as in environments that are corrosive, wet, or dirty and do not fall into the categories defined by the *NEC*®. For operation in various adverse conditions, a wide variety of specialized lighting fixtures are made. Dustproof, dust-tight, and vapor-tight fixtures are examples. Dustproof fixtures are constructed so that dust will not interfere with the fixture operation. Dust-tight fixtures are constructed so that dust is prevented from entering the enclosed case. Vapor-tight fixtures are gasketed and fully enclosed fixtures made for use in wet or dusty environments. Fixtures made for use in corrosive environments are typically constructed with stainless steel or nonmetallic housings so that they have no exposed exterior parts that can be attacked by corrosives, moisture, etc. The proper selection of fixtures for use in hazardous, non-hazardous, or adverse locations should always be done by someone with expertise in this area. Any fixture used in hazardous locations should meet *UL 844* standards and be labeled *"LISTED ELECTRIC LIGHTING FIXTURES FOR HAZARDOUS LOCATIONS."*

4.9.0 Vandal-Resistant Lighting Fixtures

Vandal-resistant lighting fixtures (*Figure 30*) can include any type of fixture (standard, emergency, security, or exit) used to light areas with high degrees of exposure to unsupervised public traffic. These areas can include public restrooms, hallways, stairwells, loading areas, transit stations, loading docks, and minimum-security institutions. Vandal-resistant fixtures are heavy-duty fixtures typically equipped with virtually unbreakable lenses that protect the lamp and socket assemblies. Many have lenses with radius corners that increase the ability of the lens to deflect blows and absorb impact. Some are also equipped with guards to protect the lenses. Vandal-resistant fixtures typically have four-point mounting bases and tamperproof hardware, making them more difficult to detach from their mounting surfaces and/or to disassemble.

4.10.0 Lighting Fixture Illumination Control

The luminance level, illumination pattern, and glare of a light fixture are determined by the type of lens panel and/or louvers used with the fixture. Lens panels, also called diffusing panels, are glass

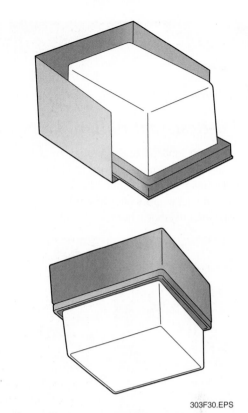

303F30.EPS

Figure 30 ◆ Vandal-resistant lighting fixtures.

or plastic translucent panels. They function to change the direction and control the distribution of the light rays to reduce the apparent brightness of the fixture at certain viewing angles. Prismatic and parabolic lens panels are two types of lens panels that have broad lighting applications, especially when used with fluorescent lighting fixtures. Another type, called a Fresnel lens, is used mainly in theater, television, and photographic lighting applications. Fresnel lenses are also used in some recessed and track lighting downlight applications. Each type of lens is distinguished by the special way in which it reflects or refracts light.

Prismatic lenses tend to be the industry standard light control media. They have good efficiency but provide limited control of light distribution. For areas where brightness and a wide light distribution pattern are required, prismatic lenses are a good choice. When compared to prismatic lenses, parabolic lenses reduce glare and eye strain. This is especially desirable when computer video displays are used in a space. Parabolic lenses are also more efficient than prismatic lenses. For example, a three-lamp fluorescent fixture with a parabolic lens produces as much light as a four-lamp fixture with a prismatic lens.

Louvers used with light fixtures consist of a series of baffles used to shield the light source from view at certain angles, to absorb or block unwanted light, or to reflect or direct light. The baffles are usually arranged in a geometric pattern. Louvers are sized so that the height of the baffles and the spacing between them create the desired cutoff angle. Louvers with flat cell walls (flat louvers) block the sightline from the viewer to the lamps. Louvers with curved cell walls (parabolic louvers) also block the sightline from the viewer to the lamps. In addition, the curved walls, redirect the rays of light downward that strike them within the cutoff angle.

Choosing the right lens or louver for a specific application is as important as choosing the lighting fixture itself. For this reason, the selection is normally made by a qualified lighting engineer and/or lighting manufacturer's representative.

5.0.0 ◆ APPLICATIONS OF LIGHTING

This section provides a brief overview of the more common commercial/industrial lighting applications. The information given here is intended to introduce you to some of the factors and considerations that are taken into account when lighting systems are designed for particular applications. Also given are some examples of the types of lighting fixtures that can be used for the various applications. A discussion of the numerous factors involved in the design of lighting systems, the criteria for selection of fixture types, and the exact methods for determining the quantity and placement of lighting fixtures in a room or area are beyond the scope of this module. These design decisions are normally made by an architect, lighting engineer, or other person with lighting system design expertise. Information specific to the design of lighting systems used for specific applications can be found in the *Lighting Handbook* referenced at the back of this module. Also, design-related information is frequently included in the technical section of many lighting fixture manufacturers' product catalogs.

5.1.0 Office Buildings

Offices and office buildings are designed for people whose work involves thinking tasks and/or the use of various forms of communication, including written, visual, telephone, and computer formats. Office lighting affects the appearance of the space and its occupants, the mood or effect, and the productivity level. Most modern office lighting uses surface-mounted or recessed fluorescent fixtures as the main light source. However, incandescent and HID fixtures are also used for some lighting applications.

5.1.1 Offices and Open Plan Areas

Two methods are commonly used to provide office lighting. One method uses general lighting to provide illumination for the various tasks. This is commonly done for private offices or other areas where task lighting is not needed. General lighting can be accomplished by using lighting fixtures that provide direct (downward) lighting, indirect (upward) lighting, or both. Several types of optical controls are available for use with these general lighting fixtures. These include diffusers, lenses, and polarizers. Diffusers scatter the light emitted by the fixture lamps before it leaves the fixture. It should be pointed out that diffusers are not recommended for use in open office areas where PCs are widely used because they can cause unacceptable reflections on the monitors.

Fixtures with lenses incorporate a series of small prisms that reduce the apparent brightness of the fixture at the near-horizontal viewing angles of 45° to 90° from vertical. Polarizers filter emitted light by transmitting it through multiple refractive layers. This helps to reduce reflected glare.

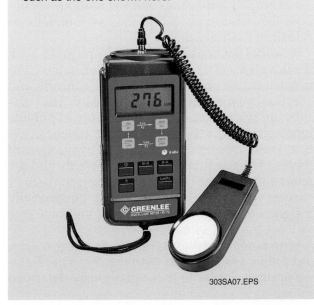
Lighting fixtures equipped with parabolic louvers are often used to precisely control the general brightness level.

The other widely used method for office lighting is to provide most of the task lighting using local fixtures in conjunction with lower levels of general lighting. This is typically done in open plan areas. An open plan area is an area that accommodates workers in a common space with few floor-to-ceiling partitions or walls. If partitions or storage cabinets over work surfaces cause shadows, local lighting fixtures can be used to provide adequate task lighting and reduce the shadows. Caution should be used when choosing downlighting provided by compact fluorescent or low-wattage HID lighting fixtures over incandescent fixtures for task lighting. This is because the point-source nature of these types of fixtures can cause harsh hand shadows or reflections.

In energy-efficient office lighting designs, low-voltage switching, programmable microprocessor control, time clocks, photocells, occupancy sensors, and/or dimmer circuits are widely used to control the lighting levels of the offices and rooms based on occupancy and/or the required levels of lighting. Some of the energy conservation measures that should be incorporated in office lighting are as follows:

- Zone the work areas in large open spaces to allow the lighting in each zone to be switched independently.
- Use multi-level electronic ballasts in fluorescent fixtures to provide for high/low switching and dimming.
- When three-lamp fluorescent fixtures are used, connect the inside lamp in each fixture to a different switch than the outside lamps. Similarly, when four-lamp fixtures are used, connect the inside pair of lamps in each fixture to a different circuit than the outside lamp pair. These circuit arrangements enable a stepped reduction in lighting during periods when full illumination is not required.
- Each separate private office should have its own control switch or switches.
- Switching lighting fixtures near windows separately from those in the interior of a room allows these fixtures to be turned off during periods of bright daylight.

5.1.2 Meeting Rooms and Common Use Areas

In addition to the work areas, office buildings normally contain other areas with different lighting requirements, including conference rooms, reception rooms, public areas, and restrooms.

- *Conference rooms* – Because of the variety of activities that occur in conference rooms, the rooms may have one or more lighting systems. Rooms with one general lighting system typically control illumination by switching and by using fluorescent fixtures with dimming capability. Some rooms have a secondary lighting system consisting of incandescent, halogen, or compact fluorescent downlighting with a dimmer control. This system is typically used to provide low-level lighting during activities involving the use of overhead or LCD projectors, etc. Other systems may use dimmer-controlled perimeter wall washing fixtures for better visual appeal and where presentation materials are mounted on the room walls.
- *Reception rooms* – These areas are designed for people who are waiting for appointments. While waiting, they typically read or talk to others. Lighting for reception rooms is typically provided by ceiling fixtures, wall fixtures, or both. Accent lighting on pictures and other objects is also commonly used. These methods of lighting are designed so as to produce a peaceful atmosphere without glare.
- *Public areas* – Public areas in a building include entrance and elevator or escalator lobbies, corridors, and stairways. These areas usually must remain lighted either continuously or for long periods of time. Lighting for entrance lobbies is typically provided by ceiling and/or wall lighting fixtures. The lighting scheme used should be attractive, but more importantly, it must provide for a safe transition from the exterior of the building to the interior. This is because the eyes must readapt from bright daylight conditions to darker indoor conditions and vice versa. Elevator lobbies are typically casual viewing areas where differences in lighting are acceptable. The important thing is to have a relatively high level of light to illuminate the elevator threshold in order to see possible height differences between the elevator cab and the floor. Corridor lighting levels should provide at least 20% of the illuminance of adjacent areas. Linear lighting fixtures orientated crosswise to the corridor can be used to make the corridor appear wider; continuous linear fixtures located adjacent to the walls generally give a feeling of spaciousness. For stairways, the stair treads must be well illuminated and the lighting fixtures should be located so as to avoid both glare and shadows cast by occupants onto the stairs. Because corridors and stairways are paths of egress, codes require that they also be provided with emergency lighting.

- *Restrooms* – Like elevator lobbies, lighting differences in restrooms are acceptable. Lighting fixtures should be placed so that they provide light in the vicinity of mirrors to illuminate the face adequately. Other lighting fixtures should be positioned so that they provide adequate lighting of the plumbing fixtures and stalls.

5.2.0 Schools

The lighting used in schools should produce a visual environment that aids in the learning process for both the students and the instructors. The lighting must be visually comfortable and amenable to the psychological and emotional needs of the students. Lighting can make schools more pleasant and attractive, reinforce a feeling of spaciousness, and be used to separate areas involving different activities or functions.

In new construction of classrooms, fluorescent fixtures equipped with higher-efficiency, rare earth phosphor lamps are preferred. These lamps also offer better color rendering. In high-ceiling spaces, suspended direct-indirect fixtures provide for reflected light from the ceiling and downlighting. For classrooms with low ceilings, surface-mounted or recessed fixtures are used. Recessed fixtures are less obtrusive than surface-mounted fixtures of the same size.

Many classrooms are used for computer training. The lighting design and fixtures used in these rooms should be the same as previously described for offices. Such designs reduce glare and reflections on the monitors. Similarly, the lighting design and fixtures used in task-oriented shop-type classrooms should be the same as those used in the related industrial or commercial environments where the same tasks are being performed. For example, a machine shop classroom should have the same type of lighting as found in an industrial machine shop.

A school gymnasium is a multi-purpose area. Besides physical education and athletics, it is commonly used for graduations, dances, concerts, and community meetings. High-intensity HID fixtures are widely used in gymnasiums to provide lighting for physical education and athletic activities. These are often supplemented by fluorescent lighting fixtures that restart more rapidly and are better suited to the other uses of the gymnasium. Fixtures in these locations must be specially designed or use guards to prevent breakage [see *NEC Section 410.10(E)*].

With some exceptions, offices, meeting rooms, and public areas located within schools use lighting designs similar to those previously described for the same areas in office buildings. The control system methods and devices used to efficiently control lighting throughout the various rooms and areas within a school complex are also similar to those previously described for office buildings. Schools typically also have emergency lighting control systems.

5.3.0 Retail Store Merchandise Areas

The function of the lighting systems used in store merchandise areas is to attract and guide the customer to the merchandise, allow the customer to visually evaluate the merchandise, and provide illumination to complete the sale and perform other administrative tasks.

The current trend for merchandising is to provide variations in illumination for different kinds of stores and for different departments within a store. Incandescent, fluorescent, and HID fixtures are all commonly used in retail lighting applications. The fixture types used depend on the particular application. Spotlights, floodlights, and wall washer fixtures are all used for merchandise highlighting. The form of objects and texture of surfaces can be revealed by directional lighting such as that produced by small, highly focused downlights. Three-dimensional merchandise is generally displayed using a mix of directional lighting to accent the form and diffuse lighting to relieve the harshness produced by dark shadows.

Retail stores can use a layout of lighting fixtures designed to provide general light throughout the sales area, with or without regard for the location of the merchandise. However, such lighting schemes produce a store environment that lacks emphasis or focal points. Other stores use a form of lighting where the fixtures are permanently located such that they emphasize the merchandise displayed in various stationary fixtures, showcases, or gondolas. Such a system is typically used in supermarkets or similar stores in which the locations of the display units and the types of merchandise being displayed do not change. Flexible lighting systems are widely used in department and similar stores in which the locations of the displays and the types of merchandise being displayed are continuously changing. These systems use a pattern of electric outlets of the continuous or individual type for use with nonpermanent lighting fixtures. This flexibility allows for a general pattern of lighting or for specific lighting. It also has the added feature of being able to change fixture types to suit the application. Such systems typically use recessed adjustable fixtures, recessed adjustable pulldown fixtures, and track lighting fixtures.

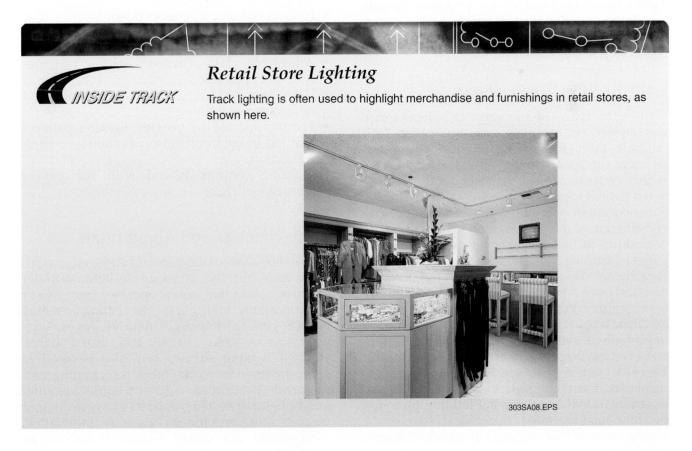

INSIDE TRACK

Retail Store Lighting

Track lighting is often used to highlight merchandise and furnishings in retail stores, as shown here.

303SA08.EPS

Merchandise within store display cases is normally lighted. Strip-type fluorescent fixtures are typically used for this purpose to provide a continuous line of sight and to minimize the heat created in such close spaces. Incandescent and miniature halogen fixtures are sometimes used when it is desirable to create more acceptable color rendition, especially when displaying items such as jewelry or glassware. Flush, surface-mounted, or suspended adjustable fixtures are commonly used for accent lighting of merchandise displayed in free-standing vertical display cases or open-front wall cases. Clothing racks located in cased walls may be lighted from above by concealed, baffled fixtures. Those located in open areas are typically lighted by flush or surface-mounted adjustable ceiling downlights.

Lighting controls are widely used within merchandising areas to decrease operating costs and increase the flexibility of the lighting system. These controls allow for selective switching and dimming of the fixtures during different operational periods. For example, all merchandise accent and display lighting can be turned off during the hours when the store is not open to the public. Areas of the store are zoned so that the general lighting level in each zone can be adjusted independently as needed during periods of security and surveillance, cleanup, and/or maintenance. Some methods of light control include low-voltage switching, programmable microprocessor control, time clocks, photocells, and occupancy sensors and dimmers.

5.4.0 Health Care Facilities

Lighting systems used in support of health care vary widely in purpose and application. For example, the lighting requirements for hospitals are more complex than those for an extended care facility or doctor's office. In hospitals, the lighting in the patient rooms must be suitable both for the patient's comfort and for the observation and examination of the patient. Lights for reading are positioned so that the light falls directly on the patient's reading material with no reflected light. Switch-controlled lighting used for the observation and/or examination of the patient typically uses compact fluorescent fixtures because of their excellent color rendition. Good color rendition is important so that the color of the patient's skin and eyes can be properly evaluated. In critical or intensive care rooms, several lighting arrangements are normally used. A typical arrangement might include indirect general lighting provided by two wall-mounted fluorescent fixtures. A ceiling-mounted or recessed fluorescent fixture

used as a downlight for reading and an incandescent night light for surveillance might be used at the nurses' station. Tungsten halogen fixtures with ellipsoidal reflectors are typically used to provide task lighting for patient examination.

General lighting at a nurses' station is normally multi-level via dimmer and/or switch control. This allows for higher illumination during the day and a lower level at night. Task lighting is designed to facilitate the reading of patient information from charts and computer monitors. Often, under-counter task lights also serve as night lights. Operating rooms, computerized tomography (CT) areas, magnetic resonance imaging (MRI) scanning rooms, and similar task areas all require very specialized lighting fixtures to provide the appropriate general and task lighting. Emergency lighting for use in case of a power outage is critical in operating rooms.

Lighting requirements for patient rooms in extended care facilities (nursing homes) are similar to those used in hospital patient rooms. However, extended care facilities are mostly populated by the elderly with a wide variety of vision problems. Lighting systems used in these facilities must take this into account in addition to the lighting needed to support patient care tasks. To compensate for the poorer vision of older patients, these systems generally incorporate:

- Indirect lighting whenever possible
- Constant floor illumination
- Higher levels of task-appropriate light
- Adjustable task-oriented lighting fixtures whenever possible
- High color rendering sources of incandescent or fluorescent lighting
- Fixtures located to provide lighting patterns that aid in environment orientation and minimize confusion
- Lighting controls placed with the patient (corded switches)

5.5.0 Theaters and Auditoriums

The seating areas of theaters and auditoriums typically have diffused, comfortable lighting under dimmer control that can be operated from two or more locations. Lighting for the seating area may include general downlight fixtures, coves, sidewall sconces, and curtain and mural lights. Selected lighting circuits may also be used in order to provide work lights for cleaning and rehearsals. In addition to general lighting requirements, lighting systems used in live performance theaters must provide for visibility, motivation, composition, and mood relative to an activity occurring onstage. Theatrical lighting design should only be done by professional designers.

Areas of the stage that must be lighted include the stage apron, acting areas upstage from the proscenium arch (the curtain and arch or framework that holds it), extension stage areas, auxiliary acting areas, and the foreground and background scenery and properties. For plays, dances, operas, and similar types of entertainment, lighting must come from the front, back, sides, and top of the stage to produce the effects needed to create the desired program mood and dimensionality. For musical programs, it is necessary to light from the front and from above and behind the stage area so that the performers can read their music.

Theater and stage lighting are provided by numerous special stage lighting fixtures, the most common of which are:

- *Parabolic (reflector) spotlights* – Fixtures that can be adjusted to produce a wide or narrow hard-edged beam of light.
- *Fresnel spotlights* – Fixtures equipped with a Fresnel lens that produce a beam of light with a smooth, soft edge. A Fresnel lens is a stepped, flat lens with a textured back. The field and beam angles are varied by changing the distance (focusing) between the lamp and lens.
- *Ellipsoidal spotlights* – Fixtures that produce a hard-edged light pattern with precise beam control. The light is focused through a gate of the unit where the beam can be shaped with the use of shutters or an iris. The shaped beam is then focused by the lens system.
- *Scoop floodlights* – Fixtures that produce diffused light with indefinite margins and poorly defined shadows. They are used to soften or fill the shadows produced by other types of fixtures. Scoops are equipped with front clips to hold a frame containing either color media or diffused media.
- *Soft lights* – Fixtures that produce a well-diffused, almost shadow-free illumination used for fill light. All the light is reflected off a matte-finish reflector before reaching the subject.

Lighting in front of the proscenium opening is accomplished in a variety of ways. Spotlights in the theater ceiling are used to frontlight the downstage and apron acting areas. These are typically installed behind a continuous slot stretching across the ceiling from side to side. Spotlights are often located on or recessed in the auditorium and proscenium side walls to supplement the ceiling spotlights. Other front lighting fixtures include a

CASE HISTORY

Lighting Installations

A ceiling contractor finished the installation on an 86'-high theater ceiling but failed to test the fixtures before the scaffolding was removed. Shortly thereafter, it was discovered that one of the fixtures was wired incorrectly. To replace the single fixture cost the contractor $6,000 for the scaffolding and rework hours.

The Bottom Line: Because of the difficulty in accessing the fixtures in theaters and auditoriums, it is very important to test and verify the operation of newly installed lighting fixtures before the scaffolding or lifts have been removed.

set of floodlights (striplights) used as footlights to soften face shadows cast by the overhead fixtures and to add general tone lighting from below. These lights are normally multicolored lights. Special spotlights called follow spotlights are located above the rear center of the audience seating area. These are used to illuminate selected performers.

Stage lighting behind the proscenium opening (upstage) is also accomplished in a variety of ways. Stage lighting fixtures are suspended above and across the stage on pipes or bridges called light pipes. Each light pipe is equipped with connector strips that provide a quick and simple method of electrically connecting the various fixtures. Typically, two or more light pipes installed parallel to each other are used to light the depth of the stage. Light pipe-mounted Fresnel and ellipsoidal fixtures are commonly used to produce a soft-edged beam that is widely variable in focus. High-intensity, narrow-beam Fresnel or parabolic spotlights or PAR fixtures positioned so that they face upstage can provide for backlighting of the artists in the main area of the stage. A border light or series of scoops can be used to provide general downlighting across the stage area and overhead illumination of hanging curtains, scenery, and so on.

Control of theater lighting systems must be very complex in order to meet the constantly changing (during each performance) lighting requirements demanded by theatrical presentations. These control systems make extensive use of memory, use accurately timed faders, and are capable of coordinating complex simultaneous operations. Most systems incorporate both manual and computer-controlled devices to accomplish the needed lighting operations.

5.6.0 Industrial Locations

The lighting used in industrial settings must provide adequate visibility for transforming raw materials into finished products. Because physical hazards exist in manufacturing processes, lighting is also a major factor in the prevention of accidents. The lighting scheme used in an industrial application depends on the specific industry and/or the tasks involved. These are too numerous to describe here. Information about lighting factors pertaining to specific industries can be found in the *Lighting Handbook* referenced in the back of this module.

Most industrial applications use either direct or semi-direct lighting. However, semi-direct fixtures with an upward light component are preferred because they contribute to visual comfort by balancing the light between the fixtures and their backgrounds. A wide variety of industrial fluorescent and HID fixtures are made for this purpose.

The interiors of many industrial buildings contain spaces commonly called bays. These are typically classified as low-bay or high-bay areas. The classifications refer to how high the bottoms of the bay lighting fixtures are mounted above the work plane (3' above the floor). As discussed earlier, low-bay fixtures are typically mounted at heights of 20' or less and high-bay fixtures at heights of 20' or more. Both industrial fluorescent and HID low-bay fixtures are widely used in low-bay areas, depending on the application. HID high-bay fixtures are used extensively in high-bay areas. High-bay fixtures use reflectors to direct light downward. Low-bay fixtures generally include a refractor to spread out the light for even light distribution and lower fixture brightness.

Fluorescent or HID fixtures are typically surface-mounted or suspended from bar joists, beams, other overhead structural members, or trolley busways to form a uniform lighting array called a ceiling plane. This arrangement provides for general lighting in the bay. Localized general lighting using a greater number of fixtures in a specific area is often used where tasks are being performed that require light levels higher than the prevailing general lighting level.

Each fixture has a spacing criterion (SC). This is the maximum ratio of spacing to mounting height that will produce uniform lighting when the fixture is used. If the spacing between adjacent fixtures exceeds the optimal spacing determined by the SC, then dark spots may occur on the work plane. For example, if the SC is 1.5, it means that for good lighting, this fixture should be spaced at a distance of no more than 1.5 times its mounting height above the work plane. For example, if it is mounted at a height of 15', then the distance between fixtures should not exceed 22½' (1.5 × 15').

Where more difficult visual tasks are performed, additional task lighting is provided to supplement the general lighting. This task lighting is used to provide a certain illuminance or color or to permit special aiming or positioning of light sources to produce or avoid highlights or shadows and best illuminate the details of the visual task. The types of fixtures used to provide task lighting in industrial situations are as varied as the individual tasks themselves.

Lighting applications frequently require the use of specialized fixtures designed for use in areas classified by the *NEC*® and/or local codes as being hazardous or in areas otherwise considered environmentally harsh. As appropriate to the environment, the fixtures used in such areas would be explosion-proof, dust-tight, dustproof, vapor-tight, or have other special features.

Many processes used in electronics and pharmaceuticals, as well as in industries such as medicine and aerospace, require the contamination-free environment provided by a clean room. Lighting fixtures used in clean rooms are specially designed for that purpose. Typically, they are designed without crevices where contaminants can accumulate. The housings are free of holes, slots, and other openings that would allow passage of air or particles into or out of the fixture. Lenses and other components are sealed to the housing with special gasketing. Some are made so that they can be used in conjunction with high-efficiency particle arrestance (HEPA) filters that allow air flow of HEPA-filtered air through the fixture and into the clean room space.

5.7.0 Outdoor Lighting

Outdoor lighting for buildings should facilitate nighttime approach and entry, whether on foot or by vehicle, provide security for the building and its contents, and enhance the architectural features of the building.

5.7.1 Security Lighting

Walks, driveways, streets, and parking lots should be lighted at night using appropriate walkway, roadway, and parking lot fixtures. Entrance and exit areas of the building should be more brightly lighted with wall-mounted fixtures or overhead lights. Building outside surfaces (facades), approaches, and outdoor activity areas should be lighted both for the activity itself and for general safety, as well as for protection against vandalism and theft. Facade lighting typically uses carefully located, vandalproof HID fixtures mounted on poles, trees, adjacent walls, or special pedestals. Many of these area fixtures are equipped with photoelectric (PE) controls that allow for automatic dusk-to-dawn operation of the fixtures.

5.7.2 Aesthetic Lighting

Some artistic or aesthetic uses of outdoor area lighting include:

- Downlighting that is mounted close to the ground to emphasize flower beds, steps, paths, etc., and to visually extend the property, making it seem more spacious
- Downlighting that is mounted high in trees, poles, etc., to floodlight a broad area
- Uplighting to highlight interesting trees, statuary, or textured walls
- Accent or spotlighting to emphasize a special object
- Spread or diffused light in areas such as patios, decks, driveways, and walkways
- Shadow lighting to provide a sense of depth and project intriguing shadows on vertical surfaces behind objects
- Silhouetting objects from behind to obscure details and emphasize their shape (the reverse of shadow lighting)
- Grazing to highlight an interesting surface to emphasize its texture
- Using multiple fixtures to provide crosslighting that creates an effect that is softer, deeper, and more pleasing

5.7.3 Roadway Lighting

Good roadway lighting is designed to promote safer conditions for drivers and pedestrians. There are several classifications of roadways, including freeways, expressways, major roadways, collector roadways, and local roadways. They can also be classified as commercial, intermediate, and residential. Roadways are designed following standards set by the Illuminating Engineering Society of North America (IESNA). The IESNA has assigned three classifications to describe the light distribution or beam pattern of a roadway fixture:

- S (short), M (medium), or L (long) indicates how far up and down a street a fixture directs light.
- C (cutoff), S (semi-cutoff), or N (non-cutoff) indicates how much light a fixture directs above 80 degrees and 90 degrees vertical. A cutoff fixture directs almost no light above 90 degrees, a semi-cutoff directs some light above 90 degrees, and a non-cutoff has no restrictions on how much light is emitted in any direction.
- Road fixture type designations I, II, III, and IV refer to fixtures that produce asymmetrical (noncircular) horizontal light distribution patterns. Simply stated, these designations indicate how far a fixture directs light across the width of the street; the higher the number, the further the light is directed across the width of the street. An IESNA Type V fixture indicates a fixture that produces a circular (symmetrical) horizontal light pattern.

5.8.0 Sports Lighting

The goal of sports lighting is to provide appropriate lighting by controlling the brightness of an object and its background so that the object will appear clear and sharp to the players, spectators, and television viewers (if applicable). For sports lighting, the level of illumination required depends on several factors, such as whether the activity is indoors or outdoors, the speed of the game, the skill of the players, and the number of

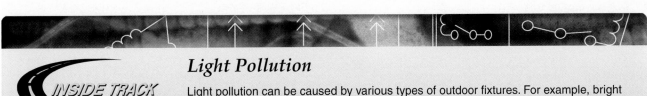

Light Pollution

INSIDE TRACK

Light pollution can be caused by various types of outdoor fixtures. For example, bright floodlighting from a business or residence can blind drivers approaching the area or using adjacent roadways.

spectators and their distance from the field of play. It also depends on whether the sport is aerial or ground-oriented. Aerial sports involve playing with an object, typically a ball, in the air as well as on the ground. Ground sports are played on the ground or a few feet above the ground so that the players and spectators do not look up in the normal course of play.

Because most sports involve seeing a fast-moving object, such as a football or ice hockey puck, sports facilities normally require very high levels of light directed and/or focused in a manner appropriate for the specific sport. Many have more than one illuminance requirement.

For example, the lighting requirements in a baseball stadium differ between the outfield and

Low-Voltage and Solar Lighting Systems

GOING GREEN

Low-voltage and solar lighting systems are two methods that are widely used for energy-efficient residential outdoor landscape and/or aesthetic lighting. Low-voltage lighting systems operate at 30VAC or less under any load conditions. They can have one or more secondary circuits, each limited to 25A maximum. A basic low-voltage outdoor lighting system operates at 12V. As shown here, it consists of an isolating power supply (step-down transformer), usually with a built-in timer, lighting fixtures, and associated equipment and wiring identified for such use. The low-voltage wires can be run over or under the ground to the low-voltage light fixtures. The length of cable runs in a low-voltage system are generally limited to distances of 100' or less from the transformer. For outdoor landscape lighting applications involving greater distances, a 120VAC line voltage system is typically used. Generally, low-voltage systems are safe and more economical to install, and the fixtures can easily be moved to make different lighting arrangements. *NEC Article 411* governs low-voltage lighting systems. Note that *NEC Chapter 3* methods must be employed for concealed lighting.

A solar outdoor lighting fixture is powered by the sun's energy using a built-in solar panel. During the daylight hours, the panel gathers and stores the sun's energy to power the fixture at night. The energy is stored in an internal rechargeable nickel-cadmium (nicad) battery. The light fixture is controlled by a built-in ambient light sensor that turns the light fixture on at dusk and off at dawn. Reliable operation of solar lighting depends on a good source of direct sunlight, so it works better in areas such as the Southwest.

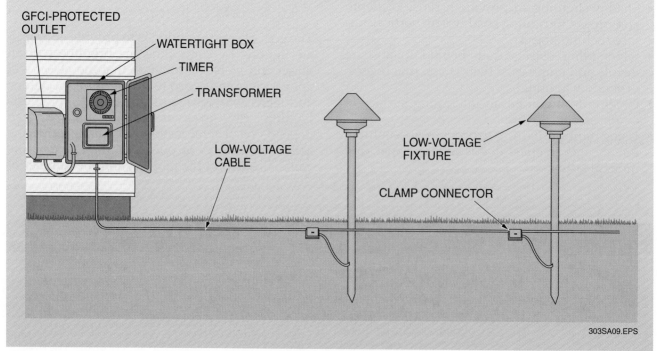

303SA09.EPS

GOING GREEN

Protecting Endangered Loggerhead Turtles

In some areas of the country, loggerhead turtle hatchlings are protected from excessive light levels to help ensure that more hatchlings make it from the nesting area to the ocean. It is thought that the young hatchlings are attracted to manmade light and move toward it instead of the natural light reflected off the ocean. To prevent this, all exterior lights, including nearby streetlights, are required to be turned off during the hatching season.

the infield. The lighting scheme used in a particular sport facility depends on the specific sport involved and the size of the playing field or arena. Information about lighting factors pertaining to specific sports can be found in the *Lighting Handbook* referenced at the back of this module. Similar information is frequently included in the technical section of lighting fixture manufacturers' product catalogs.

HID floodlight fixtures with metal halide and high-pressure sodium lamps are used extensively for sports lighting. However, incandescent or tungsten halogen floodlights are sometimes used for sports that require lower levels of light or those where the annual hours of facility usage are limited. Fluorescent fixtures are typically used indoors and/or where the fixtures are mounted less than 20' from the lighted area. The type of lighting used in indoor facilities may be totally indirect lighting for a small multi-purpose facility; it may be direct-indirect lighting for tennis courts or similar facilities; or it may be direct lighting for arenas and stadiums. Outdoor sports facilities use direct lighting systems. Normally, multiple ganged assemblies of pole-mounted or tower-mounted enclosed rectangular, enclosed round, and/or shielded floodlight fixtures are used for this purpose.

Because outdoor sports lighting is visible at distances beyond the sports facility, it must be adequately controlled to limit spill light and sky glow. Spill light is light that shines beyond the sports facility that can annoy occupants of adjacent properties. Because spill light creates annoyance problems, many municipalities have enacted ordinances to limit it. It is generally controlled by using lighting fixtures with an intensity distribution pattern that does not light areas outside the sports facility. It can also be controlled by using cutoff fixtures or high mounting poles with fixtures that have a low aiming angle. Sky glow, typically seen in urban areas, is caused by the stray or reflected light that is emitted into the atmosphere being reflected and scattered by dust, water vapor, and other pollutants in the atmosphere. This can be controlled somewhat by using fixtures that eliminate direct upward light emission.

6.0.0 ◆ SPECIAL-PURPOSE WIRING SYSTEMS USED FOR LIGHTING

There are several special-purpose wiring systems used to provide the wiring of electrical lighting fixtures in commercial and industrial applications. Some common systems are briefly described here.

6.1.0 Manufactured System Wiring

Manufactured system wiring (*Figure 31*) is commonly used as an alternative to conventional hard wiring methods in commercial and industrial applications. Manufactured system wiring components are produced by lighting fixture manufacturers for use with their lighting products. Within the lighting system, these components are interchangeable and can be reused time after time. Component labels are color-coded by voltage, and the units are keyed to prevent mismatching voltages. Lighting branch circuit wiring is run from the building panel to an outlet or junction box that serves as the starting point for the manufactured wiring run. There, a manufactured system feeder adapter is wired and mounted. From this point on, all the wiring to the lighting fixtures is manufactured using the appropriate cable adapters.

6.2.0 Lighting Trolley Busways

Lighting trolley busways (*Figure 32*) are widely used in industrial and some commercial lighting applications because of the flexibility they provide. Lighting fixtures suspended from a trolley

Manufactured System Wiring

Install manufactured wiring systems starting either at the beginning of the circuit or the end of the circuit to ensure that the wiring is oriented correctly.

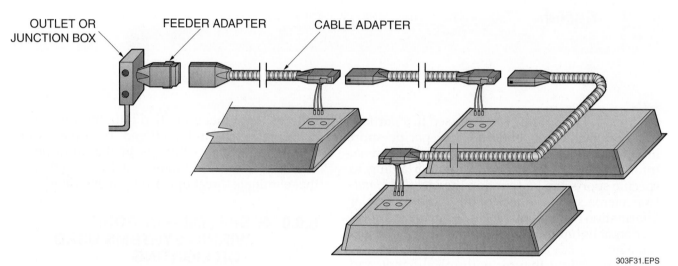

Figure 31 ◆ Manufactured wiring system.

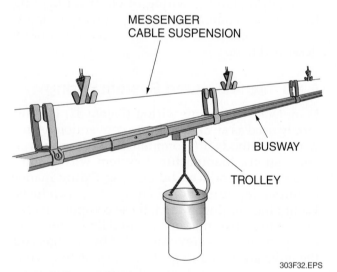

Figure 32 ◆ Lighting trolley busway.

busway can be easily shifted from one location to another along the trolley as desired. It is not necessary to take down or replace heavy conduit systems as would be required to change the location of a fixture in a fixed lighting outlet system.

Lighting trolley busway components are made in several amperage ratings and are for use in two-wire, three-wire, and four-wire systems with

a maximum voltage of 300V (AC or DC) to ground. *NEC Article 368* defines the use of busways. Generally, they can be used in exposed areas except in the following locations:

• Where they are subject to mechanical damage or corrosive vapors
• In hoistways
• In any hazardous location unless listed for such use
• Outdoors
• In wet or damp locations unless identified for such use

Trolley busways can be suspended from the building structure by various methods. Manufacturers provide a variety of hangers for surface mounting, drop-rod suspension, or use with a messenger cable. Messenger cables are stranded steel cables stretched tightly between building attachment points and adjusted by turnbuckles located at the ends of the runs. Intermediate supports for the cable are attached to the overhead structure at appropriate intervals to prevent any sagging of the lighting system.

Trolley busway runs are constructed using standard straight sections typically made in 5' and 10' lengths and nonstandard length sections as required, joined together by plain couplings.

Advanced Lighting Controls

In addition to traditional lighting control systems, there are many new home automation devices available. These include hardwired, radio frequency (RF), and power line carrier (PLC) systems. These systems can be configured to light one room, groups of rooms, or pathways of light throughout a home.

Hardwired systems use their own communications cable and must be installed when the home is built. They are the most reliable option, but are also the most expensive and cannot be easily installed in retrofit installations.

RF systems use radio signals to control selected lights in the home, both from a central location (typically a bedroom) and using remote control units, such as a handheld unit or a keychain or car visor transmitter. The lamps to be controlled are plugged into RF modules or controlled by special dimmer switches.

PLC systems operate through a home's existing wiring. These range from simple lamp timers to whole-house lighting, security, motorized window treatments, and home entertainment systems. Like RF systems, PLC systems use plug-in modules and/or special dimmer switches and a programmable controller. Since PLC systems rely on the power line as a network, they may be subject to signal problems in certain areas of the home. Booster modules are available for signal strengthening in problem areas.

Advanced X10 and similar systems (e.g., Insteon, Lutron, and others) can be configured to use both RF (or infrared) control and power line carrier signaling along with computer software to control the lighting in each room as well as outdoor lighting, security cameras, motion sensors, and home entertainment systems.

Commercial applications may soon see the use of phase cut carrier technology. This is a type of PLC system that uses special ballasts to transmit control signals from an electrical junction box to fluorescent and high intensity discharge (HID) lamps over existing wiring. This allows energy-saving options such as daylighting and demand response control. Daylighting uses photosensors to measure daylight and automatically adjust the electric lighting. Demand response control allows building managers to reduce lighting in entire facilities during periods of high use. Electricity billing is often structured to encourage the use of demand response control.

These couplings permit free passage of the trolleys along the busway. The ends of the busways are equipped with end caps that close the ends of the run and provide an entrance point for the insertion or removal of trolleys. For convenience in removing or inserting trolleys, trolley entrance couplings are often installed at other points in the run, typically at the mid-point. Several types of trolley and stationary plug-in devices, commonly called twistouts, are made for use with lighting trolley busways. These provide the electrical connection interface between the busway and the lighting fixture. Trolleys have wheels allowing them to roll freely along the busway from one position to another. Both trolleys and twistouts can be equipped with cord strain relief clamps or they can be furnished with outlet boxes with a standard receptacle installed. Heavy-duty, weight-supporting devices that also attach to the busway are used in conjunction with trolleys and twistouts when heavy fixtures are being suspended from the busway.

6.3.0 Strut-Type Channel Systems

Uni-Strut®, Power Strut®, and similar structural members are frequently used to both support and provide a wiring raceway for powering lighting fixtures in industrial/commercial lighting systems. Various support and wiring methods can be used, depending on the application. Different types of hangers and other fixture-supporting hardware used to support the fixtures from the strut sections are specially designed by the strut manufacturer for that purpose. *Figure 33* shows two typical strut-type methods for supporting and wiring fixtures.

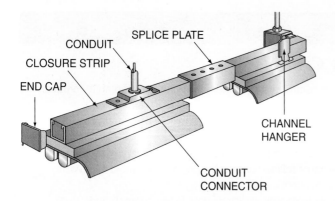

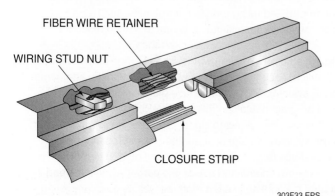

303F33.EPS

Figure 33 ◆ Typical strut-type methods for supporting and powering fixtures.

7.0.0 ◆ DIMMING

Dimming lighting fixtures is a widely used method of controlling lighting levels to meet varying illumination requirements and conserve energy.

7.1.0 Incandescent Lamps

An incandescent or tungsten halogen lighting fixture can be dimmed by simply reducing the voltage applied across the lamp filament. Dimming incandescent lamps affects the light output, life, and color temperature of the lamp. In the past, autotransformer-type dimmers were used to control large loads, but they have been replaced with solid-state dimmer banks. These dimmers are simply connected in series with the incandescent lamp load. Solid-state dimmers are used when the lighting load to be controlled is large.

Electronic dimmer controls use solid-state, silicon-controlled rectifiers (SCRs), thyristors, and/or transistor circuitry. These basically operate as high-speed switches that alternately turn the voltage applied to the lamp on and off. These dimmer controls normally include radio frequency interference (RFI) filtering to reduce interference with computer systems, intercom systems, sound

systems, and other electrical systems. Electronic incandescent dimmer controls can be of several types:

- *Touch type* – Provides even control using a touch-sensitive plate.
- *Toggle type* – Provides even control by means of a toggle switch.
- *Non-preset slide type* – Provides even control by means of a slide mechanism.
- *Preset slide type* – Provides even control by means of a slide mechanism and allows the preset lighting level position to be maintained via a separate on/off switch.
- *Rotary type* – Provides even control by means of rotary action. The dimmer may be of the preset or non-preset type.

Incandescent dimmers are made in single-pole and three-way switch designs. Some are made in master and remote designs for use in applications where the lighting must be controlled from two or more locations. *Figure 34* shows a basic circuit using multiple-location dimming with remote control.

7.2.0 Fluorescent Lamps

Dimming fluorescent lamps differs from dimming incandescent lamps in that fluorescent dimmers do not provide dimming to zero light as do incandescent dimmers. Also, the color temperature does not vary substantially over the dimming range. To dim fluorescent lamps, special dimming ballasts and dimmer controls are required. The dimmer control, working in conjunction with the dimming ballast, operates to maintain voltage to the fluorescent lamp cathodes, keeping them at the proper operating temperature. They also operate to vary the current flow in the lamp arc in order to vary the intensity of light being produced by the lamp.

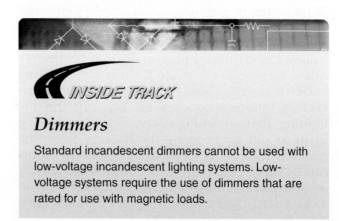

Dimmers

Standard incandescent dimmers cannot be used with low-voltage incandescent lighting systems. Low-voltage systems require the use of dimmers that are rated for use with magnetic loads.

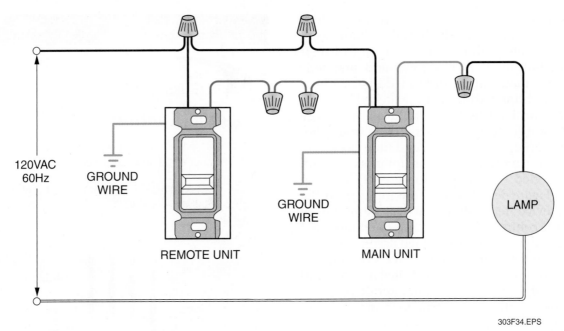

Figure 34 ◆ Simplified multiple-location incandescent lamp dimming with remote control.

303F34.EPS

Most dimming ballasts in use are of the solid-state electronic type in which a low-level signal applied from the dimmer control (*Figure 35*) causes the ballast to regulate the lamp arc current and, therefore, the lamp's light output. These all operate using high-frequency (20kHz to 50kHz) switching circuitry. The lamp or arc current is lowered and the light output dimmed by shortening the time that the current flows during each cycle. Some electronic ballasts are designed to also reduce the electrode filament voltage at full light output and restore it when the lamps are operated in the dimmed mode at low arc current. Controlling the light by low-level signaling allows electronic dimming ballast systems to be flexible. Large numbers of electronic ballasts can be controlled easily. This provides for ease of manufactured control when incorporated in an energy management system.

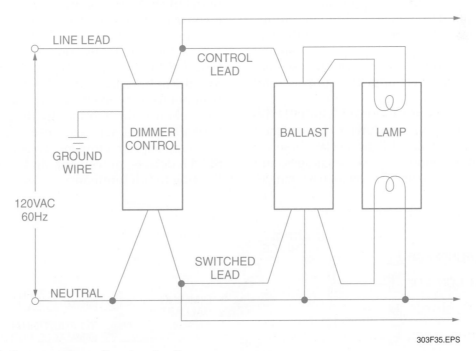

303F35.EPS

Figure 35 ◆ Basic fluorescent lamp dimming circuit.

Although not as widely used, magnetic dimming ballasts are also made that allow fluorescent lamps to be safely dimmed to about 10% of their full light output. Both electronic and magnetic ballasts are made in multi-level designs. These allow the lighting fixtures to be switched between two or more fixed light levels. In many energy management lighting schemes, inputs from photosensors, occupancy sensors, and time clocks are continuously monitored by an interface control unit connected to multiple fluorescent dimming fixtures (*Figure 36*). Based on the input from the sensor device requesting the least amount of energy, the control interface causes the fluorescent lighting in the area to be dimmed accordingly.

Always refer to the manufacturer's recommendations for lamp and ballast installation and replacement. Dimmers may not operate properly if the manufacturer's instructions are not followed. For example, many types of lamps need to run for a specified burn-in time before being turned off when they are first installed and may also need to be turned off periodically during operation.

7.3.0 HID Lamps

HID lamps are optimized to operate at full power. However, dimming circuits are sometimes used for energy conservation purposes. These dimming circuits operate in a way similar to those used with fluorescent lamps. Note that the slow warmup and hot restrike delays that are characteristic of HID fixtures also apply to dimming. HID lamps respond to changes in dimmer settings much more slowly than incandescent or fluorescent lamps. For this reason, a properly designed HID dimming circuit normally operates so that any dimming is delayed until the lamp is fully warmed up.

Multi-level ballasts are also used with HID fixtures to allow the lighting fixtures to be switched between two or more fixed light levels. These are widely used in warehouses, parking garages, and similar applications. *Figure 37* shows a basic single-

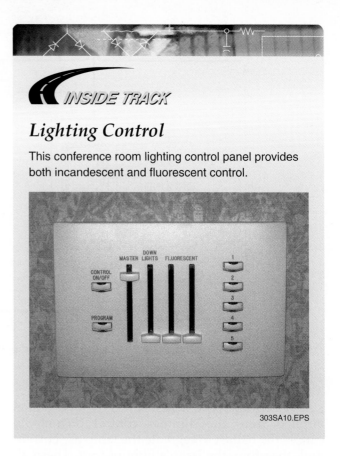

zone, two-level (high and low) HID lighting system. It is controlled via the switch level interface control (SLIC), which provides for manual selection of on/off and high/low illumination levels.

A basic automatic light dimming scheme used with HID fixtures is shown in *Figure 38*. As shown, it consists of a switch level interface panel (SLIP), a switch level occupancy detector (SLOD), and several independently controlled switch level (SL) HID lighting fixtures. This arrangement allows for HID fixtures located in different areas or zones to be controlled separately. For example, if used in a warehouse, the entrance lighting fixtures can be kept at full light, while the aisles in the warehouse areas are kept at a low level. As people enter, the SLOD detects movement and brings the aisle lighting to full luminance.

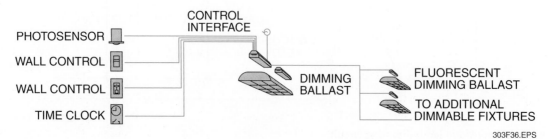

Figure 36 ◆ Basic control circuit to control dimming of multiple fluorescent fixtures.

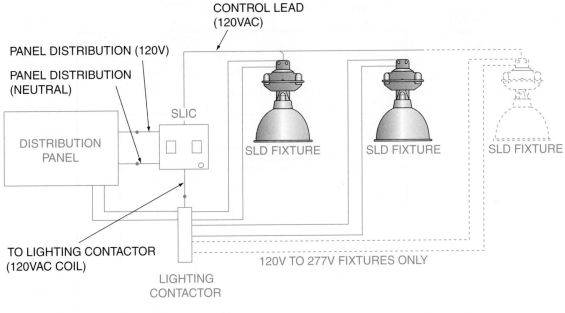

CONTROL LEAD
(120VAC)

PANEL DISTRIBUTION (120V)

PANEL DISTRIBUTION
(NEUTRAL)

SLIC

DISTRIBUTION
PANEL

SLD FIXTURE   SLD FIXTURE   SLD FIXTURE

TO LIGHTING CONTACTOR
(120VAC COIL)

120V TO 277V FIXTURES ONLY

LIGHTING
CONTACTOR

SLD = SWITCH LEVEL DIMMING
SLIC = SWITCH LEVEL INTERFACE CONTROL

303F37.EPS

Figure 37 ◆ Simplified single-zone, manually controlled HID dimming circuit.

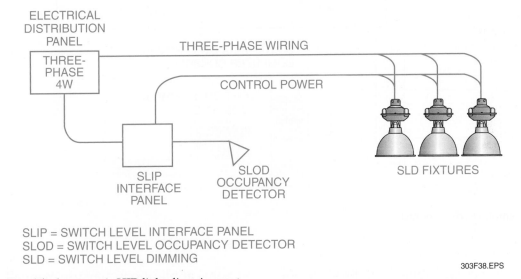

ELECTRICAL
DISTRIBUTION
PANEL

THREE-PHASE WIRING

THREE-
PHASE
4W

CONTROL POWER

SLD FIXTURES

SLIP
INTERFACE
PANEL

SLOD
OCCUPANCY
DETECTOR

SLIP = SWITCH LEVEL INTERFACE PANEL
SLOD = SWITCH LEVEL OCCUPANCY DETECTOR
SLD = SWITCH LEVEL DIMMING

303F38.EPS

Figure 38 ◆ Simplified automatic HID light dimming system.

7.4.0 Dimmer Control Racks

Some applications require the use of numerous dimmers in order to control the lighting system. Modern theater lighting control systems provide a good example because they use dimmers extensively to control the lighting to meet the constantly changing (during each performance) lighting requirements demanded by theatrical presentations. Typically, groups of electronic dimmer controls are mounted in a rack (dimmer rack) controlled from a remotely located computer console. Depending on the application, dimmer racks may house only a few dimmers or many. In large theater systems, several racks may be bussed together. Each rack is typically equipped with one dimmer device for each lighting branch circuit connected to the rack (*Figure 39*). Three-phase,

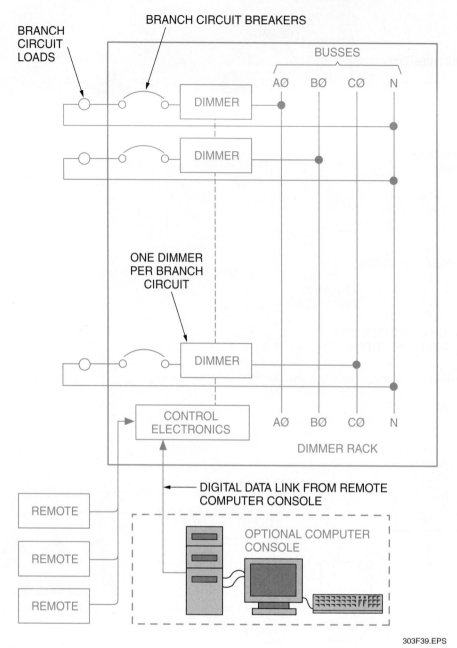

Figure 39 ◆ Simplified dimmer rack.

303F39.EPS

four-wire power is distributed via buses in the rack to all the dimmers. A digital control cable connects the remotely located computer console to the dimmer rack(s), allowing the operator to be located on or near the stage for easy viewing of the performance and the related lighting.

Some applications may use manually operated dimmers located in a dimmer rack. Manually operated racks usually contain resistance or auto-transformer-type dimmers.

1. One lumen falling on one square foot of surface produces an illumination of _____.
   a. one footcandle
   b. one candlepower
   c. one candela
   d. 12.57 candelas

2. Lighting that directs well-focused illumination to a specific surface such as a kitchen counter is classified as _____ lighting.
   a. general
   b. accent
   c. task
   d. wall wash

3. Lighting fixtures that distribute 40% to 60% of their emitted light downward and the balance upward, sometimes with a strong horizontal component, are classified as _____ lighting fixtures.
   a. direct
   b. general diffuse
   c. indirect
   d. semi-direct

4. Lighting fixtures installed in offices, stores, and public buildings are classified as being used in _____ service.
   a. industrial
   b. residential
   c. floodlighting
   d. commercial

5. Ceiling fixtures equipped with diffusers are typically used to provide _____ lighting.
   a. general diffused
   b. accent
   c. display
   d. accent or display

6. Industrial fluorescent fixtures with non-apertured reflectors typically direct _____ of their light downward toward the work surface.
   a. 10% to 20%
   b. 20% to 30%
   c. 80% to 90%
   d. 100%

7. Bollard fixtures are commonly used to light _____.
   a. parking lots
   b. roadways
   c. walkways and grounds
   d. building facades

8. *NEC Article 500* governs the application and installation of _____.
   a. emergency lighting
   b. lighting in hazardous locations
   c. recessed fluorescent lighting
   d. lighting busways

9. Two identical low-bay HID fixtures are to be mounted at a height of 18'. If these fixtures have a spacing criterion (SC) of 1.5, what is the maximum distance that should exist between them to provide proper lighting?
   a. 15½'
   b. 20'
   c. 25'
   d. 27'

10. Trolley busways can be used outdoors.
    a. True
    b. False

Summary

Proper lighting system design requires the selection of fixtures and lamps that work together to create safe, useful lighting that complements the furnishings and/or tasks being illuminated, creates the mood desired, and increases the security of the property.

Lighting fixtures can be divided into two basic groups. First, there are those meant to be seen and to be part of the overall design of the space, such as ceiling pendants, chandeliers, wall sconces, surface-mounted ceiling fixtures, and track lighting. Second, there are unobtrusive fixtures, such as cove and recessed lights, which provide light without calling undue attention to themselves. Lighting fixtures are classified by layout and location, CIE luminaire type, and type of service.

Notes

Trade Terms
Introduced in This Module

Candela (cd): The International System (SI) unit of luminous intensity. It is roughly equal to 12.57 lumens. The name has been retained from the early days of lighting when a standard candle of a fixed size and composition was used as the basis for evaluating the intensity of other light sources. See *candlepower*.

Candlepower (cp): The measure of light intensity measured in candelas (cd). It is used to indicate the intensity of a light source in a given direction. The higher intensity of light, the more candelas it represents and the higher the candlepower.

Diffused lighting: Lighting on a work surface or object that is not predominantly incident from any particular direction.

Diffuser: A part of a lighting fixture that redirects or scatters the light produced by the fixture lamp to provide diffused lighting.

Footcandle (fc): The unit used to measure how much total light is reaching a surface, such as a wall or table. One lumen falling on one square foot of surface produces an illumination level of one footcandle. One footcandle also equals 10.76 lux. Lux is the standard international unit of illuminance and is equal to one lumen per square meter.

Work plane: The distance above the floor where a task is to be performed. In offices and schools, this is usually 30". In kindergartens, it would be lower, and in some manufacturing areas, it would be much higher.

This module is intended to present thorough resources for task training. The following reference works are suggested for further study. These are optional materials for continuing education rather than for task training.

Lighting Handbook. New York, NY: Illuminating Engineering Society of North America (IESNA), 2000.

National Electrical Code® Handbook, Latest Edition. Quincy, MA: National Fire Protection Association.

NCCER makes every effort to keep these textbooks up-to-date and free of technical errors. We appreciate your help in this process. If you have an idea for improving this textbook, or if you find an error, a typographical mistake, or an inaccuracy in NCCER's Contren® textbooks, please write us, using this form or a photocopy. Be sure to include the exact module number, page number, a detailed description, and the correction, if applicable. Your input will be brought to the attention of the Technical Review Committee. Thank you for your assistance.

Instructors – If you found that additional materials were necessary in order to teach this module effectively, please let us know so that we may include them in the Equipment/Materials list in the Annotated Instructor's Guide.

Write: Product Development and Revision
National Center for Construction Education and Research
3600 NW 43rd St., Bldg. G, Gainesville, FL 32606

Fax: 352-334-0932

E-mail: curriculum@nccer.org

Craft _____ Module Name _____

Copyright Date _____ Module Number _____ Page Number(s) _____

Description _____

_____

_____

_____

(Optional) Correction _____

_____

_____

(Optional) Your Name and Address _____

_____

_____

Tower Bridge

London's Tower Bridge, opened in 1894, was originally powered by steam. In its heyday, the bridge was raised about 6,000 times a year. Since 1975, it has been powered by electric motors and now is raised about 1,000 times a year.

26304-08

26304-08
Hazardous Locations

Overview

Hazardous locations are those that contain both combustible materials and energized electrical components. The *NEC*® uses a system of classes and divisions to identify hazardous locations. Classes identify the type of combustible material, while divisions define the state or presence of the material. Class I locations contain combustible gases or vapors; combustible dust is present in Class II locations; and Class III locations contain combustible fibers or flyings.

Equipment that houses electrical components in hazardous locations must meet certain standards or guidelines depending on the division to which it is assigned. Special fittings are available for conduit systems to physically seal off the passage through the interior of the conduit in order to prevent combustible material from traveling from the hazardous area to potential ignition points. *NEC Chapter 5* contains rules and regulations that apply to hazardous locations.

Note: *NFPA 70*®, *National Electrical Code*®, and *NEC*® are registered trademarks of the National Fire Protection Association, Inc., Quincy, MA 02269. All *National Electrical Code*® and *NEC*® references in this module refer to the 2008 edition of the *National Electrical Code*®.

Objectives

When you have completed this module, you will be able to do the following:

1. Define the various classifications of hazardous locations.
2. Describe the wiring methods permitted for branch circuits and feeders in specific hazardous locations.
3. Select seals and drains for specific hazardous locations.
4. Select wiring methods for Class I, Class II, and Class III hazardous locations.
5. Follow *National Electrical Code®* (*NEC®*) requirements for installing explosionproof fittings in specific hazardous locations.

Trade Terms

Approved
Conduit
Conduit body
Equipment
Explosionproof
Explosionproof
  apparatus
Hazardous (classified)
  location
Sealing compound
Sealoff fittings

Required Trainee Materials

1. Pencil and paper
2. Appropriate personal protective equipment
3. Copy of the latest edition of the *National Electrical Code®*

Prerequisites

Before you begin this module, it is recommended that you successfully complete *Core Curriculum; Electrical Level One; Electrical Level Two; Electrical Level Three*, Modules 26301-08 through 26303-08.

This course map shows all of the modules in *Electrical Level Three*. The suggested training order begins at the bottom and proceeds up. Skill levels increase as you advance on the course map. The local Training Program Sponsor may adjust the training order.

ELECTRICAL LEVEL THREE
26311-08 Motor Controls
26310-08 Voice, Data, and Video
26309-08 Motor Calculations
26308-08 Commercial Electrical Services
26307-08 Transformers
26306-08 Distribution Equipment
26305-08 Overcurrent Protection
26304-08 Hazardous Locations
26303-08 Practical Applications of Lighting
26302-08 Conductor Selection and Calculations
26301-08 Load Calculations – Branch and Feeder Circuits
ELECTRICAL LEVEL TWO
ELECTRICAL LEVEL ONE
CORE CURRICULUM: Introductory Craft Skills

304CMAP.EPS

1.0.0 ◆ INTRODUCTION

NEC Articles 500 through 504 cover the requirements of electrical **equipment** and wiring for all voltages in locations where fire or explosion hazards may exist due to flammable gases or vapor, flammable liquids, combustible dust, or ignitable fibers or other flying materials. Locations are classified depending on the properties of the flammable vapors, liquids, gases, or combustible dusts or fibers that may be present, as well as the likelihood that a flammable or combustible concentration or quantity is present.

Any area in which the atmosphere or a material in the area is such that the arcing of operating electrical contacts, components, and equipment may cause an explosion or fire is considered a **hazardous (classified) location**. In all such cases, **explosionproof apparatus**, raceways, and fittings are used to provide an **explosionproof** wiring system.

The *NEC®* divides hazardous materials into three classes (Class I, Class II, and Class III), with two divisions for each class (Division 1 and Division 2). Of these, Class I, Division 1 represents the most hazardous location. These classes have been established on the basis of the explosive character of the atmosphere for the testing and approval of equipment for use in each class. However, it must be understood that considerable skill and judgment must be applied when deciding to what degree an area contains hazardous concentrations of vapors, combustible dusts, or easily ignitable fibers and flying materials. Furthermore, many factors, such as temperature, barometric pressure, quantity of release, humidity, ventilation, and distance from the vapor source, must be considered. When information on all factors concerned is properly evaluated, a consistent classification for the selection and location of electrical equipment can be developed.

NEC Articles 505 and 506 allow classification and application to international standards, but it will be applied only under engineering supervision. In addition, few American-made products are listed for international application. As a result, this module will not cover *NEC Articles 505 and 506* in any detail.

1.1.0 Class I Locations

Class I atmospheric hazards are divided into Divisions 1 and 2, and also into four groups (A, B, C, and D). Group A represents the greatest hazard.

Those locations in which flammable gases or vapors may be present in the air in quantities sufficient to produce explosive or ignitable mixtures are identified as Class I locations. If these gases or vapors are present during normal operation, frequent repair or maintenance operations, or where breakdown or faulty operation of process equipment might also cause simultaneous failure of electrical equipment, the area is designated as Class I, Division 1. Examples of such locations are interiors of paint spray booths where volatile, flammable solvents are used, inadequately ventilated pump rooms where flammable gas is pumped, anesthetizing locations of hospitals (to a height of 5' above floor level), and drying rooms for the evaporation of flammable solvents (see *Figure 1*).

Class I, Division 2 covers locations in which volatile flammable gases, vapors, or liquids are handled either in a closed system or confined within suitable enclosures, or where hazardous concentrations are normally prevented by positive mechanical ventilation. Areas adjacent to Division 1 locations, into which gases might occasionally flow, also belong in Division 2.

1.2.0 Class II Locations

Class II locations are those that are hazardous because of the presence of combustible dust. Class II, Division 1 locations are areas in which combustible dust may be present in the air under normal operating conditions in quantities sufficient to produce explosive or ignitable mixtures; examples are working areas of grain-handling and storage plants and rooms containing grinders or pulverizers (*Figure 2*). Class II, Division 2 locations are areas in which dangerous concentrations of suspended dust are not likely, but where dust might accumulate.

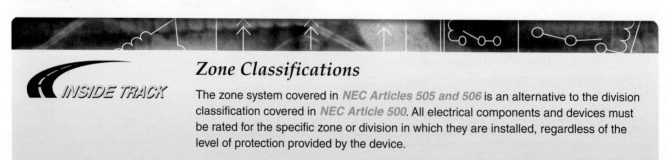

Zone Classifications

INSIDE TRACK

The zone system covered in *NEC Articles 505 and 506* is an alternative to the division classification covered in *NEC Article 500*. All electrical components and devices must be rated for the specific zone or division in which they are installed, regardless of the level of protection provided by the device.

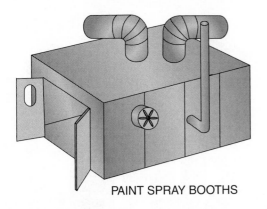

PAINT SPRAY BOOTHS

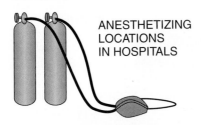

ANESTHETIZING
LOCATIONS
IN HOSPITALS

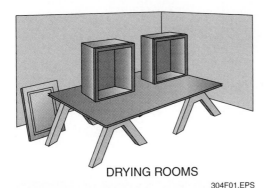

DRYING ROOMS

304F01.EPS

Figure 1 ◆ Typical *NEC*® Class I locations.

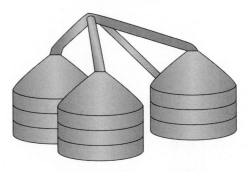

GRAIN-HANDLING AND STORAGE PLANTS

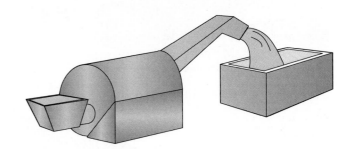

ROOMS CONTAINING GRINDERS AND PULVERIZERS

304F02.EPS

Figure 2 ◆ Typical *NEC*® Class II locations.

Besides the two divisions, Class II atmospheric hazards also cover three groups of combustible dusts (E, F, and G). The groupings are based on the electrical resistivity and ignition temperature of the dust. Group E is typically Division 1. Groups F and G may be either Division 1 or Division 2. Because the *NEC*® is considered the definitive classification tool and contains explanatory data about hazardous atmospheres, refer to *NEC Section 500.5* for exact definitions of Class II, Divisions 1 and 2.

1.3.0 Class III Locations

Class III locations are those areas that are hazardous because of the presence of easily ignitable fibers or other flying materials, but such materials are not likely to be in suspension in the air in

quantities sufficient to produce ignitable mixtures. Such locations usually include certain areas of rayon, cotton, and textile mills, clothing manufacturing plants, and woodworking plants (*Figure 3*).

1.4.0 Applications

Hazardous atmospheres are summarized in *Table 1*. For a more complete listing of flammable liquids, gases, and solids, see *Recommended Practice for the Classification of Flammable Liquids, Gases, or Vapors and of Hazardous (Classified) Locations for Electrical Installations in Chemical Process Areas*, National Fire Protection Association Publication No. 497.

Once the class of an area is determined, the conditions under which the hazardous material may be present determine the division. In Class I and Class II, Division 1 locations, the hazardous gas or dust may be present in the air under normal operating conditions in dangerous concentrations. In Division 2 locations, the hazardous material is not normally in the air, but it might be released if there is an accident or if there is faulty operation of equipment.

Tables 2 through *6* provide a summary of the various classes of hazardous locations as defined by the *NEC*®.

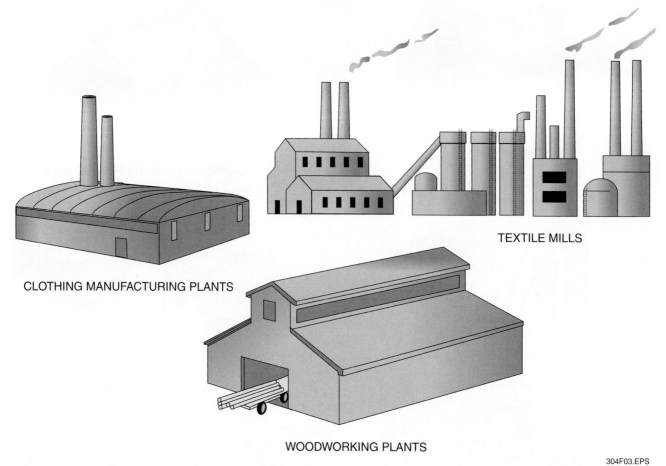

CLOTHING MANUFACTURING PLANTS

TEXTILE MILLS

WOODWORKING PLANTS

304F03.EPS

Figure 3 ◆ Typical *NEC*® Class III locations.

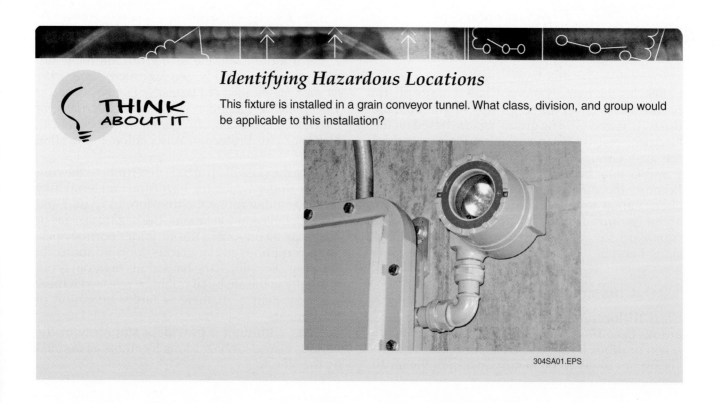

Identifying Hazardous Locations

THINK ABOUT IT

This fixture is installed in a grain conveyor tunnel. What class, division, and group would be applicable to this installation?

304SA01.EPS

Table 1 Summary of Hazardous Atmospheres

Hazardous Area	Class Subdivisions	Groups
	Class I Divisions	**Class I, Division Groups**
Class I: Material present is a flammable gas or vapor	Division 1: Locations in which hazardous concentrations of flammable gases or vapors are present normally or frequently	Group A: Atmospheres containing acetylene
	Division 2: Locations in which hazardous concentrations of flammable gases or vapors are present as a result of infrequent failure of equipment or containers	Group B: Atmospheres containing hydrogen, manufactured gases containing more than 30% hydrogen by volume, or gases or vapors of equivalent hazard
		Group C: Atmospheres containing ethylene, cyclopropane, or gases or vapors of equivalent hazard
		Group D: Atmospheres containing propane, gasoline, or gases or vapors of equivalent hazard
	Class I Zones	**Class I, Zone Groups**
	Zone 0: Locations in which combustible material is present continuously or for long periods	Group IIC: Atmospheres containing acetylene or hydrogen or other gases or vapors meeting Group IIC criteria
	Zone 1: Locations in which combustible material is likely to be present normally or frequently because of repair or maintenance operations or leakage	Group IIB: Atmospheres containing acetaldehyde, ethylene, or other gases or vapors meeting Group IIB criteria
	Zone 2: Locations in which combustible material is not likely to occur in a normal operation and, if it does occur, will exist only for a short period	Group IIA: Atmospheres containing propane, gasoline, or other gases or vapors meeting Group IIA criteria
	Class II Divisions	**Class II, Division Groups**
Class II: Material present is a combustible dust	Division 1: Locations in which hazardous concentrations of combustible dust are present normally or may exist because of equipment breakdown or where electrically conductive combustible dusts are present in hazardous quantities	Group E: Atmospheres containing combustible metal dusts including aluminum, magnesium, and other metals of similar hazards
	Division 2: Locations in which hazardous concentrations of combustible dust are not normally suspended in the air but may occur as a result of infrequent malfunction of equipment or where dust accumulation may interfere with safe dissipation of heat or may be ignitable by abnormal operation of electrical equipment	Group F: Atmospheres containing combustible carbonaceous dusts, including carbon black, charcoal, coals, or dusts that have been sensitized by other materials so that they present an explosion hazard
		Group G: Atmospheres containing combustible nonconductive dusts not included in Group E or F, including flour, grain, wood, and plastic
	Class III Divisions	
Class III: Material present is an ignitable fiber or flying	Division 1: Locations in which easily ignitable fibers or materials producing combustible flyings are handled, manufactured, or used	No Groups
	Division 2: Locations in which easily ignitable fibers are stored or handled, except in the manufacturing process	

Table 2 Application Rules for Class I, Division 1

Components	Characteristics	NEC® Reference
Boxes, fittings	Explosionproof and threaded for connection to conduit	NEC Section 501.10(A)
Wiring methods	Rigid metal conduit, steel intermediate metal conduit, Type MI cable, and, under certain conditions, ITC and MC cable	NEC Section 501.10(A)
Flexible connections	Class I, explosionproof	NEC Section 501.10(A)
Sealoffs	Approved for purpose	NEC Section 501.15(A)
Liquid-filled transformers	Installed in an approved vault	NEC Section 501.100(A)
Panelboards	Class I enclosure	NEC Section 501.115(A)
Circuit breakers	Class I enclosure	NEC Section 501.115(A)
Fuses	Class I enclosure	NEC Section 501.115(A)
Switches	Class I enclosure	NEC Section 501.115(A)
Dry-type transformers	Class I, Division 1 enclosure	NEC Section 501.120(A)
Motors	Class I, Division 1, totally enclosed or submerged	NEC Section 501.125(A)
Generators	Class I, Division 1, totally enclosed or submerged	NEC Section 501.125(A)
Lighting fixtures	Approved for Class I, Division 1	NEC Section 501.130(A)
Portable lamps	Class I, Division 1, approved as a portable assembly	NEC Section 501.130(A)
Utilization equipment	Class I, Division 1	NEC Section 501.135(A)
Receptacles	Approved for the location	NEC Section 501.145
Alarm systems	Class I, Division 1	NEC Section 501.150(A)

Table 3 Application Rules for Class I, Division 2

Components	Characteristics	NEC® Reference
Flexible connections	Class I, explosionproof	NEC Section 501.10(B)
Wiring methods	Rigid metal conduit, steel intermediate metal conduit, Types MI, MC, MV, TC, ITC, or PLTC cables, or enclosed gasketed busways or wireways	NEC Section 501.10(B)
Sealoffs	Approved for purpose	NEC Section 501.15(B)
Boxes, fittings	Do not have to be explosionproof unless current interrupting contacts are exposed	NEC Section 501.15(C)
Liquid-filled transformers	General purpose	NEC Section 501.100(B)
Fuses	Class I enclosure	NEC Section 501.105(B)(3)
Panelboards	General purpose with exceptions	NEC Section 501.115(B)
Switches	Class I enclosure	NEC Section 501.115(B)
Motor controls	Class I, Division 2	NEC Section 501.115(B)
Circuit breakers	Class I enclosure	NEC Section 501.115(B)(1)
Dry-type transformers	Class I, general purpose except switching mechanism Division 1 enclosures	NEC Section 501.120(B)
Motors	General purpose unless motor has sliding contacts, switching contacts, or integral resistance devices; if so, use Class I, Division 1	NEC Section 501.125(B)
Generators	Class I, totally enclosed or submerged	NEC Section 501.125(B)
Portable lamps	Explosionproof	NEC Section 501.130(B)
Lighting fixtures	Protected from physical damage	NEC Section 501.130(B)
Utilization equipment	Class I, Division 2	NEC Section 501.135(B)
Receptacles	Approved for the location	NEC Section 501.145
Alarm systems	Class I, Division 2	NEC Section 501.150(B)

Table 4 Application Rules for Class II, Division 1

Components	Characteristics	NEC® Reference
Wiring methods	Rigid metal conduit, steel intermediate metal conduit, or Types MI and, under certain conditions, MC cables listed for use in Class II, Division 1 locations	NEC Section 502.10(A)
Flexible connections	Extra-hard usage cord, liquid-tight, and others	NEC Section 502.10(A)(2)
Boxes, fittings	Class II boxes required when using taps, joints, or other connections; otherwise, use dust-tight boxes with no openings	NEC Section 502.10(A)(4)
Liquid-filled transformers	Install in an approved vault	NEC Section 502.100(A)
Dry-type transformers	Class II, vault	NEC Section 502.100(A)
Circuit breakers	Dust/ignitionproof enclosure	NEC Section 502.115(A)
Fuses	Dust/ignitionproof enclosure	NEC Section 502.115(A)
Switches	Dust/ignitionproof enclosure	NEC Section 502.115(A)
Motor controls	Dust/ignitionproof	NEC Section 502.115(A)
Panelboards	Dust/ignitionproof	NEC Section 502.115(A)(1)
Generators	Class II, Division 1 or totally enclosed	NEC Section 502.125(A)
Motors	Class II, Division 1 or totally enclosed	NEC Section 502.125(A)
Lighting fixtures	Class II	NEC Section 502.130(A)
Portable lamps	Class II	NEC Section 502.130(A)
Utilization equipment	Class II	NEC Section 502.135(A)
Receptacles	Class II	NEC Section 502.145(A)

Table 5 Application Rules for Class II, Division 2

Components	Characteristics	NEC® Reference
Wiring methods	Rigid metal conduit, steel intermediate metal conduit, electrical metallic tubing (EMT), Types MI, MC, TC, ITC, or PLTC cables, or enclosed dust-tight busways or wireways	NEC Section 502.10(B)
Flexible connections	Extra-hard usage cord, liquid-tight, and others	NEC Section 502.10(B)(2)
Boxes, fittings	Use tight covers to minimize entrance of dust	NEC Section 502.10(B)(4)
Liquid-filled transformers	Install in vault	NEC Section 502.100(B)
Dry-type transformers	Class II vault	NEC Section 502.100(B)
Panelboards	Dust-tight enclosure	NEC Section 502.115(B)
Circuit breakers	Dust-tight enclosure	NEC Section 502.115(B)
Fuses	Dust-tight enclosure	NEC Section 502.115(B)
Switches	Dust-tight enclosure	NEC Section 502.115(B)
Motor controls	Dust-tight enclosure	NEC Section 502.115(B)
Motors	Class II, Division 1 or totally enclosed	NEC Section 502.125(B)
Generators	Class II, Division 1 or totally enclosed	NEC Section 502.125(B)
Lighting fixtures	Class II	NEC Section 502.130(B)
Portable lamps	Class II	NEC Section 502.130(B)(1)
Utilization equipment	Class II	NEC Section 502.135(B)
Receptacles	Exposed live parts are not allowed	NEC Section 502.145(B)

Table 6 Application Rules for Class III, Divisions 1 and 2

Components	Characteristics	NEC® Reference
Wiring methods	Rigid metal conduit, steel intermediate metal conduit, EMT, Types MI and MC cables, or enclosed dust-tight busways or wireways	*NEC Section 503.10(A)*
Boxes, fittings	Dust-tight	*NEC Section 503.10(A)(1)*
Flexible connections	Extra-hard usage cord and other flexible conduit/fittings	*NEC Section 503.10(A)(2)*
Liquid-filled transformers	Install in an approved vault	*NEC Section 503.100*
Dry-type transformers	Dust-tight enclosure	*NEC Section 503.100*
Panelboards	Dust-tight enclosure	*NEC Section 503.115*
Circuit breakers	Dust-tight enclosure	*NEC Section 503.115*
Fuses	Dust-tight enclosure	*NEC Section 503.115*
Switches	Dust-tight enclosure	*NEC Section 503.115*
Motor controls	Dust-tight enclosure	*NEC Section 503.115*
Motors	Totally enclosed	*NEC Section 503.125*
Generators	Totally enclosed	*NEC Section 503.125*
Lighting fixtures	Tight enclosure with no openings	*NEC Section 503.130*
Portable lamps	Unswitched, guarded with tight enclosure for lamp	*NEC Section 503.130*
Utilization equipment	Class III	*NEC Section 503.135*
Receptacles	Minimize accumulation of fibers or flyings	*NEC Section 503.145*

Delayed Action Receptacles

The receptacle shown here is rated for Class I, Division 1 and 2 locations and features a delayed action rotating sleeve that prevents complete withdrawal of the plug in one continuous movement. This delay allows any arc-generated heat to be dissipated before the plug is released.

304SA02.EPS

2.0.0 ◆ PREVENTION OF EXTERNAL IGNITION/EXPLOSION

The main purpose of using explosionproof fittings and wiring methods in hazardous areas is to prevent ignition of flammable liquids or gases and to prevent an explosion.

2.1.0 Sources of Ignition

In certain atmospheric conditions when flammable gases or combustible dusts are mixed in the proper proportion with air, any source of energy is all that is needed to touch off an explosion.

One prime source of energy is electricity. Equipment such as switches, circuit breakers, motor starters, pushbutton stations, or plugs and receptacles can produce arcs or sparks in normal operation when contacts are opened and closed. This could easily cause ignition.

Other hazards are devices that produce heat, such as lighting fixtures and motors. In this case, the surface temperatures may exceed the safe limits of many flammable atmospheres.

Finally, many parts of the electrical system can become potential sources of ignition in the event of insulation failure. This group includes wiring (particularly splices in the wiring), transformers, impedance coils, solenoids, and other low-temperature devices without make-or-break contacts.

Non-electrical hazards such as sparking metal can also easily cause ignition. A hammer, file, or other tool that is dropped on masonry or on a ferrous surface can cause a hazard unless the tool is made of non-sparking material. For this reason, portable electrical equipment is usually made from aluminum or other material that will not produce sparks if the equipment is dropped.

Electrical safety is of crucial importance. The electrical installation must prevent accidental ignition of flammable liquids, vapors, and dusts released to the atmosphere. In addition, because much of this equipment is used outdoors or in corrosive atmospheres, the material and finish must be such that maintenance costs and shutdowns are minimized.

2.2.0 Combustion Principles

Three basic conditions must be satisfied for a fire or explosion to occur:

- A flammable liquid, vapor, or combustible dust must be present in sufficient quantity.
- The flammable liquid, vapor, or combustible dust must be mixed with air or oxygen in the proportions required to produce an explosive mixture.
- A source of ignition must be applied to the explosive mixture.

In applying these principles, the quantity of the flammable liquid or vapor that may be liberated and its physical characteristics must be recognized.

Vapors from flammable liquids also have a natural tendency to disperse into the atmosphere and rapidly become diluted to concentrations below the lower explosion limit, particularly when there is natural or mechanical ventilation.

 WARNING!

The possibility that the gas concentration may be above the upper explosion limit does not afford any degree of safety, because the concentration must first pass through the explosive range to reach the upper explosion limit.

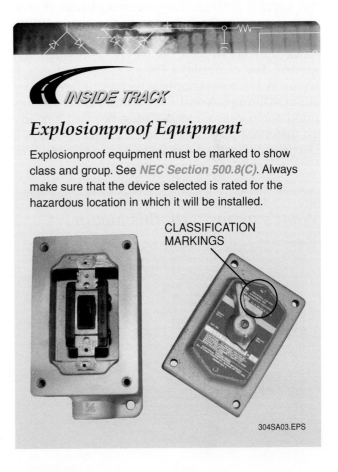

INSIDE TRACK

Explosionproof Equipment

Explosionproof equipment must be marked to show class and group. See *NEC Section 500.8(C)*. Always make sure that the device selected is rated for the hazardous location in which it will be installed.

CLASSIFICATION MARKINGS

304SA03.EPS

3.0.0 ◆ EXPLOSIONPROOF EQUIPMENT

Each area that contains gases or dusts considered hazardous must be carefully evaluated to make certain the correct electrical equipment is selected. Many hazardous atmospheres are Class I, Group D or Class II, Group G. However, certain areas may involve other groups, particularly Class I, Groups B and C. Conformity with the *NEC®* requires the use of fittings and enclosures **approved** for the specific hazardous gas or dust involved.

NOTE

Equipment is rated by both classification and temperature. See *NEC Table 500.8(C)*.

The wide assortment of explosionproof equipment makes it possible to provide adequate electrical installations under any of the various hazardous conditions. However, you must be thoroughly familiar with all *NEC®* requirements and know what fittings are available, how to install them properly, and where and when to use the various fittings. For example, some electricians are under the false impression that a fitting rated for Class I, Division 1 can be used under any hazardous conditions. However, remember the groups. For example, a fitting rated for Class I, Division 1, Group C cannot be used in areas classified as Groups A or B. On the other hand, fittings rated for use in Group A may be used for any group beneath A; fittings rated for use in Class I, Division 1, Group B can be used in areas rated as Group B areas or below, and so on.

What's wrong with this picture?

304SA04.EPS

WARNING!

Never interchange fittings or covers between one hazardous area and another. Such items must be rated for the appropriate class, division, and group.

Explosionproof fittings are rated for both classification and groups. All parts of these fittings (including covers) are rated accordingly. Therefore, if a Class I, Division 1, Group A fitting is required, a Group B (or below) fitting cover must not be used. The cover itself must be rated for Group A locations. Consequently, when working on electrical systems in hazardous locations, always make certain that fittings and their related components match the condition at hand.

3.1.0 Intrinsically Safe Equipment

Intrinsically safe equipment is incapable of releasing sufficient electrical energy under normal or abnormal conditions to cause ignition of a specific hazardous atmospheric mixture in its most easily ignited concentration. The use of intrinsically safe equipment is primarily limited to process control instrumentation because these electrical systems lend themselves to the low energy requirements.

Installation rules for intrinsically safe equipment are covered in *NEC Article 504*. In general, intrinsically safe equipment and its associated wiring must be installed so that it is positively separated from the non-intrinsically safe circuits, because induced voltages could defeat the concept of intrinsically safe circuits. Underwriters Laboratories, Inc. and FM Global list several devices in this category.

3.2.0 Explosionproof Conduit and Fittings

A typical floor plan for a hazardous area is shown in *Figure 4*.

In hazardous locations where threaded metal **conduit** is required, the conduit must be threaded with a standard conduit cutting die (*Figure 5*) that provides ¾" taper per foot. The conduit should be made up wrench-tight to prevent sparking in the event fault current flows through the raceway system per *NEC Section 500.8(E)*. All boxes, fittings, and joints shall be threaded for connection to the conduit system and shall be an approved, explosionproof type (*Figure 6*). Threaded joints must be made up with at least five threads fully engaged.

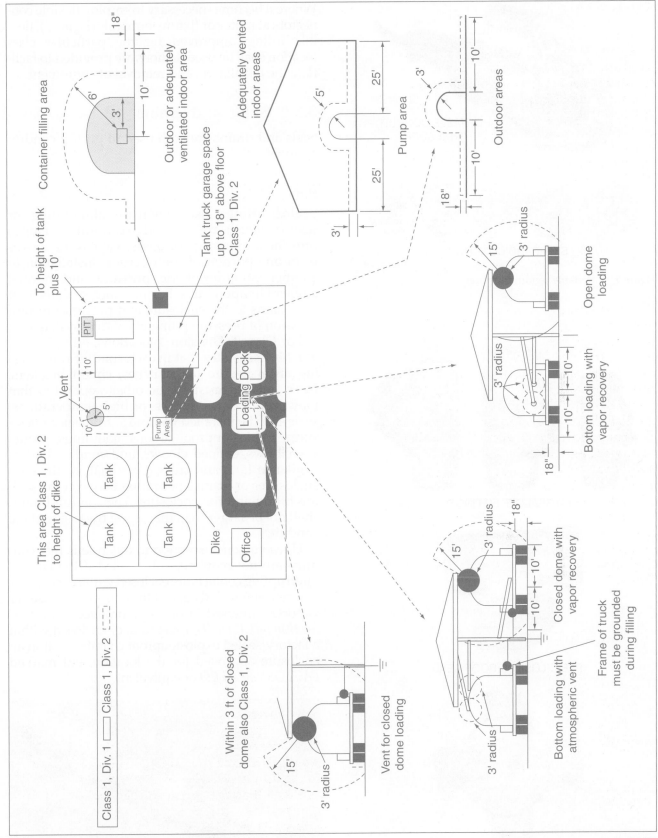

Figure 4 ◆ Floor plan of a hazardous location.

304F04.EPS

PORTABLE CONDUIT
THREADER

STANDARD CONDUIT DIES

304F05.EPS

Figure 5 ◆ Portable conduit threader.

(A) SEALOFF FITTINGS

(B) EXPLOSIONPROOF SEAL

(C) EXPLOSIONPROOF
CONDUIT BODY

304F06.EPS

Figure 6 ◆ Typical fittings approved for hazardous areas.

Where it becomes necessary to employ flexible connectors at motor or fixture terminals (*Figure 7*), flexible fittings approved for the particular class location shall be used. Unions are provided to facilitate the installation and removal of equipment.

3.3.0 Seals and Drains

Seals and drains are both used to protect conduit systems.

3.3.1 Seals

Sealoff fittings, also known as sealing fittings or seals (*Figure 8*), are required in conduit systems to prevent the passage of gases, vapors, or flames from one portion of the electrical installation to another at atmospheric pressure and normal ambient temperatures. Furthermore, sealoffs limit explosions to the enclosure and prevent precompression of pressure piling in conduit systems.

For Class I, Division 1 locations, *NEC Section 501.15(A)(1)* states that in each conduit run entering an enclosure for switches, circuit breakers, fuses, relays, resistors, or other apparatus that may produce arcs, sparks, or high temperatures, seals shall be installed within 18" from such enclosures. Explosionproof unions, couplings, reducers, elbows, capped elbows, and **conduit bodies** similar to L, T, and cross types shall be the only enclosures or fittings permitted between the sealing fitting and the enclosure. The conduit bodies shall not be larger than the largest trade size of the conduit.

However, one exception to this rule is that conduits are not required to be sealed if the current interrupting contacts are enclosed within a chamber hermetically sealed against the entrance of gases or vapors, immersed in oil in accordance with *NEC Section 501.15(A)(1), Exception*, or enclosed within a factory-sealed explosionproof chamber within an enclosure approved for the location and marked *FACTORY SEALED* or equivalent.

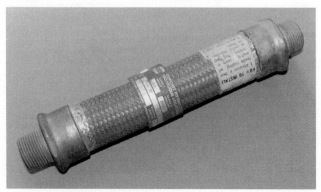

304F07.EPS

Figure 7 ◆ Explosionproof flexible connector.

Pressure Piling

Pressure piling occurs when an explosion in one section of a raceway creates expanding gases that cause the vapors to compress further down in the raceway, which will then cause a secondary explosion of a greater magnitude. This explosion will result in additional expanding gases and compressed vapors, continuing to create additional explosions, which may potentially exceed the containment capabilities of the raceway system.

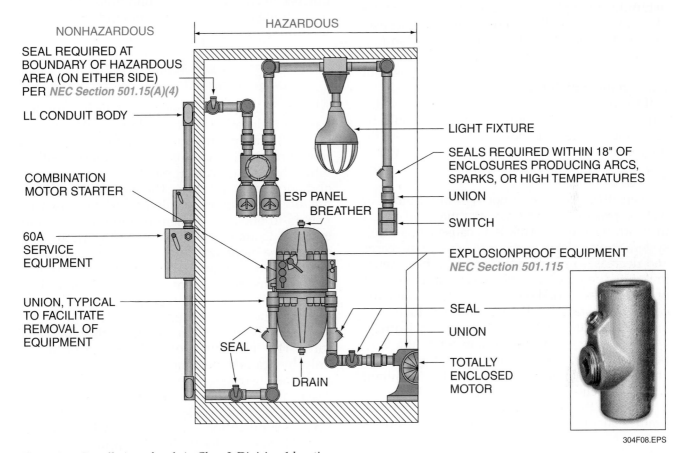

Figure 8 ◆ Installation of seals in Class I, Division 1 locations.

Seals are also required in Class II locations where a raceway provides communication between an enclosure that is required to be dust/ignitionproof and one that is not *(NEC Section 502.15)*.

A permanent and effective seal is one method of preventing the entrance of dust into the dust/ignitionproof enclosure through the raceway. A horizontal raceway, not less than 10' long, is another approved method, as is a vertical raceway not less than 5' long and extending downward from the dust/ignitionproof enclosure.

Where a raceway provides communication between an enclosure that is required to be dust/ignitionproof and an enclosure in an unclassified location, seals are not required.

Where sealing fittings are used, all must be accessible.

While it is not an *NEC®* requirement, many electrical designers sectionalize long conduit runs by inserting seals not more than 50' to 100' apart, depending on the conduit size. This is done in order to minimize the effects of pressure piling.

In general, seals are installed at the same time as the conduit system. However, the conductors are installed after the raceway system is complete and prior to packing and sealing the fittings.

Fill Requirements

THINK ABOUT IT

Why do the standards for listing sealoff fittings allow only a 25% fill for these fittings, while conduits are permitted to have a 40% fill?

3.3.2 Drains

In humid atmospheres or wet locations where it is likely that water will gain entrance to the raceway system, the raceways should be inclined so that water will not collect in enclosures or on seals, but will be led to low points where it may pass out through integral drains.

If the arrangement of raceway runs makes this impractical, special drain/seal fittings should be used, such as the type shown in *Figure 9*. These fittings prevent water from accumulating above the seal and meet the requirements of *NEC Section 501.15(F)*.

Even if the location is not typically humid or wet, surprising amounts of water may still collect in conduit systems. This is because no conduit system is completely airtight. Alternate increases and decreases in temperature and/or barometric pressure due to weather changes or to the nature of the process carried on in the location where the conduit is installed will cause the introduction of outside air. If this air carries sufficient moisture, it will condense within the system when the temperature drops. Because the internal conditions are unfavorable to evaporation, the resultant water will remain and accumulate over time.

To avoid the accumulation of moisture, install drain/seal fittings with drain covers or fittings with inspection covers. This is a recommended practice even if prevailing conditions at the time of planning or installation do not indicate a moisture problem.

3.3.3 Selection and Installation of Seals and Drains

Always select the proper sealoff fitting for the hazardous location (such as Class 1, Groups A, B, C, or D) and for the proper use in respect to its mounting position. This is particularly critical when the conduit run crosses between hazardous and nonhazardous areas. The improper positioning of a seal may permit hazardous gases or vapors to enter the system beyond the seal and escape into another portion of the hazardous area or enter a nonhazardous area. Some seals are designed to be mounted in any position; others are restricted to horizontal or vertical mounting. *Figure 10* shows various types of seals.

Install the seals on the proper side of the partition or wall, as recommended by the manufacturer. The installation of seals should be made only by trained personnel in strict compliance with the instruction sheets furnished with the seals and **sealing compound**. *NEC Section 501.15(C)(4)* prohibits splices or taps in sealoff fittings. Sealoff fittings are listed by UL for use in Class I hazardous locations with approved sealing compound only. This compound, when properly mixed and poured, hardens into a dense, strong mass that is insoluble in water, not attacked by chemicals, and not softened by heat. It is designed to withstand the pressure of the exploding trapped gases or vapors. Conductors sealed in the compound may be any approved thermoplastic or rubber insulated type. Conductors may or may not be lead-covered.

3.3.4 Sealing Compounds and Dams

Poured seals should be made only by qualified personnel following the manufacturer's instructions. Improperly poured seals serve no purpose. Sealing compound must be approved for the purpose, not be affected by the surrounding

304F09.EPS

Figure 9 ◆ Typical drain seal.

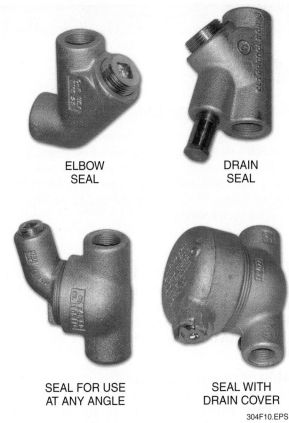

ELBOW
SEAL

DRAIN
SEAL

SEAL FOR USE
AT ANY ANGLE

SEAL WITH
DRAIN COVER

304F10.EPS

Figure 10 ◆ Various types of seals.

atmosphere or liquids, and not have a melting point of less than 200°F (93°C). The sealing compound and dams must also be approved for the type and manufacturer of the fitting. For example,

Crouse-Hinds CHICO® A sealing compound is the only sealing compound approved for use with Crouse-Hinds ECM fittings.

To pack the sealoff, remove the threaded plug or plugs from the fitting and insert the fiber supplied with the packing kit. Tamp the fiber between the wires and the hub before pouring the sealing compound into the fitting, then pour in the sealing cement and reset the threaded plug tightly. The fiber packing prevents the sealing compound from entering the conduit lines in the liquid state.

Sealing compound is poured after the conduit system and seals are installed and the conductors and packing fiber have been installed. Most sealing compound kits contain a powder in a polyethylene bag within an outer container. Remove the bag of powder, fill the outside container, pour in the powder, and mix.

 CAUTION

Always make certain that the sealing compound is compatible for use with the packing material, brand and type of fitting, and type of conductors used in the system.

In practical applications, dozens of seals may be required for a particular installation. Consequently, after the conductors are pulled, each seal in the system is first packed. To prevent the possibility of overlooking a seal, a certain color of paint is normally sprayed on the seal hub to indicate

 *INSIDE TRACK*

Sealing Conductors

On a multi-conductor cable traveling between areas where sealoff fittings are required, strip the insulation from the cable in the middle of the fitting and slightly unravel and separate (birdcage) the conductors. This ensures that the sealing compound fully surrounds each conductor to prevent the passage of gas through the seal. See *NEC Section 501.15(D)(2)*. This may require special consideration when selecting a sealoff fitting.

Pouring Vertical Seals

When pouring a vertical seal that penetrates the top of an enclosure, such as starters or panelboards, open the door while you are pouring the seal to observe whether or not the seal is leaking into the enclosure.

Anti-Seize Compound

When preparing sealing fittings, apply an approved anti-seize compound to the screw threads and interior surfaces of the caps and plugs. This will facilitate removal of the plugs for inspection purposes.

that the seal has been packed. When the sealing compound is poured, a different color paint is sprayed on the seal hub to indicate a finished job. This method permits the job supervisor to visually inspect the conduit run, and if a seal is not painted the appropriate color, he or she knows that the proper installation on this seal was not done; therefore, action can be taken to correct the situation immediately. The sealoff fittings in *Figure 11* are typical of those used. The type in *Figure 11(A)* is for vertical mounting and is provided with a threaded, plugged opening into which the sealing cement is poured. The sealoff in *Figure 11(B)* has an additional plugged opening in the lower hub to facilitate packing fiber around the conductors to form a dam for the sealing cement. *Figure 11(C)* shows a sealoff fitting along with fiber material and sealing compound.

The following guidelines should be observed when preparing sealing compound:

- Use a clean mix vessel for every batch. Particles of previous batches or dirt will spoil the seal.
- Recommended proportions are by volume—usually two parts powder to one part clean water. Slight deviations in these proportions will not affect the results.
- Do not mix more than can be poured in 15 minutes after water is added. Use cold water; warm water increases the setting speed. Stir immediately and thoroughly.
- If the batch starts to set, do not attempt to thin it by adding water or by stirring. Such a procedure will spoil the seal. Discard the partially set material and make a fresh batch. After pouring, close the opening immediately.

INSIDE TRACK

Sealing Compound

A new SpeedSeal™ compound is available that eliminates the packing material required in a typical seal filling. To use this compound, you simply pump the special applicator to mix the compound, and then inject it into the fitting. This greatly reduces the time required to complete a seal.

- Do not pour compound in sub-freezing temperatures or when these temperatures are likely to occur during curing.
- Ensure that the compound level is in accordance with the instruction sheet for the specific fitting.

Most other explosionproof fittings are provided with threaded hubs for securing the conduit as described previously. Typical fittings include switch and junction boxes, conduit bodies, unions and connectors, flexible couplings, explosionproof lighting fixtures, receptacles, and panelboard and motor starter enclosures. A practical representation of these and other fittings is shown in *Figures 12* through *14*.

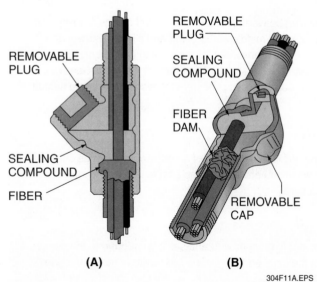

(A)   (B)

304F11A.EPS

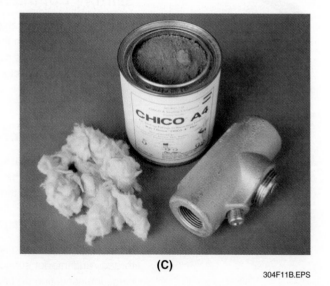

(C)

304F11B.EPS

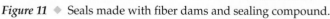

Figure 11 ◆ Seals made with fiber dams and sealing compound.

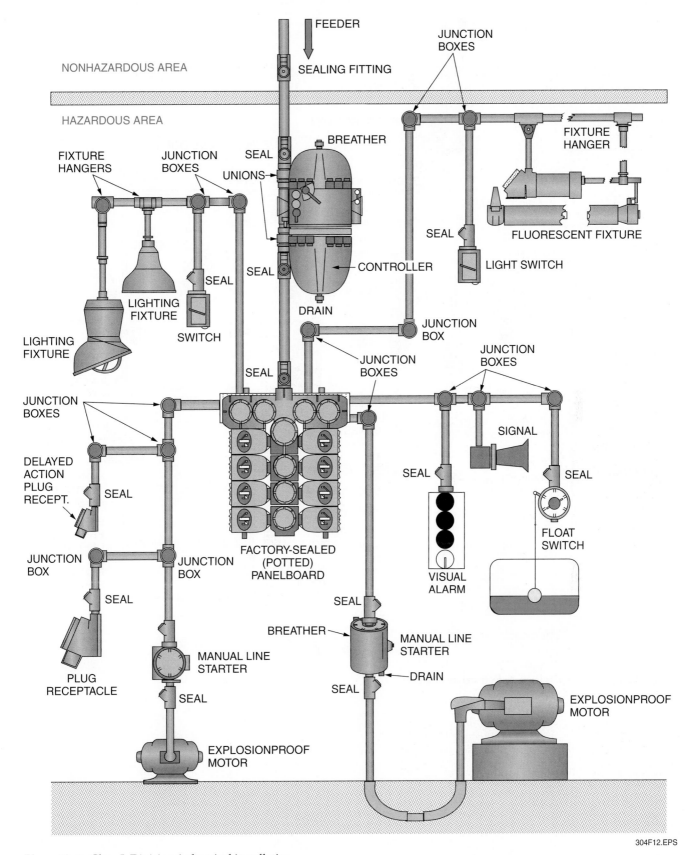

Figure 12 ◆ Class I, Division 1 electrical installation.

304F12.EPS

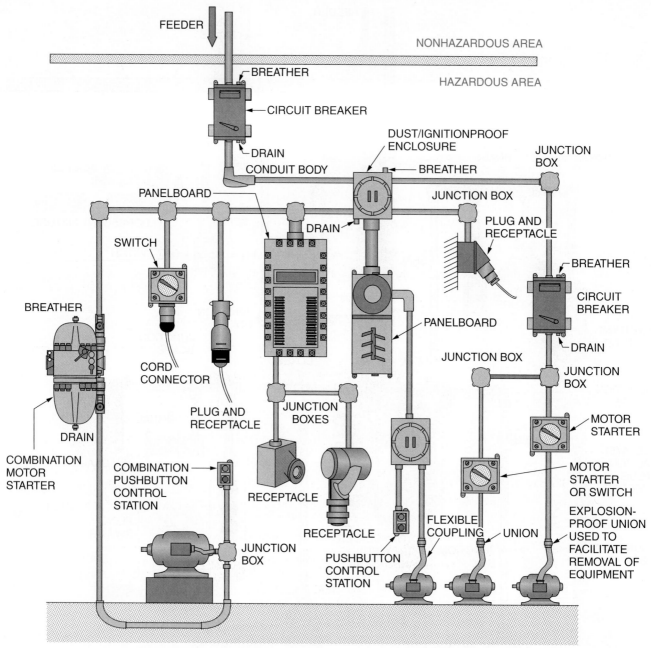

Figure 13 ◆ Class I, Division 2 electrical installation.

304F13.EPS

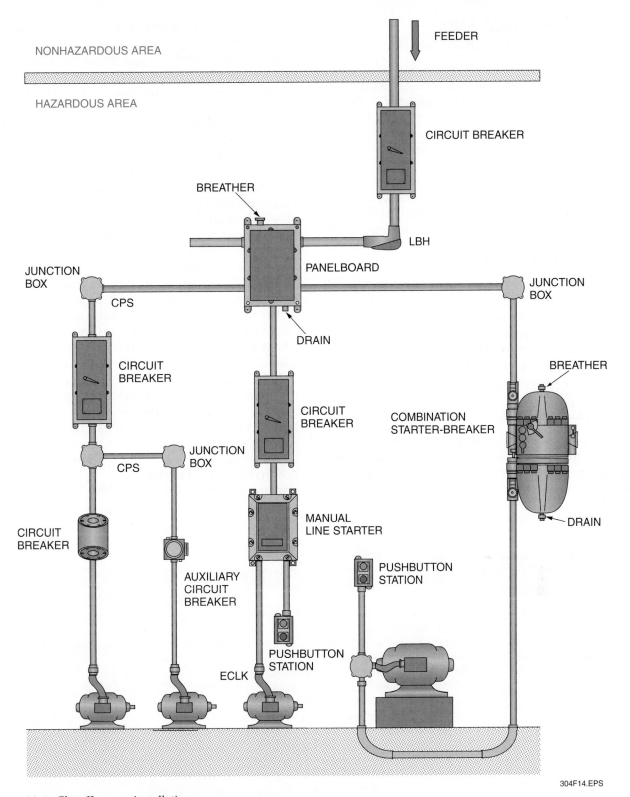

Figure 14 ◆ Class II power installation.

4.0.0 ◆ GARAGES AND SIMILAR LOCATIONS

Garages and similar locations where volatile or flammable liquids are handled or used as fuel in self-propelled vehicles (including automobiles, buses, trucks, tractors, and marina repair garages) are not usually considered critically hazardous locations. However, the entire area up to a level 18" above the floor is considered a Class I, Division 2 location, and certain precautionary measures are required by the *NEC*®. Likewise, any pit or depression below floor level shall be considered a Class I, Division 2 location, and the pit or depression may be judged as a Class I, Division 1 location if it is unvented.

NOTE

In major repair garages that service vehicles using lighter-than-air fuels such as hydrogen and compressed natural gas, the code also classifies the area within 18" of the ceiling if the area is not ventilated. See *NEC Section 511.3(C)(2)*.

Normal raceway (conduit) and wiring may be used for the wiring method above this hazardous level, except where conditions indicate that the area concerned is more hazardous than usual. In this case, the applicable type of explosionproof wiring may be required.

Approved sealoff fittings should be used on all conduit passing from hazardous areas to nonhazardous areas. The requirements set forth in *NEC Article 501* apply to horizontal as well as vertical boundaries of the defined hazardous areas. Raceways embedded in a masonry floor or buried beneath a floor are considered to be within the hazardous area above the floor if any connections or extensions lead into or through such an area. However, conduit systems terminating to an open raceway in an outdoor unclassified area shall not be required to be sealed between the point at which the conduit leaves the classified location and enters the open raceway.

Figure 15 shows a typical automotive service station with applicable *NEC*® requirements. The space in the immediate vicinity of the gasoline dispensing island is denoted as Class I, Division 1. The surrounding area, within a radius of 20' of the island, falls under Class I, Division 2 to a height of 18" above grade. Bulk storage plants for gasoline are subject to comparable restrictions.

NEC Article 514 covers gasoline dispensing and service stations. *NEC Article 511* covers commercial garages.

A summary of *NEC*® rules governing the installation of electrical wiring at and near gasoline dispensing pumps is shown in *Table 7*.

Table 7   *NEC*® Application Rules for Service Stations

Application	*NEC*® Regulation	*NEC*® Reference
Equipment in hazardous locations	All wiring and components must conform to the rules for Class I locations.	*NEC Section 514.4*
Equipment above hazardous locations	All wiring must conform to the rules for such equipment in commercial garages.	*NEC Section 514.4*
Gasoline dispenser	A disconnecting means must be provided for each circuit leading to or through a dispensing pump to disconnect all voltage sources, including feedback, during periods of service and maintenance. An approved seal (sealoff) is required in each conduit entering or leaving a dispenser.	*NEC Sections 514.7 and 514.13*
Grounding	Metal portions of all noncurrent-carrying parts of dispensers must be effectively grounded and bonded.	*NEC Section 514.16*
Underground wiring	Underground wiring installed within 2' of ground level shall be in threaded rigid metal conduit or IMC. If underground wiring is buried 2' or more, rigid nonmetallic conduit may be used along with the types mentioned above; Type MI cable may also be used in some cases.	*NEC Section 514.8*

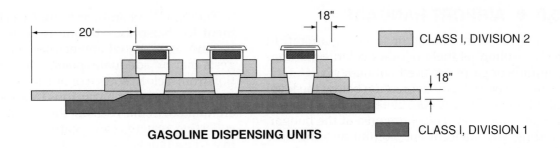

GASOLINE DISPENSING UNITS

CLASS I, DIVISION 2

CLASS I, DIVISION 1

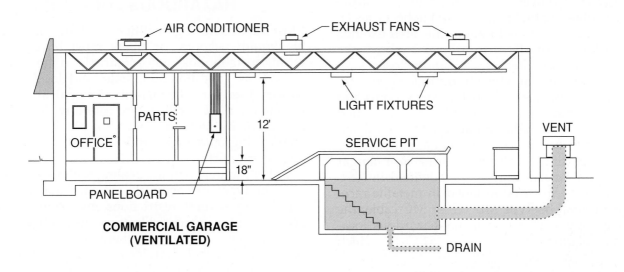

AIR CONDITIONER

EXHAUST FANS

LIGHT FIXTURES

PARTS

OFFICE

12'

VENT

SERVICE PIT

18'

PANELBOARD

COMMERCIAL GARAGE (VENTILATED)

DRAIN

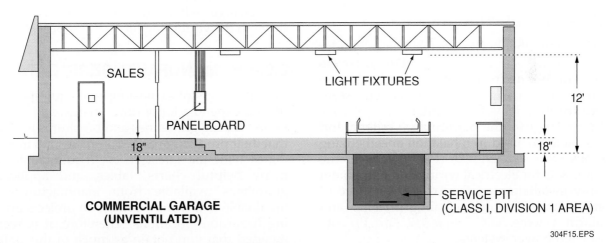

SALES

LIGHT FIXTURES

12'

PANELBOARD

18'

18'

COMMERCIAL GARAGE (UNVENTILATED)

SERVICE PIT (CLASS I, DIVISION 1 AREA)

304F15.EPS

Figure 15 ◆ Commercial service station and garage classifications.

5.0.0 ◆ AIRPORT HANGARS

Buildings used for storing or servicing aircraft in which gasoline, jet fuels, or other volatile flammable liquids or gases are used fall under *NEC Article 513*. In general, any depression below the level of the hangar floor is considered to be a Class I, Division 1 location. The entire area of the hangar, including any adjacent and communicating area not suitably cut off from the hangar, is considered to be a Class I, Division 2 location up to a level of 18" above the floor. The area within 5' horizontally from aircraft power plants, fuel tanks, or structures containing fuel is considered to be a Class I, Division 2 hazardous location; this area extends upward from the floor to a level 5' above the upper surface of wings and engine enclosures.

Adjacent areas in which hazardous vapors are not likely to be released, such as stockrooms and electrical control rooms, should not be classified as hazardous when they are adequately ventilated and effectively cut off from the hangar itself by walls or partitions. All fixed wiring in a hangar not within a hazardous area as defined in *NEC Section 513.3* must be installed in metallic raceways or shall be Type MI, TC, or MC cable; the only exception is wiring in nonhazardous locations as defined in *NEC Section 513.3(D)*, which may be of any type recognized in *NEC Chapter 3*. *Figure 16* summarizes the *NEC®* requirements for airport hangars.

6.0.0 ◆ HOSPITALS

Hospitals and other healthcare facilities fall under *NEC Article 517*. *NEC Article 517, Part II* covers the general wiring in patient areas of healthcare facilities. *NEC Article 517, Part III* covers essential electrical systems for hospitals. *NEC Article 517, Part IV* gives the performance criteria and wiring methods used in inhalation anesthetizing locations. *NEC Article 517, Part V* covers the requirements for electrical wiring and equipment in X-ray installations. *NEC Article 517, Part VI* covers communications, signaling systems, and fire alarm systems. *NEC Article 517, Part VII* covers isolated power systems.

Anesthetizing locations of hospitals are considered Class I, Division 1 to a height of 5' above the floor. Gas storage rooms are designated as Class I, Division 1 throughout. Most of the wiring in these areas, however, can be limited to lighting fixtures only by locating all switches and other devices outside of the hazardous area.

The *NEC®* recommends that electrical equipment for hazardous locations be located in less hazardous areas wherever possible. It also suggests that by adequate, positive-pressure ventilation from a clean source of outside air, the hazards may be reduced or hazardous locations limited or eliminated. In many cases, the installation of dust-collecting systems can greatly reduce the hazards in a Class II area.

7.0.0 ◆ PETROCHEMICAL HAZARDOUS LOCATIONS

Most manufacturing facilities involving flammable liquids, vapors, or fibers must have their wiring installations conform strictly to the *NEC®* as well as governmental, state, and local ordinances. Therefore, the majority of electrical installations for these facilities are carefully designed by experts in the field—either the plant in-house engineering staff or an independent consulting engineering firm.

Industrial installations dealing with petroleum or some types of chemicals are particularly susceptible to several restrictions involving many governmental agencies. Electrical installations for petrochemical plants will therefore have many pages of electrical drawings and specifications, which require approval from all the agencies involved. Once approved, these drawings and specifications must be followed exactly, because any change whatsoever must once again go through the various agencies for approval.

8.0.0 ◆ MANUFACTURERS' DATA

Manufacturers of explosionproof equipment and fittings expend a lot of time, energy, and expense in developing guidelines and brochures to ensure that their products are used correctly and in accordance with the latest *NEC®* requirements. The many helpful charts, tables, and application guidelines available from manufacturers are invaluable to anyone working on projects involving hazardous locations. Therefore, it is recommended that you obtain as much of this data as possible. Once obtained, study this data thoroughly. Doing so will enhance your qualifications for working in hazardous locations of any type. Manufacturers' data is usually available to qualified personnel at little or no cost and can be obtained from local distributors or directly from the manufacturer.

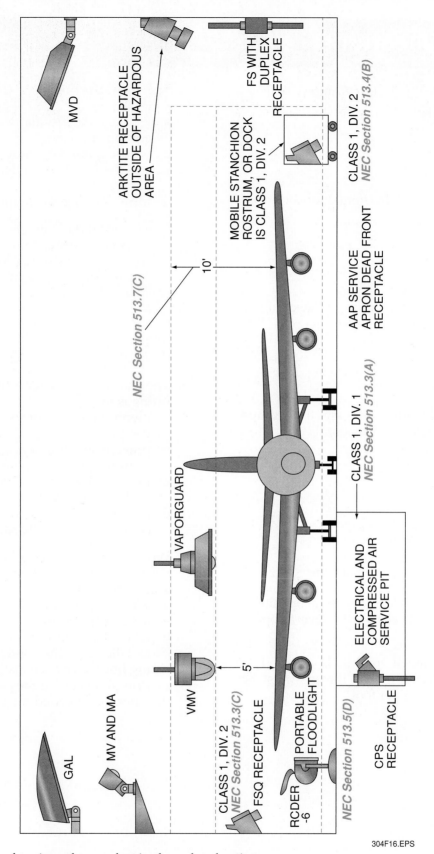

Figure 16 ◆ Sections of an airport hangar showing hazardous locations.

304F16.EPS

1. The *NEC*® lists _____ classification(s) of hazardous atmospheres.
   a. one
   b. two
   c. three
   d. four

2. There is/are _____ division(s) for each classification.
   a. one
   b. two
   c. three
   d. four

3. There is/are _____ group(s) listed under Class I, Division 1.
   a. one
   b. two
   c. three
   d. four

4. Which of the following are the groups listed under Class II, Division 1?
   a. A, B, C, and D
   b. E, F, G
   c. H, I, J
   d. L, M, N

5. When installing circuit breakers in Class II, Division 1 hazardous locations, a _____ must be used.
   a. dust/ignitionproof enclosure
   b. dust-tight enclosure
   c. tight metal enclosure with no openings
   d. standard enclosure with several openings

6. When rigid metal conduit is required in hazardous locations, the threads must be cut at _____ taper per foot.
   a. ½"
   b. ¾"
   c. 1"
   d. 1¼"

7. The main purpose of a union in conduit runs is to _____.
   a. facilitate the installation and removal of equipment
   b. ensure grounding continuity
   c. seal the system from flammable gases or vapors
   d. form tighter joints in the system

8. When installing switches or other arc-producing apparatus in Class I, Division 1 locations, within what distance of the switch (or other apparatus) must sealing fittings be installed?
   a. 6"
   b. 12"
   c. 18"
   d. 24"

9. The purpose of packing fiber in a sealing fitting is to _____.
   a. identify the type of seal and the sealing compound to use
   b. prevent any liquids, gases, or vapors from passing through the fitting
   c. prevent flammable vapors from mixing with the sealing compound
   d. provide a dam to contain the sealing compound until it hardens

10. Which of the following best describes the time to pour in the sealing compound in a sealing fitting?
    a. Within five minutes after it is installed in a raceway system
    b. After the conduit system and seals are installed and the conductors and packing fiber have been installed
    c. Prior to installing the packing fiber
    d. After the threaded plug is set tight

Summary

Any area in which the atmosphere or a material in the area is such that the arcing of operating electrical contacts, components, and equipment may cause an explosion or fire is considered a hazardous location. In all such cases, explosionproof equipment, raceways, and fittings are used to provide an explosionproof wiring system.

The wide assortment of explosionproof equipment now available makes it possible to provide adequate electrical installations under any of the various hazardous conditions. However, you must be thoroughly familiar with all *NEC®* requirements and know what fittings are available, how to install them properly, and where and when to use the various fittings.

Notes

Trade Terms
Introduced in This Module

Approved: Acceptable to the authority having jurisdiction.

Conduit: A tubular raceway such as electrical metallic tubing (EMT); rigid metal conduit, rigid nonmetallic conduit, etc.

Conduit body: A separate portion of a conduit or tubing system that provides access through removable covers to the interior of the system at a junction of two or more sections of the system or at a terminal point of the system.

Equipment: A general term including material, fittings, devices, appliances, fixtures, apparatus, and the like used as a part of (or in connection with) an electrical installation.

Explosionproof: Designed and constructed to withstand an internal explosion without creating an external explosion or fire.

Explosionproof apparatus: Apparatus enclosed in a case that is capable of withstanding an explosion of a specified gas or vapor that may occur within it; also capable of preventing the ignition of a specified gas or vapor surrounding the enclosure by sparks, flashes, or explosion of the gas or vapor within; which operates at such an external temperature that a surrounding flammable atmosphere will not be ignited thereby.

Hazardous (classified) location: A location in which ignitable vapors, dust, or fibers may cause a fire or explosion.

Sealing compound: The material poured into an electrical fitting to seal and minimize the passage of vapors.

Sealoff fittings: Fittings required in conduit systems to prevent the passage of gases, vapors, or flames from one portion of the electrical installation to another through the conduit. Also referred to as sealing fittings or seals.

Additional Resources

This module is intended to present thorough resources for task training. The following reference works are suggested for further study. These are optional materials for continued education rather than for task training.

Code Digest. Latest Edition. Syracuse, NY: Crouse-Hinds.

National Electrical Code® Handbook. Latest Edition. Quincy, MA: National Fire Protection Association.

NCCER makes every effort to keep these textbooks up-to-date and free of technical errors. We appreciate your help in this process. If you have an idea for improving this textbook, or if you find an error, a typographical mistake, or an inaccuracy in NCCER's Contren® textbooks, please write us, using this form or a photocopy. Be sure to include the exact module number, page number, a detailed description, and the correction, if applicable. Your input will be brought to the attention of the Technical Review Committee. Thank you for your assistance.

Instructors – If you found that additional materials were necessary in order to teach this module effectively, please let us know so that we may include them in the Equipment/Materials list in the Annotated Instructor's Guide.

Write: Product Development and Revision
National Center for Construction Education and Research
3600 NW 43rd St., Bldg. G, Gainesville, FL 32606

Fax: 352-334-0932

E-mail: curriculum@nccer.org

Craft   Module Name

Copyright Date   Module Number   Page Number(s)

Description

(Optional) Correction

(Optional) Your Name and Address

Overcurrent Protection

Lightning

Lightning is nature's electrical charge. An average strike can easily release 250 kilowatt-hours of energy, enough to operate a 100-watt light bulb continuously for more than three months. At 54,032 degrees Fahrenheit lightning is five times as hot as the surface of the sun. All of this energy is contained in a bolt about the width of a human thumb. Researchers are working to find ways to harvest lightning as an alternative source of energy.

26305-08

26305-08
Overcurrent Protection

Topics to be presented in this module include:

Overview

Overcurrent protective devices are used primarily to protect the conductors from excessive heat caused by overcurrent. However, some overcurrent protective devices also function to protect equipment. There are three categories of overcurrent: overloads, short circuits, and ground faults.

An overload occurs when the load in amperes exceeds the normal operating or intended current level. A short circuit occurs when an unintended path develops from one current-carrying circuit conductor to another, either bypassing the load entirely or in some cases, partially bypassing the load. Ground faults only occur in grounded electrical systems in which one current-carrying conductor is intentionally grounded. In a ground fault situation, an ungrounded current-carrying conductor makes unintended contact with earth ground or a conductive part that is grounded. This condition provides a path for the electrons to get back to the source by using the earth ground as a conductive path back to the point at which the system is grounded.

Overcurrent protective devices must be selected based on the voltage to which they will be exposed, the trip ampere rating for the circuit, and the interrupting rating (in amperes).

Objectives

When you have completed this module, you will be able to do the following:

1. Apply the key *National Electrical Code®* (*NEC®*) requirements regarding overcurrent protection.
2. Check specific applications for conformance to *NEC®* sections that cover short circuit current, fault currents, interrupting ratings, and other sections relating to overcurrent protection.
3. Determine let-through current values (peak and rms) when current-limiting overcurrent devices are used.
4. Select and size overcurrent protection for specific applications.

Trade Terms

Ampere rating
Ampere squared seconds (I^2t)
Amperes interrupting capacity (AIC)
Arcing time
Clearing time
Current-limiting device
Fast-acting fuse
Inductive load
Melting time
NEC® dimensions
Overload
Peak let-through (Ip)
Root-mean-square (rms)
Semiconductor fuse
Short circuit current
Single phasing
UL classes
Voltage rating

Required Trainee Materials

1. Pencil and paper
2. Appropriate personal protective equipment
3. Copy of the latest edition of the *National Electrical Code®*

Prerequisites

Before you begin this module, it is recommended that you successfully complete *Core Curriculum; Electrical Level One; Electrical Level Two; Electrical Level Three*, Modules 26301-08 through 26304-08.

This course map shows all of the modules in *Electrical Level Three*. The suggested training order begins at the bottom and proceeds up. Skill levels increase as you advance on the course map. The local Training Program Sponsor may adjust the training order.

ELECTRICAL LEVEL THREE

26311-08
Motor Controls

26310-08
Voice, Data, and Video

26309-08
Motor Calculations

26308-08
Commercial Electrical Services

26307-08
Transformers

26306-08
Distribution Equipment

26305-08
Overcurrent Protection

26304-08
Hazardous Locations

26303-08 Practical Applications of Lighting

26302-08 Conductor Selection and Calculations

26301-08 Load Calculations – Branch and Feeder Circuits

ELECTRICAL LEVEL TWO

ELECTRICAL LEVEL ONE

CORE CURRICULUM:
Introductory Craft Skills

305CMAP.EPS

1.0.0 ◆ INTRODUCTION

Electrical distribution systems are often quite complicated. They cannot be absolutely fail-safe. Circuits are subject to destructive overcurrents. Factors that contribute to the occurrence of such overcurrents are harsh environments, general deterioration, accidental damage or damage from natural causes, excessive expansion, or electrical distribution system overload. Reliable protective devices prevent or minimize costly damage to transformers, conductors, motors, and many other components and loads that make up the complete distribution system. Reliable circuit protection is essential to avoid the severe monetary losses that can result from power blackouts and prolonged downtime of facilities. It is the need for reliable protection, safety, and freedom from fire hazards that has made overcurrent protective devices absolutely necessary in all electrical systems, both large and small.

Overcurrent protection of electrical circuits is so important that the *NEC®* devotes an entire article to this subject. *NEC Article 240* provides the general requirements for overcurrent protection and overcurrent protective devices. *NEC Article 240, Parts I through VIII* cover systems 600V (nominal) and under; *NEC Article 240, Part IX* covers overcurrent protection over 600V (nominal). This entire article will be covered in this module, along with practical examples.

All conductors must be protected against overcurrents in accordance with their ampacities as set forth in *NEC Section 240.4*. They must also be protected against **short circuit current** damage as required by *NEC Sections 110.10 and 240.1*. The two basic types of overcurrent protective devices that are in common use are fuses and circuit breakers.

1.1.0 Fault Currents

The function of overcurrent protection is to protect electrical conductors and equipment from damage due to fault currents. This is done by opening the circuit as quickly as necessary but not so fast as to cause nuisance trips from normal surges such as those that occur during motor starting. Types of fault current include the following:

- Overloads
- Short circuits
- Ground faults
- Arc faults

1.1.1 Overloads

Overloads are most often between one and six times the normal current level. Usually, they are caused by temporary surge currents that occur when motors are started up or transformers are energized. Such overload currents or transients are normal occurrences. Since they are of brief duration, any temperature rise is momentary and has no harmful effect on the circuit components, so long as the system is designed to accommodate them.

Continuous overloads can result from defective motors (such as worn motor bearings), overloaded equipment, or too many loads on one circuit. Such sustained overloads are destructive and must be cut off by protective devices before they damage the distribution system or system loads. However, since they are of relatively low magnitude compared to short circuit currents, removal of the overload current within a few seconds will generally prevent equipment damage. A sustained overload current results in overheating of conductors and other components and will cause deterioration of insulation, which may eventually result in severe damage and short circuits if not interrupted.

1.1.2 Short Circuits

A short circuit is a conducting connection, whether intentional or accidental, between any of the conductors of an electrical system, either line-to-line or line-to-ground.

The **amperes interrupting capacity (AIC)** rating of a circuit breaker or fuse is the maximum fault current that the breaker will safely interrupt. This AIC rating is at rated voltage and frequency.

Whereas overload currents occur at rather modest levels, a short circuit or fault current can be many hundreds of times larger than the normal operating current. A high-level fault may be 50,000A or larger). If not cut off within a matter of a few thousandths of a second, damage and destruction can become rampant—there can be severe insulation damage, melting of conductors, vaporization of metal, ionization of gases, arcing, and fires. Simultaneously, high-level short circuit currents can develop huge magnetic field stresses. The magnetic forces between busbars and other conductors can be many hundreds of pounds per lineal foot; even heavy bracing may not be adequate to keep them from being warped or distorted beyond repair.

NEC Section 110.9 states that equipment intended to interrupt current at fault levels (fuses and circuit breakers) must have an interrupting rating sufficient for the nominal circuit voltage and the current that is available at the line terminals of the equipment.

Equipment intended to interrupt current at other than fault levels must have an interrupting rating at nominal circuit voltage sufficient for the current that must be interrupted.

These *NEC*® statements mean that fuses and circuit breakers (and their related components) designed to interrupt fault or operating currents (open the circuit) must have a rating sufficient to withstand such currents. This section emphasizes the difference between clearing fault level currents and clearing operating currents. Protective devices such as fuses and circuit breakers are designed to clear fault currents and therefore must have short circuit interrupting ratings sufficient for fault levels. Equipment such as contactors and safety switches have interrupting ratings for currents at other than fault levels. Thus, the interrupting rating of electrical equipment is now divided into two parts:

- Current at fault (short circuit) levels
- Current at operating levels

Most people are familiar with the normal current-carrying **ampere rating** of fuses and circuit breakers. For example, if an overcurrent protective device is designed to open a circuit when the circuit load exceeds 20A for a given time period, as the current approaches 20A, the overcurrent protective device begins to overheat. If the current barely exceeds 20A, the circuit breaker will open normally or a fuse link will melt after a given period of time with little, if any, arcing. For example, if 40A of current were instantaneously applied to the circuit, the overcurrent protective device would open, but again with very little arcing. However, if a ground fault occurs on the circuit that ran the amperage up to 5,000A, for example, an explosion effect would occur within the protective device. One simple indication of this is the blackened windows of plug fuses.

If this fault current exceeds the interrupting rating of a fuse or circuit breaker, the protective device can be damaged or destroyed; such current can also cause severe damage to equipment and injure personnel. Therefore, selecting overcurrent protective devices with the proper interrupting capacity is extremely important in all electrical systems.

There are several factors that must be considered when calculating the required interrupting capacity of an overcurrent protective device. *NEC*

Section 110.10 states that the overcurrent protective devices, total impedance, component short circuit current ratings, and other characteristics of the circuit to be protected shall be selected and coordinated to permit the circuit protective devices used to clear a fault to do so without extensive damage to the electrical components of the circuit. This fault shall be assumed to be either between two or more of the circuit conductors, or between any circuit conductor and the grounding conductor or enclosing metal raceway.

The component short circuit rating is a current rating given to conductors, switches, circuit breakers, and other electrical components, which, if exceeded by fault currents, will result in extensive damage to the component. The rating is expressed in terms of time intervals and/or current values. Short circuit damage can be the result of heat generated or the electromechanical force of a high-intensity magnetic field.

The *NEC*®'s intent is that the design of a system must be such that short circuit currents cannot exceed the short circuit current ratings of the components selected as part of the system. Given specific system components and the level of available short circuit currents that could occur, overcurrent protective devices (mainly fuses and/or circuit breakers) must be used that will limit the energy let-through of fault currents to levels within the withstand ratings of the system components.

1.1.3 Ground Faults

A ground fault is an unintentional conducting connection between any of the conductors of an electrical system and the conducting material that encloses the conductors or any conducting material that is grounded or may become grounded. A low-resistance ground fault behaves in the same way as a short circuit and the protective device operates to clear the fault. A ground fault of one conductor of the circuit will also be responded to by the normal circuit protective response. Ground faults often progress to line-to-line and line-to-ground faults and can be very destructive.

In many power distribution systems, it is common to use protective relays to detect and respond to levels of ground fault below full-load current values. A dedicated ground fault relay may be applied. There are electronic trip units available for circuit breakers that add ground fault protection to the breaker. A typical response to a ground fault in a distribution system is to trip immediately when low-level ground faults occur or after a time delay for high-level ground faults. *NEC Section 230.95* discusses the requirements for ground fault protection of solidly grounded, wye-connected systems not over 600V for service disconnects rated at 1,000A and above.

1.1.4 Arc Faults

While faults of any type causing trips are often called shorts, an arcing fault is the result of an intermittent or higher resistance fault than a short circuit. Most arcing faults begin as ground faults and often grow to line-to-line and line-to-ground faults.

Arc faults can be catastrophic and expose the worker to the hazards of electrical arc and blast. This includes thermal burns, flash burns to the eyes, inhalation of molten metal vapor, injuries caused by shrapnel and molten metal splatter, the concussive effects of blast forces to the body, and hearing damage.

In systems 1,000V and above, arc fault currents are near the magnitude of short circuit currents and usually detected and cleared in the same way. For 480V circuits, however, the arc current may be as low as 38% of the available bolted fault current and seldom exceeds 60%. Very often the arc fault current in systems at or below 600V is seen as an overload instead of a fault, and will not act on the instantaneous trip or clearing value of the protective device.

2.0.0 ◆ FUSES

The fuse is a reliable overcurrent protective device. It consists of a fusible link or links encapsulated in a tube and connected to contact terminals. The electrical resistance of the link is so low that it simply acts as a conductor. However, when destructive currents occur, the link quickly melts and opens the circuit to protect conductors and other circuit components and loads. Fuse characteristics are stable and do not require periodic maintenance or testing.

2.1.0 Types of Fuses

NEC Article 240, Part V contains the requirements associated with plug fuses, plug fuseholders, and adapters; *NEC Article 240, Part VI* lists those requirements applying to cartridge fuses and cartridge fuseholders. Plug fuses are normally used in general-purpose branch circuits. Cartridge fuses may be found in branch circuits, feeder circuits, motor circuits, and many other special applications.

Per *NEC Section 240.50(A)*, plug fuses are only permitted in circuits not exceeding 125V between conductors or in circuits supplied by a system having a grounded neutral where the line-to-neutral voltage does not exceed 150V.

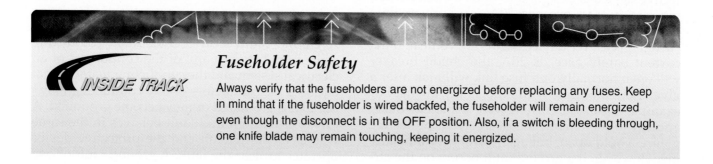

Fuseholder Safety

Always verify that the fuseholders are not energized before replacing any fuses. Keep in mind that if the fuseholder is wired backfed, the fuseholder will remain energized even though the disconnect is in the OFF position. Also, if a switch is bleeding through, one knife blade may remain touching, keeping it energized.

Plug fuses rated at 15A or less have hexagonal windows or caps so that they may easily be distinguished from fuses of higher ampere ratings. This is a requirement of *NEC Section 240.50(C)*. Plug fuses with an ampere rating above 15A have a round window.

There are two screw-in configurations associated with plug fuses, Edison-base and Type S. An adapter is available to convert an Edison-base socket to a Type S socket. The Edison-base socket is a standard screw-in base, which allows any ampere-rated Edison fuse to be installed. The Type S socket is matched to a specific-rated fuse and will not accept any other fuse except that rating.

NEC Section 240.52 prohibits the use of Edison-base fuseholders in new installations except where they are made to accept Type S fuses using the adapter method.

Per *NEC Section 240.60(A)*, cartridge fuses and fuseholders rated at 300V are only permitted to be used in circuits not exceeding 300V between conductors, or in single-phase, line-to-neutral circuits supplied from a three-phase, four-wire, solidly grounded, neutral source where the line-to-neutral voltage does not exceed 300V.

2.2.0 Voltage Rating

Most low-voltage power distribution fuses have 250V or 600V ratings. The **voltage rating** of a fuse must be at least equal to or greater than the circuit voltage. It can be higher, but never lower. For example, a 600V fuse can be used in a 240V circuit.

The voltage rating of a fuse depends upon its capability to open a circuit under an overcurrent condition. Specifically, the voltage rating determines the ability of the fuse to suppress the internal arcing that occurs after a fuse link melts and an arc is produced (**arcing time**). If a fuse is used with a voltage rating lower than the circuit voltage, arc suppression will be impaired and, under some fault current conditions, the fuse may not safely clear the overcurrent. Special consideration is necessary for **semiconductor fuse** applications in which a fuse of a certain voltage rating is used on a lower-voltage circuit.

2.3.0 Ampere Rating

Every fuse has a specific ampere rating. When selecting the ampere rating, consideration must be given to the size and type of load and *NEC®* requirements. The ampere rating of a fuse should normally not exceed the current-carrying capacity of the circuit. For instance, if a conductor is rated to carry 20A, a 20A fuse is the largest that should be used. However, there are specific circumstances in which the ampere rating is permitted to be greater than the current-carrying capacity of the circuit. A typical example is the motor circuit; dual-element fuses are generally permitted to be sized up to 175% and nontime-delay fuses up to 300% of the motor full-load amperes. Generally, the ampere rating of a fuse and switch combination should be selected at 125% of the continuous load current (this usually corresponds to the circuit capacity, which is also selected at 125% of the load current). There are exceptions, such as when the fuse-switch combination is approved for continuous operation at 100% of its rating.

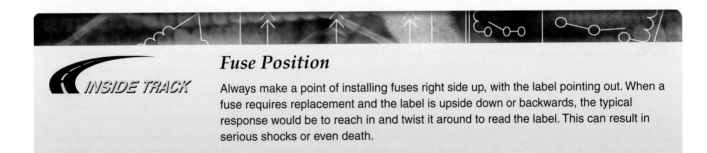

Fuse Position

Always make a point of installing fuses right side up, with the label pointing out. When a fuse requires replacement and the label is upside down or backwards, the typical response would be to reach in and twist it around to read the label. This can result in serious shocks or even death.

2.4.0 Interrupting Rating

A protective device must be able to withstand the destructive energy of short circuit currents. If a fault current exceeds a level beyond the capability of the protective device, the device may actually rupture, causing additional damage. Therefore, it is important to choose a protective device that can sustain the largest potential short circuit currents. The rating that defines the capacity of a protective device to maintain its integrity when reacting to fault currents is called its interrupting rating.

NEC Section 110.9 requires equipment intended to interrupt current at fault levels to have an interrupting rating sufficient for the current that must be interrupted. Interrupting rating and interrupting capacity were covered in your Level Two training; more advanced material is presented in this module.

2.5.0 Selective Coordination

The coordination of protective devices prevents system power outages or blackouts caused by overcurrent conditions. When only the protective device nearest a faulted circuit opens and larger upstream fuses remain closed, the protective devices are selectively coordinated (they discriminate). The word selective is used to denote total coordination (isolation of a faulted circuit by the opening of only the localized protective device).

Figure 1 shows the minimum ratios of ampere rating of low peak fuses that are required to provide selective coordination of upstream and downstream fuses.

2.6.0 Current Limitation

If a protective device cuts off a short circuit current in less than one half cycle, before it reaches its total available (and highly destructive) value, the

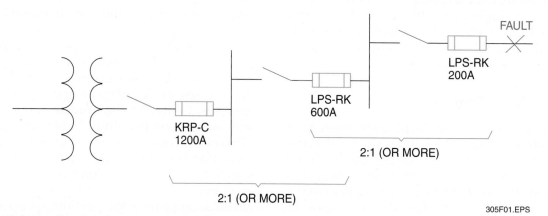

305F01.EPS

Figure 1 ◆ Selective coordination.

Fault Current Protection

INSIDE TRACK

In a fully rated system, each overcurrent protective device has an interrupting rating that is equal to or greater than the maximum available fault current. Fully rated systems are approved for use in all locations.

In a series rated system, the main upstream breaker has an interrupting capacity equal to or greater than the maximum available fault current, but the downstream protective devices can be rated at lower values and therefore, will have a lower installed cost. Series rated systems have their place in the industry in areas where the tripping of a main breaker would cause a nuisance but would not result in an unsafe condition or the interruption of a critical process. For this reason, series rated systems are not approved for use in all locations. (See also *NEC Section 110.22.*)

device is a **current-limiting device**. Most modern fuses are current-limiting devices. They restrict fault currents to such low values that a high degree of protection is given to circuit components against even very high short circuit currents. They permit breakers with lower interrupting ratings to be used and can reduce bracing of bus structures. They also minimize the need for other components to have high short circuit current ratings. If not limited, short circuit currents can reach levels of 30,000A or 40,000A or higher in the first half cycle (0.008 second at 60Hz) after the start of a short circuit. The heat that can be produced in circuit components by the immense energy of short circuit currents can cause severe insulation damage or even an explosion. At the same time, the huge magnetic forces developed between conductors can crack insulators and distort and destroy bracing structures. Thus, it is extremely important that a protective device limit fault currents before they can reach their full potential level.

A noncurrent-limiting protective device, by permitting a short circuit current to build up to its full value, can let an immense amount of destructive short circuit heat energy through before opening the circuit, as shown in *Figure 2*. On the other hand, a current-limiting device (such as a current-limiting fuse) has such a high speed of response that it cuts off a short circuit before it can build up to its full peak value, as shown in *Figure 3*.

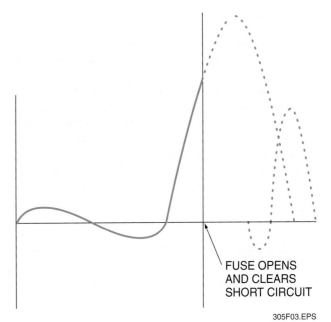

FUSE OPENS
AND CLEARS
SHORT CIRCUIT

305F03.EPS

Figure 3 ◆ Characteristics of a current-limiting fuse.

3.0.0 ◆ OPERATING PRINCIPLES OF FUSES

This section describes the operating characteristics of various types of cartridge fuses. These fuses include the nontime-delay and the dual-element, time-delay fuses.

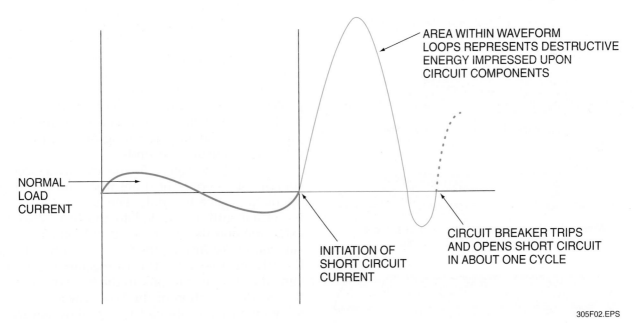

AREA WITHIN WAVEFORM
LOOPS REPRESENTS DESTRUCTIVE
ENERGY IMPRESSED UPON
CIRCUIT COMPONENTS

NORMAL
LOAD
CURRENT

INITIATION OF
SHORT CIRCUIT
CURRENT

CIRCUIT BREAKER TRIPS
AND OPENS SHORT CIRCUIT
IN ABOUT ONE CYCLE

305F02.EPS

Figure 2 ◆ Characteristics of a noncurrent-limiting protective device.

3.1.0 Nontime-Delay Fuses

The basic component of a fuse is the link. Depending upon the ampere rating of the fuse, the single-element, nontime-delay fuse may have one or more links. They are electrically connected to the end blades (or ferrules) and enclosed in a tube or cartridge surrounded by an arc-quenching filler material.

Under normal operation, when the fuse is operating at or near its ampere rating, it simply functions as a conductor. However, as illustrated in *Figure 4*, if an overload current occurs and persists for more than a short interval of time, the temperature of the link eventually reaches a level that causes a restricted segment of the link to melt; as a result, a gap is formed and an electric arc established. As the arc causes the link metal to burn back, the gap becomes progressively larger. The

electrical resistance of the arc eventually reaches such a high level that the arc cannot be sustained and is extinguished; the fuse will have then completely cut off all current flow in the circuit. Suppression or quenching of the arc is accelerated by the filler material.

Overload current normally falls within the region of between one and six times normal current, resulting in currents that are quite high. Consequently, a fuse may be subjected to short circuit currents of 30,000A to 40,000A or higher. The response of current-limiting fuses to such currents is extremely fast. The restricted sections of the fuse link will simultaneously melt within a matter of two-thousandths or three-thousandths of a second in the event of a high-level fault current.

The high resistance of the multiple arcs, together with the quenching effects of the filler particles, results in rapid arc suppression and clearing of the circuit. Again, refer to *Figure 3*. The short circuit current is cut off in less than one half cycle, long before the short circuit current can reach its full value.

3.2.0 Dual-Element, Time-Delay Fuses

Unlike single-element fuses, the dual-element, time-delay fuse can be applied in circuits subject to temporary motor overloads and surge currents to provide both high-performance short circuit and overload protection. Oversizing to prevent nuisance openings is not necessary with this type of fuse. The dual-element, time-delay fuse contains two distinctly separate types of elements. Electrically, the two elements are connected in series. The fuse links (similar to those used in the nontime-delay fuse) perform the short circuit protection function; the overload element provides protection against low-level overcurrents or overloads and will hold an overload that is five times greater than the ampere rating of the fuse for a minimum time of 10 seconds.

As shown in *Figure 5*, the overload section consists of a copper heat absorber and a spring-operated trigger assembly. The heat absorber bar is permanently connected to the heat absorber extension and the short circuit link on the opposite end of the fuse by the S-shaped connector of the trigger assembly. The connector electrically joins the short circuit link to the heat absorber in the overload section of the fuse. These elements are joined by a calibrated fusing alloy. An overload current causes heating of the short circuit link connected to the trigger assembly. The transfer of heat from the short circuit link to the heat absorber begins to raise the temperature of the

LINK

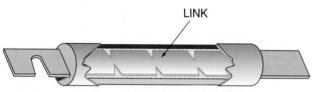

Cut-away view of single-element fuse.

Under sustained overload, a section of the link melts and an arc is established.

The open single-element fuse after opening a circuit overload.

When subjected to a short circuit, several sections of the fuse link melt almost instantly.

The appearance of an open single-element fuse after opening a short circuit.

305F04.EPS

Figure 4 ◆ Characteristics of a single-element fuse.

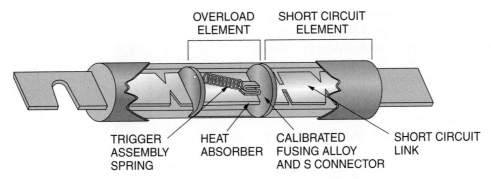

OVERLOAD ELEMENT

SHORT CIRCUIT ELEMENT

TRIGGER ASSEMBLY SPRING

HEAT ABSORBER

CALIBRATED FUSING ALLOY AND S CONNECTOR

SHORT CIRCUIT LINK

The true dual-element fuse has distinct and separate overload and short circuit elements.

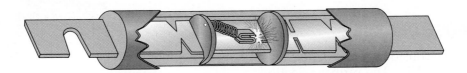

Under sustained overload conditions, the trigger spring fractures the calibrated fusing alloy and releases the connector.

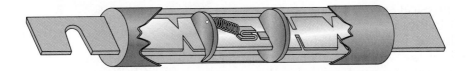

The dual-element fuse after opening under an overload.

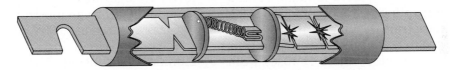

Like the single-element fuse, a short circuit current causes the restricted portions of the short circuit elements to melt and arcing to burn back the resulting gaps until the arcs are suppressed by the arc-quenching material and increased arc resistance.

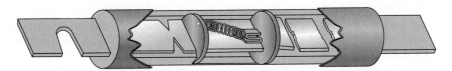

The dual-element fuse after opening under a short circuit condition.

305F05.EPS

Figure 5 ◆ Characteristics of a dual-element, time-delay fuse.

heat absorber. If the overload is sustained, the temperature of the heat absorber eventually reaches a level that permits the trigger spring to fracture the calibrated fusing alloy and pull the connector free of the short circuit link and the heat absorber. As a result, the short circuit link is electrically disconnected from the heat absorber, the conducting path through the fuse is opened, and the overload current is interrupted. A critical aspect of the fusing alloy is that it retains its original characteristics after repeated temporary overloads without degradation.

The advantages of dual-element fuses are:

• Provide motor overload, ground fault, and short circuit protection
• Permit the use of smaller and less costly switches
• Give a higher degree of short circuit protection (greater current limitation) in circuits in which surge currents or temporary overloads occur
• Simplify and improve blackout prevention (selective coordination)

4.0.0 ◆ UL FUSE CLASSES

Safety is the primary consideration of Underwriters Laboratories, Inc. (UL). The proper selection, overall functional performance, and reliability of a product are factors that are not within the basic scope of UL activities. However, to develop its safety test procedures, UL does develop the basic performance and physical specifications of standards of a product. In the case of fuses, these standards have culminated in the establishment of distinct UL classes of low-voltage (600V or less) fuses. Various UL fuse classes are described in the following paragraphs.

UL Class R (rejection) fuses are high-performance ¹⁄₁₀A to 600A units, 250V and 600V, having a high degree of current limitation and a short circuit interrupting rating of up to 200,000A (root-mean-square [rms] symmetrical). This type of fuse is designed to be mounted in rejection-type fuse clips to prevent older Class H fuses from being installed. Since Class H fuses are not current-limiting devices and are recognized by UL as having only a 10,000A interrupting rating, serious damage could result if a Class H fuse were inserted in a system designed for Class R fuses. Consequently, *NEC Section 240.60(B)* requires fuseholders for current-limiting fuses to reject noncurrent-limiting fuses.

Figure 6 shows standard Class H and Class R cartridge fuses. A grooved ring in one ferrule of the Class R fuse provides the rejection feature of the Class R fuse in contrast to the lower interrupting capacity, nonrejection type. *Figure 7* shows fuse rejection clips for Class R fuses.

Class CC fuses are 600V, 200,000A interrupting rating, branch circuit fuses with overall dimensions of ¹⁵⁄₃₂" × 1½". Their design incorporates rejection features that allow them to be inserted into rejection fuseholders and fuse blocks that reject all

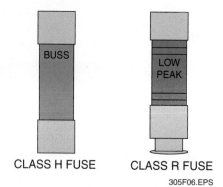

CLASS H FUSE   CLASS R FUSE

305F06.EPS

Figure 6 ◆ Comparison of Class H and Class R fuses.

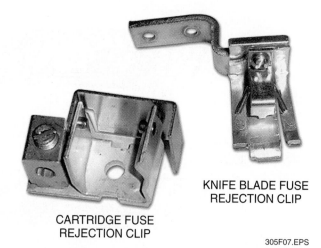

KNIFE BLADE FUSE REJECTION CLIP

CARTRIDGE FUSE REJECTION CLIP

305F07.EPS

Figure 7 ◆ Fuse rejection clips.

lower voltage, lower interrupting rating ¹⁵⁄₃₂" × 1½" fuses. They are available from ¹⁄₁₀A through 30A.

Class G fuses are 300V, 100,000A interrupting rating branch circuit fuses that are size rejecting to eliminate overfusing. The fuse diameter is ¹³⁄₃₂", while the length varies from 1⁵⁄₁₆" to 2¼". They are available in ratings from 1A through 60A.

Class H fuses are 250V and 600V, 10,000A interrupting rating branch circuit fuses that may be renewable or nonrenewable. They are available in ampere ratings of 1A through 600A.

Class J fuses are rated to interrupt 200,000A. They are UL labeled as current limiting, are rated for 600VAC, and are not interchangeable with other classes.

Class K fuses are listed by UL as K-1, K-5, or K-9. Each subclass has designated **ampere squared seconds (I^2t)** and **peak let-through (Ip)** maximums. These are dimensionally the same as Class H fuses (*NEC®* dimensions), and they can have interrupting ratings of 50,000A, 100,000A, or 200,000A. These fuses are current-limiting devices; however, they are not marked current limiting since they do not have a rejection feature.

Class L fuses are available in ampere ratings of 601A through 6,000A, and are rated to interrupt 200,000A. They are labeled current limiting and are rated for 600VAC. They are intended to be bolted into their mountings and are not normally used in clips. Some Class L fuses have time-delay features for all-purpose use.

Class T fuses are 300V and 600V, with ampere ratings from 1A through 1,200A. They are physically very small and can be applied where space is at a premium. They are fast-acting fuses with an interrupting rating of 200,000A.

4.1.0 Branch Circuit Listed Fuses

Branch circuit listed fuses are designed to prevent the installation of fuses that cannot provide a comparable level of protection to equipment. The characteristics of branch circuit fuses are as follows:

- They must have a minimum interrupting rating of 10,000A.
- They must have a minimum voltage rating of 125V.
- They must be size rejecting such that a fuse of a lower voltage rating cannot be installed in the circuit.
- They must be size rejecting such that a fuse with a current rating higher than the fuseholder rating cannot be installed in the circuit.

4.2.0 Medium-Voltage Fuses

As defined in *ANSI/IEEE 40-1981*, fuses above 600V are classified as general-purpose, current-limiting; backup current-limiting; or expulsion types.

- *General-purpose current-limiting fuses* – Fuses that are capable of interrupting all currents from the rated interrupting current down to the current that causes melting of the fusible element in one hour.
- *Backup current-limiting fuses* – Fuses that are capable of interrupting all currents from the maximum rated interrupting current down to the rated minimum interrupting current.
- *Expulsion fuses* – Vented fuses in which the expulsion effect of gases produced by the arc and lining of the fuseholder, either alone or aided by a spring, extinguishes the arc.

In the definitions just given, the fuses are defined as either expulsion or current-limiting types. A current-limiting fuse is a sealed, non-venting fuse that, when melted by a current within its interrupting rating, produces arc voltages exceeding the system voltage, which in turn forces the current to zero. The arc voltages are produced by introducing a series of high resistance arcs within the fuse. The result is a fuse that typically interrupts high fault currents within the first half cycle of the fault.

In contrast, an expulsion fuse depends on one arc to initiate the interruption process. The arc acts as a catalyst, causing the generation of de-ionizing gas from its housing. The arc is then elongated either by the force of the gases created or a spring. At some point, the arc elongates far enough to prevent a restrike after passing through a current zero. Therefore, it is not atypical for an expulsion fuse to take many cycles to clear.

Knife Blade Fuses

In a Class R knife blade fuse, a notch in one blade provides the rejection feature. This picture shows both Class H and Class R knife blade fuses.

CLASS H FUSE

CLASS R FUSE

305SA01.EPS

Many of the rules for applying expulsion fuses and current-limiting fuses are the same, but because the current-limiting fuse operates much faster on high fault currents, some additional rules must be applied.

Three basic factors must be considered when applying any fuse: voltage, continuous current-carrying capacity, and interrupting rating.

- *Voltage* – The fuse must have a voltage rating that is equal to or greater than the normal frequency recovery voltage that will be seen across the fuse under all conditions. On three-phase systems, it is a good rule of thumb that the voltage rating of the fuse be greater than or equal to the line-to-line voltage of the system.
- *Continuous current-carrying capacity* – Continuous current values that are shown on the fuse represent the level of current the fuse can carry continuously without exceeding the temperature rises as specified in *ANSI C37.46*. An application that exposes the fuse to a current slightly above its continuous rating but below its minimum interrupting rating may damage the fuse due to excessive heat. This is the main reason overload relays are used in series with backup current-limiting fuses for motor protection.
- *Interrupting rating* – All fuses are given a maximum interrupting rating. This rating is the maximum level of fault current that the fuse can safely interrupt. Backup current-limiting fuses are also given a minimum interrupting rating. When using backup current-limiting fuses, it is important that other protective devices are used to interrupt currents below this level.

When choosing a fuse, it is important that the fuse be properly coordinated with other protective devices located upstream and downstream. To accomplish this, one must consider the **melting time** and **clearing time** characteristics of the devices. Two curves, the minimum melting curve and the total clearing curve, provide this information. To ensure proper coordination, the following rules should be observed:

- The total clearing curve of any downstream protective device must be below a curve representing 75% of the minimum melting curve of the fuse being applied.
- The total clearing curve of the fuse being applied must lie below a curve representing 75% of the minimum melting curve for any upstream protective device.

4.3.0 Current-Limiting Fuses

To ensure proper application of a current-limiting fuse, it is important that the following additional rules be applied:

- Current-limiting fuses produce arc voltages that exceed the system voltage. Care must be taken to ensure that the peak voltages do not exceed the insulation level of the system. If the fuse voltage rating is not permitted to exceed 140% of the system voltage, there should not be a problem. This does not mean that a higher rated fuse cannot be used, but one must be assured that the system insulation level (BIL) will handle the peak arc voltage produced. BIL stands for basic impulse level, which is the reference impulse insulation strength of an electrical system.
- As with the expulsion fuse, current-limiting fuses must be properly coordinated with other protective devices on the system. For this to happen, the rules for applying an expulsion fuse must be used at all currents that cause the fuse to interrupt in 0.01 second or greater.

When other current-limiting protective devices are on the system, it becomes necessary to use I^2t values for coordination at currents causing the fuse to interrupt in less than 0.01 second. These values may be supplied as minimum and maximum values or minimum melting and total clearing I^2t curves. In either case, the following rules should be followed:

- The minimum melting I^2t of the fuse should be greater than the total clearing I^2t of the downstream current-limiting device.
- The total clearing I^2t of the fuse should be less than the minimum melting I^2t of the upstream current-limiting device.

The fuse selection chart in *Figure 8* should serve as a guide for selecting fuses on circuits of 600V or less. Other valuable information may be found in catalogs furnished by manufacturers of overcurrent protective devices. These are usually obtainable from electrical supply houses or from manufacturers' representatives. You may also write the various manufacturers for a complete list (and price, if any) for all reference materials offered by them.

4.4.0 Fuses for Selective Coordination

The larger the upstream fuse is relative to a downstream fuse (feeder to branch, etc.), the less possibility there is of an overcurrent in the downstream circuit causing both fuses to open. Fast action,

Circuit	Load	Ampere Rating	Fuse Type	Symbol	Voltage Rating (ac)	UL Class	Interrupting Rating (K)	Remarks
Main, Feeder, and Branch (Conventional dimensions)	All load types (optimum overcurrent protection)	0 to 600A	LOW PEAK® (dual-element time-delay)	LPN-RK	250V	RK1	200	All-purpose fuses. Unequaled for combined short circuit and overload protection.
				LPS-RK	600V			
		601 to 6,000A	LOW PEAK® time-delay	KRP-C	600V	L	200	
	Motors, welders, transformers, capacitor banks (circuits with heavy inrush currents)	0 to 600A	FUSETRON® (dual-element time-delay)	FRN-R	250V	RK5	200	Moderate degree of current limitation. Time-delay passes surge currents.
				FRS-R	600V			
		601 to 4,000A	LIMITRON® (time-delay)	KLU	600V	L	200	All-purpose fuse. Time-delay passes surge currents.
	Non-motor loads (circuits with no heavy inrush currents)	0 to 600A	LIMITRON® (fast-acting)	KTN-R	250V	RK1	200	Same short circuit protection as LOW PEAK® fuses but must be sized larger for circuits with surge currents: i.e., up to 300%.
				KTS-R	600V			
	LIMITRON® fuses particularly suited for circuit breaker protection	601 to 6,000A	LIMITRON® (fast-acting)	KTU	600V	L	200	A fast-acting, high-performance fuse.
	All load types (optimum overcurrent protection)	0 to 600A	LOW PEAK® (dual-element, time-delay)	LPJ	600V	J	200	All-purpose fuses. Unequaled for combined short circuit and overload protection.
	Non-motor loads (circuits with no heavy inrush currents)	0 to 600A	LIMITRON® (quick-acting)	JKS	600V	J	200	Very similar to KTS-R LIMITRON®, but smaller.
		0 to 1,200A	T-TRON™	JJN	300V	T	200	The space saver (1/3 the size of KTN-R/KTS-R).
				JJS	600V			

305F08.EPS

Figure 8 ◆ Fuse selection chart (600V or less).

nontime-delay fuses require at least a 3:1 ratio between the ampere rating of a large upstream, line-side, time-delay fuse to that of the downstream, load-side fuse in order to be selectively coordinated. In contrast, the minimum selective coordination ratio necessary for dual-element fuses is only 2:1 when used with low peak load-side fuses (*Figure 9*).

The use of dual-element, time-delay fuses affords easy selective coordination, which hardly requires anything more than a routine check of a tabulation of required selectivity ratios. As shown in *Figure 10*, close sizing of dual-element fuses in the branch circuit for motor overload protection provides a large difference (ratio) in the ampere ratings between the feeder fuse and the branch fuse compared to the single-element, nontime-delay fuse.

4.5.0 Fuse Time-Current Curves

When a low-level overcurrent occurs, a long interval of time will be required for a fuse to open (melt) and clear the fault. On the other hand, if the overcurrent is large, the fuse will open very quickly. The opening time is a function of the magnitude of the level of overcurrent. Overcurrent levels and the corresponding intervals of opening times are logarithmically plotted in graph form, as shown in *Figure 11*. Levels of overcurrent are scaled on the horizontal axis, with time intervals on the vertical axis. The curve is therefore called a time-current curve.

The plot in *Figure 11* reflects the characteristics of a 200A, 600V, dual-element fuse. Note that at the 1,000A overload level, the time interval that is required for the fuse to open is 10 seconds. Yet, at approximately the 2,200A overcurrent level, the opening (melt) time of the fuse is only 0.01 second. It is apparent that the time intervals become shorter and shorter as the overcurrent levels become larger. This relationship is called an inverse time-to-current characteristic. Time-current curves are published or are available on most commonly used fuses showing minimum melt, average melt, and/or total clear characteristics. Although upstream and downstream fuses are easily coordinated by adhering to simple ampere ratios, these time-current curves permit close or critical analysis of coordination.

4.5.1 Peak Let-Through Charts

Peak let-through charts enable you to determine both the peak let-through current and the apparent prospective rms symmetrical let-through

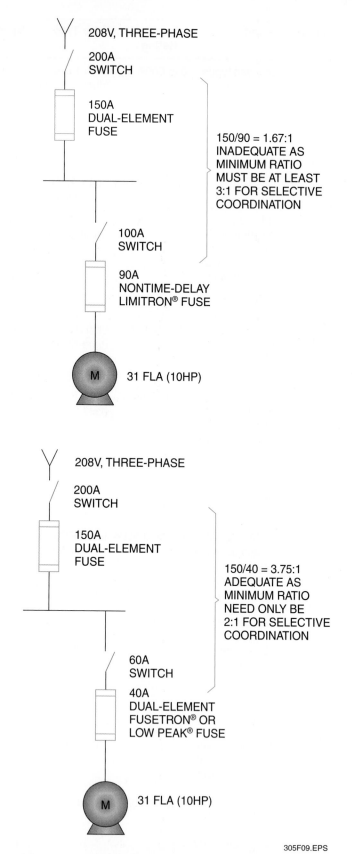

305F09.EPS

Figure 9 ◆ Fuses used for selective coordination.

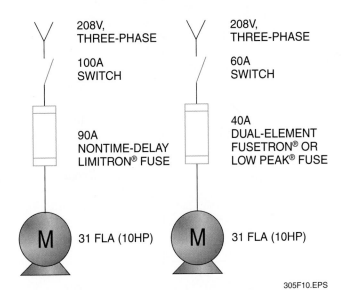

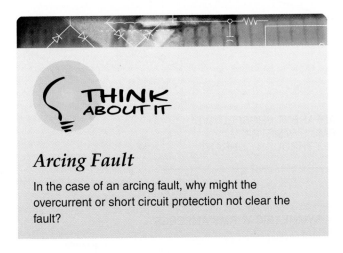

Figure 10 ◆ Comparison of dual-element fuse and single-element, nontime-delay fuse.

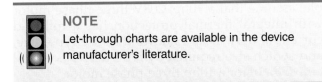

THINK ABOUT IT

Arcing Fault

In the case of an arcing fault, why might the overcurrent or short circuit protection not clear the fault?

current. Such charts are commonly referred to as current limitation curves. *Figure 12* shows a simplified chart with explanations of the various functions.

NOTE

Let-through charts are available in the device manufacturer's literature.

See *Figure 12*. Point 1 shows the system available rms short circuit current of 40,000A. Upward from point 1 is the intersection with the fuse size of 100A (point 2). Follow this left to point 3, the peak let-through current (10,400A). This line represents the point at which the fuse becomes

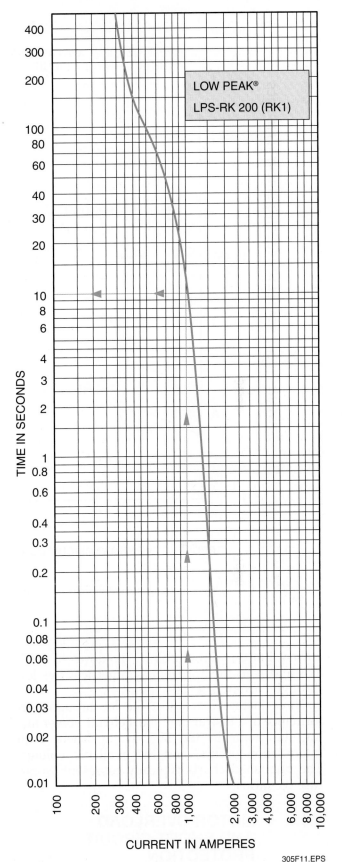

Figure 11 ◆ Typical time-current curve of a fuse.

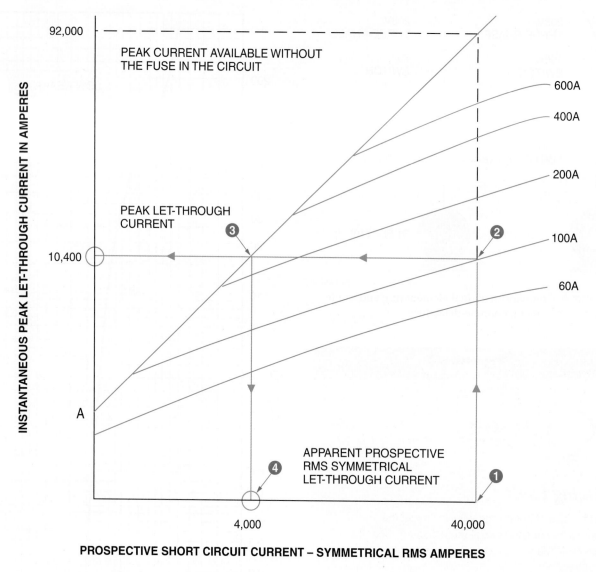

Figure 12 ♦ Principles of forming current limitation curves.

305F12.EPS

current limiting. From point 3, go down to the X axis and read the apparent symmetrical rms let-through current of 4,000A (point 4).

Three factors that affect the let-through performance of a fuse are the short circuit power factor, the point in the sine wave where fault interruption occurs, and the available voltage. The published charts typically represent worst-case scenarios.

5.0.0 ♦ MOTOR OVERLOAD AND SHORT CIRCUIT PROTECTION

When used in circuits with surge currents such as those caused by motors, transformers, and other inductive components, dual-element, time-delay

fuses can be sized close to full-load amperes to give maximum overcurrent protection. For example, assume that a 10hp, 208V three-phase motor with integral thermal protection has a full-load current of 31A. *Table 1* shows the fuse type, size, and switch size required by *NEC Sections 430.52 and 430.110* for a 10hp three-phase motor.

Table 1 shows that a 45A, dual-element fuse will protect the 31A motor compared to the much larger 100A, single-element fuse necessary. It is apparent that if a sustained, harmful overload of 300% occurred in the motor circuit, the 100A, single-element fuse would never open and the motor could be damaged. The nontime-delay fuse provides only ground fault and short circuit protection—requiring separate overload protection as per the *NEC®*.

Table 1 Fuse and Switch Size for a 10-Horsepower Motor (208V, 3Ø, 31 FLA)

Fuse Type	Maximum Fuse Size (Amps)	Required Switch Size (Amps)
Dual-element, time-delay	45A	60A
Single-element, nontime-delay	100A	100A

In contrast, the 45A, dual-element fuse provides ground fault and short circuit protection plus a high degree of backup protection against motor burnout from overload or **single phasing** should other overload protective devices fail. If thermal overloads, relays, or contacts should fail to operate, the dual-element fuses will act independently to protect the motor.

Aside from providing only short circuit protection, the single-element fuse also makes it necessary to use larger size switches since a switch rating must be equal to or larger than the ampere rating of the fuse; as a result, the larger switch may cost two or three times more than would be necessary if a dual-element fuse were used (*Figure 13*).

When secondary single phasing occurs, the current in the remaining phases increases to a value of 170% to 200% of the rated full-load current. When primary single phasing occurs, unbalanced voltages that occur in the motor circuit cause excessive current. Dual-element fuses sized for motor overload protection can protect motors against the overload damage caused by single phasing.

The nontime-delay, **fast-acting fuse** must be oversized in circuits in which surge or temporary overload currents occur (**inductive load**). The response of the oversized fuse to short circuit cur-

rents is slower. Current builds up to a high level before the fuse opens, causing the current-limiting action of the oversized fuse to be less than a fuse whose ampere rating is closer to the normal full-load current of the circuit. Consequently, oversizing sacrifices some component protection and although it is permitted by the *NEC*®, the practice is not recommended.

In actual practice, dual-element fuses used to protect motors keep short circuit currents to approximately half the value of the nontime-delay fuses, since the nontime-delay fuses must be oversized to carry the temporary starting current of a motor per *NEC Table 430.52.*

While dual-element, time delay fuses can provide motor overload protection, they are nonrenewable and must be replaced after they have blown. In most motor branch circuits, a motor starter, also called a motor controller, usually containing a resettable overload relay is used for the overload protection of a motor. However, a branch circuit supplying a motor starter requires short-circuit protection in the form of a backup overload/overcurrent device or instantaneous-trip overcurrent device. From this device, the motor branch circuit is routed to the motor starter. In some cases, the motor starter and overcurrent devices are contained in a combination motor starter. Branch circuit power, switched by the motor starter, is applied to the motor through the overload relay, which is selected specifically for the motor. Like dual-element time delay fuses, the overload relay provides protection to the motor for overload conditions due to low line voltage, single phasing, or excessive loading of the motor. Overload relays can be thermal, magnetic, or solid-state. Thermal overload relays are commonly used because they can be matched to the heating curve of the motor for maximum motor protection. Overload relays are designed to withstand numerous trip and reset cycles without replacement.

When the overload relay in a manual motor starter is actuated by an overload, a trip lever opens the motor starter switch contacts supplying power through the overload relay to the motor. In a magnetic contactor motor starter, the hold-in (seal-in) circuit for the contactor is interrupted.

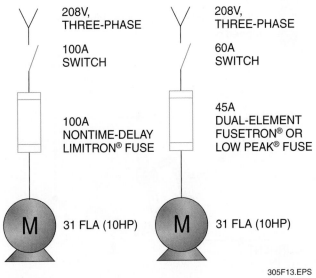

208V, THREE-PHASE

100A SWITCH

100A NONTIME-DELAY LIMITRON® FUSE

M 31 FLA (10HP)

208V, THREE-PHASE

60A SWITCH

45A DUAL-ELEMENT FUSETRON® OR LOW PEAK® FUSE

M 31 FLA (10HP)

305F13.EPS

Figure 13 ◆ Dual-element fuses permit the use of smaller and less costly switches.

This causes the contactor to deenergize and open contacts supplying power through the overload relay to the motor. Once the overload problem has been resolved, the overload relay can be reset and the motor restarted. In certain specific applications, motor starters and overload relays with an automatic reset function are used. Motor starters and thermal/magnetic overload relays as well as motor branch circuit calculations are covered in more detail in later modules.

6.0.0 ◆ CIRCUIT BREAKERS

Circuit breakers were covered in your Level Two training. However, some of the more important points are worth repeating here before we cover practical applications of both fuses and circuit breakers.

Basically, a circuit breaker is a device for closing and interrupting a circuit between separable contacts under both normal and abnormal conditions. This is done manually (normal condition) by using its handle to switch it to the ON or OFF positions. However, the circuit breaker is also designed to open a circuit automatically on a predetermined overload or ground fault current without damage to itself or its associated equipment. As long as a circuit breaker is applied within its rating, it will automatically interrupt any fault and is therefore classified as an inherently safe overcurrent protective device.

The internal arrangement of a circuit breaker is shown in *Figure 14*, while its external operating characteristics are shown in *Figure 15*. Note that the handle on a circuit breaker resembles an ordinary toggle switch. On an overload, the circuit breaker opens itself or trips. In a tripped position, the handle jumps to the middle position (*Figure 15*). To reset it, turn the handle to the OFF position and then turn it as far as it will go beyond this position (RESET position); finally, turn it to the ON position.

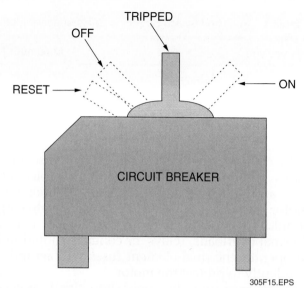

Figure 15 ◆ External characteristics of a circuit breaker.

A standard molded case circuit breaker usually contains:

- A set of contacts
- A magnetic trip element
- A thermal trip element
- Line and load terminals
- Busing used to connect these individual parts
- An enclosing housing of insulating material

The circuit breaker handle manually opens and closes the contacts and resets the automatic trip units after an interruption. Some circuit breakers also contain a manually operated push-to-trip testing mechanism.

 CAUTION

When a breaker trips, always determine why it tripped before resetting it.

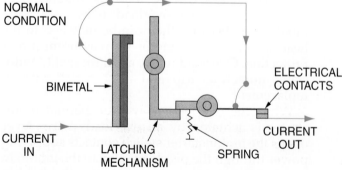

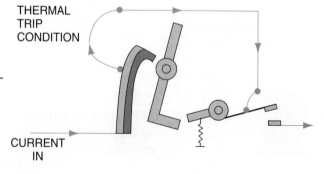

Figure 14 ◆ Internal arrangement of a circuit breaker.

Circuit breakers are grouped for identification according to given current ranges. Each group is classified by the largest ampere rating of its range. These groups are as follows:

- 15A–100A
- 125A–225A
- 250A–400A
- 500A–1,000A
- 1,200A–2,000A

Therefore, they are classified as 100A, 225A, 400A, 1,000A, and 2,000A frames. These numbers are commonly referred to as frame classifications or frame sizes and are terms applied to groups of molded case circuit breakers that are physically interchangeable with each other.

6.1.0 Interrupting Capacity Rating

In most large commercial and industrial installations, it is necessary to calculate available short circuit currents at various points in a system to determine if the equipment meets the requirements of *NEC Sections 110.9 and 110.10.* There are a number of methods used to determine the short circuit requirements in an electrical system. Some give approximate values; others require extensive computations and are quite exacting. A recommended method is given in a later section.

The breaker interrupting capacity is based on tests to which the breaker is subjected. There are two such tests; one is set up by UL and the other by NEMA. The NEMA tests are self-certification.

UL tests are certified by unbiased witnesses. UL tests have been limited to a maximum of 10,000A in the past, so the emphasis was placed on NEMA tests with higher ratings. UL tests now include the NEMA tests plus other ratings. Consequently, the emphasis is now being placed on UL tests.

Arc Flash

A serious injury resulted when a journeyman electrician on an industrial service call attempted to install a three-phase, 400A bolt-in circuit breaker into an energized panel. He miscalculated the positioning of the bolt-in lugs of the breaker with the breaker handle in the ON position, causing one phase terminal to contact an energized busbar, while contacting the grounded cabinet with another phase terminal. The circuit breaker exploded in his face, in addition to releasing a toxic vapor due to a phenomenon known as arc flash. He survived but sustained severe injuries to his eyes and lungs.

The Bottom Line: Working with energized equipment requires the use of properly rated personal protective equipment, including a full-body flash suit, face shield, and glasses.

The interrupting capacity of a circuit breaker is based on its rated voltage. Where the circuit breaker can be used on more than one voltage, the interrupting capacity will be shown for each voltage level. For example, the LA-type circuit breaker has 42,000A symmetrical interrupting capacity at 240V, 30,000A symmetrical at 480V, and 22,000A symmetrical at 600V.

7.0.0 ◆ CIRCUIT PROTECTION

All conductors in a circuit are to be protected against overcurrents in accordance with their ampacities as set forth in *NEC Section 240.4*. They must also be protected against short circuit current damage as required.

Ampere ratings of overcurrent protective devices must not be greater than the ampacity of the conductor. There is, however, an exception. *NEC Section 240.4(B)* states that if the conductor rating does not correspond to a standard size overcurrent protective device, the next larger size overcurrent protective device may be used, provided its rating does not exceed 800A. Likewise, the conductor cannot be part of a multi-outlet branch circuit supplying receptacles for cord- and plug-connected portable loads. When the ampacity of a busway or cablebus does not correspond to a standard overcurrent protective device, the next larger standard rating may be used only if the rating does not exceed 800A per *NEC Sections 368.17(A) and 370.5*.

7.1.0 Lighting/Appliance Branch Circuits

The branch circuit rating must be classified in accordance with the rating of the overcurrent protective device. Classifications for those branch circuits other than individual loads must be 15A, 20A, 30A, 40A, and 50A, as specified in *NEC Section 210.3*.

Branch circuit conductors and equipment must be protected by an overcurrent protective device with ampere ratings that conform to *NEC Section 210.20*. Basically, the branch circuit conductor and overcurrent protective device must be sized for the actual noncontinuous load plus 125% of the continuous load. The overcurrent protection size must not be greater than the conductor ampacity. Branch circuits rated 15A through 50A with two or more outlets (other than receptacle circuits) must be protected at their rating and the branch circuit conductor sized according to *NEC Table 210.24*.

INSIDE TRACK

Surge Protection

In addition to the plug-in surge strips that are typically used to protect electronic equipment, some manufacturers are now offering residential surge protection devices that mount directly on the main breaker panel. These devices provide comprehensive protection for AC power in use throughout the residence, and are used in conjunction with surge strips. Surge strips are not rated to handle the higher surges caused by induced lightning, and the comprehensive device reduces the voltage surge to a level that can be handled by the surge protector.

This protection is normally installed at the point where the conductors receive their supply.

7.1.1 Feeder Circuits

The feeder overcurrent protective device ampere rating and feeder conductor ampacity must be as follows:

- *Feeder circuit with no motor load* – The overcurrent protective device size must be at least 125% of the continuous load plus 100% of the noncontinuous load.
- *Feeder circuit with all motor loads* – Size the overcurrent protective device at 100% of the largest branch circuit protective device plus the sum of the full-load current of all other motors.
- *Feeder circuit with mixed loads* – Size the overcurrent protective device at 100% of the largest branch circuit protective device plus the sum of the full-load current of all other motors, plus 125% of the continuous, nonmotor load, plus 100% of the noncontinuous, nonmotor load.

7.1.2 Service Equipment

Each ungrounded service entrance conductor must have an overcurrent device in series with a rating not higher than the ampacity of the conductor. The service overcurrent devices shall be part of the service disconnecting means or be located immediately adjacent to it (*NEC Section 230.91*).

Service disconnecting means can consist of one to six switches or circuit breakers for each service or for each set of service-entrance conductors permitted in *NEC Section 230.2*. When more than one switch is used, the switches must be grouped together (*NEC Section 230.72*).

7.1.3 Transformer Secondary Circuits

Field installations indicate nearly 50% of transformers installed do not have secondary protection. The *NEC*® requires overcurrent protection for lighting and appliance panelboards and recommends that secondary conductors be protected from damage by applying the proper overcurrent protective device. For example, the primary overcurrent device protecting a three-wire transformer cannot offer protection to the secondary conductors. *NEC Section 240.4(F)* and *NEC Section 240.21(C)* discuss protection of transformer secondary conductors.

7.1.4 Motor Circuit Protection

Motors and motor circuits have unique operating characteristics and circuit components. Therefore, these circuits must be dealt with differently from other types of loads. Generally, two levels of overcurrent protection are required for motor branch circuits:

- *Overload protection* – Motor running overload protection is intended to protect the system components and motor from damaging overload currents.
- *Short circuit protection (includes ground fault protection)* – Short circuit protection is intended to protect the motor circuit components such as the conductors, switches, controllers, overload relays, motor, etc., against short circuit currents or grounds. This level of protection is commonly referred to as motor branch circuit protection. Dual-element fuses are designed to provide this protection, as long as they are sized correctly.

There are a variety of ways to protect a motor circuit, depending upon the user's objective. The ampere rating of an overcurrent protective device selected for motor protection depends on whether the overcurrent protective device is of the dual-element, time-delay type or the nontime-delay type.

NOTE
Design B energy-efficient motors require special circuit protection consideration.

In general, nontime-delay fuses can be sized at 300% of the motor full-load current for ordinary motors so that the normal motor starting current does not affect the fuse (*NEC Table 430.52*). Dual-element, time-delay fuses are able to withstand normal motor starting current and can be sized closer to the actual motor rating than nontime-delay fuses.

A summary of *NEC*® regulations governing overcurrent protection is covered in *Table 2*, while the table in *Figure 16* gives generalized fuse application guidelines for motor branch circuits. *Table 3* may be used to select dual-element fuses for motor protection.

NOTE
In many cases, the overcurrent device size is less than that allowed by the *NEC*®. This is an additional safeguard adopted by many manufacturers and installers.

8.0.0 ◆ SHORT CIRCUIT CALCULATIONS

Adequate interrupting capacity and protection of electrical components are two essential aspects required by *NEC Sections 110.9, 110.10, and 240.1*. The first step in ensuring that the system protective devices have the proper interrupting rating and provide component protection is to determine the available short circuit currents. Although several methods for determining short circuit currents are in use, the point-to-point method of short circuit calculation is normally the recommended method. The application of the point-to-point method permits the determination of available short circuit currents with a reasonable degree of accuracy at various points for either single-phase or three-phase electrical distribution systems. This method assumes unlimited primary short circuit current (infinite bus) for a facility transformer.

Type of Motor	Dual-Element, Time-Delay Fuses			Nontime-Delay Fuses
	Desired Level of Protection			
	Motor Overload and Short Circuit	Backup Overload and Short Circuit	Short Circuit Only (Based on *NEC Tables 430.247 through 430.250* current ratings)	Short Circuit Only (Based on *NEC Tables 430.247 through 430.250* current ratings)
Service Factor 1.15 or Greater or 40°C Temp. Rise or Less	125% or less of motor nameplate current	125% or next standard size (not to exceed 140%)	150% to 175%	150% to 300%
Service Factor Less Than 1.15 or Greater Than 40°C Temp. Rise	115% or less of motor nameplate current	115% or next standard size (not to exceed 130%)	150% to 175%	150% to 300%

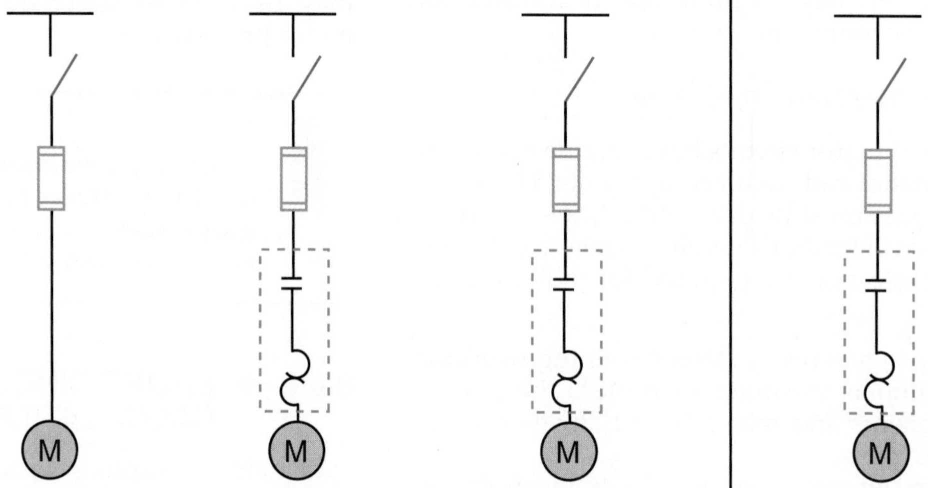

Fuses give overload and short circuit protection.

Overload relay gives overload protection and fuses provide backup overload protection.

Overload relay provides overload protection and fuses provide only short circuit protection.

Overload relay provides overload protection and fuses provide only short circuit protection.

305F16.EPS

Figure 16 ◆ Fuse application guidelines for motor branch circuits.

Table 2 *NEC*® Regulations for Overcurrent Protection

Application	Rule	*NEC*® Reference
Scope	Overcurrent protection for conductors and equipment is provided to open the circuit if the current reaches a value that will cause an excessive or dangerous temperature in conductors or conductor insulation. See also *NEC Sections 110.9 and 110.10* for requirements for interrupting capacity and protection against fault currents.	*NEC Section 240.1 FPN*
Protection required	Each ungrounded service-entrance conductor must have overcurrent protection in series with each ungrounded conductor.	*NEC Section 230.90(A)*
Number of devices	Up to six circuit breakers or sets of fuses may be considered as the overcurrent device.	*NEC Section 230.90(A), Exception No. 3*
Location in building	The overcurrent device must be part of the service disconnecting means or be located immediately adjacent to it.	*NEC Section 230.91*
Accessibility	In a property comprising more than one building under single management, the ungrounded conductors supplying each building served shall be protected by overcurrent devices, which may be located in the building served or in another building on the same property, provided they are accessible to the occupants of the building served. In a multiple-occupancy building, each occupant shall have access to the overcurrent protective devices.	*NEC Sections 230.92 and 230.72(C), Exception*
Location in circuit	The overcurrent device must protect all circuits and devices, except equipment which may be connected on the supply side, including: (1) Service switch; (2) Special equipment, such as surge arrestors; (3) Circuits for emergency supply and load management (where separately protected); (4) Circuits for fire alarms or fire pump equipment (where separately protected); (5) Meters with all metal housing grounded (600V or less); (6) Control circuits for automatic service equipment, if suitable overcurrent protection and disconnecting means are provided.	*NEC Section 230.94 plus Exceptions*
Installation and use	Listed or labelled equipment shall be used and installed in accordance with any instructions included in the listing or labeling.	*NEC Section 110.3(B)*
Interrupting rating	Equipment intended to interrupt current at fault levels shall have an interrupting rating sufficient for the system voltage and the current which is available at the line terminals of the equipment.	*NEC Section 110.9*
Circuit impedance and other characteristics	The overcurrent protective devices, total impedance, component short circuit current ratings, and other characteristics of the circuit to be protected shall be so selected and coordinated as to permit the circuit protective devices used to clear a fault without the occurrence of extensive damage to the electrical components of the circuit.	*NEC Section 110.10*
General	Bonding of normally noncurrent-carrying electrically conductive materials that could become energized shall be provided where necessary to ensure electrical continuity and the capacity to safely conduct any fault current likely to be imposed.	*NEC Section 250.4(A)(4)*
Bonding other enclosures	Metal raceways, cable trays, cable armor, cable sheath, enclosures, frames, fittings, and other metal noncurrent-carrying parts that are to serve as grounding conductors, with or without the use of supplementary equipment grounding conductors, shall be effectively bonded where necessary to ensure electrical continuity and the capacity to safely conduct any fault current likely to be imposed on them. Any nonconductive paint, enamel, or similar coating shall be removed at threads, contact points, and contact surfaces or be connected by means of fittings so designed as to make such removal unnecessary.	*NEC Section 250.96(A)*

Table 3 Selection of Fuses for Motor Protection (1 of 3)

Dual-Element Fuse Size	Motor Protection (Used without properly sized relays) Motor Full-Load Amps		Backup Motor Protection (Used with properly sized overload relays) Motor Full-Load Amps	
	Motor service factor of 1.15 or greater or with temperature rise not over 40°C	Motor service factor less than 1.15 or with temperature rise not over 40°C	Motor service factor of 1.15 or greater or with temperature rise not over 40°C	Motor service factor of less than 1.15 or with temperature rise not over 40°C
$^1/_{10}$	0.08–0.09	0.09–0.10	0–0.08	0–0.09
$^1/_8$	0.10–0.11	0.11–0.15	0.09–0.10	0.10–0.11
$^5/_{100}$	0.12–0.15	0.14–0.15	0.11–0.12	0.12–0.13
$^2/_{10}$	0.16–0.19	0.18–0.20	0.13–0.16	0.14–0.17
$^1/_4$	0.20–0.23	0.22–0.25	0.17–0.20	0.18–0.22
$^3/_{10}$	0.24–0.30	0.27–0.30	0.21–0.24	0.23–0.26
$^4/_{10}$	0.32–0.39	0.35–0.40	0.25–0.32	0.27–0.35
$^1/_2$	0.40–0.47	0.44–0.50	0.33–0.40	0.36–0.43
$^6/_{10}$	0.48–0.60	0.53–0.60	0.41–0.48	0.44–0.52
$^8/_{10}$	0.64–0.79	0.70–0.80	0.49–0.64	0.53–0.70
1	0.80–0.89	0.87–0.97	0.65–0.80	0.71–0.87
$1^1/_8$	0.90–0.99	0.98–1.08	0.81–0.90	0.88–0.98
$1^1/_4$	1.00–1.11	1.09–1.21	0.91–1.00	0.99–1.09
$1^4/_{10}$	1.12–1.19	1.22–1.30	1.01–1.12	1.10–1.22
$1^1/_2$	1.20–1.27	1.31–1.39	1.13–1.20	1.23–1.30
$1^6/_{10}$	1.28–1.43	1.40–1.56	1.21–1.28	1.31–1.39
$1^8/_{10}$	1.44–1.59	1.57–1.73	1.29–1.44	1.40–1.57
2	1.60–1.79	1.74–1.95	1.45–1.60	1.58–1.74
$2^1/_4$	1.80–1.99	1.96–2.17	1.61–1.80	1.75–1.96
$2^1/_2$	2.00–2.23	2.18–2.43	1.81–2.00	1.97–2.17

305T01A.EPS

Table 3 Selection of Fuses for Motor Protection (2 of 3)

Dual-Element Fuse Size	Motor Protection (Used without properly sized relays) Motor Full-Load Amps		Backup Motor Protection (Used with properly sized overload relays) Motor Full-Load Amps	
	Motor service factor of 1.15 or greater or with temperature rise not over 40°C	Motor service factor less than 1.15 or with temperature rise not over 40°C	Motor service factor of 1.15 or greater or with temperature rise not over 40°C	Motor service factor of less than 1.15 or with temperature rise not over 40°C
$2^6/_{10}$	2.24–2.39	2.44–2.60	2.01–2.24	2.18–2.43
3	2.40–2.55	2.61–2.78	2.25–2.40	2.44–2.60
$3^2/_{10}$	2.56–2.79	2.79–3.04	2.41–2.56	2.61–2.78
$3^1/_2$	2.80–3.19	3.05–3.47	2.57–2.80	2.79–3.04
4	3.20–3.59	3.48–3.91	2.81–3.20	3.05–3.48
$4^1/_2$	3.60–3.99	3.92–4.34	3.21–3.60	3.49–3.91
5	4.00–4.47	4.35–4.86	3.61–4.00	3.92–4.35
$5^6/_{10}$	4.48–4.79	4.87–5.21	4.01–4.48	4.36–4.87
6	4.80–4.99	5.22–5.43	4.49–4.80	4.88–5.22
$6^1/_4$	5.00–5.59	5.44–6.08	4.81–5.00	5.23–5.43
7	5.60–5.99	6.09–6.52	5.01–5.60	5.44–6.09
$7^1/_2$	6.00–6.39	6.53–6.95	5.61–6.00	6.10–6.52
8	6.40–7.19	6.96–7.82	6.01–6.40	6.53–6.96
9	7.20–7.99	7.83–8.69	6.41–7.20	6.97–7.83
10	8.00–9.59	8.70–10.00	7.21–8.00	7.84–8.70
12	9.60–11.99	10.44–12.00	8.01–9.60	8.71–10.43
15	12.00–13.99	13.05–15.00	9.61–12.00	10.44–13.04
$17^1/_2$	14.00–15.99	15.22–17.39	12.01–14.00	13.05–15.21
20	16.00–19.99	17.40–20.00	14.01–16.00	15.22–17.39
25	20.00–23.99	21.74–25.00	16.01–20.00	17.40–21.74
30	24.00–27.99	26.09–30.00	20.01–24.00	21.75–26.09
35	28.00–31.99	30.44–34.78	24.01–28.00	26.10–30.43

305T01B.EPS

Table 3   Selection of Fuses for Motor Protection (3 of 3)

Dual-Element Fuse Size	Motor Protection (Used without properly sized relays) Motor Full-Load Amps		Backup Motor Protection (Used with properly sized overload relays) Motor Full-Load Amps	
	Motor service factor of 1.15 or greater or with temperature rise not over 40°C	Motor service factor less than 1.15 or with temperature rise not over 40°C	Motor service factor of 1.15 or greater or with temperature rise not over 40°C	Motor service factor of less than 1.15 or with temperature rise not over 40°C
40	32.00–35.99	34.79–39.12	28.01–32.00	30.44–37.78
45	36.00–39.00	39.13–43.47	32.01–36.00	37.79–39.13
50	40.00–47.99	43.48–50.00	36.01–40.00	39.14–43.48
60	48.00–55.99	52.17–60.00	40.01–48.00	43.49–52.17
70	56.00–59.99	60.87–65.21	48.01–56.00	52.18–60.87
75	60.00–63.99	65.22–69.56	56.01–60.00	60.88–65.22
80	64.00–71.99	69.57–78.25	60.01–64.00	65.23–69.57
90	72.00–79.99	78.26–86.95	64.01–72.00	69.58–78.26
100	80.00–87.99	86.96–95.64	72.01–80.00	78.27–86.96
110	88.00–99.00	95.65–108.69	80.01–88.00	86.97–95.65
125	100.00–119.00	108.70–125.00	88.01–100.00	95.66–108.70
150	120.00–139.99	131.30–150.00	100.01–120.00	108.71–130.43
175	140.00–159.99	152.17–173.90	120.01–140.00	130.44–152.17
200	160.00–179.99	173.91–195.64	140.01–160.00	152.18–173.91
225	180.00–199.99	195.65–217.38	160.01–180.00	173.92–195.62
250	200.00–239.99	217.39–250.00	180.01–200.00	195.63–217.39
300	240.00–279.99	260.87–300.00	200.01–240.00	217.40–260.87
350	280.00–319.99	304.35–347.82	240.01–280.00	260.88–304.35
400	320.00–359.99	347.83–391.29	280.01–320.00	304.36–347.83
450	360.00–399.99	391.30–434.77	320.01–360.00	347.84–391.30
500	400.00–479.99	434.78–500.00	360.01–400.00	391.31–434.78
600	480.00–600.00	521.74–600.00	400.01–480.00	434.79–521.74

305T01C.EPS

8.1.0 Basic Short Circuit Calculation Procedure

Step 1 Determine the transformer full-load amperes (I_{FLA}) from the nameplate, *Table 4*, or the following equation. (E_{L-L} = line-to-line voltage.)

$$I_{FLA}(3\phi) = \frac{kVA \times 1,000}{E_{L-L} \times 1.73}$$

$$I_{FLA}(1\phi) = \frac{kVA \times 1,000}{E_{L-L}}$$

Step 2 Find the transformer multiplier using the following equation:

$$\text{Multiplier} = \frac{100}{\text{transformer \%Z}}$$

Step 3 Determine the transformer let-through short circuit current (I_{SCA}). Use *Table 5* or the following equation:

$$I_{SCA} = \text{transformer}_{FLA} \times \text{multiplier}$$

Table 4 Three-Phase Transformer—Full-Load Current Rating in Amperes

Voltage (Line-to-Line)	Transformer kVA Rating								
	150	167	225	300	500	750	1,000	1,500	2,000
208	417	464	625	834	1,388	2,080	2,776	4,164	5,552
220	394	439	592	788	1,315	1,970	2,630	3,940	5,260
240	362	402	542	722	1,203	1,804	2,406	3,609	4,812
440	197	219	296	394	657	985	1,315	1,970	2,630
460	189	209	284	378	630	945	1,260	1,890	2,520
480	181	201	271	361	601	902	1,203	1,804	2,406

Table 5 Short Circuit Currents Available from Various Size Transformers

Voltage and Phase	kVA	Full-Load Amps	% Impedance	Short Circuit Amps
120/240V Single phase	25	104	1.6	10,300
	37½	156	1.6	15,280
	50	209	1.7	19,050
	75	313	1.6	29,540
	100	417	1.6	38,540
	167	695	1.8	54,900
120/208V Three phase	150	417	2.0	20,850
	225	625	2.0	31,250
	300	834	2.0	41,700
	500	1,388	2.0	69,444
	750	2,080	3.5	59,426
	1,000	2,776	3.5	79,310
	1,500	4,164	3.5	118,965
	2,000	5,552	5.0	111,040
	2,500	6,950	5.0	139,000
277/480V Three phase	112½	135	1.0	13,500
	150	181	1.2	15,083
	225	271	1.2	22,583
	300	361	1.2	30,083
	500	601	1.3	46,230
	750	902	3.5	25,770
	1,000	1,203	3.5	34,370
	1,500	1,804	3.5	51,540
	2,000	2,406	5.0	48,120
	2,500	3,077	5.0	60,140

NOTE

The motor short circuit contribution, if significant, may be added to the transformer secondary short circuit current value as determined here. Proceed with the adjusted figure through Steps 4, 5, and 6 to follow. A practical estimate of motor short circuit contribution is to multiply the total load current in amperes by four.

Step 4 Calculate the f factor using the following equations along with *Table 6*.

For parallel runs, multiply the C value by the number of conductors per phase:

$$3\phi \text{ faults} = \frac{1.73 \times L \times I}{C \times E_{L-L}}$$

1ϕ line-to-line (L-L) faults on 1ϕ, center-tapped transformers:

$$f = \frac{2 \times L \times I}{C \times E_{L-L}}$$

1ϕ line-to-neutral (L-N) faults on 1ϕ, center-tapped transformers:

$$f = \frac{2 \times L \times I}{C \times E_{L-N}}$$

Where:

L = length (in feet) of the circuit to the fault

C = constant (from *Table 6*)

I = available short circuit current in amperes at the beginning of the circuit

Step 5 Calculate the multiplier (M) using the following equation, or use *Table 7*:

$$M = \frac{1}{1 + f}$$

Step 6 Compute the available short circuit current (symmetrical) at the fault using the following equation:

$$I_{SCA} \text{ at fault} = I_{SCA} \text{ at beginning of circuit} \times M$$

Table 6 C Values for Conductors and Busways

| AWG or kcmil | Copper, Three Single Conductors | | | | Copper, Three-Conductor Cable | | | |
| | Steel Conduit | | Nonmagnetic Conduit | | Steel Conduit | | Nonmagnetic Conduit | |
	600V	5kV	600V	5kV	600V	5kV	600V	5kV
14	389	389	389	389	389	389	389	389
12	617	617	617	617	617	617	617	617
10	981	981	981	981	981	981	981	981
8	1,557	1,551	1,558	1,555	1,559	1,557	1,559	1,558
6	2,425	2,406	2,430	2,417	2,431	2,424	2,433	2,428
4	3,806	3,750	3,825	3,789	3,830	3,811	3,837	3,823
3	4,760	4,760	4,802	4,802	4,760	4,790	4,802	4,802
2	5,906	5,736	6,044	5,926	5,989	5,929	6,087	6,022
1	7,292	7,029	7,493	7,306	7,454	7,364	7,579	7,507
1/0	8,924	8,543	9,317	9,033	9,209	9,086	9,472	9,372
2/0	10,755	10,061	11,423	10,877	11,244	11,045	11,703	11,528
3/0	12,843	11,804	13,923	13,048	13,656	13,333	14,410	14,118
4/0	15,082	13,605	16,673	15,351	16,391	15,890	17,482	17,019
250	16,483	14,924	18,593	17,120	18,310	17,850	19,779	19,352
300	18,176	16,292	20,867	18,975	20,617	20,051	22,524	21,938
350	19,703	17,385	22,736	20,526	22,646	21,914	24,904	24,126
400	20,565	18,235	24,296	21,786	24,253	23,371	26,915	26,044
500	22,185	19,172	26,706	23,277	26,980	25,449	30,028	28,712
600	22,965	20,567	28,033	25,203	28,752	27,974	32,236	31,258
750	24,136	21,386	28,303	25,430	31,050	30,024	32,404	31,338
1,000	25,278	22,539	31,490	28,083	33,864	32,688	37,197	35,748

Note: The C values of other conductors/busways can be found in the manufacturer's literature for the device being sized.

Table 7	Multipliers		
f	**M**	**f**	**M**
0.01	0.99	1.50	0.40
0.02	0.98	1.75	0.36
0.03	0.97	2.00	0.33
0.04	0.96	2.50	0.29
0.06	0.94	3.50	0.22
0.07	0.93	4.00	0.20
0.08	0.93	5.00	0.17
0.09	0.92	6.00	0.14
0.10	0.91	7.00	0.13
0.15	0.87	8.00	0.11
0.20	0.83	9.00	0.10
0.25	0.80	10.00	0.09
0.30	0.77	15.00	0.06
0.35	0.74	20.00	0.05
0.40	0.71	30.00	0.03
0.50	0.67	40.00	0.02
0.60	0.63	50.00	0.02
0.70	0.59	60.00	0.02
0.80	0.55	70.00	0.01
0.90	0.53	80.00	0.01
1.00	0.50	90.00	0.01
1.20	0.45	100.00	0.01

NOTE

The line-to-neutral current is higher than the line-to-line fault current at the secondary terminals of a single-phase, center-tapped transformer. The short circuit current available (I) for Step 4 above should be adjusted at the transformer terminals as follows:

I = 1.5 × L-L short circuit amperes
at transformer terminals

At some distance from the terminals, depending on wire size, the L-N fault current is lower than the L-L fault current. The 1.5 multiplier is an approximation and will theoretically vary from 1.33 to 1.67. These figures are based on the change in turns ratio between the primary and the secondary, infinite source available, zero feet from the terminals of the transformer, and 1.2 × %X and 1.5 × %R for L-N vs. L-L resistance and reactance values. Begin L-N calculations at the transformer secondary terminals, then proceed point-to-point.

8.2.0 Practical Application

To demonstrate how the point-to-point method of short circuit calculation is used, find the total short circuit amperes at Fault #1 (I_{SCA1}) and Fault #2 (I_{SCA2}) in the circuit shown in *Figure 17*.

Step 1 $I_{FLA} = \dfrac{kVA \times 1,000}{E_{L-L} \times 1.73} = \dfrac{300 \times 1,000}{208 \times 1.73} = 834A$

Step 2 Multiplier $= \dfrac{100}{transformer\ \%Z} = \dfrac{100}{2} = 50$

Step 3 I_{SCA} at transformer secondary
$= 834 \times 50 = 41,700A$

Step 4 $f_1 = \dfrac{1.73 \times L_1 \times I_{SCA}}{C_1 \times E_{L-L}}$

$= \dfrac{1.73 \times 20 \times 41,700}{22,185 \times 208} = 0.313$

Step 5 $M_1 = \dfrac{1}{1 + f_1} = \dfrac{1}{1 + 0.313} = 0.762$

Step 6 $I_{SCA1} = I_{SCA} \times M = I_{SCA1}$
$= 41,700 \times 0.762 = 31,775A$ at Fault #1

To find the available short circuit current at Fault #2 (I_{SCA2}), use the current at Fault #1 as the available short circuit current.

Step 7 $f_2 = \dfrac{1.73 \times L \times I_{SCA1}}{C_2 \times E_{L-L}}$

$= \dfrac{1.73 \times 20 \times 31,775}{5,906 \times 208} = 0.895$

Step 8 $M_2 = \dfrac{1}{1 + f_2} = \dfrac{1}{1 + 0.895} = 0.528$

Step 9 $I_{SCA2} = I_{SCA} \times M_2$
$= I_{SCA2} = 31,775 \times 0.528 = 16,777A$
at Fault #2

8.3.0 Peak Let-Through Charts

Peak let-through charts (*Figure 18*) allow you to determine both the peak let-through current and the apparent prospective root-mean-square (rms) symmetrical let-through current. Such charts are commonly referred to as current limitation curves. These charts vary with the type of overcurrent protective device and are shown here only to familiarize you with their appearance.

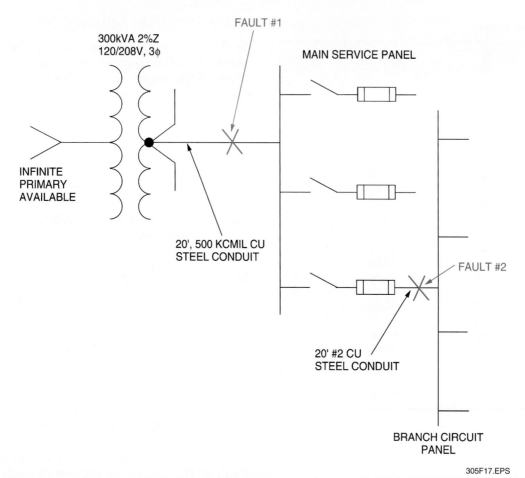

Figure 17 ◆ Sample one-line diagram for short circuit calculation.

INSIDE TRACK

Current-Limiting Circuit Breakers

Most circuit breakers have current-limiting characteristics; however, they do not limit current at the mid- to low-current levels required by UL to qualify as current-limiting breakers. Only those listed as current-limiting have published peak let-through charts.

SYMBOL INDICATES A NONADJUSTABLE (FIXED) CURRENT-LIMITING BREAKER

305SA02.EPS

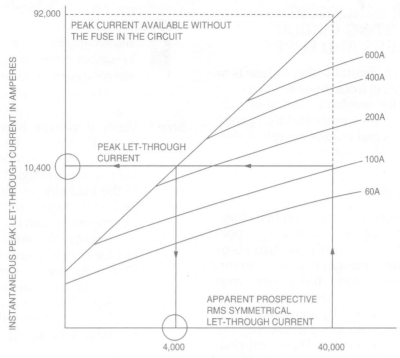

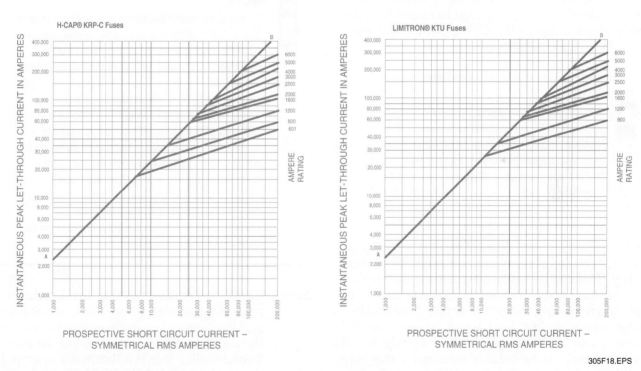

Figure 18 ◆ Peak let-through charts.

305F18.EPS

9.0.0 ◆ TROUBLESHOOTING AND TESTING CIRCUIT BREAKERS AND FUSES

The main function of a circuit breaker or fuse is to protect the circuit wiring from damage caused by the heat generated by overloads and short circuits. While both perform similar functions, they do so in different ways and require slightly different procedures to troubleshoot them.

9.1.0 Circuit Breakers

A standard circuit breaker operates on both thermal and magnetic principles. The magnetic portion provides an instantaneous trip for currents of high levels such as those present in a short circuit. The thermal portion protects the circuit from excessive heat damage generated by lower level currents such as those caused by an overload.

The first step in troubleshooting is to verify the position of the breaker handle. When tripped, most breaker handles move to a position halfway between ON and OFF. Some have visual indicators, while others do not. To reset a circuit breaker, move the handle to the OFF position and then move it back to the ON position.

CAUTION

Whenever re-energizing a breaker, stand to one side and look away from the breaker. If the breaker trips immediately when reset, stop. Do not re-energize the circuit until the fault is located or the load is disconnected from the breaker.

An instant trip or failure to reset is a good sign that a short circuit is causing the magnetic portion to trip. This can be verified by turning the breaker off, checking to ensure there is no presence of voltage, and removing the load wires from the breaker. After removing the load wires, re-energize the breaker and verify the presence of voltage on all load terminals. If the breaker remains on, this may indicate a short circuit in the wiring or in a wiring device.

If the breaker stays on for a limited time, the problem is likely an overload condition, a high impedance or arcing fault, or a short circuit condition. Use the following procedure to troubleshoot the problem:

WARNING!

The following procedure may only be performed by qualified individuals operating under the appropriate safe work plan or permit.

Step 1 Verify if voltage is present using a voltmeter. First verify the operation of the meter by testing it on a known voltage source. Then test for voltage by applying probe one to the load side of the circuit breaker and probe two to ground. Continue the test by moving probe two to other phases that may be present in the panel. If voltage is present, test for current using a clamp-on ammeter. Clamp the meter around the wire to be tested. In the case of multi-pole circuit breakers, this test will need to be performed on each wire. Verify that the current does not exceed the rating of the breaker.

NOTE

Thermal magnetic breakers are designed not to trip at 100% of the handle rating under normal operating conditions. At slightly over 100% (between 105% and 120%) of the handle rating, the breaker must trip between 2,000 and 10,000 seconds. At 150% of the handle rating, the same molded case breaker must trip between 200 and 900 seconds. This type of breaker will actually sustain a load in excess of its rating for a short time, depending upon the size of the load. If the current is less than the handle rating of the breaker but the breaker keeps tripping, it may be due to other factors such as unmeasured harmonic currents and/or excess heat or a defective breaker.

Step 2 Check for excess heat. If the breaker is remaining on for a limited amount of time and the current levels are within tolerance, the problem may be heat related. As mentioned earlier, this is a thermally operated device. It senses the heat produced due to current and trips if that heat is exceeded. Outside heat sources may also affect the operation. If the breaker is located in an area where there is a high ambient temperature, the breaker may trip at currents less than the handle rating of the breaker.

Other factors that may cause this are loose or corroded connections, such as where the breaker is attached to the busbar in the panel, or loose connections where the load wire is attached to the breaker. If an infrared heat detector is available, it may be used to pinpoint the source of heat causing the breaker to trip or you may check by using your own senses. You may also check for heat in the panel by lightly placing your hand on top of the suspected breaker and the adjacent breakers. The breaker may be cool to slightly warm, but should not feel hot.

WARNING!
When testing for this condition, be aware of the voltages present and take the appropriate precautionary measures before proceeding.

Step 3 Listen for any crackling sounds that may indicate arcing. Electrical arcs generate ozone, which has a distinct sharp odor. You may also detect the odor of burned plastic.

WARNING!
If you suspect arcing, immediately disconnect the power.

Step 4 Remove the breaker to inspect the connection between the breaker and the busbar. Follow the manufacturer's recommendations for removing the breaker. Inspect the breaker at the termination points. If excess heat has been present, the terminals may be discolored. Check the wire near the point of attachment for discoloration or charring of the insulation.

Step 5 Replace the breaker.

Breakers are usually stable, reliable, and long-lasting. However, excess current and tripping may damage the internal components and lead to premature wear or unreliable operation. If there is any doubt as to the operation or condition of a breaker, replace it immediately and inspect the connecting bus and nearby components for potential problems.

9.2.0 Fuses

There are two types of fuses: single-element fuses and dual-element fuses. A single-element fuse is designed to act immediately upon overcurrent, which may be either a short circuit or an overload. A dual-element fuse has two fuse elements: a quick-acting element similar to that of a single-element fuse and a thermal operation time-delay element. The quick-acting element is sized to react to short circuit conditions while allowing smaller overcurrents to pass. The time-delay element will protect against small overcurrents over a given time, similar to that of a thermal breaker.

WARNING!
Since fuses usually have live exposed parts within the fuseholder, extra care must be taken when troubleshooting.

To troubleshoot a fuse using an ohmmeter, proceed as follows:

WARNING!
The following procedure may only be performed by qualified individuals operating under the appropriate safe work plan or permit.

Step 1 Disconnect the power to the fuse. This usually involves turning off a local disconnect switch. If the fuseholder is not adjacent to the power switch, follow appropriate lockout/tagout procedures.

Step 2 Verify that the voltage has been removed from all fuses associated with the circuit. Use a voltmeter to take readings from the top of the fuseholder to ground and other phases, and then repeat the procedure from the bottom of the fuseholder. Another check between the top and bottom of the fuseholder will verify that no voltage is present.

Step 3 After checking for voltage, remove the fuse. Using an ohmmeter or continuity tester, place one probe at one end of the fuse and the other probe on the opposite end. If the reading shows no resistance, indicating continuity, the fuse is good. Otherwise, the fuse is bad.

Step 4 To locate a blown fuse in an energized circuit, use a voltmeter to check for voltage first at the top of the fuse, checking to ground or another phase, and then at the bottom of the fuse. If voltage is present at the top but not the bottom, the fuse is bad. Disconnect the power and verify voltage is not present before replacing the fuse. If voltage-to-ground or another phase is present at both the top and bottom of the fuse and a bad fuse is still suspected, conduct further voltage checks from the bottom of the fuse to all phases (line-to-line in a 240V single-phase circuit or phase-to-phase and ground). If voltage is not obtained when checking phase-to-phase, a final check across the fuse from top to bottom should also be done. If voltage is present across the fuse, the fuse is bad.

NOTE

Fuses have varying time curves between types and classes and therefore vary in the amount of time they may sustain an overload. They are still affected by the same external problems that affect a circuit breaker, including unmeasured harmonic currents and excess heat.

Step 5 To check for excess heat, disconnect all power, verify that the voltage has been removed, and proceed to do the visual and other tests. In this case, a visual test is easy to perform. Look for signs of discoloration on the fuse, fuseholders, and terminals. If fuseholder clips are discolored from heat, they may need to be replaced. The fuse clip material will lose its spring temper, and squeezing it with pliers won't fix it.

Check the wire for insulation damage or discoloration that may result from excessive heat. Since the voltage is connected, the sound of arcing should not be present, but the smell may exist.

Step 6 Disconnect the power, verify that voltage has been removed, and replace the fuse. Note that voltage detected across a fuse (end-to-end) in a 240V or three-phase circuit is indicative of an open link.

Step 7 Once the new fuse is installed, verification should follow. After installing the new fuse, re-energize the circuit using the proper safety techniques. Verify the presence of voltage on all load terminals by checking phase-to-phase and phase-to-ground using a clamp-on ammeter to verify the current on the circuit.

WARNING!

Verify that all voltage has been removed prior to touching or replacing any fuse.

Fuses are stable devices that seldom fail due to problems within the fuses themselves. However, a fuse may be stressed over time from excessive currents and eventually fail. More often there is a problem within the circuit that causes the failure. Since fuses are good for one time only, it is sometimes more difficult to determine the problem that caused a fuse to fail. If replacement of one fuse does not correct the problem, and the visual inspections have been completed, it may be necessary to troubleshoot the circuit using known methods for that circuit. Do not continue to replace fuses hoping to eventually find one that holds. Do not increase the fuse size or change fuse types without investigating the impact on the protection of the equipment. Be aware of the limitations of single-element fuses and the limited use with motor circuits.

1. Which of the following best describes the maximum short circuit current that a fuse or circuit breaker will safely interrupt?
   a. CIA
   b. AIC
   c. SOL
   d. ICA

2. A plug fuse is typically used in a _____ circuit.
   a. branch
   b. feeder
   c. motor
   d. high-voltage

3. A 600V fuse can be used in a 240V circuit.
   a. True
   b. False

4. A single-element fuse can be used in circuits subject to temporary motor overloads.
   a. True
   b. False

5. The minimum interrupting rating of branch circuit listed fuses is _____.
   a. 5,000A
   b. 10,000A
   c. 15,000A
   d. 20,000A

6. The minimum voltage rating of branch circuit fuses is _____.
   a. 24V
   b. 120V
   c. 125V
   d. 240V

7. All of the following are classifications of medium-voltage fuses *except* _____.
   a. general-purpose current-limiting fuses
   b. backup current-limiting fuses
   c. expulsion fuses
   d. proportion fuses

8. An expulsion fuse is best described as a _____.
   a. fuse capable of interrupting all currents from the rated interrupting current down to the current that causes melting of the fusible element in an hour
   b. fuse capable of interrupting all currents from the maximum rated interrupting current down to the rated minimum interrupting current
   c. strap-mounted device
   d. vented fuse in which the expulsion effect of gases produced by the arc and lining of the fuseholder, either alone or aided by a spring, extinguishes the arc

9. Each of the following is a basic factor to consider when applying any fuse *except* _____.
   a. voltage
   b. continuous current-carrying capacity
   c. interrupting rating
   d. manufacturer's brand name

10. The total clearing time of any downstream protective device must be below a curve representing _____ of the minimum melting curve of the fuse being applied.
    a. 10%
    b. 20%
    c. 50%
    d. 75%

11. When a common circuit breaker trips, the handle is in the _____ position.
    a. OFF
    b. ON
    c. middle
    d. RESET

12. Circuit breakers are grouped for identification according to given current ranges. Each group is classified by the _____.
    a. largest ampere rating of its range
    b. smallest ampere rating of its range
    c. absolute lowest ampere rating of its range
    d. overall average ampere rating of its range

13. When circuit breakers are classified as 100A through 2,000A frames, these numbers are normally referred to as the _____.
    a. frame size
    b. overload protective current
    c. maximum voltage allowed on the circuit
    d. physical size of the circuit breaker

14. The full-load current rating of a 150kVA, three-phase transformer operating at 240V is _____.
    a. 296A
    b. 324A
    c. 362A
    d. 397A

15. A circuit breaker that trips instantly or fails to reset is most likely the result of _____.
    a. corroded connections
    b. a high ambient temperature
    c. a short circuit
    d. loose connections

Summary

Reliable overcurrent protective devices prevent or minimize costly damage to transformers, conductors, motors, and the many other components and electrical loads that make up the complete electrical distribution system. Consequently, reliable circuit protection is essential to avoid the severe monetary losses that can result from power blackouts and prolonged downtime of various types of facilities. The *NEC*® has set forth various minimum requirements dealing with overcurrent devices and how they should be installed in various types of electrical circuits.

Notes

Trade Terms
Introduced in This Module

Ampere rating: The current-carrying capacity of an overcurrent protective device. The fuse or circuit breaker is subjected to a current above its ampere rating; it will open the circuit after a predetermined period of time.

Ampere squared seconds (I^2t): The measure of heat energy developed within a circuit during the fuse's clearing. It can be expressed as melting I^2t, arcing I^2t, or the sum of them as clearing I^2t. I stands for effective let-through current (rms), which is squared, and t stands for time of opening in seconds.

Amperes interrupting capacity (AIC): The maximum short circuit current that a circuit breaker or fuse can safely interrupt.

Arcing time: The amount of time from the instant the fuse link has melted until the overcurrent is interrupted or cleared.

Clearing time: The total time between the beginning of the overcurrent and the final opening of the circuit at rated voltage by an overcurrent protective device. Clearing time is the total of the melting time and the arcing time.

Current-limiting device: A device that will clear a short circuit in less than one half cycle. Also, it will limit the instantaneous peak let-through current to a value substantially less than that obtainable in the same circuit if that device were replaced with a solid conductor of equal impedance.

Fast-acting fuse: A fuse that opens on overloads and short circuits very quickly. This type of fuse is not designed to withstand temporary overload currents associated with some electrical loads (inductive loads).

Inductive load: An electrical load that pulls a large amount of current—an inrush current—when first energized. After a few cycles or seconds, the current declines to the load current.

Melting time: The amount of time required to melt a fuse link during a specified overcurrent.

NEC® dimensions: These are dimensions once referenced in the *National Electrical Code®*. They are common to Class H and K fuses and provide interchangeability between manufacturers for fuses and fusible equipment of given ampere and voltage ratings.

Overload: Can be classified as an overcurrent that exceeds the normal full-load current of a circuit.

Peak let-through (Ip): The instantaneous value of peak current let-through by a current-limiting fuse when it operates in its current-limiting range.

Root-mean-square (rms): The effective value of an AC sine wave, which is calculated as the square root of the average of the squares of all the instantaneous values of the current throughout one cycle. Alternating current rms is that value of an alternating current that produces the same heating effect as a given DC value.

Semiconductor fuse: Fuse used to protect solid-state devices.

Short circuit current: Can be classified as an overcurrent that exceeds the normal full-load current of a circuit by a factor many times greater than normal. Also characteristic of this type of overcurrent is that it leaves the normal current-carrying path of the circuit—it takes a shortcut around the load and back to the source.

Single phasing: The condition that occurs when one phase of a three-phase system opens, either in a low-voltage or high-voltage distribution system. Primary or secondary single phasing can be caused by any number of events. This condition results in unbalanced loads in polyphase motors and unless protective measures are taken, it will cause overheating and failure.

UL classes: Underwriters Laboratories has developed basic physical specifications and electrical performance requirements for fuses with voltage ratings of 600V or less. These are known as UL standards. If a type of fuse meets with the requirements of a standard, it can fall into that UL class. Typical UL classes are R, K, G, L, H, T, CC, and J.

Voltage rating: The maximum value of system voltage in which a fuse can be used, yet safely interrupt an overcurrent. Exceeding the voltage rating of a fuse impairs its ability to safely clear an overload or short circuit.

This module is intended to present thorough resources for task training. The following reference work is suggested for further study. This is optional material for continued education rather than for task training.

National Electrical Code® Handbook, Latest Edition. Quincy, MA: National Fire Protection Association.

CONTREN® LEARNING SERIES – USER UPDATE

NCCER makes every effort to keep these textbooks up-to-date and free of technical errors. We appreciate your help in this process. If you have an idea for improving this textbook, or if you find an error, a typographical mistake, or an inaccuracy in NCCER's Contren® textbooks, please write us, using this form or a photocopy. Be sure to include the exact module number, page number, a detailed description, and the correction, if applicable. Your input will be brought to the attention of the Technical Review Committee. Thank you for your assistance.

Instructors – If you found that additional materials were necessary in order to teach this module effectively, please let us know so that we may include them in the Equipment/Materials list in the Annotated Instructor's Guide.

Write: Product Development and Revision
National Center for Construction Education and Research
3600 NW 43rd St., Bldg. G, Gainesville, FL 32606

Fax: 352-334-0932

E-mail: curriculum@nccer.org

Craft _____ Module Name _____

Copyright Date _____ Module Number _____ Page Number(s) _____

Description _____

_____

_____

_____

(Optional) Correction _____

_____

_____

(Optional) Your Name and Address _____

_____

_____

Distribution Equipment

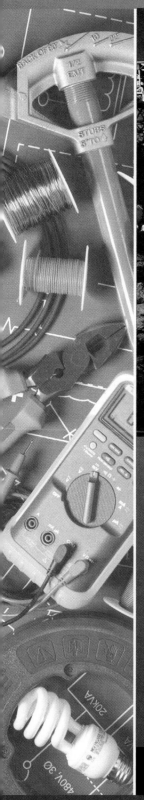

Hoover Dam

Upon completion in 1935, Hoover Dam was the world's largest electric power-producing facility and its largest concrete structure. Today, this National Historic Landmark is the 34th largest hydroelectric generating station on the globe, producing an average 4.4 billion kilowatt-hours per year.

26306-08

26306-08
Distribution Equipment

Topics to be presented in this module include:

Overview

Electrical power is generated at power plants. The voltage is stepped up to compensate for voltage drops that may occur when sending the power across long runs of transmission lines. Once the voltage arrives in the general area of usage, it must be stepped down to usable voltage levels and distributed to consumers. Stepping down and distributing the power is accomplished by distribution equipment.

Switchboards make up a large portion of distribution equipment. They are used to control the routing of power to large areas of usage or directly to high-end consumers such as industrial or manufacturing facilities. Technicians who maintain switchboard equipment must be trained in the proper inspection, testing, and maintenance of this type of equipment. The *NEC*® regulates the construction, installation, and accessories used with switchboard equipment.

Distribution systems are typically illustrated using one-line electrical drawings. These types of drawings provide a traceable path from the incoming power in a substation, through step-up or step-down transformers, fuses, circuit breakers, switches, and eventually out to the consumer or individual loads.

Objectives

When you have completed this module, you will be able to do the following:

1. Describe the purpose of switchgear.
2. Describe the four general classifications of circuit breakers and list the major circuit breaker ratings.
3. Describe switchgear construction, metering layouts, wiring requirements, and maintenance.
4. List *National Electrical Code®* (*NEC®*) requirements pertaining to switchgear.
5. Describe the visual and mechanical inspections and electrical tests associated with low-voltage and medium-voltage cables, metal-enclosed busways, and metering and instrumentation.
6. Describe a ground fault relay system and explain how to test it.

Trade Terms

Air circuit breaker
Basic impulse insulation level (BIL)
Branch circuit
Bus
Bushing
Capacity
Current transformer (CT)
Distribution system equipment

Distribution transformer
Feeder
Metal-enclosed switchgear
Potential transformer (PT)
Service-entrance equipment
Switchboard
Switchgear

Required Trainee Materials

1. Pencil and paper
2. Appropriate personal protective equipment
3. Copy of the latest edition of the *National Electrical Code®*

Prerequisites

Before you begin this module, it is recommended that you successfully complete *Core Curriculum; Electrical Level One; Electrical Level Two; Electrical Level Three*, Modules 26301-08 through 26305-08.

This course map shows all of the modules in *Electrical Level Three*. The suggested training order begins at the bottom and proceeds up. Skill levels increase as you advance on the course map. The local Training Program Sponsor may adjust the training order.

ELECTRICAL LEVEL THREE

26311-08 Motor Controls

26310-08 Voice, Data, and Video

26309-08 Motor Calculations

26308-08 Commercial Electrical Services

26307-08 Transformers

26306-08 Distribution Equipment

26305-08 Overcurrent Protection

26304-08 Hazardous Locations

26303-08 Practical Applications of Lighting

26302-08 Conductor Selection and Calculations

26301-08 Load Calculations – Branch and Feeder Circuits

ELECTRICAL LEVEL TWO

ELECTRICAL LEVEL ONE

CORE CURRICULUM: Introductory Craft Skills

306CMAP.EPS

1.0.0 ◆ INTRODUCTION

An electrical power system consists of several subsystems on both the utility (supply) side and the customer (user) side. Electricity generated in power plants is stepped up to transmission voltage and fed into a nationwide grid of transmission lines. This power is then bought, sold, and dispatched as needed. Local utility companies take power from the grid and reduce the voltage to levels suitable for subtransmission and distribution through various substations to the customer. This may range from the common 200A, 120/240V residential service to thousands of amps at voltages from 480V to 69kV in an industrial facility.

From the point of service, customers must control, distribute, and manage the power to supply their electrical needs. This module will discuss how this is done using a typical industrial facility as an example. We will discuss the various components of the distribution system and their interdependence. An understanding of single-line diagrams will allow analysis of a facility's distribution system.

NOTE

The voltage conventions used in this module are industry standards for distribution systems.

2.0.0 ◆ VOLTAGE CLASSIFICATIONS

While electrical systems and equipment are often classified by voltage rating, switchgear is classified first by the type of construction and secondly by voltage rating. It is important to note that there is no official industry-wide voltage classification system. For example, the *NEC®* considers anything above 600V as high voltage, while the transmission sector considers anything below 72,500V (72.5kV) as low voltage. In industrial applications, the term low-voltage refers to systems rated up to 1,000V, while medium voltage refers to systems rated above 1,000V and up to 38,000V (38kV). This is the range in which metal-clad switchgear and circuit breakers are manufactured in standard configurations. This is also the voltage range in which pre-molded and shrink-on termination kits are readily available for shielded cable terminations. Above this voltage level, cable is run on overhead power lines rather than in raceway or cable tray.

Low-voltage power circuit breaker switchgear, for example, may be rated up to 1,000VAC or 3,200VDC. Metal-clad or metal-enclosed switchgear is applied at voltages over 1,000VAC up to a maximum of 38,000VAC.

3.0.0 ◆ SWITCHBOARDS

According to the *National Electrical Code®*, the term **switchboard** may be defined as a large single panel, frame, or assembly of panels on which switches, overcurrent and other protective devices, **buses**, and instruments may be mounted, either on the face or back or both. Switchboards are generally accessible from both the rear and from the front and are not intended to be installed in cabinets.

3.1.0 Applications

Switchboards are used in modern distribution systems to subdivide large blocks of electrical power. One location for switchboards is typically where the main power enters the building. In this location, the switchboard is referred to as **service-entrance equipment**. The other location common for switchboards is downstream from the service-entrance equipment. In the downstream location, the switchboard is commonly referred to as **distribution system equipment**.

3.2.0 General Description

A switchboard consists of a stationary structure that includes one or more freestanding units of uniform height that are mechanically and electrically joined to make a single coordinated installation. These cubicles contain circuit-interrupting devices. They take up less space in a plant, have more eye appeal, and eliminate the need for a separate room to protect personnel from contact with lethal voltages.

The main portion of the switchboard is formed from heavy-gauge steel welded with members across the top and bottom to provide a rigid enclosure. Most switchboard enclosures are divided into three sections: the front section, the bus section, and the cable section. These three sections are physically separated from one another by metal partitions. This confines any damage that may occur to any one section and keeps it from affecting the other sections.

Typical switchboard components include:

- Circuit breakers
- Fuses
- Motor starters
- Ground fault systems
- Instrument transformers
- Switchboard metering
- Control power transformers
- Busbars

Electrical ratings include three-phase, three-wire and three-phase, four-wire systems with voltage ratings up to 600V and current ratings up to 4,000A.

A switchboard enclosure is described as a dead front panel, which means that no live parts are exposed on the opening side of the equipment; however, it contains energized breakers. Busbars can be a standard size or customized. Standard sizes are usually made of silver-plated or tin-plated copper or tin-plated aluminum. Conventional bus sizing is 0.25" × 2" through 0.375" × 7". Copper provides an ampacity of 1,000A/sq. in. of cross-sectional area. When using aluminum, the ampacity is 750A/sq. in.

When two busbars are bolted together using Grade S hardware with the proper torque, the ampacity of the connection is 200A/sq. in. of the lapped portion for aluminum or copper bussing. Bussing joints must be bolted together to the specified torque and include Belleville washers or Keps nuts. Aluminum busbars must be tin-plated, and copper busbars over 600A must be plated with tin or silver.

3.3.0 Switchboard Frame Heating

Table 1 shows guidelines that should be observed in order to keep heat losses in the iron switchboard frame members to a safe minimum. The dimensions are recommended values and should be adhered to whenever possible.

NOTE

Some switchboard frames are engineered differently and will have values other than those shown in *Table 1*.

3.4.0 Low-Voltage Spacing Requirements

To minimize tracking or arcing from energized parts to ground, switchboard construction includes spacing requirements. These spacing requirements are measured between live parts of opposite polarity and between live parts and grounded metal parts. *Figure 1* illustrates typical switchboard spacing requirements.

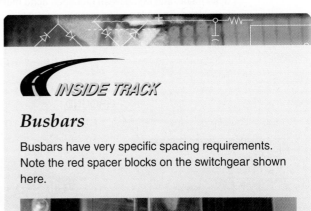

INSIDE TRACK

Busbars

Busbars have very specific spacing requirements. Note the red spacer blocks on the switchgear shown here.

306SA01.EPS

Table 1 Switchboard Frame Heating Guidelines

Amperes	Minimum Distance from Phase Bus to Closest Steel Member	Minimum Distance from Neutral Bus to Closest Steel Member
3,000	4"	2"
4,000	6"	3"
5,000 and over	12"	see below
5,000 to 6,000	An aluminum or nonmagnetic material should be used in place of steel frame sections. Wherever possible, you must maintain 12" to steel members and 6" to aluminum or nonmagnetic members. Neutral spacing can be 6" and 3", respectively. If the main bus is tapered, it is permissible (at 4,000A and below) to use steel frames for those sections containing the tapered bus.	
6,000 and over	You must use an aluminum or nonmagnetic material for frame sections and maintain 12" to steel members and 6" to aluminum or nonmagnetic members. Neutral spacing can be 6" and 3", respectively. The use of any steel frame members is discouraged. If the main bus is tapered, it is permissible (at 4,000A and below) to use steel frames for those sections containing the tapered bus.	

Note: For amperages above 8,000A, the neutral spacing must be 12" wherever possible.

VOLTAGE INVOLVED		MINIMUM SPACING BETWEEN LIVE PARTS OF OPPOSITE POLARITY		MINIMUM SPACING THROUGH AIR AND OVER SURFACE BETWEEN LIVE PARTS AND GROUNDED METAL PARTS
GREATER THAN	MAX.	THROUGH AIR	OVER SURFACE	BOTH THROUGH AIR AND OVER SURFACE
0 – 125		½"	¾"	½"
125 – 250		¾"	1¼"	½"
250 – 600		1"	2"	*1"

* A through air spacing of not less than ½" is acceptable (1) at a molded-case circuit breaker or a switch other than a snap switch, (2) between uninsulated live parts of a meter mounting or grounded dead metal, and (3) between grounded dead metal and the neutral of a 480Y/277V, three-phase, four-wire switchgear section.

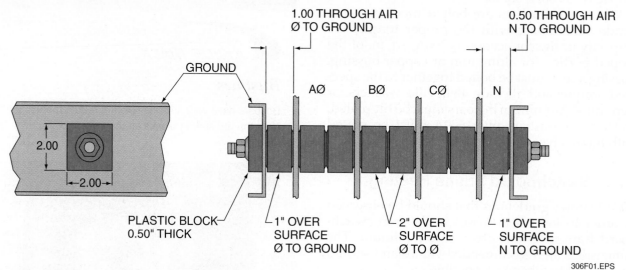

Figure 1 ◆ Typical busbar spacing requirements.

306F01.EPS

An isolated dead metal part, such as a screwhead or washer, interposed between uninsulated live parts of opposite polarity or between an uninsulated live part and grounded dead metal is considered to reduce the spacing by an amount equal to the dimension of the interposed part along the path of measurement.

When measuring over-surface spacing, any slot, groove, and the like that is 0.013" (0.33 mm) wide or less and in the contour of the insulating material is to be disregarded.

When measuring spacing, an air space of 0.013" or less between a live part and an insulating surface is to be disregarded, and the live part is to be considered in contact with the insulating material. A pressure wire connector shall be prevented from any turning motion that would result in less than the minimum acceptable spacings. The means used to ensure turn prevention must be reliable, such as a shoulder or boss. A lock washer alone is not acceptable.

A means of turn prevention need not be provided if spacings are not less than the following minimum accepted values:

- When the connector and any connector of opposite polarity have each been turned 30 degrees toward the other
- When the connector has been turned 30 degrees toward other live parts of opposite polarity and toward grounded dead metal parts

3.5.0 Cable Bracing

All construction using conductors and having a short circuit current rating greater than 50,000 rms symmetrical amperes requires a cable brace positioned as close to the supply lugs as possible. The cable brace is intended to be mounted in the same area that is allotted for wire bending. It is not necessary to provide additional mounting height to accommodate the cable brace.

The cable brace requirement does not apply to load-side cables, main breakers, or switches. It only applies when cables are connected directly to an unprotected line-side bus. The bus restrictions for a line-side bus are as follows:

- There can be no splice in edgewise bus mounting of 2,100A or less rated at 50,000 rms symmetrical amperes.
- There can be no splice in flatwise bus mounting of 600A or less rated over 50,000 rms symmetrical amperes.

> **NOTE**
> This does not apply to connections made from the through bus to a switch or circuit breaker.

The cable restrictions for a line-side bus include:

- Busing of 600A or less that is rated over 50,000 rms symmetrical amperes cannot use cables; it must be bus connected.
- If cabling is required, 800A minimum busing must be used.

Cable bracing requirements may be excluded if the busing is able to fully withstand the total available short circuit current.

4.0.0 ◆ SWITCHGEAR

Switchgear is a general term used to describe switching and interrupting devices and assemblies of those devices containing control, metering, protective, and regulatory equipment, along with the associated interconnections and supporting structures. Switchgear performs two basic functions:

- Provides a means of switching or disconnecting power system apparatus
- Provides power system protection by automatically isolating faulty components

Switchgear can be classified as:

- Metal-enclosed switchgear (low voltage)
- Metal-clad switchgear (low and medium voltage)
- Metal-enclosed interrupters
- Unit substations

The low-voltage and medium-voltage switchgear assemblies are completely enclosed on all sides and topped with sheet metal, except for ventilating openings and inspection windows. They contain primary power circuit switching or interrupting devices, buses, connections, and control and auxiliary devices. *Figure 2* shows typical low-voltage, metal-clad switchgear.

The station-type cubicle switchgear consists of indoor and outdoor types with power circuit breakers rated from 14.4kV to 34.5kV, 1,200A to 5,000A, and 1,500kVA to 2,500kVA interrupting

306F02.EPS

Figure 2 ◆ Typical low-voltage, metal-clad switchgear.

capacity. Equipment can be special ordered and built at higher kVA ratings.

4.1.0 Switchgear Construction

Switchgear consists of a stationary structure that includes one or more freestanding units of uniform height that are mechanically and electrically joined to make a single coordinated installation. These units, commonly referred to as cubicles, contain circuit-interrupting devices such as circuit breakers.

Switchgear enclosures are formed from heavy-gauge sheet steel that has been welded or bolted together. Structural members across the top, sides, and bottom provide a rigid enclosure. Metal-clad switchgear enclosures are divided into three sections: the front section, the bus section, and the cable or termination section.

These three sections are physically separated from one another by metal partitions. This confines any damage that may occur to any one section and keeps it from affecting the other sections. It also separates power between the sections for ease and safety of maintenance.

The rigid enclosure provides the primary structural strength of the switchgear assembly and the means by which the switchgear is fastened to its foundation. The strength of the enclosure and its mounting system will vary depending on its intended use. For example, switchgear used in a nuclear application must meet certain seismic qualifications.

The enclosure also provides the required supports and mounts for items to be located in the switchgear and provides for the necessary interconnections between the switchgear and other plant systems. The number of sections and physical makeup of switchgear varies depending on the voltage and current ratings, project specifications, and specific manufacturer.

Figures 3 and *4* show external and internal views, respectively, of medium-voltage, metal-clad switchgear. This equipment is available in voltages from 4.76kV to 27kV and current ranges from 1,200A through 3,000A.

4.2.0 Control and Metering Safety Standards

There is a tendency among some people in the industry to use the terms "switchboard" and "switchgear" interchangeably. However, they are not the same. Switchgear is manufactured and tested to more exacting standards and is configured differently than switchboards. For example, in switchgear there are physical barriers between breakers, and between the breakers and the bus. Switchgear is more durable and fault resistant, and is commonly selected for larger applications where low-voltage power circuit breakers and selective coordination are applied, such as computer data centers, manufacturing, and process facilities.

306F03.EPS

Figure 3 ◆ Medium-voltage, metal-clad switchgear (exterior view).

306F04.EPS

Figure 4 ◆ Medium-voltage, metal-clad switchgear (interior view).

4.3.0 Wiring System

The *NEC*® requires wiring to be supported mechanically to keep the wiring in place. Wire harnessing is generally used within the switchboard with the following restrictions:

- Each bundle or cable of wires must be run in a vertical or horizontal direction, securing the harness by means of plastic cable ties or cable clips.
- Plastic wire cable clamps shall be placed at strategic locations along the harnessing to hold the harness firmly in place to prevent interference with the control components' required electrical, mechanical, and arcing clearances.
- Apply wire ties to the harnessed wiring every 3" to 4" with self-adhesive cable ties spaced at every 12".
- Some precautions to be observed when wiring the switchboard electrical components are:
  - Keep control wires at least ½" from moving parts.
  - Avoid running wires across sharp metal edges. To protect the wiring from mechanical damage, use approved cable protectors, such as a nylon clip cable guard, a wire guard for edge protection, or special edge protection molding.
  - Wires must not touch exposed bare electrical parts of opposite polarity.
  - Wires must not interfere with the adjustment or replacement of components.
  - Wires should be as straight and as short as possible.
  - Wires shall not be spliced.
  - To eliminate possible strain on the control wire, a certain amount of slack should be given to the individual or harnessed conductor terminated at a component connection.
  - The equipment ground busbar shall not be used as a portion of the control or metering circuits.
  - Do not use pliers for bending control wiring. Use your hands or an approved wire bending device.

4.3.1 Door-Mounted Wiring Restrictions

No incoming wiring connections may be made directly to the door-mounted devices. Wires from the door-mounted equipment to the panel terminal block should be a minimum of 19-strand wire.

Wires from the door must be neatly cabled so that the door can be opened easily without placing excessive strain on the wire terminal connections. In some cases, the cable must be separated into two bundles to accomplish this. Insulated sleeving, tubing, or vinyl tape must be used to bundle and protect the flexible wires.

4.3.2 Terminal Connections

All control or metering wiring entering or leaving the switchboard should terminate at terminal blocks, leaving one side of the terminal block free for the user's connections. No factory connections are allowed on the user's terminal connection point. For factory wiring, allow a maximum of two control wires on the same side of a terminal block. No more than three connections are allowed on terminals of control transformers, meters, meter selector switches, and metering equipment.

Since bolted pressure switches or any 100% current-rated, molded-case circuit breaker's line and load power terminals are allowed a higher maximum operating temperature than the recommended insulated conductor's operating temperature, the control wires cannot be placed directly on the 100%-rated disconnect device's line and load connections.

In all cases, control wires cannot touch any exposed part of opposite electrical polarity.

4.4.0 Metering Current and Potential Transformers

Ground connections on a **potential transformer (PT)** or **current transformer (CT)** secondary terminal must be connected to the ground bus. CT secondary terminals must be shorted if no metering equipment is connected to the current transformer.

PTs are required to have primary and secondary fusing. If protective circuits, such as ground fault or phase failure protective systems, are placed in the secondary circuit of the potential transformer, no secondary fusing is required.

Metering circuit connections made directly to the incoming bus must be provided with current-limiting fuses that are equal in rating to the available interrupting capacity.

NOTE

CTs and PTs will be discussed later.

4.5.0 Switchgear Handling, Storage, and Installation

The following is a basic guideline for the handling of switchgear. It is important to note that these recommendations only supplement the manufacturer's instructions. Manufacturers include instruction books and drawings with their equipment. It is absolutely imperative that you read and understand these documents before handling any equipment.

- *Switchgear handling* – Immediately upon receipt of switchgear, an inspection for damage during transit should be performed. If any damage is noted, the transportation company should be notified immediately.
- *Switchgear rigging* – Instructions for switchgear should be found in the manufacturer's instruction books and drawings. Verify that the rigging is suitable for the size and weight of the equipment.
- *Switchgear storage* – Indoor switchgear that is not being installed right away should be stored in a clean, dry location. The equipment should be level and protected from the environment if construction is proceeding. The longer equipment is in storage, the more care is required for protection of the equipment. If a temporary cover is used to protect the equipment, this cover should not prevent air circulation. If the building is not heated or temperature controlled, heaters should be used to prevent moisture/condensation buildup. Outdoor switchgear that cannot be installed immediately must be provided with temporary power. This power will allow operation of the space heaters provided with the equipment.
- *Bus connections* – The main bus that is usually removed during shipping should be reconnected. Ensure that the contact surfaces are clean and pressure is applied in the correct manner. The conductivity of the joints is dependent on the applied pressure at the contact points. The manufacturer's torque instructions should be referenced.
- *Cable connections* – When making cable connections, verify the phasing of each cable. This procedure is done in accordance with the connection diagrams and the cable tags. When forming and mounting cables, ensure that the cables are tightened per the manufacturer's instructions.
- *Grounding* – Any sections of ground bus that were previously disconnected for shipping should be reconnected when the units are installed. In addition, the system must be bonded at this time. The ground bus should be connected to the system ground with as direct a connection as possible. If the system ground is to be run in metal conduit, bonding to the conduit is required. The ground connection is

necessary for all switchgear and should be sized per the *NEC*®.

5.0.0 ◆ TESTING AND MAINTENANCE

This section covers general testing and maintenance procedures.

WARNING!

When working on switchgear or any piece of electrical equipment, you must always be aware of and follow all applicable safety procedures. You must also understand the construction and operation of the equipment. You must be specifically trained and qualified to work on or near energized electrical circuits and equipment. National consensus standards such as *NFPA 70E* and *70B* provide specific guidance for achieving an electrically safe work condition. *NFPA 70E, Standard for Electrical Safety in the Workplace, Article 120*, provides a step-by-step procedure for achieving an electrically safe work condition.

Chapter 7 in *NFPA 70B, Recommended Practice for Electrical Equipment Maintenance*, provides personnel safety for qualified electrical workers, while other chapters provide specific direction for maintenance and troubleshooting of various types of equipment.

NOTE

Test values will differ depending on whether you are performing an acceptance test or a maintenance test.

5.1.0 General Maintenance Guidelines

To perform a visual inspection:

Step 1 Check the exterior for the proper fit of doors and covers, paint, etc.

Step 2 Check the interior, particularly the current-carrying parts, including:
- Inspect the busbars for dirt, corrosion, and/or overheating.
- If necessary, perform an infrared or thermographic test. Note any discoloration that would represent a poor bus joint.
- Check the busbar supports for cracks.
- Check for correct electrical spacing.
- Verify the integrity of all bolted connections.

Protective Grounding

Even after a circuit has been isolated, de-energized, locked out, and verified without voltage, it still may not be safe to work on. This is because there is still a possibility that a circuit or conductor may be inadvertently re-energized through any one of the following means:

- Induced voltages from other energized conductors
- Static buildup from wind on outdoor conductors
- High voltage from lightning strikes
- Any condition that might bring an energized conductor into contact with the de-energized circuit
- Switching errors causing re-energizing of the circuit
- Capacitive charges in equipment or conductors

When any of these conditions are possible, *NFPA 70B, Recommended Practice for Electrical Equipment Maintenance*, requires that temporary grounds be applied before the circuit or equipment is considered safe. In fact, standard practice in overhead line construction and within open substations is that any conductor without a temporary ground connection is considered energized. While the terms temporary ground, safety ground, and protective ground are often used interchangeably, temporary grounds cover both personal protective grounds and static grounds. Personal protective grounds consist of cable connected to de-energized lines and equipment by jumpering and bonding with appropriate clamps, to limit the voltage difference between accessible points at a work site to safe values if the lines or equipment are accidentally re-energized. Protective grounds are sized to carry the maximum available fault current at the work site for the expected fault duration. Static grounds include any grounding cable or bonding jumper (including clamps) that has an ampacity less than the maximum available fault current at the work site, or is smaller than No. 2 AWG copper equivalent. Static grounds are used for potential equalizing between conductive parts in grounding configurations that cannot subject them to significant current. Therefore, smaller wire that provides adequate mechanical strength is sufficient (e.g., No. 12 AWG).

Low-voltage equipment with only a single source of supply usually does not require temporary grounding for safety. Low-voltage equipment with dual supply and medium-voltage equipment should be grounded at the bus.

ASTM International Standard F855-04, Temporary Protective Grounds to Be Used on De-energized Electric Power Lines and Equipment, is the national consensus standard covering the equipment making up the temporary grounding system. This standard addresses the parts of a temporary grounding system, which include the clamps, ferrules, cables, or a complete protective ground assembly of clamps, ferrules, and cables. These components work together and must be capable of conducting the maximum available fault current that could occur at a work location if lines or equipment become re-energized from any source, and for the expected duration of the fault. Because the circuit is NOT safe until grounding is applied, placing and removing temporary grounds is considered work on live parts, and appropriate PPE and safe work practices must be followed.

This picture shows a temporary protective ground cluster on incoming medium-voltage feeders at equipment. Not shown is the connection to the permanent system and feeder grounding conductors. This arrangement will provide safety for the connected equipment bus. Notice the phase arrangement is from left to right at the front of the equipment (the back of the equipment is shown here).

306SA02.EPS

Keyed Interlocks

Keyed interlocks, such as the one shown here, ensure that qualified personnel perform operations in the required sequence by preventing or allowing the operation of one part only when another part is locked in a predetermined position. These devices can be used for a variety of safety applications, such as preventing personnel from accessing a high-voltage compartment before opening the disconnect switch.

306SA03.EPS

To clean the switchboard:

Step 1 Vacuum the interior (do not use compressed air).

Step 2 Wipe down the interior using a clean, lint-free cloth. Use nonconductive, nonresidue solution, such as contact cleaner or denatured alcohol.

To check equipment operation:

Step 1 Manually open and close circuit breakers and switches.

Step 2 Electrically operate all components, such as ground fault detectors, sure trip metering, current transformers, test blocks, ground lights, blown main fuse detectors, and phase failure detectors.

To perform a megger test:

Step 1 Isolate the bus by opening all circuit breakers and switches.

Step 2 Disconnect any devices, such as relays and transformers, that may be connected to the busbars.

Step 3 Make sure all personnel are clear of the switchboard.

Step 4 Use a 1,000V megger to check the phase-to-phase and phase-to-ground resistance. Megger readings should reflect the values listed in the equipment manufacturer's instructions. Typical values are shown in *Table 2*.

5.2.0 Test Guidelines

This section provides typical guidelines for performing various tests on distribution equipment.

 WARNING!

This test is performed while the equipment is energized and the covers are removed. This test may only be performed by qualified personnel under the appropriate safe work plan or permit.

Table 2 Typical Insulation Resistance Tests on Electrical Apparatus and Systems at 68°F

Minimum Voltage Rating of Equipment	Minimum Test Voltage (VDC)	Recommended Minimum Insulation Resistance (in Megohms)
2–250V	500	50
251–600V	1,000	100
601–5,000V	2,500	1,000
5,001–15,000V	2,500	5,000
15,001–39,000V	5,000	20,000

Meggers

To test for potential insulation breakdown, phase-to-phase shorts, or phase-to-ground shorts in switchgear, you need to apply a much higher potential than that supplied by the battery of an ohmmeter. A megohmmeter, or megger, is commonly used for these tests. The megger is a portable instrument consisting of a hand-driven DC generator, which supplies the level of voltage for making the measurement, and the instrument portion, which indicates the value of the resistance being measured.

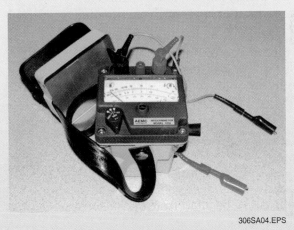

306SA04.EPS

306F05.EPS

Figure 5 ◆ Infrared imager used in thermographic surveys.

5.2.1 Thermographic Survey

A thermographic survey (*Figure 5*) involves checking switches, busways, open buses, switchgear, cable and bus connections, circuit breakers, rotating equipment, and load tap changers.

Infrared surveys should be performed during periods of maximum possible loading and not at less than 40% of the rated load of the electrical equipment being inspected. Negative test results include:

- Temperature gradients of 1°C to 3°C indicate a possible deficiency and require investigation.
- Temperature gradients of 4°C to 15°C indicate a deficiency. Repair as time permits.
- Temperature gradients of 16°C and above indicate a major deficiency. Secure power and repair as soon as possible.

WARNING!

You must be certified and authorized to perform these tests. Ensure that there is no voltage present prior to testing.

To perform a visual and mechanical inspection:

Step 1 Inspect the physical, electrical, and mechanical condition of the equipment.

Step 2 Compare the equipment nameplate information with the latest single-line diagram, and report any discrepancies.

Step 3 Check for proper anchorage, required area clearances, physical damage, and proper alignment.

Step 4 Inspect all doors, panels, and sections for missing paint, dents, scratches, fit, and missing hardware.

Step 5 Inspect all bus connections for high resistance. Use a low-resistance ohmmeter or check tightness of bolted bus joints using a calibrated torque wrench.

Step 6 Test all electrical and mechanical interlock systems for proper operation and sequencing:
- A closure attempt must be made on all locked-open devices. An opening attempt must be made on all locked-closed devices.
- A key exchange must be made with all devices operated in normally off positions.

Step 7 Clean the entire switchgear using the manufacturer's approved methods and materials.

Step 8 Inspect insulators for evidence of physical damage or contaminated surfaces.

Step 9 Inspect the lubrication:
- Verify appropriate contact lubricant on moving current-carrying parts.
- Verify appropriate lubrication of moving and sliding surfaces.
- Exercise all active components.
- Inspect all indicating devices for proper operation.

WARNING!

Electrical testing may produce hazardous voltages and may only be performed by qualified personnel under the appropriate safe work plan or permit. Prepare the area to avoid any accidental contact with the system under test, and wear appropriate personal protective equipment.

To perform electrical testing:

Step 1 Perform ratio and polarity tests on all current and voltage transformers.

Step 2 Perform ground resistance tests.

Step 3 Perform insulation resistance tests on each bus section (phase-to-phase and phase-to-ground) for one minute. Refer to the specific manufacturer's guidelines, an example of which is shown in *Table 2*.

Step 4 Perform an overpotential test on each bus section (phase-to-ground) for one minute. Refer to specific manufacturer's guidelines, an example of which is shown in *Table 3*.

NOTE

The values shown in *Tables 2* and *3* are typical acceptance values. Maintenance values will vary by manufacturer.

Step 5 Perform an insulation resistance test on the control wiring. Do not perform this test on wiring connected to solid-state components.

Step 6 Perform a phasing check on double-ended switchgear to ensure proper bus phasing from each source.

Table 3	Overpotential DC Test Voltages for Electrical Apparatus Other Than Inductive Equipment	
	DC Test Voltage Max.	
Nominal Voltage Class	**New**	**Used**
250V	2,500VDC	1,500VDC
600V	3,500VDC	2,000VDC
5,000V	18,000VDC	11,000VDC
15,000V	50,000VDC	30,000VDC

Any values of insulation resistance less than those listed in the manufacturer's literature should be investigated. Overpotential tests should not proceed until insulation resistance levels are raised above minimum values.

Overpotential test voltages must be applied in accordance with the manufacturer's literature. Test results are evaluated on a go/no-go basis by slowly raising the test voltage to the required value. The final test voltage is applied for one minute.

5.2.3 Low-Voltage Cables (600V Maximum)

To perform a visual and mechanical inspection:

Step 1 Inspect cables for physical damage and proper connection in accordance with the single-line diagram.

Step 2 Verify the integrity of all bolted connections.

Step 3 Check color-coded cable against the applicable engineer's specifications and *NEC*® standards.

To perform electrical testing:

Step 1 Perform an insulation resistance test on each conductor with respect to ground and adjacent conductors. The applied potential should be 1,000VDC for one minute.

Step 2 Perform a continuity test to ensure proper cable connection. The minimum insulation resistance values must not be less than two megohms.

5.2.4 Medium-Voltage Cables (15kV Maximum)

To perform a visual and mechanical inspection:

Step 1 Inspect exposed sections for physical damage.

Step 2 Inspect for shield grounding, cable support, and termination.

Step 3 Inspect for proper fireproofing in common cable areas.

Step 4 If cables are terminated through window-type CTs, make an inspection to verify that neutrals and grounds are properly terminated for normal operation of the protective devices.

Step 5 Visually inspect the jacket and insulation condition.

Step 6 Inspect for proper phase identification and arrangement.

5.2.5 Metal-Enclosed Busways

To perform a visual and mechanical inspection:

Step 1 Inspect the bus for physical damage.

Step 2 Inspect for proper bracing, suspension, alignment, and enclosure.

Step 3 Check the tightness of bolted joints using a calibrated torque wrench.

Step 4 Check for proper physical orientation per the manufacturer's labels to ensure proper cooling. Perform continuity tests on each conductor to verify that proper phase relationships exist.

Step 5 Check outdoor busways for removal of weep-hole plugs if applicable and also for the proper installation of a joint shield.

To perform electrical testing:

Step 1 Perform an insulation resistance test. Measure the insulation resistance on each bus run (phase-to-phase and phase-to-ground) for one minute.

Step 2 Perform AC or DC overpotential tests on each bus run, both phase-to-phase and phase-to-ground.

Manufacturer's Data

Never assume anything when it comes to equipment operation, testing, or maintenance. Always refer to the manufacturer's installation, operating, and maintenance instructions for the equipment in use. These materials provide important data that explain the warranty requirements, appropriate test procedures, and specific maintenance and test points.

Step 3 Perform a contact resistance test on each connection point of the uninsulated bus. On an insulated bus, measure the resistance of the bus section and compare values with adjacent phases.

Step 4 Insulation resistance test voltages and resistance values must be in accordance with the manufacturer's specifications.

Step 5 Apply overpotential test voltages in accordance with the manufacturer's specifications.

5.2.6 Metering and Instrumentation

To perform a visual and mechanical inspection:

Step 1 Examine all devices for broken parts, indication of shipping damage, and wire connection tightness.

Step 2 Verify that meter connections are in accordance with appropriate diagrams.

To perform electrical testing:

Step 1 Check the calibration of meters at all cardinal points.

Step 2 Calibrate watt-hour meters to one-half of one percent (0.5%).

Step 3 Verify all instrument multipliers.

6.0.0 ◆ NEC® REQUIREMENTS

This section is designed to provide a brief description of the NEC® articles that are applicable to switchboard construction, installation, and accessories.

6.1.0 Requirements for Electrical Installations

NEC® requirements for electrical installations include the following:

- *Interrupting rating* – The interrupting rating is the maximum current a device is intended to interrupt under standard test conditions. *NEC Section 110.9* defines the equipment interrupting rating as sufficient to interrupt the current that is available at the line-side terminals of the equipment.
- *Deteriorating agents* – *NEC Section 110.11* provides for the protection of equipment and conductors from environments that could cause deterioration, such as gases, vapors, liquids, or moisture, unless specifically designed for such environments.

- *Mechanical execution of work* – *NEC Section 110.12* states that electrical equipment is to be installed in a neat and professional manner. Any openings provided by the equipment manufacturer or at the time of installation that are not being used must be sealed equivalent to the structure wall. This section also forbids the use of electrical equipment with damaged parts that may affect the safe operation or mechanical strength of the equipment.
- *Mounting and cooling* – *NEC Section 110.13* states that electrical equipment shall be securely fastened to its mounting surface by mechanical fasteners, excluding wooden plugs driven into concrete, masonry, plaster, or similar materials. Equipment shall be located so as not to restrict air flow required for convection or forced-air cooling.
- *Electrical connections* – Due to the resistive oxidation created when dissimilar metals are connected, splicing devices and pressure connectors must be identified for the conductor material with which they are to be used (*NEC Section 110.14*). Dissimilar metal conductors may not be mixed in terminations or splices. Antioxidation compounds must be suitable for use and must not adversely affect conductors, installation, or equipment. Terminals for use with more than one conductor or aluminum must be identified as such.
- *Markings* – The manufacturer's trademark or logo and system ratings, including voltage, current, and wattage, must be permanently attached to the equipment (*NEC Section 110.21*).
- *Disconnect identification* – Each disconnecting means, such as circuit breakers, fused switches, **feeders**, or unfused disconnects, must be clearly marked as to its purpose at its point of origin unless located in such a manner that its purpose is evident (*NEC Section 110.22*).
- *Working space* – Suitable access and working space shall be maintained around electrical equipment to permit safe operation and maintenance (*NEC Section 110.26*). Minimum clearances in front of all electrical enclosures must conform to those specified in *NEC Section 110.26*; in all cases, space must be adequate to allow doors or hinged parts to open to a 90-degree angle. In differing conditions, the distances in *NEC Table 110.26(A)(1)* must be adhered to. Storage of any kind is not permitted within the clearance area. In accordance with *NEC Section 110.26(C)(1)*, at least one entrance of ample size must be provided to enter and exit the work area. In cases of services over 1,200A and over 6' wide, two entrances are required. The work space must be adequately illuminated.

- *Flash protection* – *NEC Section 110.16* states that electrical equipment such as switchboards, panelboards, industrial control panels, meter socket enclosures, and motor control centers in other-than-dwelling occupancies that are likely to require examination, adjustment, servicing, or maintenance while energized shall be field marked to warn qualified persons of potential electric arc flash hazards. The marking shall be located so as to be clearly visible to qualified persons before examination, adjustment, servicing, or maintenance of the equipment.

6.2.0 Requirements for Conductors

NEC Section 200.6 covers requirements associated with identifying grounded conductors. It includes the following:

- *Neutrals* – Grounded conductors (neutrals) size No. 6 AWG and smaller are color coded with a solid white or gray marking for the entire length of the conductor. Conductors size No. 6 and larger may be color coded with a solid white marking tape at termination points at the time of installation. Where different electrical systems are run together, each system's grounded conductor must be distinctively identified *[NEC Section 200.6(D)]*.

- *Protection* – Branch circuit conductors must be protected by overcurrent devices, as specified in *NEC Section 240.4*.

- *Loading* – *NEC Section 210.19(A)* states that protective device calculations for continuous duty circuits are calculated at 125% of the continuous load. This equates to an 80% loading factor on the branch circuit.

- *Tap rules* – Tap conductors are conductors that are tapped onto the line-side bus of the switchboard to feed control circuits, control power transformers, and metering devices. Overcurrent devices (typically fuses) are connected where the conductor to be protected receives its supply. Per *NEC Section 240.21*, tap conductors do not require protection if the following conditions are met:
  - The length of the conductor is not over 10'.
  - The ampacity of the conductor is not less than the combined loads supplied by the conductor.
  - The conductors do not extend beyond the switchboard.
  - The conductors are enclosed in a raceway except at the point of connection to the bus.
  - For field installations where the tap conductors leave the enclosure or vault in which the tap is made, the rating of the overcurrent device on the line side of the tap conductors does not exceed 10 times the tap conductor's ampacity.

- *Markings* – All conductors and cables shall be permanently marked to indicate the manufacturer, voltage, AWG size, and insulation type *(NEC Section 310.11)*.
  - Grounded conductors (neutrals) size No. 6 and smaller shall have a continuous marking of white or gray for the entire length of the conductor. Larger conductors may be marked at each termination with white marking tape.
  - Grounding conductors (equipment grounding wires) shall be permitted to be bare wire. In cases of insulated grounding conductors, the conductor will have a continuous marking of green for the entire length of the conductor. Larger conductors may be marked at each end and every point where the conductor is accessible.

- Ungrounded conductors (phase wires) must be distinguishable from grounded or grounding conductors with colors other than white, gray, or green. Typical ungrounded conductor identification colors are black, red, blue, brown, orange, and yellow. Conductors size No. 6 or smaller must have a continuous marking. Larger cables may be marked at each termination.
- In switchboards fed by a four-wire, delta system in which one phase is grounded at its midpoint, the phase having the higher voltage must be marked with an orange color according to *NEC Section 110.15.*
- *Ampacities* – The ampacities of cable are determined by the tables referenced in *NEC Section 310.15(B)* or with engineering support per *NEC Section 310.15(C).*

6.3.0 Grounding

NEC® grounding requirements include the following:

- *Grounding* – *NEC Section 250.20(B)* states that AC systems between 50V and 1,000V must be grounded when any of the following conditions are met:
  - Where the system can be grounded in such a way that the maximum phase-to-ground voltage does not exceed 150V
  - When the system is three-phase, four-wire, wye-connected and the neutral is used as a circuit conductor
  - When the system is three-phase, four-wire, delta-connected and the midpoint of a phase is used as a conductor (developed neutral)
- *Grounding electrode conductor* – *NEC Sections 250.24 and 250.66* cover the requirements of grounding electrode conductors, including proper sizing of the equipment grounding conductors to the service equipment enclosures. *NEC Section 250.24* states that for grounded systems (delta or wye), an unspliced main bonding jumper in the service equipment must be used to connect the grounding conductor and the service disconnect enclosure to the grounded conductor of the system within the enclosure.
- *Electrodes* – *NEC Sections 250.52, 250.53, and 250.56* require that when rod or pipe electrodes are used, they must extend a minimum of 8' into the soil. The electrode must be no less than ¾" in diameter for pipe and ⅝" in diameter for

rods. It must be galvanized metal or copper-coated to resist corrosion. Underground structures, such as water piping systems, may also be used as an electrode. Underground gas piping systems must not be used. Aluminum electrodes are not permitted. Rod, pipe, or plate electrodes must maintain a resistance of no more than 25Ω to ground. If the resistance is above 25Ω, an additional electrode is required to maintain the minimum resistance.

- *Grounding of ground wire conduits* – *NEC Section 250.64(E)* states that a grounding conductor or its enclosure must be securely mounted to the surface along which it runs. In cases where the conductor is enclosed, the enclosure must be electrically continuous and firmly grounded.
- *Ground connection surfaces* – Nonconducting coatings, such as paint, enamel, or insulating materials, must be thoroughly removed at any point where a grounding connection is made (*NEC Section 250.12).*

6.4.0 Switchboards and Panelboards

NEC® requirements for switchboards and panelboards include the following:

- *Dedicated space* – *NEC Section 110.26(F)* states that panelboards and switchboards may only be installed in spaces specifically designed for such purposes. No other piping, ducts, or devices may be installed or pass through such areas, except equipment that is necessary to the operation of the electrical equipment.
- *Inductive heating* – *NEC Section 408.3(B)* states that busbars and conductors must be arranged so as to avoid overheating due to inductive forces.
- *Phasing* – *NEC Section 408.3(E)* states that phasing in switchboards must be arranged A, B, C from front to back, top to bottom, and left to right, respectively, when facing the front of the switchboard. In systems containing a high leg, the B phase must be the phase conductor having a higher voltage to ground.
- *Wire bending space* – *NEC Section 408.3(G)* states that the wire bending space must be in accordance with *NEC Table 312.6.*
- *Minimum spacing* – *NEC Section 408.56* states that the spacing between bare metal parts and conductors must be as specified in *NEC Table 408.56.* Conductors entering the bottom of switchboards must have the clearances specified in *NEC Table 408.5.*

- *Conductor insulation* – Insulated conductors within switchboards must be listed as flame-retardant and rated at not less than the voltage applied to them or any adjacent conductors they may come in contact with *(NEC Section 408.19)*.

7.0.0 ◆ GROUND FAULTS

Ground faults exist when an unintended current path is established between an ungrounded conductor and ground on a solidly grounded service. These faults can occur due to deteriorated insulation, moisture, dirt, rodents, foreign objects, such as tools, and careless installation.

Ground faults are usually high arcing and low level in nature, which conventional breakers will not detect. Ground fault protection is used to protect equipment and cables against these low-level faults.

Ground fault protection is required per the *NEC®* on solidly-grounded wye services of more than 150V to ground but not exceeding a phase-to-phase voltage of 600V with each service disconnecting means of 1,000A or more.

7.1.0 Ground Fault Systems

The three basic methods of sensing ground faults include:

- Ground-return method
- Residual method
- Zero sequence method

The ground-return method incorporates a sensing coil around the grounding electrode conductor. The residual method uses three individual sensing coils to monitor the current on each phase conductor. The zero sequence method requires a single, specially designed sensor to monitor all the phases and the neutral conductor of a system at the same time, as shown in *Figure 6*.

INSIDE TRACK

Transformer Grounding

This is the secondary termination compartment of a 2.5MVA padmount transformer that shows the connection and arrangement of parallel feeders serving downstream 480V switchgear. This is a solidly grounded wye transformer.

SYSTEM BONDING JUMPER

306SA06.EPS

7.2.0 Sensing Operation

When circuit conditions are normal, the currents from all the phase and neutral (if used) conductors add up to zero, and the sensor current transformer produces no signal. When any ground fault occurs, the currents add up to equal the ground fault current, and the sensor produces a signal proportional to the ground fault. This signal provides power to the ground fault relay, which trips the circuit breaker.

A ground fault lasting for less than the time-delay period will not pick up the ground trip coil, thus eliminating nuisance tripping of self-clearing faults.

The ground fault relay is a high-reliability device due to its solid-state construction. The use of redundant, self-protecting, and high-reliability components further improves the performance. Self-protection against failure is provided through an internal fuse that will blow and result in a tripping function if the solid-state circuitry fails during a ground fault situation.

7.3.0 Zero Sequencing Sensor Mounting

The sensor current transformer (sensor) should be mounted so that all phase and neutral (if used) conductors pass through the core window once. The ground conductor (if used) must not pass through the core window. The neutral conductors must be free of all grounds after passing through the core window (see *Figure 6*).

When so specified by the system design engineer, the sensor may be mounted so that only the

conductor connecting the neutral to ground at the service equipment passes through the core window. In such cases, the sensor must provide power to the particular ground fault relay that is associated with the main circuit breaker.

Maintain at least two inches of clearance from the iron core of the sensor to the nearest busbar or cable to avoid false tripping. Cable conductors should be bundled securely and braced to hold them at the center of the core window. The sensor should be mounted within an enclosure and protected from mechanical damage.

7.4.0 Relay Mounting

The ground fault relay should be mounted in a vertical position within an enclosure with the terminal block at the lower end. The location of the relay should be such that the trip setting knob is accessible without exposing the operator to contact with live parts or arcing from disconnect operations.

7.5.0 Connections

Connections for standard applications should be made in accordance with the wiring diagrams in the manufacturer's literature. An example of one circuit is shown in *Figure 7*. Wires from the sensor to the ground fault relay should be no longer than 25' and no smaller than No. 14 AWG wire. Wires from the ground fault relay to the trip coil should be no longer than 50' and no smaller than No. 14 AWG wire. All wires should be protected from arcing fault and physical damage by barriers, conduit, armor, or location in an equipment enclosure. Do not disconnect or short circuit wires to the circuit breaker trip coil at any time when the power is turned on.

7.6.0 Relay Settings

The ground fault relay has an adjustable trip setting. The amount of time delay is factory set and is available in nominal time delays of 0.1, 0.2, 0.3, and 0.5 second. When ground fault protection is used in downstream steps, the feeder should have the next lower time-delay curve than the main, the branch the next lower curve than the feeder, and so on.

High trip settings on main and feeder circuits are desirable to avoid nuisance tripping. High settings usually do not reduce the effectiveness of the protection if the ground path impedance is reasonably low. Ground faults usually quickly reach a value of 40% or more of the available short circuit current in the ground path circuit.

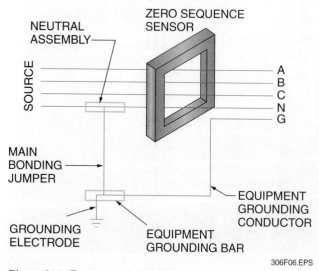

Figure 6 ◆ Zero sequencing diagram.

306F06.EPS

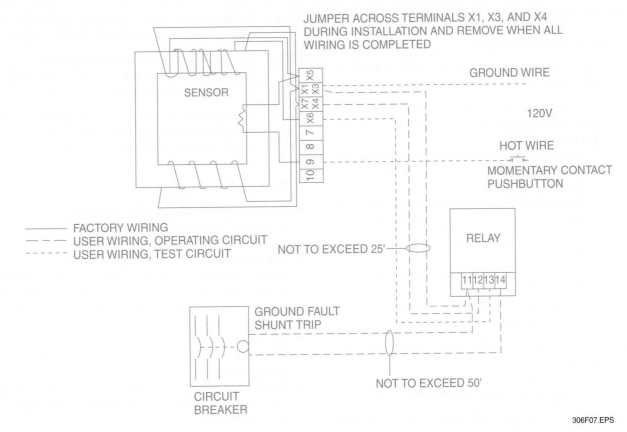

JUMPER ACROSS TERMINALS X1, X3, AND X4
DURING INSTALLATION AND REMOVE WHEN ALL
WIRING IS COMPLETED

————— FACTORY WIRING
— — — USER WIRING, OPERATING CIRCUIT
- - - - - USER WIRING, TEST CIRCUIT

306F07.EPS

Figure 7 ◆ Typical wiring diagram.

7.6.1 Coordination with Downstream Circuit Breakers

It is recommended that the magnetic trips of any downstream circuit breakers that are not equipped with ground fault protection be set as low as possible. Likewise, the ground fault relay trip settings for main or feeder circuits should be higher than the magnetic trip settings for unprotected downstream breakers where possible. This will minimize nuisance tripping of the main or feeder breaker for ground faults occurring on downstream circuits.

7.6.2 Instantaneous Trip Feature

Standard ground-powered ground fault relays have a built-in instantaneous trip feature. This instantaneous trip has a fixed time delay of approximately 1½ cycles, and the fixed trip setting is higher than found on most feeder or branch breakers to avoid nuisance tripping. Its purpose is to interrupt very high-current ground faults on main disconnects as quickly as possible and to protect the ground fault relay components.

7.7.0 Ground Fault System Test

This section provides an overview of a generic visual inspection and electrical test for ground faults. Always follow the procedures specified by the equipment manufacturer for the system being tested.

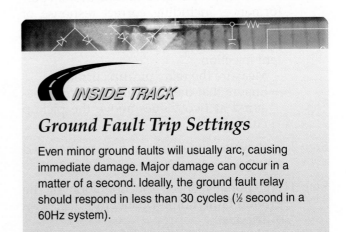

INSIDE TRACK

Ground Fault Trip Settings

Even minor ground faults will usually arc, causing immediate damage. Major damage can occur in a matter of a second. Ideally, the ground fault relay should respond in less than 30 cycles (½ second in a 60Hz system).

7.7.1 *Procedures*

Perform a visual inspection:

Step 1 Inspect the components for physical damage.

Step 2 Determine if a ground sensor was located properly around the appropriate conductor(s):
- Zero sequence sensing requires all phases and the neutral to be encircled by the sensor(s).
- Ground return sensing requires the sensor to encircle the main bonding jumper.

Step 3 Inspect the main bonding jumper to ensure:
- Proper size
- Termination on the line side of the neutral disconnect link
- Termination on the line side of the sensor on zero sequence systems

Step 4 Inspect the grounding electrode conductor to ensure:
- Proper size
- Correct switchboard termination

Step 5 Inspect the ground fault control power transformer for proper installation and size. When the control transformer is supplied from the line side of the ground fault protection circuit interrupting device, overcurrent protection and a circuit disconnecting means must be provided.

Step 6 Visually inspect the switchboard neutral bus downstream of the neutral disconnect line to verify the absence of ground connections.

Perform electrical tests as required by *NEC Section 230.95(C)*:

Step 1 Check for proper ground fault system performance, including correct response of the circuit interrupting device confirmed by primary/secondary ground sensor current injection:
- Measure the relay pickup current.
- Ensure that the relay time delay is measured at two values above the pickup current.

Step 2 Test system operation at 57% of the rated voltage.

Step 3 Functionally check the operation of the ground fault monitor panel for:
- Trip test
- No-trip test
- Nonautomatic reset

Step 4 Verify proper sensor polarity on the phase and neutral sensors for residual systems.

Step 5 Measure the system neutral insulation resistance downstream of the neutral disconnect link to verify the absence of grounds.

Step 6 Test systems (zone interlock/time coordinates) by simultaneous ground sensor current injection, and monitor for the proper response.

Test result evaluation:

- The system neutral insulation resistance should be above 100Ω and preferably one megohm or greater.
- The maximum pickup setting of the ground fault protection must be 1,200A and the maximum time delay must be one second for ground fault currents equal to or greater than 3,000A, according to *NEC Section 230.95(A)*.
- The relay pickup current should be within 10% of the manufacturer's calibration marks or fixed setting.
- The relay timing should be in accordance with the manufacturer's published time-current characteristics.

8.0.0 ◆ HVL SWITCH

Figure 8 shows the general appearance of an HVL (high-voltage limiting) switch. The HVL switch is a switching device for primary circuits up to the full interrupting current of the switch. The switches are single-throw devices designed for use on 2.4kV to 34.5kV systems.

HVL switches may provide both switching and overcurrent protection. HVL switches are commonly used as a service disconnect in unit substations and for sectionalizing medium-voltage feeder systems. The HVL switch is designed to conform to ANSI standards for metal-enclosed switchgear.

8.1.0 Ratings

Switch ratings are as follows:

- *Switch kV* – The design voltage for the switch. Of course, nominal system voltage is the normal application method; thus, a 5kV switch may be used for nominal system voltages of 2.4kV or 4.16kV, etc.
- *Basic impulse insulation level, or BIL (kV)* – The maximum voltage pulse that the equipment will withstand.
- *Frequency (Hertz)* – All HVL switches may be used in either 50Hz or 60Hz power systems.

Figure 8 ◆ High-voltage limiting switch.

306F08.EPS

- *Withstand (kV)* – The maximum 60Hz voltage that can be applied to the switch for one minute without causing insulation failure.
- *Capacitor switching (kVAR)* – The maximum capacitance expressed in kVAR that can be switched with the HVL.
- *Fault close* – The maximum, fully offset fault current that the switch can be closed into without sustaining damage. The term fully offset means that the fault current will have a delaying DC component in addition to the AC component.
- *Short time current* – The amount of current that the switch will carry for 10 seconds without sustaining any damage.
- *Continuous current (amps)* – The amount of current that the switch will carry continuously.
- *Interrupting current (amps)* – The maximum amount of current that the switch will safely interrupt.

8.2.0 Variations

There are six main types of switches:

- *Upright* – The upright switch design is the most common type. The upright construction of the service entry, jaws, and arc chutes are located near the top of the cubicle. The hinge point is below the jaws and arc chutes.
- *Inverted* – The inverted switch design has the terminals, jaws, and arc chutes located near the bottom of the cubicle. The hinge point is above

the jaws and arc chutes. This type of switch is used primarily as a main switch to a lineup of other switches. Its handle operation is identical to that of an upright switch; to close the switch, the handle is moved up, and to open it, the handle is moved down.

- *Fused/unfused* – HVL switches are available in both fused and unfused models. If equipped with fuses, the entire HVL switch has the fault interrupting capacity of the fuse and therefore provides fault protection. Either current-limiting or boric acid fuses may be used in the HVL switch.
- *Duplex* – A duplex switch is actually two switches, each in its own bay. The bays are mechanically connected and the switches are electrically connected on the load side. This switch may be used to supply power to a single load from two different sources.
- *Selector* – A selector allows an HVL switch to have double-throw characteristics. The selector switch is a single switch with a load connected to the moving or switch mechanism. Throwing the switch to one side connects the load to one source, while throwing it the other way connects it to a second source. The selector switch will be interlocked with another switch to prevent the selector switch from interrupting current flow. The selector serves a purpose similar to the duplex switch. However, the selector switch is not an interrupter; it is a disconnect.
- *Motor-operated* – This type of switch is most commonly used as the major component in an automatic transfer scheme. It can also be used when open and close functions are to be initiated from remote locations.

8.3.0 Opening Operation

In the closed position, the main switch blade is engaged on the stationary interrupting contacts. The circuit current flows through the main blades.

As the switch operating handle is moved toward the open position, the stored energy springs are charged. After the springs become fully charged, they toggle over the dead center position, discharging force to the switch operating mechanism.

The action of the switch operating mechanism forces the movable main blade off the stationary main contacts while the interrupting contacts are held closed, momentarily carrying all the current without arcing. Once the main contacts have separated well beyond the striking distance, the interrupting blade contact that was held captive has charged the interrupter blade hub spring, and the interrupter blade is suddenly forced free and flips open.

The resulting arc drawn between the stationary and movable interrupting contacts is elongated and cooled as the plastic arc chute absorbs heat and generates an arc-extinguishing gas to break up and blow out the arc. The combination of arc stretching, arc cooling, and extinguishing gas causes a quick interruption with only minor erosion of the contacts and arc chutes. The movable main and interrupting contacts continue to the fully open position and are maintained there by spring pressure.

8.4.0 Closing

When the switch operating handle is moved toward the closed position, the stored energy springs are being charged and the main blades begin to move. As the main and interrupter blades approach the arc chute, the stored energy springs become fully charged and toggle over the dead center position.

When the main and movable blades approach the main stationary contacts, a high-voltage arc leaps across the diminishing air gap in an attempt to complete the circuit. The arc occurs between the tip of the stationary main contacts and a remote corner of the movable main blades. This arc is short and brief because the fast-closing blades minimize the arcing time.

The spring pressure and momentum of the fast-moving main blades completely close the contacts. The force is great enough to cause the contacts to close even against repelling short circuit magnetic forces if a fault exists. At the same time, the interrupter blade tip is driven through the twin stationary interrupting contacts, definitely latching and preparing them for an interrupting operation when the switch is opened.

 WARNING!
Maintenance and testing may only be performed by qualified personnel under the appropriate safe work plan or permit.

8.5.0 Maintenance

Maintenance tasks for an HVL switch include the following:

Step 1 The HVL switch should be operated several times. Observe the mechanism and check for binding.

Step 2 Inspect the interrupting and main blades every 100 operations for excessive wear or damage; replace as necessary. Also, inspect the arc chutes for damage.

Step 3 Clean the switch and its compartment thoroughly. Use a clean cloth and avoid solvents.

Step 4 Lubricate the switch. The pivot points on the switch should be greased. The switch contacts should also be lubricated with a light film of grease after being cleaned.

Step 5 Final maintenance checks include phase-to-ground and phase-to-phase megger testing. If the results are satisfactory, then a DC high-potential test is performed.

8.6.0 Sluggish Operation

A switch that is operating sluggishly hesitates on the opening cycle. This contrasts with the normal snapping action. Observing the interrupter blade during the opening operation is the proper way to determine sluggish operation. Sluggishness must be repaired to prevent the switch from locking up completely. Perform the following procedure:

Step 1 Tease the switch closed and then open again while watching the interrupter blades closely. Sluggishness on close is shown by the main blade's being engaged behind the contacts of the arc chute. On opening, the interrupter blades may hesitate momentarily.

Step 2 Disconnect the links from the operating shaft. Never operate the switch with the links off as this may break the handle crank casting. This is because the main spring energy is absorbed by the handle crank rather than the main blades.

Step 3 Rotate the handle approximately 45 degrees, and hold it in this position while trying to operate the switch by hand. Excessive binding will prevent rotation of the shaft.

Step 4 Check the contact adjustment at the jaw and hinge.

Step 5 Check for binding between the interrupter blade and the arc chute.

Step 6 Remove the front panel over the operating mechanism and disconnect the spring yoke from the cam. Check for binding between the spring pivot and the sides of the operator. Check the spring for breaks.

9.0.0 ◆ BOLTED PRESSURE SWITCHES

Bolted pressure switches (*Figure 9*) are used frequently on service-entrance feeders in switchgear such as that shown in *Figure 10*. They are often used in lieu of circuit breakers because they are inexpensive. Bolted pressure switches can be manually operated or motor operated. However, unlike a circuit breaker, they can only be automatically tripped by three events: a ground fault, a phase failure, or a blown main fuse detector.

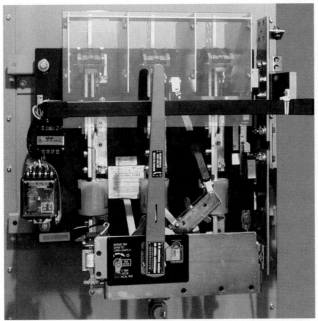

306F09.EPS

Figure 9 ◆ Bolted pressure switch.

306F10.EPS

Figure 10 ◆ Switchgear.

9.1.0 Ground Fault

Under normal conditions, the currents in all conductors surrounded by the ground fault CT equal zero. When a ground fault occurs, this sensed current increases, eventually reaching the ground fault relay pickup point and causing the bolted pressure switch to trip.

The ground fault system may also be tested. By pressing the test button, a green test light will illuminate, indicating correct circuit operation. To actually test the switch, press the TEST and RESET buttons simultaneously. This sends an actual trip signal through the current sensor, thus tripping the switch. Whenever a bolted pressure switch is tripped, a red light or a red flag will trip. Additionally, the ground fault relay must be reset before the switch can be reclosed.

9.2.0 Phase Failure

If a phase failure relay is installed, it will cause a trip of the bolted pressure switch if a phase is lost. This could occur when a tree limb knocks a line down. Under this condition, the phase failure relay will sense the lost phase and trip the bolted pressure switch, preventing a single-phasing condition.

9.3.0 Blown Main Fuse Detector

If one of the in-line main fuses were to blow, the blown main fuse detector would detect it and cause a trip of the bolted pressure switch. The trip signal generated comes from a capacitor trip unit. This ensures that power is always available to trip the switch.

9.4.0 Maintenance

These switches have a high failure rate due to lack of maintenance. All manufacturers of bolted pressure switches recommend annual maintenance. Lack of annual maintenance will eventually result in a switch that is stuck shut. Since these switches are often used as service-entrance equipment, a stuck switch can pose immediate personnel safety hazards, as well as equipment failures.

 WARNING!
When performing any maintenance, always follow the safety procedures of your company.

High-Resistance Grounding

High-resistance grounding (HRG) is increasingly applied in both medium- and low-voltage distribution systems to limit ground fault energy. Medium-voltage systems have long used low-impedance grounding systems to limit ground fault current. Limiting ground fault current to values of 25A or less increased system reliability by allowing ground faults to be detected and selectively cleared. These systems used either a large resistor or inductance (transformer primary with a shorted secondary or a relay coil in the secondary) connected between the neutral bushing of the service transformer and the system bonding jumper. Detection of a ground fault at certain levels caused a protective relay to open the circuit and clear the fault. The resistors used were mounted in wire cages next to the supply transformers. The resistors were rated at maximum ground fault power for the duration of the fault before it cleared. If the resistors burned up due to failure to clear the fault, the system became an ungrounded wye and extremely dangerous.

In practice, low-resistance grounding for 480V or 600V systems was seldom applied due to the requirements for large resistors and space for enclosures and heat dissipation. The use of high-resistance grounding for low-voltage, high-current systems is becoming common. The high fault currents available on large 480V or 600V systems present a significant arc flash hazard while at the same time the arcing fault may be seen as only an overload by the protective device. High-resistance ground systems allow detection of ground faults and facilitate location of the faulted circuit.

The most common types of faults in power systems are:

- Three-phase (balanced) fault
- Phase to phase
- Phase to phase to ground
- Phase to ground

In industrial facilities it is estimated that 98% or more of all faults begin as a ground fault. If the arcing ground fault current is high enough, the fault develops into a phase-to-phase or three-phase fault. High-resistance grounding in 600V systems limits ground fault current to less than 10A; 5A is a common limit. This level of fault current is too low to present an arc flash hazard or to sustain an arc by itself. This makes development of phase-to-phase faults unlikely. The low fault current also allows continuity of service for a period of time with little risk of equipment damage. Good practice requires that a fault be located and cleared as quickly as possible. The current allowed in the ground fault must be greater than the capacitive coupled charging current in the system in order to avoid false alarms. In low-voltage systems, a sensed ground fault is usually indicated and alarmed but does not send a trip signal to controlling equipment. This is done to allow location of the fault using various features available in HRG panels and cabinets.

In the event of a ground fault in an HRG 480V system, the two non-faulted phases will be at 480V (line potential) to ground and to the faulted phase. If a second phase should fault to ground before an existing fault is cleared, the second fault will NOT be limited and fault currents greater than encountered on a solidly grounded system should be expected. In performing arc hazard analysis we are aware that there is a much lower risk of our actions causing an uncontrolled fault but we cannot reduce the evaluated incident energy levels or PPE. This is because the risk is not eliminated and a line-to-line or three-phase fault is still possible. Surveys indicate that human error is responsible for most faults that start as line to line or three phase. It should be noted that resistance or reactance grounded systems may NOT supply line-to-neutral loads, but may supply line-to-line loads.

Due to the high interrupting capacity of the switch when operated under load, the grease that is used on the movable blades deteriorates over time and eventually turns into an adhesive. Even when the switch is not operated on a recurring basis, the grease still deteriorates due to the high temperatures associated with the current drawn by the phase. The deterioration of this grease has been shown to cause the switch to stick shut. The grease must be cleaned off yearly with denatured alcohol and replaced.

 CAUTION
Regular electrical grease cannot be used; use only the grease specifically recommended by the switch manufacturer.

Additionally, infrared scanning of in-service bolted pressure switches has revealed a marked heating concern in switches. A digital low-resistance ohmmeter (DLRO) is used to ensure that all three phases carry similar current loads. DLRO readings should never be greater than 75 microhms, and there should not be more than a 5% difference between the phases.

Typical annual maintenance includes:

Step 1 De-energize the switch, lock and tag it, and perform a preliminary operational check.

Step 2 Record pre-maintenance DLRO readings.

Step 3 With the switch open, disassemble the crossbar to free all three phases.

Step 4 Clean off all old grease with denatured alcohol or a similar solvent.

Step 5 Inspect the arc tips and arc chutes for damage.

Step 6 Adjust all pivotal connections on each blade to within the manufacturer's recommended tolerances.

Step 7 Apply an appropriate grease to the movable blades and the area where the blades come in contact with the stationary assembly.

Step 8 Check the pullout torque on each individual blade prior to crossbar reassembly. It should be in accordance with the manufacturer's prescribed limits. Too much torque will result in a switch that will be unable to open under load.

Step 9 Record the DLRO readings.

Step 10 Reassemble the crossbar assembly.

Step 11 Close and open the switch manually several times. Ensure that no phases hang up on the arc chute assembly.

Step 12 Megger the switch.

Step 13 Energize and test all accessories, such as the ground fault detector, phase failure detector, and blown main fuse detector.

Remember, if the switch is physically stuck shut, de-energize the switch from the incoming power supply, and take extra precautions when trying to unstick the switch. It may be necessary to pry the blades open, but beware of the excessive outward force that will result from a charged opening spring. To alleviate this, discharge the opening spring before commencing any work on the switch.

10.0.0 ◆ TRANSFORMERS

Transformers are used to step voltage up and down in the power transmission and distribution system.

The reason for such high transmission voltages is twofold. First, as a transformer increases transmission voltage, the required current decreases in the same proportion; therefore, larger amounts of power can be transmitted and line losses reduced. Second, to send large amounts of power over long distances at a high current and a low voltage requires a very large diameter wire. The reduction in current reduces the conductor size, which results in a cost reduction.

A transformer is an electrical device that uses the process of electromagnetic induction to change the levels of voltage and current in an AC circuit without changing the frequency and with very little loss of power.

10.1.0 Transformer Theory

As current flows through a conductor, a magnetic field is produced around the conductor. This magnetic field begins to form at the instant current begins to flow and expands outward from the conductor as the current increases in magnitude.

When the current reaches its peak value, the magnetic field is also at its peak value. When the current decreases, the magnetic field also decreases.

Alternating current (AC) changes direction twice per cycle. These changes in direction or alternation create an expanding and collapsing magnetic field around the conductor.

If the conductor is wound into a coil, the magnetic field expanding from each turn of the coil cuts across other turns of the coil. When the source current starts to reverse direction, the magnetic field collapses, and again the field cuts across the other turns of the coil.

The result in both cases is the same as if a conductor is passed through a magnetic field. An electromotive force (EMF) is induced in the conductor. This EMF is called a self-induced EMF because it is induced in the conductor carrying the current.

The direction of this induced EMF is always opposite the direction of the EMF that caused the current to flow initially. This principle is known as Lenz's law:

- An induced EMF always has such a direction as to oppose the action that produced it.
- For this reason, the EMF induced is also known as a counter-electromotive force (CEMF).

The counter-electromotive force reaches a value nearly equal to the applied voltage; thus, the primary current is limited when the secondary is open circuited.

10.1.1 No-Load Operation

The operation of a transformer is based on the principle that electrical energy can be transferred efficiently by mutual induction from one winding to another. When the primary winding is energized from an AC source, an alternating magnetic flux is established in the transformer core. This flux links the turns of the primary with the secondary, thereby inducing a voltage in them. Since the same flux cuts both windings, the same voltage is induced in each turn of both windings. Whenever the secondary of a transformer is left disconnected (or open), there is no current drawn by the secondary winding. The primary winding draws the amount of current required to supply the magnetomotive force, which produces the transformer core flux. This current is called the exciting or magnetizing current.

The exciting current is limited by the CEMF in the primary and a small amount of resistance, which cannot be avoided in any current-carrying conductor.

10.1.2 Load Operation

When a load is connected to the secondary winding of a transformer, the secondary current flowing through the secondary turns produces a counter-magnetomotive force. According to Lenz's law, this magnetomotive force is in a direction that opposes the flux that produced it. This opposition tends to reduce the transformer flux and is accompanied by a reduction in the CEMF in the primary. Since the primary current is limited by the internal impedance of the primary winding and the CEMF in the winding, whenever the CEMF is reduced, the primary current continues to increase until the original transformer flux reaches a state of equilibrium.

10.2.0 Transformer Types

Transformers can be divided into two main categories: power transformers and distribution transformers. Power transformers handle large amounts of power and step down from transmission voltages to distribution voltages. Distribution transformers are designed to handle larger currents at lower voltage levels. Distribution transformers have smaller kVA ratings and are physically much smaller than power transformers. Power transformers often have an auxiliary means of cooling, such as fans and radiators. Distribution transformers are usually self-cooled, using no fans or other cooling methods. Whereas distribution transformers may be pole-mounted or pad-mounted, power transformers are always freestanding.

Although there is some overlap between power and distribution transformers, a transformer that is rated at more than 500kVA and/or 34.5kV is generally a power transformer. A transformer rated below these values can be considered a distribution transformer. Remember, there is an overlap in kVA capacity and voltage depending on the system and power requirements.

10.3.0 Dry Transformers (Air-Cooled)

Many transformers do not use an insulating liquid to immerse the core and windings. Dry or air-cooled transformers are used for many jobs where small, low-kVA transformers are required. Large distribution transformers are usually oil filled for better cooling and insulating. However, for installations in buildings and other locations where the oil in oil-filled transformers would be a serious fire hazard, dry transformers are used. These transformers are generally of the core form. The core and coils are similar to those of other transformers. A three-phase, dry-type transformer is shown in *Figure 11*.

The case is made of sheet metal and provided with ventilating louvers for the circulation of cooling air. To increase the output, fans can be installed to draw cooling air through the coils at a faster rate than is possible with natural circulation.

Figure 11 ◆ Dry-type transformer.

306F11.EPS

Either Class B or Class H insulation is used for the windings. Class B insulation may be operated safely at a hot-spot temperature of 130°C. Class H insulation may be operated safely at a hot-spot temperature of 180°C. The use of these materials makes it possible to manufacture smaller transformers. Both Class B and Class H insulation consist of mica, asbestos, fiberglass, and similar inorganic material. Temperature-resistant organic varnishes are used as the binder for Class B insulation. Silicone or fluorine compounds or similar materials are used as the binder for Class H insulation. Such transformers use high-temperature insulation only in locations where the high temperature requires such insulation.

10.4.0 Sealed Dry Transformers

Hermetically sealed dry transformers are constructed in large sizes for voltages above 15kV. They are used for installations in buildings and other locations where oil-filled transformers would be a serious fire hazard, but they may also be used for lower voltages and kVA ratings and for water-submersible transformers in locations subject to floods. Nitrogen is typically used for the insulation and cooling of sealed dry transformers.

10.5.0 Transformer Nameplate Data

Transformer nameplate data includes the following:

- *Electrical ratings* – The information relating to the transformer electrical parameters can be found on the nameplate.
- *Voltage ratings* – The voltage rating identifies the nominal root mean square (rms) voltage value at which the transformer is designed to operate. A transformer can operate within a ±5% range of its rated primary voltage. If the primary voltage is increased to more than +5%, the windings of the transformer can overheat. Operation of the transformer at more than −5% decreases its power output proportionally to the percent voltage reduction. Transformer windings are rated as follows:
  - Phase-to-phase and phase-to-neutral for wye windings, such as 480Y/277VAC
  - Phase-to-phase for delta windings, such as 480VAC
  - Dual-voltage windings, such as 480VAC × 240VAC

When transformers are equipped with a tap changer, the voltage ratings in the nameplate indicate the nominal voltages.

- *BIL* – This identifies the maximum impulse voltage the winding insulation can withstand without failure.
- *Phase* – The phase information indicates the number of phase windings contained in a transformer tank.
- *Frequency* – The frequency rating of a transformer is the normal operating system frequency. When a transformer is operated at a lower frequency, the reactance of the primary winding decreases. This causes a higher exciting current and an increase in flux density. In addition, there is an increase in core loss, which results in overall heating.
- *Class* – Transformers are classified by the type of cooling they employ.
- *Temperature rise* – The temperature rise rating is the maximum elevation above ambient temperature that can be tolerated without causing insulation damage.
- *Capacity* – The capacity of a transformer to transfer energy is related to its ability to dissipate the heat produced in the windings. The capacity rating is the product of the rated voltage and the current that can be carried at that voltage without exceeding the temperature rise limitation.

- *Impedance* – Impedance identifies the opposition of a transformer to the passage of short circuit current.
- *Phasor diagrams* – Phasor diagrams show phase and polarity relationships of the high and low windings. They can be used with the schematic connection diagram to provide test connection points and to provide proper external system connections.

10.6.0 Transformer Case Inspections

When inspecting the inside of a dry-type, air-cooled transformer case, look for the following:

- Temporary shipping supports or guards
- Bent, broken, or loose parts
- Debris on the floor or in the coils
- Corrosion of any part
- Worn or frayed insulation
- Shifted core members
- Damaged tap changer mounts or mechanisms
- Misaligned core spacers and loose coil elements
- Broken or loose blocking

Upon the completion of the inspection, replace the covers and bolt securely. All information should be recorded on appropriate inspection sheets.

10.7.0 Transformer Tests

The following tests are the recommended minimum tests that should be included as part of a maintenance program. These tests are conducted to determine and evaluate the present condition of the transformer. From the results of these tests, a determination is made as to whether the transformer is suitable for service. All tests should be performed using the standards and procedures provided by the transformer manufacturer.

- *Continuity and winding resistance test* – There should be a continuity check of all windings. If possible, measure the winding resistance and compare it to the factory test values. An increase of more than 10% could indicate loose internal connections.
- *Insulation resistance test* – To ensure that no grounding of the windings exists, a 1,000V insulation resistance test should be made.
- *Ratio test* – A turns ratio test should be made to ensure proper transformer ratios and to ensure that all connections were made. If equipped with a tap changer, all positions should be checked.

- *Core ground* – This test is performed in the same way as the insulation resistance test, except the measurement is made from the core to the frame and ground bus. Remove the core ground strap before the test.
- *Heat scanning* – After the transformer is energized, a heat scan test should be done to detect loose connections. This test is performed using an infrared scanning device that shows or indicates hot spots.

11.0.0 ◆ INSTRUMENT TRANSFORMERS

For all practical purposes, the voltages and currents used in the primary circuits of substations are much too large to be used to provide operating quantities to relaying or metering circuits. In order to reduce voltage and currents to usable levels, instrument transformers are employed. Instrument transformers are used to:

- Protect personnel and equipment from the high voltages and/or currents used in electric power transmission and distribution
- Provide reasonable use of insulation levels and current-carrying capacity in relay and metering systems and other control devices
- Provide a means to combine voltage and/or current phasors to simplify relaying or metering

Instrument transformers are manufactured with a multitude of different ratios to provide a standard output for the many different system primary voltage levels and load currents. There are two types of instrument transformers, potential transformers and current transformers. In general, a potential transformer (*Figure 12*) is used to a supply voltage signal to devices such as voltmeters, frequency meters, power factor meters, watt-hour meters, and protective relays. The voltage is proportional to the primary voltage, but it is small enough to be safe for the test instrument. The secondary of a potential transformer may be designed for several different voltages, but most are designed for 120V. The potential transformer is primarily a distribution transformer especially designed for voltage regulation so that the secondary voltage (under all conditions) will be as close as possible to a specified percentage of the primary voltage.

A current transformer is used to supply current to an instrument connected to its secondary with the current being proportional to the primary current but small enough to be safe for the instrument. The secondary of a current transformer is usually designed for a rated current of 5A.

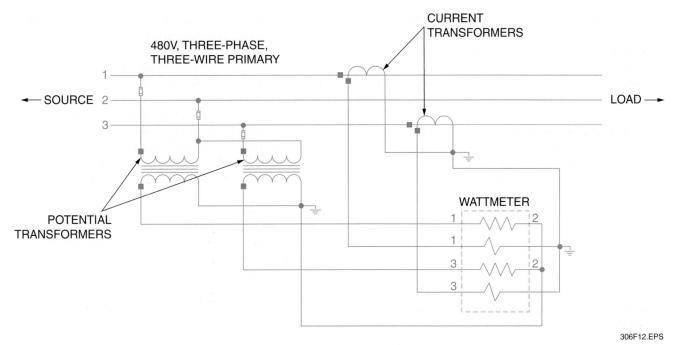

306F12.EPS

Figure 12 ◆ Current and potential transformers connected for power metering of a three-phase circuit.

A current transformer operates in the same way as any other transformer in that the same relationship exists between the primary current and the secondary current. A current transformer uses the circuit conductors as its primary winding. The secondary of the current transformer is connected to current devices such as ammeters, wattmeters, watt-hour meters, power factor meters, some forms of relays, and the trip coils of some types of circuit breakers.

When no instruments or other devices are connected to the secondary of the current transformer, a short circuit device or shunt is placed across the secondary to prevent the secondary circuit from being opened while the primary winding is carrying current.

 WARNING!

If the secondary circuit is open, there will be no secondary ampere turns to balance the primary ampere turns, so the total primary current becomes exciting current and magnetizes the core to a high flux density. This produces a high voltage across both the primary and secondary windings and endangers the life of anyone coming in contact with the meters or leads. This is why current transformers should never be fused. A current transformer is the only transformer that may be short-circuited on the secondary while energized.

11.1.0 Potential Transformers

Potential transformers are designed to reduce primary system voltages down to usable levels for metering and are often referred to as voltage transformers or VTs. Potential transformers are often used where the system's primary voltage exceeds 600V and sometimes on 240V and 480V systems.

The standard secondary circuit voltage level for a potential transformer circuit is 120V for circuits below 25kV and 115V for circuits above 25kV at the potential transformer's rated primary voltage. These voltages correspond to typical transformation ratios of standard transmission voltages. The current flowing in the secondary of the potential transformer circuit is very low under normal operating conditions, typically less than one ampere.

Potential transformers are constructed to be lightly loaded with the design emphasis on winding ratio accuracy rather than current rating. Potential transformer construction can be air-insulated dry, case epoxy-insulated, oil-filled, or SF_6-insulated, depending upon the primary circuit voltage level.

The standard output voltage of potential transformers is either 120V or 69.3V, depending on whether its primary winding uses phase-to-phase or phase-to-neutral connections. Understanding the operation of a potential or voltage transformer is simplified by the inspection of its equivalent circuit.

Potential transformers must have their secondary circuits grounded for safety reasons in the event that a short circuit develops between the primary and secondary windings and to negate the effects of parasitic capacitance between the primary and the secondary. *Figure 13* shows the connection of an ideal potential transformer circuit.

11.2.0 Current Transformers

A current transformer is designed to reduce high primary system currents down to usable levels. Current transformers are used whenever system primary voltage isolation is required. The standard secondary circuit current for a current transformer circuit is 5A with full-rated current flowing in the primary circuit.

⚠ WARNING!

The voltage level across a current transformer's secondary terminals can rise to a very dangerous level if the secondary circuit opens while the primary circuit is energized.

The primary considerations in current transformer design are the current-carrying capability and saturation characteristics. Insulation systems are of the same generic types as potential transformers; however, SF_6 insulation is infrequently used in current transformer construction.

Current transformers are manufactured in four basic types: oil-filled (for example, donut type), bar, window, and bushing. The bushing-type transformer is normally applied on circuit breakers or power transformers. The other types are used for the remaining indoor and outdoor installations. *Figure 14* illustrates some common types of current transformer construction.

The major criteria for the selection of the current transformer for relaying are its primary current rating, maximum burden, and saturation characteristics. Saturation is particularly important in relaying due to the fact that many relays are called upon to operate only under fault conditions.

Current transformer circuits operate at a very low voltage. Connected loads (burdens) range from 0.2Ω to 2Ω. These small impedances, together with a maximum continuous current of up to 5A, keep these circuits at low potentials. The voltage can become high momentarily during faults when large secondary currents flow. This voltage is a function of the current, burden, and transformer VA capability.

As with potential transformers, current transformers must also have their secondary windings grounded in the event of an insulation breakdown between the primary and secondary and to negate the effects of parasitic capacitance.

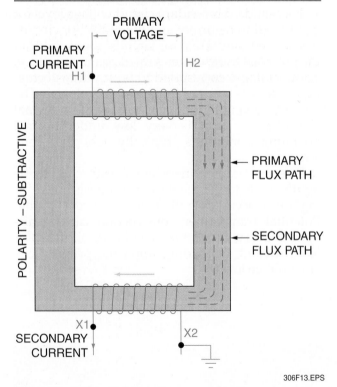

306F13.EPS

Figure 13 ◆ Potential transformer construction.

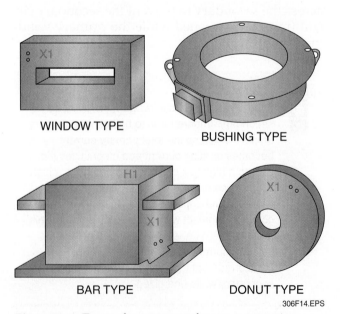

306F14.EPS

Figure 14 ◆ Types of current transformer construction.

SF_6 Insulation

Sulfur hexafluoride (SF_6) is a colorless, odorless, nontoxic, nonflammable gas that is used as an insulating gas in electrical equipment. SF_6 is used as a gaseous dielectric for transformers, condensers, and circuit breakers, often replacing harmful PCBs.

11.3.0 Instrument Transformer Maintenance

Instrument transformers require regular inspection and maintenance. A typical inspection includes:

Step 1 Inspect for physical damage, and check the nameplate information for compliance with instructions and specification requirements.

Step 2 Verify the proper connection of transformers against the system requirements.

Step 3 Verify the tightness of all bolted connections, and ensure that adequate clearances exist between the primary circuits and the secondary circuit wiring.

Step 4 Verify that all required grounding and shorting connections provide good contact.

Step 5 Test for proper operation of the potential transformer isolation (PT tip out) compartment and grounding operation when applicable.

12.0.0 ◆ CIRCUIT BREAKERS

Circuit breakers are the only circuit interrupting devices that combine a full fault current interruption rating and the ability to be manually or automatically opened or closed. A circuit breaker is defined as a mechanical switching device that is capable of making, carrying, and breaking currents under normal circuit conditions and also making, carrying (for a specified time), and breaking currents under specified abnormal circuit conditions, such as a short circuit (according to IEEE). The four general classifications of circuit breakers are:

- Air circuit breakers (ACBs)
- Oil circuit breakers (OCBs)
- Vacuum circuit breakers (VCBs)
- Gas circuit breakers (GCBs)

Circuit breakers may conveniently be divided into low-voltage, medium-voltage, and high-voltage classes. Although there is considerable overlap among these classes, each one has certain characteristic features.

12.1.0 Circuit Breaker Ratings

Circuit breaker ratings are given on the breaker nameplate. The information from the nameplate should be reviewed when considering any breaker selection problem. The same rating information should be included in any documentation for breaker applications. The rating information includes:

- *Rated voltage* – The rated voltage is the maximum voltage for which the circuit breaker is designed.
- *Rated current* – This is the continuous current that the circuit breaker can carry without exceeding a standard temperature rise (usually 55°C).

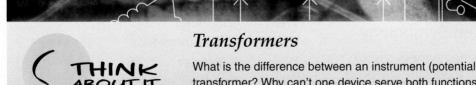

Transformers

What is the difference between an instrument (potential) transformer and a control transformer? Why can't one device serve both functions?

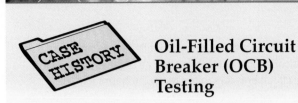

Oil-Filled Circuit Breaker (OCB) Testing

CASE HISTORY

An assistant engineer and an electrician had the responsibility of testing the insulation values of the bushings on an OCB. The assistant engineer pulled down the handle and opened the breaker, visually determining that the contacts were open and safe. He instructed the electrician to move the test leads from one bushing to the next as the assistant engineer read the values. On the third and last bushing, the electrician was electrocuted as he placed the test leads on the bushing. The last contact had not opened due to a broken porcelain insulator skirt, which allowed the cap to remain stationary while the rest of the mechanism rotated as the handle was pulled down. This was not noticeable from the assistant engineer's eye level when he determined all contacts were opened.

The Bottom Line: *Never* come in contact with switchgear such as this without first testing each and every contact point with the proper meter, and *never* rely on someone else's word that the power is off.

• *Interrupting rating* – This is the maximum value of current at rated voltage that the circuit breaker is required to successfully interrupt for a limited number of operations under specified conditions. The term is usually applied to abnormal or emergency conditions.

13.0.0 ◆ ELECTRICAL DRAWING IDENTIFICATION

Before looking at actual plant diagrams, it is necessary to understand the symbology used to condense electrical drawings. The designer uses symbols and abbreviations as a type of shorthand. This section will present the standard symbols, abbreviations, and device numbers that make up the designer's shorthand. For IEEE device numbers, see *Appendix A*.

13.1.0 Electrical Diagram Symbology

It is imperative that every line, symbol, figure, and letter in a diagram have a specific purpose and that the information be presented in its most concise form. For example, when the rating of a current transformer is given, a transformer symbol is shown, and an abbreviation such as CT is not needed; the information is implied by the symbol itself. Writing the unit of measure (amp) in this case is also unnecessary, since a current transformer is always rated in amperes. Thus, the numerical rating and the transformer symbol are sufficient. The key to reading and interpreting electrical diagrams is to understand and use the electrical legend. The legend shows the symbols used in the diagram and also contains general notes and other important information. Most electrical legends are very similar; however, there are some variations between the legends developed by different companies. Only the legend specifically designed for a given set of drawings should be used for those drawings.

The legend prevents the necessity of memorizing all the symbols presented on a diagram and can be used as a reference for unfamiliar symbols. Typically, the legend will be found in the bottom right corner of a print or on a separate drawing. In addition to symbols, abbreviations are an important part of the designer's shorthand. For example, a circle can be used to symbolize a meter, relay, motor, or indicating light. A circle's application can generally be distinguished by its location in the circuit; however, the designer uses a set of standard abbreviations to make the distinction clear. The following abbreviations are used to represent meters:

A	Ammeter
AH	Ampere-hour meter
CRO	Oscilloscope
DM	Demand meter
F	Frequency meter
GD	Ground detector
OHM	Ohmmeter
OSC	Oscillograph
PF	Power factor meter
PH	Phase meter
SYN	Synchroscope
TD	Transducer
V	Voltmeter
VA	Volt-ammeter
VAR	VAR meter
VARH	VAR hour meter
W	Wattmeter
WH	Watt-hour meter

As mentioned earlier, indicating lamps may also be represented by a circle. The following abbreviations are used to represent indicating lamps:

A   Amber
B   Blue
C   Clear
G   Green
R   Red
W   White

Relays are another component commonly represented by a circle. The following abbreviations are used for relays:

CC   Closing coil
CR   Closing/control relay
TC   Trip coil
TR   Trip relay
TD   Time-delay relay
TDE   Time-delay energize
TDD   Time-delay de-energize
X   Auxiliary relay

Still another component that is commonly represented by a circle is the motor. Motors usually have the horsepower rating in or near the circle representing them. The abbreviation for horsepower is hp. Any other piece of equipment represented by a circle will be identified in the legend, notes, or spelled out on the diagram itself.

Contacts and switches are also identified using standard abbreviations. The following is a list of these abbreviations:

A   Breaker A contact
B   Breaker B contact
BAS   Bell alarm switch
BLPB Backlighted pushbutton
CS   Control switch
FS   Flow switch
LS   Limit switch
PB   Pushbutton
PS   Pressure switch
PSD   Differential pressure switch
TDO   Time-delay open
TDC   Time-delay closed
TS   Temperature switch
XS   Auxiliary switch

The following figures illustrate examples of these abbreviations and symbols. *Figure 15* shows A and B contacts in their normally de-energized

state. If relay CR is de-energized, contact A is open and contact B is shut. When relay CR is energized, contact A is shut and contact B is open.

Figure 16 illustrates a control switch and its associated contacts. Contacts 1 through 4 open and close as a result of the operation of control switch 1 (CS1).

In the stop position, contact 2 is shut, and the red indicating lamp is lit. In the start position, contacts 3 and 4 are shut, energizing the M coil and the amber indicating lamp, respectively. When the switch handle is released, the spring returns to the run position, and contact 4 opens, de-energizing the amber lamp and closing contact 1 to energize the green lamp.

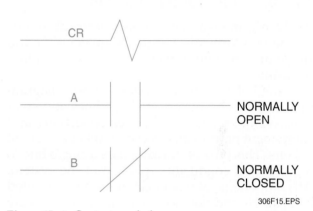

Figure 15 ◆ Contact symbols.

306F15.EPS

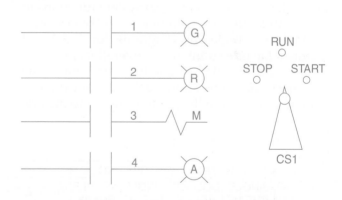

Figure 16 ◆ Switch development.

306F16.EPS

CONTACTS		STOP	RUN	START	
1			X		
2		X			
3			X	X	CS1
4				X	
SPRING RETURN			←		

X-SHUT

There are many abbreviations used on electrical drawings. The designer makes an effort to use standard abbreviations; however, you will encounter nonstandard abbreviations. Nonstandard abbreviations will typically be defined in the diagram notes or legend. *Figure 17* defines abbreviations commonly used in wiring prints and specifications. The symbols further illustrate descriptions of the abbreviations.

14.0.0 ◆ ELECTRICAL PRINTS

This section will cover the specific types of electrical prints with which you need to be familiar in order to install and maintain electrical systems.

14.1.0 Single-Line Diagrams

Analyzing and reading complex electrical circuits can be very difficult. Diagrams are simplified to single-line (one-line) diagrams to aid in reading the prints.

A single-line diagram is defined as a diagram that indicates by means of single lines and standard symbology the paths, interconnections, and component parts of an electric circuit or system of circuits. This type of drawing uses a single line to represent all conductors (phases) of the system. All components of power circuits are represented by symbols and notations. One-line diagrams are valuable tools for system visualization during planning, installation, operation, and maintenance, and they provide a basic understanding of how a portion of the electrical system functions in terms of the physical components of the circuit.

Types of single-line diagrams for industrial facilities include summary, or overall facility, diagrams and detailed single-line diagrams. The summary diagrams show each bus and disconnecting device from the point of supply to the line side connection of motor control centers (MCCs). The drawing(s) typically include all voltage levels and power transformers down to the voltage level for three-phase power usage (575V or 480V). The detailed single-line diagrams will identify the components and disconnect devices all the way to downstream users. Individual motors or panelboards fed from the MCC will be shown, as well as lighting transformers and lighting panels. The single line of branch circuits and end devices from panelboards is usually called the panel schedule. An example of a summary single-line diagram is shown in *Figure 18*.

14.2.0 Elementary Diagrams

An elementary diagram is a drawing that falls between one-line diagrams and schematics in terms of complexity. An elementary diagram is a wiring diagram showing how each individual conductor is connected. *Figure 19* is an example of an elementary schematic diagram with the circuit powered by the phase voltage between L1 and L2.

Elementary diagrams, interconnection diagrams, and connection diagrams all illustrate individual conductors. Elementary diagrams are used to show the wiring of instrument and electrical control devices in an elementary ladder or schematic form. The elementary diagram reflects the control wiring required to achieve the operation and sequence of operations described in the logic diagram. When presented in ladder form, the vertical lines in each ladder diagram represent the control power wires of a control circuit. If a number of schemes are connected to the same control circuit, the vertical lines are continuous from the top to the bottom of the ladder. The power source wire is always shown on the left side of the ladder and the ground or neutral wire on the right. A ground symbol is normally not shown on the neutral wire.

SPST NO		SPST NC		SPDT		TERMS	
SINGLE BREAK	DOUBLE BREAK	SINGLE BREAK	DOUBLE BREAK	SINGLE BREAK	DOUBLE BREAK	SPST	SINGLE-POLE SINGLE-THROW
○—○	○‾‾○	○—○	○__○			SPDT	SINGLE-POLE DOUBLE-THROW
DPST NO		DPST NC		DPDT		DPST	DOUBLE-POLE SINGLE-THROW
SINGLE BREAK	DOUBLE BREAK	SINGLE BREAK	DOUBLE BREAK	SINGLE BREAK	DOUBLE BREAK	DPDT	DOUBLE-POLE DOUBLE-THROW
						NO	NORMALLY OPEN
						NC	NORMALLY CLOSED

306F17.EPS

Figure 17 ◆ Supplementary contact symbols.

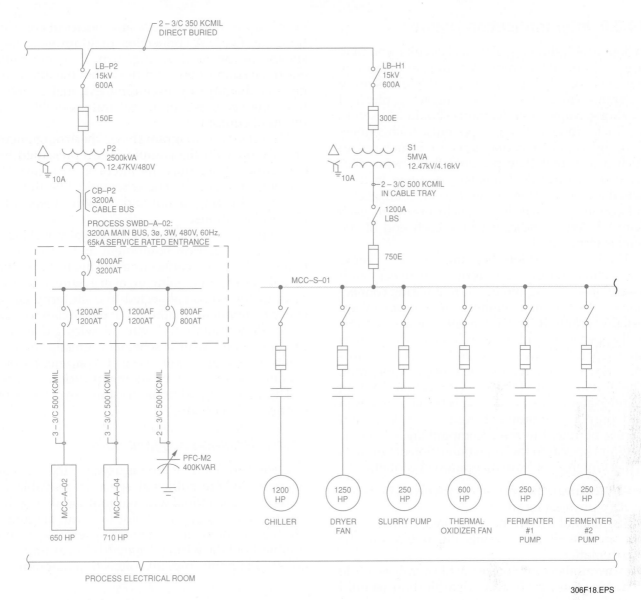

Figure 18 ◆ Single-line diagram.

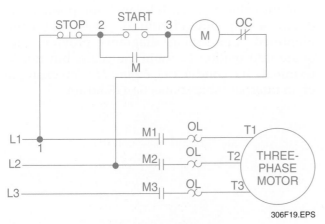

Figure 19 ◆ Elementary schematic diagram.

The control power wire numbers are shown at the top of the vertical lines on each ladder diagram. The circuit identification number and source of the control circuit are also shown at the top center of each ladder. If two or more control circuits are represented in a single ladder, the vertical lines are broken, and the wire numbers and circuit identification are entered at the top of each ladder segment. Each horizontal line in a ladder diagram represents a circuit path. All devices shown on a single horizontal line represent a series circuit path; parallel circuit paths are shown on two or more horizontal lines.

14.3.0 Interconnection Diagrams

When troubleshooting electrical circuits, you may use an elementary circuit diagram to determine the cause of a failure; however, since elementary diagrams are drawn without regard to physical locations, connection diagrams should be used to aid in locating faulty components. Interconnection and connection diagrams are structured in such a way that they present all the wires that were shown in the elementary drawing in their actual locations. These drawings show all electrical connections within an enclosure, with each wire labeled to indicate where each end of the wire is terminated.

The interconnection diagram is made to show the actual wiring connections between unit assemblies or equipment. Internal wiring connections within unit assemblies or equipment are usually omitted. The interconnection diagrams will appear adjacent to the schematic diagram or on a separate drawing, depending upon the format chosen when making the schematic diagram. The development of the interconnection diagram is integrated with that of the schematic diagram and only the equipment, terminal blocks, and wiring pertinent to the accompanying schematic diagram appear in the interconnection diagram. A typical interconnection diagram will contain:

- An outline of the equipment involved in its relative physical location
- Terminal blocks in the equipment that are concerned with the wiring illustrated on the schematic
- Wire numbers, cable sizes, cable numbers, cable routing, and cable tray identification (should not be repeated on the interconnection diagram except where necessary)
- Wiring between equipment (normally shown as individual cables but may be combined on complex drawings)
- Equipment identification information

14.4.0 Connection Diagrams

The connection diagram shows the internal wiring connections between the parts that make up an apparatus. It will contain as much detail as necessary to make or trace any electrical connections involved. A connection diagram generally shows the physical arrangement of component electrical connections. It differs from the interconnection diagram by excluding external connections between two or more unit assemblies or pieces of equipment.

The schematic diagram shows the arrangement of a circuit with the components represented by conventional symbols. Its intent is to show the function of a circuit. The schematic, like the elementary drawing, is not laid out with respect to physical locations.

A wiring diagram also shows the physical locations of all electrical equipment and/or components with all interconnecting wiring. It shows the actual connection point of every wire and the color of the wires connected to each terminal of every component. It allows the electrician to easily locate terminals and wires. A wiring diagram in conjunction with a schematic greatly aids in troubleshooting a given piece of equipment. Connection diagrams can be shown in various forms.

The following sections illustrate two types of connection diagrams.

14.4.1 Point-to-Point Method

The point-to-point method is used for the simpler diagrams where sufficient space is available to show each individual wire without sacrificing the clarity of the diagram. Point-to-point diagrams provide accurate information to terminate and troubleshoot the wiring. *Figure 20* is an example of an internal point-to-point connection diagram.

14.4.2 Cable Method

In complex diagrams, only individual cables are shown between devices or terminal strips. Lines from the cable go to each termination point at the device or strip, but individual connections are not identified. Connection diagrams provide adequate information to route the cable, but not to terminate the conductors. *Figure 21* is an example of an internal cable connection diagram.

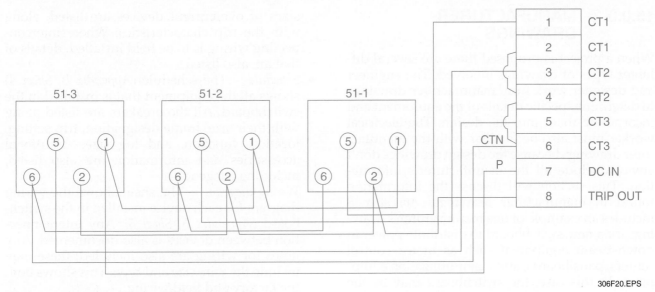

Figure 20 ◆ Point-to-point connection diagram.

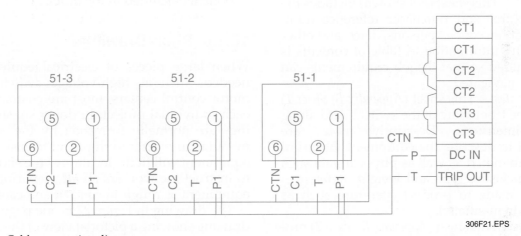

Figure 21 ◆ Cable connection diagram.

15.0.0 ◆ MANUFACTURER DRAWINGS

When a project is proposed there are several different types of drawings involved. The engineer and designer work from manufacturer drawings to determine specific details of the equipment and incorporate them into the design. The electrical worker must also be familiar with the manufacturer drawings because the design drawings don't always include all the manufacturer's information. This section will discuss the information found in manufacturer drawings. *Appendix B* includes an example of one manufacturer's drawings for a new switchboard to distribute power to downstream equipment such as motor control centers, panelboards, and major utilization equipment. In this case, the switchboard may be the service equipment or may be fed from the facility power transformer. A brief description of each sheet of the drawing follows:

- *Title sheet* – This sheet (not shown) includes the manufacturer and purchaser reference numbers, location, contact persons, dates, and other identifying information. A table of contents is also included so that the job requirements can be easily navigated.
- *General notes* – This sheet (*Appendix B, Sheet 1*) shows the front view of the switchboard along with dimensions. Descriptive notes are included to describe the equipment. Electrical ratings are noted as well as physical data such as the enclosure type and weight. References are also made to product literature such as instruction manual(s).
- *Floor plan* – This sheet (*Appendix B, Sheet 2*) provides a top view, side view, and floor plan view. This gives the installer a footprint of the space needed for the installation. The installer must make sure to include the minimum working space around the equipment as required by the *NEC*.
- *One-line diagram* – This sheet (*Appendix B, Sheet 3*) shows that the project consists of a three section switchboard. The one-line (single-line) diagram for each section is shown, along with how they are interconnected with the main bus. The

sizes of overcurrent devices are listed, along with the trip characteristics. Where interconnecting wiring is to be field installed, details of that are also listed.

- *Schedule* – The schedule (*Appendix B, Sheet 4*) shows all the equipment that is installed in the switchboard. All the breakers are listed along with their size, frame designation, trip setting, location, function, and lug size. Additional accessories and information are also listed, including a legend.
- *Wiring diagram* – The sheet shows the wiring diagram for each device installed in the switchboard (*Appendix B, Sheet 5*). Any interconnection between devices is also documented. Any notes for wiring are also included; these may include the wire size and type. This shows both the factory and field wiring.
- *Catalog sheet* – This sheet (*Appendix B, Sheet 6*), sometimes called a Bill of Materials, gives catalog numbers and additional details on the components included in the order.

15.1.0 Shop Drawings

When large pieces of electrical equipment are needed, such as high-voltage switchgear and motor control centers, most are custom built for each individual project. In doing so, shop drawings are normally furnished by the equipment manufacturer prior to shipment to ensure that the equipment will fit the location at the shop site and to instruct the workers about preparing for such equipment as rough-in conduit and cable trays.

The drawing in *Figure 22* is one page of a shop drawing showing a pictorial view of the enclosure.

Shop drawings will also usually include connection diagrams for all components that must be field wired or connected.

As-built drawings, including detailed factory-wired connection diagrams, are also included to assist workers and maintenance personnel in making the final connections and then in troubleshooting problems once the system is in operation.

Typical drawings are shown in *Figures 23* and *24*.

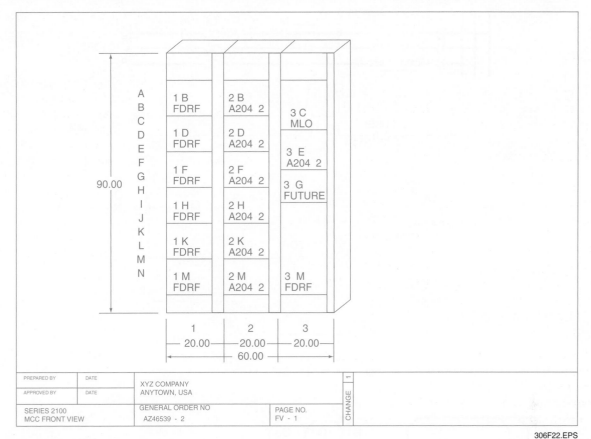

Figure 22 ◆ View of a motor control center.

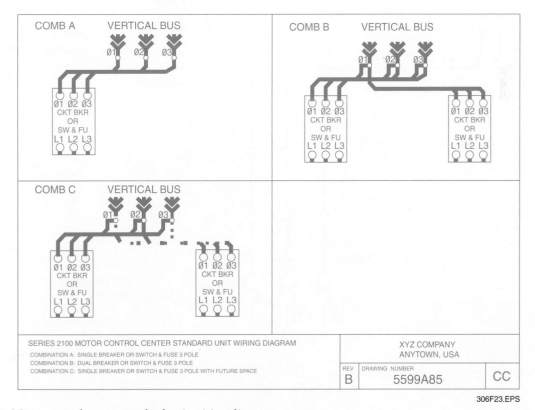

Figure 23 ◆ Motor control center standard unit wiring diagram.

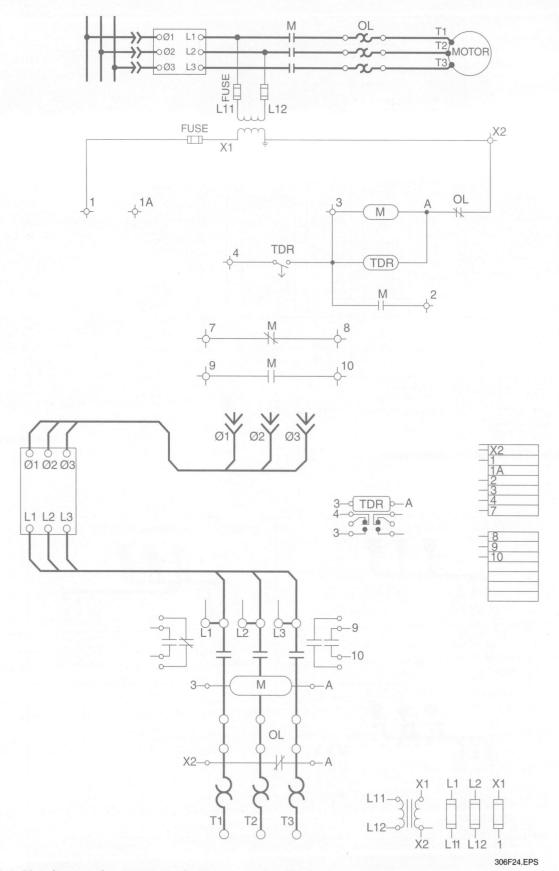

Figure 24 ◆ Unit diagrams for motor control center.

306F24.EPS

16.0.0 ◆ PANELBOARDS

This section covers panelboard construction and protective devices.

16.1.0 Lighting and Power Panelboards

Circuit control and overcurrent protection must be provided for all circuits and the power-consuming devices connected to these circuits. Lighting and power panels located throughout large buildings being supplied with electrical energy provide this control and protection. *Figure 25* shows a schedule of fifteen panelboards provided in a typical industrial building to feed electrical energy to the various circuits.

16.2.0 Panelboard Construction

In general, panelboards are constructed so that the main feed busbars run the height of the panelboard. The buses to the branch circuit protective devices are connected to the alternate main buses. In an arrangement of this type, the connections directly across from each other are on the same phase, and the adjacent connections on each side are on different phases. As a result, multiple protective devices can be installed to serve the 208V equipment. An example of a panelboard is shown in *Figure 26*.

16.2.1 Identification of Conductors

The ungrounded conductors may be any color except green (or green with a yellow stripe), which is reserved for grounding purposes only, or white or gray, which are reserved for the grounded circuit conductor. See *NEC Section 200.6.*

NEC Section 210.5(C) requires that where different voltages exist in a building, the ungrounded conductors for each system must be identified at each accessible location. Identification may be by color-coding, marking, tape, tagging, or other approved means. The means of identification must be permanently posted at each branch circuit panelboard, or readily available.

For example, this situation may occur when the building is served with 277/480V and step-down transformers are used to provide 120/208V for lighting and receptacle outlets. Examples of panelboard wiring connections are shown in *Figures 27, 28, 29,* and *30.*

16.2.2 Number of Circuits

The number of overcurrent devices in a panelboard is determined by the needs of the area being served. Using the bakery panelboard in *Figure 29* as an example, there are 13 single-pole circuits and five three-pole circuits. This is a total of 28 poles. When using a three-phase supply, the incremental number is six (a pole for each of the three phases on both sides of the panelboard). The minimum number of poles that could be specified for the bakery is 30. This would limit the power available for growth and would not permit the addition of a three-pole lead. The reasonable choice is to go to 36 poles, which provides flexibility for growth loads.

16.3.0 Panelboard Protective Devices

The main protective device for a panelboard may be either a fuse or a circuit breaker. This section concentrates on the use of circuit breakers. The selection of the circuit breaker should be based on the necessity to:

- Provide the proper overload protection
- Ensure a suitable voltage rating
- Provide a sufficient interrupting current rating
- Provide short circuit protection
- Coordinate the breaker(s) with other protective devices

The choice of the overload protection is based on the rating of the panelboard. The trip rating of the circuit breaker cannot exceed the amperage capacity of the busbars in the panelboard. The number of branch circuit breakers is generally not a factor in the selection of the main protective device except in a practical sense. It is a common practice to have the total amperage of the branch breakers greatly exceed the rating of the main breaker; however, it makes little sense for a single branch circuit breaker to be the same size as, or larger than, the main breaker.

The voltage rating of the breaker must be higher than that of the system. Breakers are usually rated at 250V to 600V.

The importance of the proper interrupting rating cannot be overstressed. You should recall that if there is ever any question as to the exact value of the short circuit current available at a point, the circuit breaker with the higher interrupting rating is to be installed.

PANEL NO.	LOCATION	MAINS	VOLTAGE RATING	NO. OF CIRCUITS	BREAKER RATINGS	POLES	PURPOSE
P-1	BASEMENT N. CORRIDOR	BREAKER 100A	208/120V 3Ø, 4W	19 2 5	20A 20A 20A	1 2 1	LIGHTING AND RECEPTACLES SPARES
P-2	BASEMENT N. CORRIDOR	BREAKER 100A	208/120V 3Ø, 4W	24 2 0	20A 20A	1 2	LIGHTING AND RECEPTACLES SPARES
P-3	2ND FLOOR N. CORRIDOR	BREAKER 100A	208/120V 3Ø, 4W	24 2 0	20A 20A	1 2	LIGHTING AND RECEPTACLES SPARES
P-4	BASEMENT S. CORRIDOR	BREAKER 100A	208/120V 3Ø, 4W	24 2 0	20A 20A	1 2 1	LIGHTING AND RECEPTACLES SPARES
P-5	1ST FLOOR S. CORRIDOR	BREAKER 100A	208/120V 3Ø, 4W	23 2 1	20A 20A 20A	1 2 1	LIGHTING AND RECEPTACLES SPARES
P-6	2ND FLOOR S. CORRIDOR	BREAKER 100A	208/120V 3Ø, 4W	22 2 2	20A 20A 20A	1 2 1	LIGHTING AND RECEPTACLES SPARES
P-7	MFG. AREA S. WALL E.	BREAKER 100A	208/120V 3Ø, 4W	5 7 2	20A 20A 20A	1 1 1	LIGHTING AND RECEPTACLES SPARES
P-8	MFG. AREA S. WALL W.	BREAKER 100A	208/120V 3Ø, 4W	5 7 2	20A 20A 20A	1 1 1	LIGHTING AND RECEPTACLES SPARES
P-9	MFG. AREA S. WALL E.	BREAKER 100A	208/120V 3Ø, 4W	5 7 2	50A 20A 20A	1 1 1	LIGHTING AND RECEPTACLES SPARES
P-10	MFG. AREA S. WALL W.	BREAKER 100A	208/120V 3Ø, 4W	5 7 2	50A 20A 20A	1 1 1	LIGHTING AND RECEPTACLES SPARES
P-11	MFG. AREA EAST WALL	LUGS ONLY 225A	208/120V 3Ø, 4W	6	20A	3	BLOWERS AND VENTILATORS
P-12	BOILER ROOM	BREAKER 100A	208/120V 3Ø, 4W	10 4	20A 20A	1 1	LIGHTING AND RECEPTACLES SPARES
P-13	BOILER ROOM	LUGS ONLY 225A	208/120V 3Ø, 4W	6	20A	3	OIL BURNERS AND PUMPS
P-14	MFG. AREA EAST WALL	LUGS ONLY 400A	208/120V 3Ø, 4W	3 2 1	175A 70A 40A	3 3 3	CHILLERS FAN COIL UNITS FAN COIL UNITS
P-15	MFG. AREA WEST WALL	LUGS ONLY 600A	208/120V 3Ø, 4W	5	100A	3	TROLLEY BUSWAY AND ELEVATOR

306F25.EPS

Figure 25 ◆ Schedule of electric panelboards for an industrial building.

Figure 26 ◆ Typical panelboard.

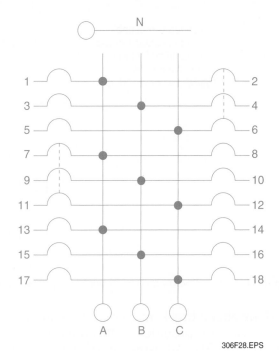

306F28.EPS

Figure 28 ◆ Lighting and appliance branch circuit panelboard—three-phase, four-wire connections.

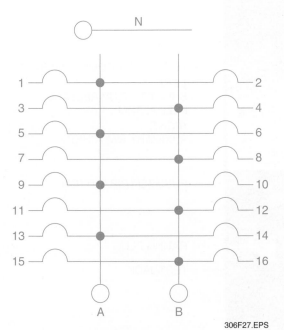

306F27.EPS

Figure 27 ◆ Lighting and appliance branch circuit panelboard—single-phase, three-wire connections.

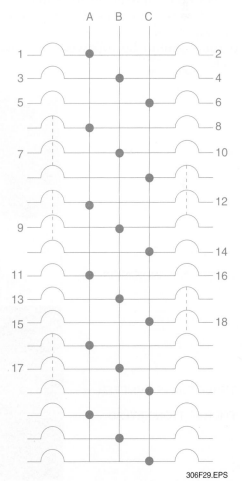

306F29.EPS

Figure 29 ◆ Bakery panelboard circuit showing alternate numbering scheme.

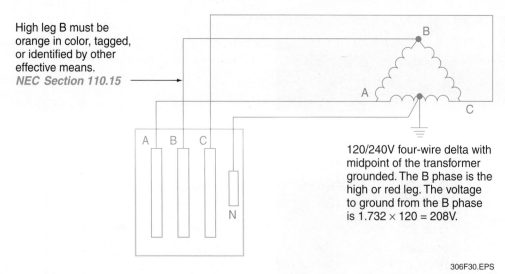

High leg B must be orange in color, tagged, or identified by other effective means.
NEC Section 110.15

120/240V four-wire delta with midpoint of the transformer grounded. The B phase is the high or red leg. The voltage to ground from the B phase is 1.732 × 120 = 208V.

306F30.EPS

Figure 30 ◆ Panelboards and switchboards supplied by four-wire, delta-connected system.

Many circuit breakers used as the main protective device are provided with an electronic trip unit (*Figure 31*). Adjustments of this trip determine the degree of protection provided by the circuit breaker if a short circuit occurs. The manufacturer of this device provides exact information about the adjustments to be made. In general, a low setting may be 10 or 12 times the overload trip rating.

Two rules should be followed whenever the trip is set:

- The trip must be set to the minimum practical setting.
- The setting must be lower than the value of the short circuit current available at that point.

If subfeed lugs are used, ensure that the lugs are suitable for making multiple breaker connections, as required by *NEC Section 110.14(A)*. In general, this means that a separate lug is to be provided for each conductor being connected.

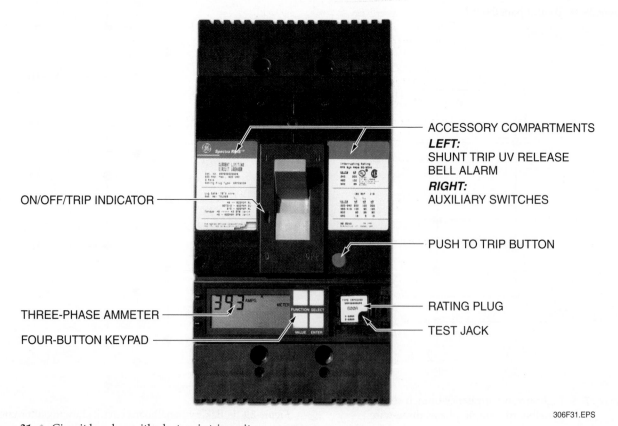

ON/OFF/TRIP INDICATOR

THREE-PHASE AMMETER

FOUR-BUTTON KEYPAD

ACCESSORY COMPARTMENTS
LEFT:
SHUNT TRIP UV RELEASE
BELL ALARM
RIGHT:
AUXILIARY SWITCHES

PUSH TO TRIP BUTTON

RATING PLUG

TEST JACK

306F31.EPS

Figure 31 ◆ Circuit breaker with electronic trip unit.

If taps are made to the subfeeder, they can be reduced in size according to *NEC Section 240.21*. This specification is very useful in cases such as that of panel P-12 in *Figure 25*. For this panel, a 100A main breaker is fed by a 350MCM conductor. Within the distances given in *NEC Section 240.21(B)(1)*, a conductor with a 100A rating may be tapped to the subfeeder and connected to the 100A main breaker in the panel.

Per *NEC Section 110.14(C)*, the temperature rating of conductors must be selected and coordinated so as not to exceed the lowest temperature rating of any connected termination, conductor, or device.

16.4.0 Branch Circuit Protective Devices

The schedule of panelboards for the industrial building (*Figure 25*) shows that lighting panels P-1 through P-6 have 20A circuit breakers, including double-pole breakers to supply special receptacle outlets. A double-pole breaker requires the same installation space as two single-pole breakers. Breakers are shown in *Figure 32*.

306F32.EPS

Figure 32 ◆ Branch circuit protective devices.

1. In industrial applications, medium-voltage may refer to systems rated over _____.
   a. 50V up to 480V
   b. 600V up to 120,000V
   c. 1,000V up to 38,000V
   d. 2,000V up to 69,000V

2. The term interrupting rating refers to the _____.
   a. trip setting of a circuit breaker
   b. voltage rating of a fuse
   c. highest voltage level a device can withstand
   d. maximum current a device will safely interrupt at rated voltage

3. Which of the following is a color that can be used to designate an ungrounded conductor?
   a. Green
   b. White
   c. Gray
   d. Red

4. When an unintended path is established between an ungrounded conductor and ground, it is called a(n) _____.
   a. phase fault
   b. open circuit
   c. ground fault
   d. overload

5. A device that is specifically designed to protect equipment from ground faults through the use of sensors is a _____.
   a. molded-case circuit breaker
   b. dual-element fuse
   c. ground fault relay
   d. ground fault circuit interrupter

6. The maximum voltage that a piece of equipment can withstand is known as its _____.
   a. interrupting capacity
   b. basic impulse insulation level (BIL)
   c. current limit
   d. frequency

7. A transformer rated at more than 500kVA is considered a(n) _____ transformer.
   a. power
   b. control
   c. distribution
   d. isolation

8. The term capacity on a transformer nameplate refers to _____.
   a. its voltage rating
   b. its ability to transfer energy
   c. the voltage produced by the secondary
   d. the number of secondary windings

9. The term class on a transformer nameplate refers to _____.
   a. its use, such as control or power
   b. the type of cooling it uses
   c. whether it is step-up or step-down
   d. its range of operating frequencies

10. The secondary of a current transformer is usually designed for a rated current of _____.
    a. 5A
    b. 10A
    c. 15A
    d. 20A

11. Of the following types of current transformers, which is normally used with circuit breakers or power transformers?
    a. Bar
    b. Bushing
    c. Oil filled
    d. Window

12. Which of the following is *not* a general classification of circuit breakers?
    a. Fuse circuit breaker
    b. Air circuit breaker
    c. Oil circuit breaker
    d. Gas circuit breaker

13. What type of diagram shows the actual wiring connections between unit assemblies or equipment with each wire labeled to indicate where to terminate it?
    a. Schematic diagram
    b. Front panel diagram
    c. Interconnection diagram
    d. Block diagram

14. The trip rating of a circuit breaker used as the main protective device in a panelboard *cannot* exceed _____.
    a. the total amperage of the branch breakers
    b. the amperage capacity of the busbars in the panelboard
    c. the amperage of the individual branch fuses
    d. 250V

15. According to the IEEE Identification System found in *Appendix A* of this module, if device No. 51 is indicated on an electrical print, it would be a(n) _____.
    a. circuit breaker
    b. reverse power relay
    c. field circuit breaker
    d. AC time overcurrent relay

Summary

This module explained the purpose of switchgear. Switchgear construction, metering layouts, wiring requirements, and maintenance were discussed. It also explained *NEC*® requirements for these systems and provided a basic understanding of how to apply them. Circuit breakers, their four general classifications, and the major circuit breaker ratings were also addressed. Additionally, ground fault relay systems and the testing of such systems were explained. This module also covered visual and mechanical inspections and electrical tests associated with low-voltage and medium-voltage cables, metal-enclosed busways, and metering and instrumentation.

Notes

Trade Terms
Introduced in This Module

Air circuit breaker: A circuit breaker in which the interruption occurs in air.

Basic impulse insulation level (BIL): The maximum impulse voltage the winding insulation can withstand without failure.

Branch circuit: A set of conductors that extends beyond the last overcurrent device in the low-voltage system of a given building.

Bus: A conductor or group of conductors that serves as a common connection for two or more circuits in a switchgear assembly.

Bushing: An insulating structure including a through conductor, or providing a passageway for such a conductor, for the purpose of insulating the conductor from the barrier and conducting from one side of the barrier to the other.

Capacity: The rated load-carrying ability, expressed in kilovolt-amperes or kilowatts, of generating equipment or other electric apparatus.

Current transformer (CT): A single-phase instrument transformer connected in series in a line that carries the full-load current. The turns ratio is designed to produce a reduced current in the secondary suitable for the current coil of standard measuring instruments and in proportion to the load current.

Distribution system equipment: Switchboard equipment that is downstream from the service-entrance equipment.

Distribution transformer: A transformer that is used for transferring electric energy from a primary distribution circuit to a secondary distribution circuit. Distribution transformers are usually rated between 5kVA and 500kVA.

Feeder: A set of conductors originating at a main distribution center that supply one or more secondary distribution centers, one or more branch circuit distribution centers, or any combination of these two types of load.

Metal-enclosed switchgear: Switchgear that is primarily used in indoor applications up to 600V.

Potential transformer (PT): A special transformer designed for use in measuring high voltage; normally, the secondary voltage is 120V.

Service-entrance equipment: Equipment located at the service entrance of a given building that provides overcurrent protection to the feeder and service conductors and also provides a means of disconnecting the feeders from the energized service equipment.

Switchboard: A large single panel, frame, or assembly of panels on which switches, fuses, buses, and instruments are mounted.

Switchgear: A general term covering switching or interrupting devices and any combination thereof with associated control, instrumentation, metering, protective, and regulating devices.

IEEE Identification System

The devices in switching equipment are referred to by numbers with appropriate suffix letters when necessary, according to the functions they perform.

These numbers are based on a system adopted as standard for automatic switchgear by IEEE and incorporated in *American Standard C37.2-1970.* This system is used in connection diagrams, instruction books, and specifications.

Device Number	Definition and Function

1   **Master Element** – The initiating device, such as a control switch, voltage relay, float switch, that serves either directly, or through such permissive devices as protective and time-delay relays, to place equipment in or out of operation.

2   **Time-Delay Starting or Closing Relay** – A device that functions to give a desired amount of time delay before or after any point of operation in a switching sequence or protective relay system, except as specifically provided by device functions 48, 62, and 79 described later.

3   **Checking or Interlocking Relay** – A device that operates in response to the position of a number of other devices (or to a number of predetermined conditions) in equipment to allow an operating sequence to proceed, to stop, or to provide a check of the position of these devices or of these conditions for any purpose.

4   **Master Contactor** – A device, generally controlled by device No. 1 or equivalent, and the required permissive and protective devices, that serves to make and break the necessary control circuits to place equipment into operation under the desired conditions and to take it out of operation under other or abnormal conditions.

5   **Stopping Device** – A control device used primarily to shut down equipment and hold it out of operation. [This device may be manually or electrically actuated, but excludes the function of electrical lockout (see device function 86) on abnormal conditions.]

6   **Starting Circuit Breaker** – A device whose principal function is to connect a machine to its source of starting voltage.

7   **Anode Circuit Breaker** – A device used in the anode circuits of a power rectifier for the primary purpose of interrupting the rectifier circuit if an arc-back should occur.

8   **Control Power Disconnecting Device** – A disconnective device, such as a knife switch, circuit breaker, or pullout fuse block, that is used for the purpose of connecting and disconnecting the source of control power to and from the control bus or equipment.

    Note: Control power is considered to include auxiliary power which supplies such apparatus as small motors and heaters.

9   **Reversing Device** – A device used for the purpose of reversing a machine field or for performing any other reversing functions.

10   **Unit Sequence Switch** – A device used to change the sequence in which units may be placed in and out of service in multiple-unit equipment.

11   Reserved for future application.

12   **Over-Speed Device** – Usually a direct-connected speed switch that functions on machine over-speed.

13   **Synchronous-Speed Device** – A device such as a centrifugal-speed switch, slip-frequency relay, voltage relay, or undercurrent relay, that operates at approximately the synchronous speed of a machine.

14   **Under-Speed Device** – A device that functions when the speed of a machine falls below a predetermined value.

15   **Speed- or Frequency-Matching Device** – A device that functions to match and hold the speed or frequency of a machine or of a system equal to, or approximately equal to, that of another machine, source, or system.

16   Reserved for future application.

17   **Shunting or Discharge Switch** – A device that serves to open or close a shunting circuit around any piece of apparatus (except a resistor), such as machine field, machine armature, capacitor, or reactor.

Note: This excludes devices that perform such shunting operations as may be necessary in the process of starting a machine by devices 6 or 42, or their equivalent, and also excludes the device 73 function, which serves for the switching of resistors.

18   **Accelerating or Decelerating Device** – A device used to close or to cause the closing of circuits that are used to increase or decrease the speed of a machine.

19   **Starting-to-Running Transition Contactor** – A device that operates to initiate or cause the automatic transfer of a machine from the starting to the running power connection.

20   **Electrically Operated Valve** – An electrically operated, controlled, or monitored valve in a fluid line.

Note: The function of the valve may be indicated by the use of suffixes.

21   **Distance Relay** – A device that functions when the circuit admittance, impedance, or reactance increases or decreases beyond predetermined limits.

22   **Equalizer Circuit Breaker** – A breaker that serves to control or to make and break the equalizer or the current-balancing connections for a field, or for regulating equipment, in a multiple-unit installation.

23   **Temperature Control Device** – A device that functions to raise or lower the temperature of a machine or other apparatus, or of any medium, when its temperature falls below or rises above a predetermined value.

Note: An example is a thermostat that switches on a space heater in a switchgear assembly when the temperature falls to a desired value as distinguished from a device that is used to provide automatic temperature regulation between close limits and would be designated as 90T.

24   Reserved for future application.

25   **Synchronizing or Synchronism-Check Device** – A device that operates when two AC circuits are within the desired limits of frequency, phase angle, or voltage to permit or to cause the paralleling of these two circuits.

26   **Apparatus Thermal Device** – A device that functions when the temperature of the shunt field or the armortisseur winding of a machine or that of a load limiting or load shifting resistor or of a liquid or other medium exceeds a predetermined value; it also functions if the temperature of the protected apparatus, such as a power rectifier, or of any medium decreases below a predetermined value.

27   **Undervoltage Relay** – A device that functions on a given value of undervoltage.

28   **Flame Detector** – A device that monitors the presence of the pilot or main flame in such apparatus as a gas turbine or steam boiler.

29   **Isolating Contactor** – A device used expressly for disconnecting one circuit from another for the purposes of emergency operation, maintenance, or testing.

30   **Annunciator Relay** – A nonautomatic reset device that gives a number of separate visual indications upon the functioning of protective devices and that may also be arranged to perform a lockout function.

31   **Separate Excitation Device** – A device that connects a circuit, such as the shunt field of a synchronous converter, to a source of separate excitation during the starting sequence or one that energizes the excitation and ignition circuits of a power rectifier.

306A02.EPS

32   **Directional Power Relay** – A device that functions on a desired value of power flow in a given direction or upon reverse power resulting from arc-back in the anode or cathode circuits of a power rectifier.

33   **Position Switch** – A device that makes or breaks its contacts when the main device or piece of apparatus that has no device function number reaches a given position.

34   **Master Sequence Device** – A device, such as a motor-operated multi-contact switch or the equivalent, or a programming device, such as a computer, that establishes or determines the operating sequence of the major devices in equipment during starting and stopping or during other sequential switching operations.

35   **Brush-Operating or Slip-Ring Short-Circuiting Device** – A device used for raising, lowering, or shifting the brushes of a machine; for short-circuiting its slip rings, or for engaging or disengaging the contacts of a mechanical rectifier.

36   **Polarity or Polarizing Voltage Device** – A device that operates or permits the operation of another device on a predetermined polarity only, or one that verifies the presence of a polarizing voltage in equipment.

37   **Undercurrent or Underpower Relay** – A device that functions when the current or power flow decreases below a predetermined value.

38   **Bearing Protective Device** – A device that functions on excessive bearing temperature or on other abnormal mechanical conditions, such as undue wear, that may eventually result in excessive bearing temperature.

39   **Mechanical Condition Monitor** – A device that functions upon the occurrence of an abnormal mechanical condition (except that associated with bearings as covered under device function 38), such as excessive vibration, eccentricity, expansion, shock, tilting, or seal failure.

40   **Field Relay** – A device that functions on a given or abnormally low value or failure of machine field current or on an excessive value of the reactive component of armature current in an AC machine indicating abnormally low field excitation.

41   **Field Circuit Breaker** – A device that functions to apply or remove the field excitation of a machine.

42   **Running Circuit Breaker** – A device whose principal function is to connect a machine to its source of running or operating voltage. This function may also be used for a device, such as a contactor, that is used in series with a circuit breaker or other fault protecting means, primarily for frequent opening and closing of the circuit.

43   **Manual Transfer or Selector Device** – A device that transfers the control circuits so as to modify the plan of operation of the switching equipment or of some of the devices.

44   **Unit Sequence Starting Relay** – A device that functions to start the next available unit in multiple-unit equipment on the failure or non-availability of the normally preceding unit.

45   **Atmospheric Condition Monitor** – A device that functions upon the occurrence of an abnormal atmospheric condition, such as damaging fumes, explosive mixtures, smoke, or fire.

46   **Reverse-Phase or Phase-Balance Current Relay** – A device that functions when the polyphase currents are of reverse-phase sequence or when the polyphase currents are unbalanced or contain negative phase-sequence components above a given amount.

47   **Phase-Sequence Voltage Relay** – A relay that functions upon a predetermined value of polyphase voltage in the desired phase sequence.

48   **Incomplete Sequence Relay** – A relay that generally returns the equipment to the normal or off position and locks it out if the normal starting, operating, or stopping sequence is not properly completed within a predetermined time. If the device is used for alarm purposes only, it should preferably be designated as 48A (alarm).

306A03.EPS

49   **Machine or Transformer Thermal Relay** – A relay that functions when the temperature of a machine armature, or other load-carrying winding or element of a machine or the temperature of a power rectifier or power transformer (including a power rectifier transformer) exceeds a predetermined value.

50   **Instantaneous Overcurrent or Rate-of-Rise Relay** – A relay that functions instantaneously on an excessive value of current or on an excessive rate of current rise, indicating a fault in the apparatus or circuit being protected.

51   **AC Time Overcurrent Relay** – A relay with either a definite or inverse time characteristic that functions when the current in an AC circuit exceeds a predetermined value.

52   **AC Circuit Breaker** – A device that is used to close and interrupt an AC power circuit under normal conditions or to interrupt this circuit under fault or emergency conditions.

53   **Exciter or DC Generator Relay** – A relay that forces the DC machine field excitation to build up during starting or which functions when the machine voltage has built up to a given value.

54   Reserved for future application.

55   **Power Factor Relay** – A relay that operates when the power factor in an AC circuit rises above or below a predetermined value.

56   **Field Application Relay** – A relay that automatically controls the application of the field excitation to an AC motor at some predetermined point in the slip cycle.

57   **Short-Circuiting or Grounding Device** – A primary circuit switching device that functions to short-circuit or ground a circuit in response to automatic or manual means.

58   **Rectification Failure Relay** – A device that functions if one or more anodes of a power rectifier fail to fire, to detect an arc-back, or on failure of a diode to conduct or block properly.

59   **Overvoltage Relay** – A relay that functions on a given value of overvoltage.

60   **Voltage or Current Balance Relay** – A relay that operates on a given difference in voltage or current input or output of two circuits.

61   Reserved for future application.

62   **Time-Delay Stopping or Opening Relay** – A time-delay relay that serves in conjunction with the device that initiates the shutdown, stopping, or opening operation in an automatic sequence.

63   **Pressure Switch** – A switch that operates on given values or on a given rate of change of pressure.

64   **Ground Protective Relay** – A relay that functions on failure of the insulation of a machine, transformer, or other apparatus to ground or on flashover of a DC machine to ground.

    Note: This function is assigned only to a relay that detects the flow of current from the frame of a machine or enclosing case or structure of a piece of apparatus to ground or one that detects a ground on a normally ungrounded winding or circuit. It is not applied to a device connected in the secondary circuit or secondary neutral of a current transformer connected in the power circuit of a normally grounded system.

65   **Governor** – The assembly of fluid, electrical, or mechanical control equipment used for regulating the flow of water, steam, or other medium to the prime mover for such purposes as starting, holding speed or load, or stopping.

306A04.EPS

66   **Notching or Jogging Device** – A device that functions to allow only a specified number of operations of a given device or equipment or a specified number of successive operations within a given time of each other. It also functions to energize a circuit periodically or for fractions of specified time intervals or that is used to permit intermittent acceleration or jogging of a machine at low speeds for mechanical positioning.

67   **AC Directional Overcurrent Relay** – A relay that functions on a desired value of AC overcurrent flowing in a predetermined direction.

68   **Blocking Relay** – A relay that initiates a pilot signal for blocking of tripping on external faults in a transmission line or in other apparatus under predetermined conditions, or a relay cooperates with other devices to block tripping or to block reclosing on an out-of-step condition or on power swings.

69   **Permissive Control Device** – Generally a two-position, manually operated switch that in one position permits the closing of a circuit breaker or the placing of equipment into operation and in the other position prevents the circuit breaker or the equipment from being operated.

70   **Rheostat** – A variable resistance device used in an electric circuit that is electrically operated or has other electrical accessories, such as auxiliary, position, or limit switches.

71   **Level Switch** – A switch that operates on given values or on a given rate of change of level.

72   **DC Circuit Breaker** – A circuit breaker used to close and interrupt a DC power circuit under normal conditions or to interrupt this circuit under fault or emergency conditions.

73   **Load-Resistor Contactor** – A contactor used to shunt or insert a step of load limiting, shifting, or indicating resistance in a power circuit, to switch a space heater in a circuit, or to switch a light or regenerative load resistor of a power rectifier or other machine in and out of a circuit.

74   **Alarm Relay** – A device other than an annunciator, as covered under device No. 30, that is used to operate or to operate in connection with a visual or audible alarm.

75   **Position Changing Mechanism** – A mechanism that is used for moving a main device from one position to another in equipment (for example, shifting a removable circuit breaker unit to and from the connected, disconnected, and test positions).

76   **DC Overcurrent Relay** – A relay that functions when the current in a DC circuit exceeds a given value.

77   **Pulse Transmitter** – A device used to generate and transmit pulses over a telemetering or pilot-wire circuit to remove the indicating or receiving device.

78   **Phase Angle Measuring or Out-of-Step Protective Relay** – A relay that functions at a predetermined phase angle between two voltages, between two currents, or between voltage and current.

79   **AC Reclosing Relay** – A relay that controls the automatic reclosing and locking out of an AC circuit interrupter.

80   **Flow Switch** – A switch that operates on given values, or a given rate of change of flow.

81   **Frequency Relay** – A relay that functions on a predetermined value of frequency, either under, over, or on normal system frequency or rate of change of frequency.

82   **DC Reclosing Relay** – A relay that controls the automatic closing and reclosing of a DC circuit interrupter, generally in response to load circuit conditions.

83   **Automatic Selective Control or Transfer Relay** – A relay that operates to select automatically between certain sources or conditions in equipment or that performs a transfer operation automatically.

306A05.EPS

84 **Operating Mechanism** – The complete electrical mechanism or servo-mechanism, including the operating motor, solenoids, position switches, and for a tap changer, induction regulator, or any similar piece of apparatus that has no device function number.

85 **Carrier or Pilot-Wire Receiver Relay** – A relay that is operated or restrained by a signal used in connection with carrier-current or DC pilot-wire fault directional relaying.

86 **Locking-Out Relay** – An electrically operated relay that functions to shut down and hold equipment out of service on the occurrence of abnormal conditions. It may be reset either manually or electrically.

87 **Differential Protective Relay** – A protective relay that functions on a percentage of phase angle or other quantitative difference of two currents or of some other electrical quantities.

88 **Auxiliary Motor or Motor Generator** – A device used for operating auxiliary equipment, such as pumps, blowers, exciters, and rotating magnetic amplifiers.

89 **Line Switch** – A switch used as a disconnecting load-interrupter or isolating switch in an AC or DC power circuit when this device is electrically operated or has electrical accessories, such as an auxiliary switch or magnetic lock.

90 **Regulating Device** – A device that functions to regulate a quantity, or quantities, such as voltage, current, power, speed, frequency, temperature, and load, at a certain value or between certain (generally close) limits for machines, tie lines, or other apparatus.

91 **Voltage Directional Relay** – A relay that operates when the voltage across an open circuit breaker or contactor exceeds a given value in a given direction.

92 **Voltage and Power Directional Relay** – A relay that permits or causes the connection of two circuits when the voltage difference between them exceeds a given value in a predetermined direction and causes these two circuits to be disconnected from each other when the power flowing between them exceeds a given value in the opposite direction.

93 **Field Changing Contactor** – A device that functions to increase or decrease in one step the value of field excitation on a machine.

94 **Tripping or Trip-Free Relay** – A device that functions to trip a circuit breaker, contactor, or equipment, to permit immediate tripping by other devices, or to prevent immediate reclosure of a circuit interrupter in case it should open automatically even though its closing circuit is maintained closed.

95
96 } Used only for specific applications on individual installations where none of the assigned numbered functions
97   from 1 to 94 is suitable.

306A06.EPS

Typical Manufacturer Drawings

This module is intended to present thorough resources for task training. The following reference works are suggested for further study. These are optional materials for continued education rather than for task training.

American Electrician's Handbook, 1996. Terrell Croft and Wilfred I. Summers. New York, NY: McGraw-Hill.

National Electrical Code® Handbook, Latest Edition. Quincy, MA: National Fire Protection Association.

CONTREN® LEARNING SERIES – USER UPDATE

NCCER makes every effort to keep these textbooks up-to-date and free of technical errors. We appreciate your help in this process. If you have an idea for improving this textbook, or if you find an error, a typographical mistake, or an inaccuracy in NCCER's Contren® textbooks, please write us, using this form or a photocopy. Be sure to include the exact module number, page number, a detailed description, and the correction, if applicable. Your input will be brought to the attention of the Technical Review Committee. Thank you for your assistance.

Instructors – If you found that additional materials were necessary in order to teach this module effectively, please let us know so that we may include them in the Equipment/Materials list in the Annotated Instructor's Guide.

Write: Product Development and Revision
National Center for Construction Education and Research
3600 NW 43rd St., Bldg. G, Gainesville, FL 32606

Fax: 352-334-0932

E-mail: curriculum@nccer.org

Craft _____ Module Name _____

Copyright Date _____ Module Number _____ Page Number(s) _____

Description _____

_____

_____

_____

(Optional) Correction _____

_____

_____

(Optional) Your Name and Address _____

_____

_____

Transformers

Growth of wind farming

The U.S. Department of Energy says that wind power may be a major source of energy by the year 2030. The DOE says up to 20% of U.S. power needs could be met by wind power, reducing pollution and cutting use of water, natural gas, and other resources, while creating new jobs.

26307-08

26307-08
Transformers

Topics to be presented in this module include:

Overview

Voltage levels on power transmission can exceed 800kV. These high levels of voltage are necessary to transmit the generated power over long distances to the usage areas. Once the power is received at distribution substations, it must be stepped down and regulated to a usable level. This is the work performed by distribution system power transformers.

Keep in mind the power formula of $P = EI$ when thinking about transformers. If you increase the voltage from one side of a transformer to the other, you decrease the available current. This is the reason that step-up transformers have limited use; the more you increase the voltage, the less current is available to operate a load.

Power transformer windings may be wound and tapped to provide various levels of voltage on the secondary side, or selective connections based on available voltages on the primary side. Varying the point at which connections are made on transformer primary or secondary windings varies the voltage available by changing the turns ratio between the windings.

Control and metering circuits in distribution substations require small transformers called control transformers. These transformers can be used to regulate the voltage supplies to control and metering circuits, or they may be used to reduce monitored high current levels to user-safe metering levels. The *NEC*® regulates overcurrent protection, installations, and grounding of all types of distribution transformers.

Objectives

When you have completed this module, you will be able to do the following:

1. Describe transformer operation.
2. Explain the principle of mutual induction.
3. Describe the operating characteristics of various types of transformers.
4. Connect a multi-tap transformer for the required secondary voltage.
5. Explain *National Electrical Code® (NEC®)* requirements governing the installation of transformers.
6. Compute transformer sizes for various applications.
7. Connect a control transformer for a given application.
8. Describe how current transformers are used in conjunction with watt-hour meters.

Trade Terms

Ampere turn
Autotransformer
Capacitance
Flux
Induction
Kilovolt-amperes (kVA)
Loss
Magnetic field

Magnetic induction
Mutual induction
Power transformer
Reactance
Rectifiers
Transformer
Turn
Turns ratio

Required Trainee Materials

1. Pencil and paper
2. Appropriate personal protective equipment
3. Copy of the latest edition of the *National Electrical Code®*

Prerequisites

Before you begin this module, it is recommended that you successfully complete *Core Curriculum; Electrical Level One; Electrical Level Two; Electrical Level Three*, Modules 26301-08 through 26306-08.

This course map shows all of the modules in *Electrical Level Three*. The suggested training order begins at the bottom and proceeds up. Skill levels increase as you advance on the course map. The local Training Program Sponsor may adjust the training order.

ELECTRICAL LEVEL THREE

26311-08
Motor Controls

26310-08
Voice, Data, and Video

26309-08
Motor Calculations

26308-08
Commercial Electrical Services

26307-08
Transformers

26306-08
Distribution Equipment

26305-08
Overcurrent Protection

26304-08
Hazardous Locations

26303-08 Practical
Applications of Lighting

26302-08 Conductor
Selection and Calculations

26301-08 Load Calculations –
Branch and Feeder Circuits

ELECTRICAL LEVEL TWO

ELECTRICAL LEVEL ONE

CORE CURRICULUM:
Introductory Craft Skills

307CMAP.EPS

1.0.0 ◆ INTRODUCTION

The electric power produced by alternators in a generating station is transmitted to locations where it is utilized and distributed to users. Many different types of **transformers** play an important role in the distribution of electricity. The main purpose of a transformer is to change the output voltage. **Power transformers** are located at generating stations to step up the voltage for more economical transmission. Substations with additional power transformers and distribution equipment are installed along the transmission line. Finally, distribution transformers are used to step down the voltage to a level suitable for utilization.

Transformers are also used quite extensively in all types of control work to raise and lower AC voltage on control circuits. They are also used in 480Y/277V systems to reduce the voltage for operating 208Y/120V lighting and other electrically operated equipment. Buck-and-boost transformers are used for maintaining appropriate voltage levels in certain electrical systems.

It is important for anyone working with electricity to become familiar with all aspects of transformer operation—how they work, how they are connected into circuits, their practical applications, and precautions to take during the installation or while working on them. This module is designed to cover these items, as well as overcurrent protection and grounding. Other subjects include correcting power factor with capacitors and the application of **rectifiers.**

2.0.0 ◆ TRANSFORMER BASICS

A very basic transformer consists of two coils, or windings, formed on a single magnetic core, as shown in *Figure 1*. Such an arrangement will allow transforming a large alternating current at a low voltage into a small alternating current at a high voltage, or vice versa.

2.1.0 Mutual Induction

The term **mutual induction** refers to the condition in which two circuits are sharing the energy of one of the circuits. It means that energy is being transferred from one circuit to the other. Induction can only occur when there is a change in the magnetic force, which can happen in either a DC or an AC circuit.

Consider the diagram in *Figure 2*. Coil A is the primary circuit that obtains energy from the power source. In a DC circuit, when the switch is closed and a voltage is applied to coil A, the current starts to flow and a **magnetic field** expands out of coil A. Coil A then changes the electrical energy of the power source into the magnetic energy (**induction**) of a magnetic field. When the field of coil A is changing, it cuts across coil B, the secondary circuit, inducing a voltage in coil B. The indicator (a galvanometer) in the secondary circuit is deflected and shows that a current, developed by the induced voltage, is flowing in the circuit. Once the voltage applied to coil A reaches its peak value, no changes occur in the field and the induction to coil B ceases, the field collapses, and no voltage is produced in coil B. The induced voltage may be generated again by moving coil B through the flux of coil A. As long as motion or change occurs in coil B or coil A and the field changes, voltage can be induced in coil B. Opening and closing the switch will create a change or motion in the field as well.

In a DC circuit, the induced voltage may be generated by moving coil B through the **flux** of coil A. However, in an AC transformer this voltage is induced without moving coil B. When the switch in the primary circuit is open, coil A has no current and no field. As soon as the switch is closed, current passes through the coil, and the magnetic field is generated. With each half cycle, this expanding field moves or cuts across the wires of coil B, thus inducing a voltage without the movement of coil B.

The magnetic field expands to its maximum strength and remains constant as long as full current flows. Flux lines stop their cutting action across the **turns** of coil B because the expansion of the field has ceased. At this point, the indicator

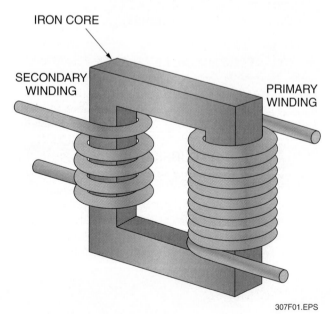

IRON CORE

SECONDARY WINDING

PRIMARY WINDING

307F01.EPS

Figure 1 ◆ Basic components of a transformer.

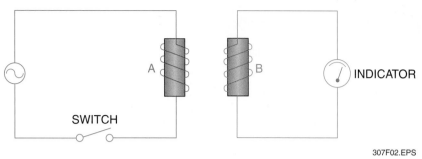

Figure 2 ◆ Mutual induction circuits.

needle on the meter reads zero because the induced voltage no longer exists. If the switch is opened, the field collapses back to the wires of coil A. As it does so, the changing flux cuts across the wires of coil B, but in the opposite direction. The current present in the coil causes the indicator needle to deflect, showing this new direction. Therefore, the indicator shows current flow only when the field is changing, either building up or collapsing. In effect, the changing field produces an induced voltage in the same way as a magnetic field moving across a conductor. This principle of inducing voltage by holding the coils steady and forcing the field to change is used in innumerable applications. The transformer is particularly suitable for operation by mutual induction. Transformers are the ideal components for transferring and changing AC voltages as needed.

Transformers are generally composed of two coils placed close to each other but not connected. Refer once more to *Figure 1*. The coil that receives energy from the line voltage source is called the primary, and the coil that delivers energy to a load is called the secondary. Even though the coils are not physically connected, they manage to convert and transfer energy as required by a process known as mutual induction.

Transformers, therefore, enable changing or converting power from one voltage to another. For example, generators that produce moderately large alternating currents at moderately high voltages use transformers to convert the power to a very high voltage and proportionately small current in transmission lines, permitting the use of smaller cable and producing less power loss.

When AC flows through a coil, an alternating magnetic field is generated around the coil. This alternating magnetic field expands outward from the center of the coil and collapses into the coil as the AC through the coil varies from zero to a maximum and back to zero again. Since the alternating magnetic field must cut through the turns of

the coil, a self-inducing voltage occurs in the coil, which opposes the change in current flow.

If the alternating magnetic field generated by one coil cuts through the turns of a second coil, voltage will be generated in this second coil just as voltage is induced in a coil that is cut by its own magnetic field. The induced voltage in the second coil is called the voltage of mutual induction, and the action of generating this voltage is called transformer action. In transformer action, electrical energy is transferred from one coil (the primary) to another (the secondary) by means of a varying magnetic field.

2.2.0 Induction in Transformers

As stated previously, a simple transformer consists of two coils located very close together and electrically insulated from each other. The primary coil generates a magnetic field that cuts through the turns of the secondary coil and generates a voltage in it. The coils are magnetically coupled to each other, and consequently, a transformer transfers electrical power from one coil to another by means of an alternating magnetic field.

Assuming that all the magnetic lines of force from the primary cut through all the turns of the secondary, the voltage induced in the secondary will depend on the ratio of the number of turns in the primary to the number of turns in the secondary. For example, if there are 100 turns in the primary and only 10 turns in the secondary, the voltage in the primary will be 10 times the voltage in the secondary. Since there are more turns in the primary than there are in the secondary, the transformer is called a step-down transformer. Transformers are rated in kilovolt-amperes (kVA) because they are independent of power factor. *Figure 3* shows a diagram of a step-down transformer with a turns ratio of 100:10, or 10:1.

$$\frac{10\ turns}{100\ turns} = 0.10 = 0.10 \times 120V = 12V$$

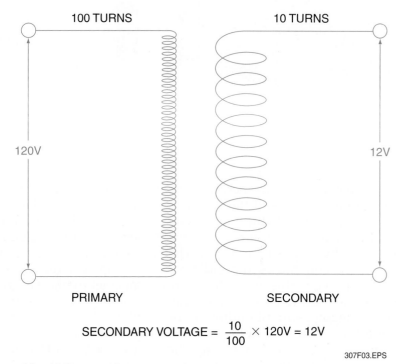

$$\frac{10 \text{ TURNS}}{100 \text{ TURNS}} = 0.10 = 0.10 \times 120V = 12V$$

100 TURNS   10 TURNS

120V   12V

PRIMARY   SECONDARY

$$\text{SECONDARY VOLTAGE} = \frac{10}{100} \times 120V = 12V$$

307F03.EPS

Figure 3 ◆ Step-down transformer with a 10:1 turns ratio.

Assuming that all the primary magnetic lines of force cut through all the turns of the secondary, the amount of induced voltage in the secondary will vary with the ratio of the number of turns in the secondary to the number of turns in the primary.

If there are more turns in the secondary winding than in the primary winding, the secondary voltage will be higher than that in the primary and by the same proportion as the number of turns in the winding. The secondary current, in turn, will be proportionately smaller than the primary current. With fewer turns in the secondary than in the primary, the secondary voltage will be proportionately lower than that in the primary, and the secondary current will be proportionately larger. Since alternating current continually increases and decreases in value, every change in the primary winding of the transformer produces a similar change of flux in the core. Every change of flux in the core and every corresponding movement of the magnetic field around the core produce a similarly changing voltage in the secondary winding, causing an alternating current to flow in the circuit that is connected to the secondary.

For example, if there are 100 turns in the secondary and only 10 turns in the primary, the voltage induced in the secondary will be 10 times the voltage applied to the primary. See *Figure 4*. Since there are more turns in the secondary than in the

primary, the transformer is called a step-up transformer.

$$\frac{100}{10} = 10 \times 12V = 120V$$

NOTE

A transformer does not generate electric power. It simply transfers electric power from one coil to another by **magnetic induction**. Transformers are rated in either volt-amperes (VA) or kilovolt-amperes (kVA).

2.3.0 Magnetic Flux in Transformers

Figure 5 shows a cross section of what is known as a high-leakage flux transformer. In these transformers, if no load were connected to the secondary or output winding, a voltmeter would indicate a specific voltage reading across the secondary terminals. If a load were applied, the voltage would drop, and if the terminals were shorted, the voltage would drop to zero. During these circuit changes, the flux in the core of the transformer would also change; it is forced out of the transformer core and is known as leakage flux. Leaking flux can actually be demonstrated with iron filings placed close to the transformer core.

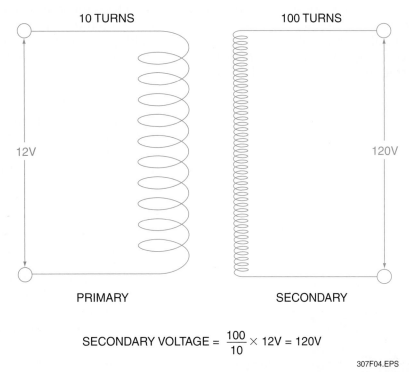

$$\frac{100}{10} = 10 \times 12V = 120V$$

10 TURNS   100 TURNS

12V   120V

PRIMARY   SECONDARY

$$\text{SECONDARY VOLTAGE} = \frac{100}{10} \times 12V = 120V$$

307F04.EPS

Figure 4 ◆ Step-up transformer with a 1:10 turns ratio.

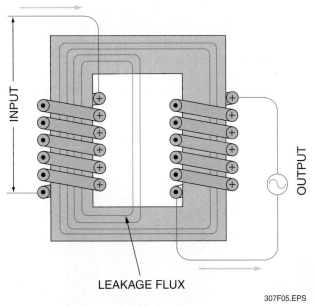

LEAKAGE FLUX

307F05.EPS

Figure 5 ◆ Transformer with high-leakage flux.

As the changes take place, the filings will shift their position, clearly showing the change in the flux pattern.

What actually happens is that as the current flows in the secondary, it tries to create its own magnetic field, which is in opposition to the original flux field. This action, like a valve in a water system, restricts the flux flow, which forces the excess flux to find another path, either through the air or in adjacent structural steel, such as transformer housings or supporting clamps.

Note that the coils in *Figure 5* are wrapped on the same iron core but are separated from each other, while the transformer in *Figure 6* has its coils wrapped around each other, which results in a low-leakage transformer design.

3.0.0 ◆ TRANSFORMER CONSTRUCTION

Transformers that are designed to operate on low frequencies have their coils, called windings, wound on iron cores. Since iron offers little resistance to magnetic lines, nearly all the magnetic field of the primary flows through the iron core and cuts the secondary.

Iron cores of transformers are constructed in three basic types: the open core, the closed core, and the shell type. See *Figure 7*. The open core is the least expensive to manufacture because the primary and secondary are wound on one cylindrical core. The magnetic path, as shown in *Figure 7*, is partially through the core and partially through the surrounding air. The air path opposes the magnetic field so that the magnetic interaction or linkage is weakened. Therefore, the open core transformer is highly inefficient.

The closed core improves the transformer efficiency by offering more iron paths and a reduced

air path for the magnetic field. The shell-type core further increases the magnetic coupling, and therefore, the transformer efficiency is greater due to two parallel magnetic paths for the magnetic field, providing maximum coupling between the primary and the secondary.

3.1.0 Cores

Special core steel is used to provide a controlled path for the flow of magnetic flux generated in a transformer. In most practical applications, the transformer core is not a solid bar of steel but is constructed of many layers of thin sheet steel called laminations. Although the specifications of the core steel are primarily of interest to the transformer design engineer, the electrical worker should at least have a conversational knowledge of the materials used.

The steel used for transformer core laminations will vary with the manufacturer, but a popular size is 0.014" thick and is called 29-gauge steel. It is processed from silicon iron alloys containing approximately 3¼% silicon. The addition of silicon to the iron increases its ability to be magnetized and also renders it essentially non-aging.

The most important characteristic of electrical steel is core loss. It is measured in watts per pound at a specified frequency and flux density. The core loss is responsible for the heating in the transformer and also contributes to the heating of the windings. Much of the core loss is a result of eddy currents that are induced in the laminations when the core is energized. To hold this loss to a minimum, adjacent laminations are coated with an inorganic varnish.

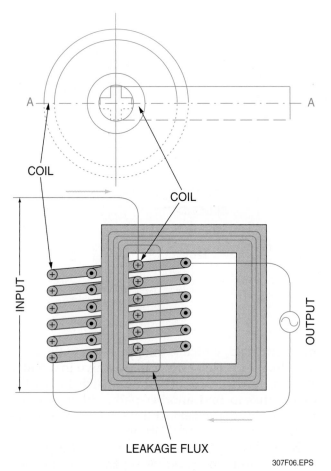

307F06.EPS

Figure 6 ◆ Low-leakage transformer.

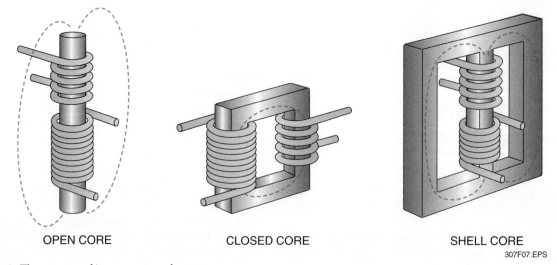

OPEN CORE   CLOSED CORE   SHELL CORE

307F07.EPS

Figure 7 ◆ Three types of iron core transformers.

Cores may either be of the core type, as shown in *Figure 8*, or the shell type, as shown in *Figure 9*. Of the two, the core type is favored for dry-type transformers for the following reasons:

- Only three core legs require stacking, which reduces cost.
- Steel does not encircle the two outer coils; this provides better cooling.
- The required floor space is reduced.

3.2.0 Types of Cores

Transformer cores are normally available in three types:

- Butt
- Wound
- Mitered cores

The butt-and-lap core is shown in *Figure 10*. Only two sizes of core steel are needed in this type of core due to the lap construction shown at the top and right side. For ease of understanding, the core strips are shown much thicker than the 0.014" thickness mentioned earlier. Each strip is carefully cut so that the air gap indicated in the lower left corner is as small as possible. The permeability of steel to the passage of flux is about 10,000 times as effective as air, hence, the air gap must be held to the barest minimum to reduce the **ampere turns** necessary to achieve adequate flux density. Also, the amount of sound produced by a transformer is a function of the flux density, which produces a difference between this construction and other types.

Another phenomenon in core steel is that the flux flows more easily in the direction in which the steel was rolled. This also varies between hot-rolled steel and cold-rolled steel. For example, the core loss due to flux passing at right angles to the rolling direction is almost 1½ times as great in hot-rolled steel and 2½ times as great in cold-rolled steel when compared with the core loss in the direction of rolling. The difference in exciting current is more dramatic, with ratios of two to one in hot-rolled steel and almost 40 to 1 in cold-rolled steel. These are primarily the designer's concern, but you should know that there is a difference.

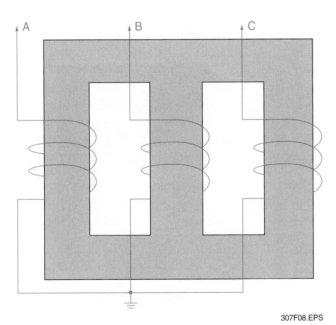

307F08.EPS

Figure 8 ◆ Core-type transformer construction.

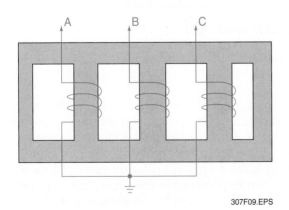

307F09.EPS

Figure 9 ◆ Shell-type transformer core.

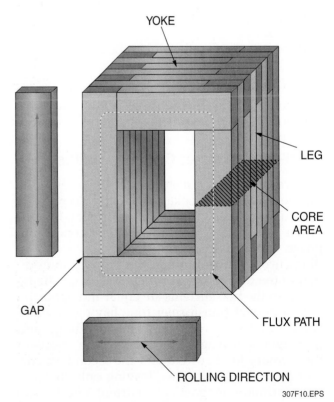

307F10.EPS

Figure 10 ◆ Butt-and-lap transformer core.

Eddy currents are restricted from passage from one lamination to another due to the inorganic insulating coating. However, the magnetic lines of flux easily transfer at adjacent laminations in the lap area, but in so doing, they are forced to cross at an angle to the preferred direction.

3.3.0 Wound Cores

Because of the unique characteristics of core steel, some core designs are made to take advantage of these differences. One such type is shown in *Figure 11*. The core loops are cut to predetermined lengths so that the gap locations do not coincide. These cuts permit assembling the core around a prewound coil that passes through both openings. Another design, now discontinued because of unfavorable cost, used a continuous core with no cuts. Separate coils had to be wound on each of the vertical legs of the completed core. You may encounter transformers of this type in existing installations.

3.4.0 Mitered Cores

Figure 12 shows a mitered core design. It is basically a butt-lap core with the joints made at 45-degree angles.

There are two benefits derived from this type of joint:

- It eliminates all cross grain flux, thereby improving the core loss and exciting current values.
- It reduces the flux density in the air gap, resulting in lower sound levels.

This type of core is normally used only with cold-rolled, grain-oriented steel and permits this steel to be used to its fullest capability.

3.5.0 Transformer Characteristics

In a well-designed transformer, there is very little magnetic leakage. Leakage causes a decrease in secondary voltage when the transformer is loaded. When a current flows through the secondary in phase with the secondary voltage, a corresponding current flows through the primary in addition to the magnetizing current. The magnetizing effects of the two currents are equal and opposite.

In a perfect transformer (one having no eddy current losses, no resistance in its windings, and no magnetic leakage), the magnetizing effects of the primary load current and the secondary current neutralize each other, leaving only the constant primary magnetizing current effective in setting up the constant flux. If supplied with a

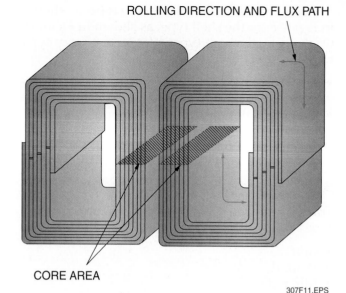

Figure 11 ◆ Wound transformer coil.

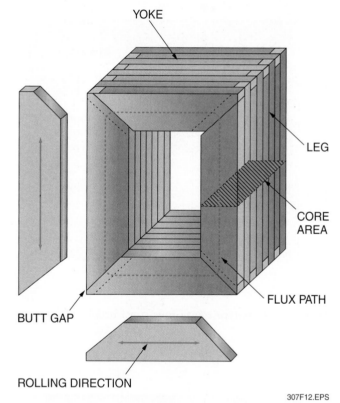

Figure 12 ◆ Mitered transformer core.

constant primary pressure, such a transformer would maintain constant secondary pressure at all loads. Obviously, the perfect transformer has yet to be built; the best transformers available today have a very small eddy current loss where the drop in pressure in the secondary windings is not more than 1% to 3%, depending on the size of the transformer.

4.0.0 ◆ TRANSFORMER TAPS

If the exact rated voltage could be delivered at every transformer location, transformer taps would be unnecessary. However, this is not possible, so taps are provided to either increase or decrease the secondary voltage.

Generally, if a load is very close to a substation or power plant, the voltage will consistently be above normal. Near the end of the line, the voltage may be below normal.

In large transformers, it would naturally be very inconvenient to move the thick, well-insulated primary leads to different tap positions when changes in source voltage levels make this necessary. Therefore, taps are used, such as those shown in the wiring diagram in *Figure 13*. In this transformer, the permanent high-voltage leads would be connected to H_1 and H_2, and the secondary leads, in their normal fashion, to X_1 and X_2, and X_3 and X_4. Note, however, the tap arrangements available at taps 2 through 7. Until a pair of these taps is interconnected with a jumper wire, the primary circuit is not completed. If this were a typical 7,200V primary, the transformer would normally have 1,620 turns. Assume 810 of these turns are between H_1 and H_6 and another 810 between H_3 and H_2. Then, if taps 6 and 3 are connected with a flexible jumper on which lugs have already been installed, the primary circuit is completed, and we have a normal ratio transformer that could deliver 120/240V from the secondary.

Between taps 6 and either 5 or 7, 40 turns of wire exist. Similarly, between taps 3 and either 2 or 4, 40 turns are present. Changing the jumper from 3 to 6 to 3 to 7 removes 40 turns from the left half of the primary. The same condition would apply on the right half of the winding if the jumper were between taps 6 and 2. Either connection would boost secondary voltage by 2½%. Had taps 2 and 7 been connected, 80 turns would have been omitted, and a 5% boost would result. Placing the jumper between taps 6 and 4 or 3 and 5 would reduce the output voltage by 5%.

5.0.0 ◆ BASIC TRANSFORMER CONNECTIONS

Transformer connections are many, and space does not permit the description of all of them here. However, an understanding of a few connection types will give the basic requirements and make it possible to use manufacturer's data for others should the need arise.

5.1.0 Single-Phase Light and Power Systems

Figure 14 is a single-phase transformer line diagram showing a connection used primarily for residential and small commercial applications. It is the most common single-phase distribution transformer in use today. It is known as

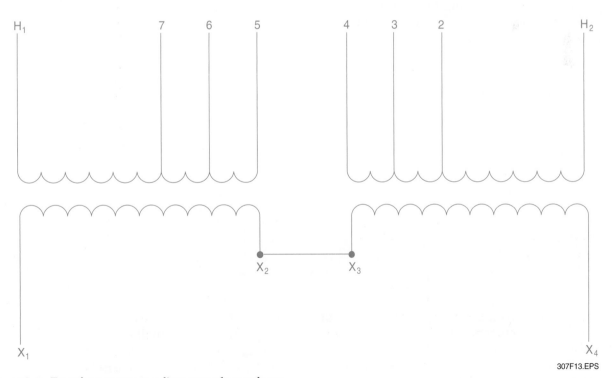

307F13.EPS

Figure 13 ◆ Transformer taps to adjust secondary voltage.

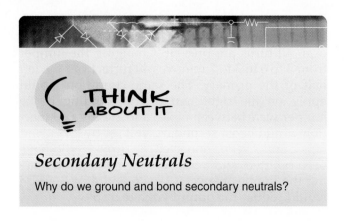
a three-wire, 120/240V, single-phase system. Because of its configuration, it is easy to balance the load between the two coils. The kVA of a single-phase transformer is calculated by dividing the total VA by 1,000.

5.2.0 Three-Phase Power Systems

The following factors need to be considered when choosing a three-phase power system:

- The load(s) to be supplied
- The voltages needed for the application
- Future expansion

There are advantages and disadvantages associated with the different types of three-phase systems. The wye system can be used for both single-phase and three-phase loads. Because of its configuration, it is easy to balance the single-phase loads while still having the ability to use it for three-phase loads. Transformers installed in

this type of system are sized by first dividing the total single-phase load in kVA by three, then taking this result and adding it to the total three-phase load in kVA divided by three. The result of these two loads added together identifies the required kVA rating of the transformer. The nominal voltage levels provided by this type of system are 120/208V or 277/480V. Note that 240V is not an option, and this may prove to be a disadvantage in many applications since 240V is a common operating voltage.

5.2.1 Delta-Wye Transformers

One of the most common transformer systems found in today's commercial and industrial settings is the delta primary, wye secondary transformer system (*Figure 15*). This is a three-phase, four-wire system that has the advantage of both providing three-phase power and also allowing lighting to be connected between any of the secondary phases and the neutral.

In *Figure 15*, the system is basically made up of three typical single-phase, step-down transformers with the interconnections between each individual transformer's primary and secondary coils determining the output voltage on the secondary side. However, this does not mean that any three single-phase transformers can be developed into a functional three-phase, delta-wye system for power and lighting. The coils must be rated for the loads to be served, as well as the primary voltage level to be connected. Power and light delta-wye transformers are universally found in both

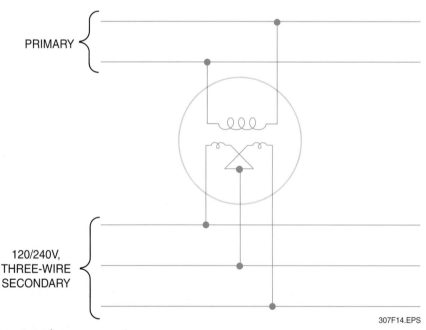

PRIMARY

120/240V,
THREE-WIRE
SECONDARY

307F14.EPS

Figure 14 ◆ Single-phase transformer connection.

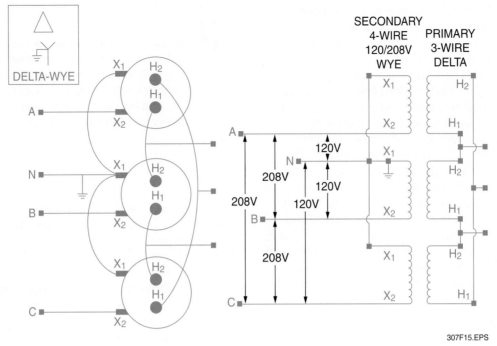

Figure 15 ◆ Delta-wye transformer system.

307F15.EPS

commercial and industrial installations as floor-mounted, dry-type transformers.

Although this transformer provides the convenience of three-phase and single-phase power, as well as each phase of the 208V system sharing the neutral to supply three legs of 120V lighting circuits, remember that the three-phase and single-phase power availability is at the 208V level and not 240V. Should a power requirement specifically call for 240V and not permit a 208V supply, this type of transformer is not the best selection.

5.2.2 Delta-Delta Transformers

The delta-delta system in *Figure 16* operates a little differently from the delta-wye system. Whereas the wye-connected system is formed by connecting one terminal from each of three equal voltage

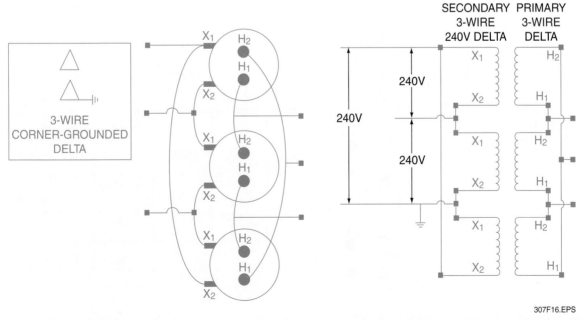

Figure 16 ◆ Delta-connected secondary.

307F16.EPS

transformer windings together to make a common terminal, the delta-connected system has its windings connected in series, forming a triangle, or the Greek symbol delta (Δ). In *Figure 17*, a center-tapped terminal is used on one winding to ground the system. A 120/240V system has 120V between the center-tapped terminal and each ungrounded terminal on either side such as phases A and C, and 240V across the full winding of each phase.

Refer to *Figure 17* and note the high leg. This is also known as the wild leg. This high leg has a higher voltage to ground than the other two phases. The voltage of the high leg can be determined by multiplying the voltage to ground of either of the other two legs by the square root of 3, which we round to a value of 1.732. Therefore, if the voltage between phase A to ground is 120V, the voltage between phase B to ground may be determined as follows:

$$120V \times 1.732 = 207.84V = 208V$$

From this, it should be obvious that no single-pole breakers should be connected to the high leg of a center-tapped, four-wire, delta-connected system. In fact, *NEC Section 110.15* requires that the phase busbar or conductor having the higher voltage to ground be durably and permanently marked by an outer finish that is orange in color, or by other effective means. This prevents future workers from connecting 120V single-phase loads to this high leg, which would probably damage any equipment connected to the circuit. Remember the color orange; no line-to-neutral loads are to be connected to this phase.

WARNING!

Always use caution when working on a center-tapped, four-wire, delta-connected system. Phase B has a higher voltage to ground than phases A and C. Never connect 120V circuits to the high leg. Doing so will result in damage to the circuits and equipment.

5.2.3 Open Delta Transformers

Three-phase, delta-connected systems may be connected so that only two transformers are used; this arrangement is known as an open delta system, as shown in *Figure 18*. It is frequently used on a delta system when one of the three transformers becomes damaged. The damaged transformer is disconnected from the circuit, and the remaining two transformers carry the load. In doing so, the three-phase load carried by the open delta bank is only 86.6% of the combined rating of the remaining two equally sized units. It is only 57.7% of the normal full-load capability of a full bank of transformers. In an emergency, however, this capability permits single-phase and three-phase power at a location where one unit burned out and a replacement was not readily available. The total load must be curtailed to avoid another burnout.

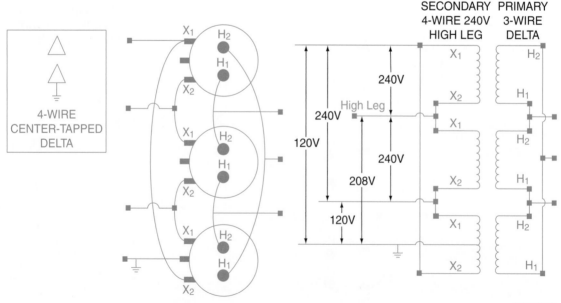

307F17.EPS

Figure 17 ◆ Characteristics of a center-tapped, delta-connected system.

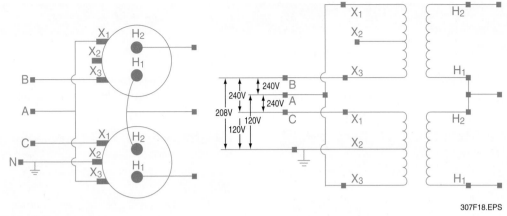

Figure 18 ◆ Open delta system.

5.3.0 Parallel Operation of Transformers

Transformers may operate in parallel under certain conditions (*Figure 19*). Generally, satisfactory operation occurs only when applying transformers with equivalent winding configurations, voltages, and impedances. Small differences between transformers can cause circulating currents to flow, resulting in excess heat and loss of efficiency.

If transformers of unequal kVA ratings must be paralleled they must also have equal XR ratios (impedance to reactance) in order to divide the load proportionately between the transformers.

5.3.1 Impedance in Parallel-Operated Transformers

Impedance plays an important role in the successful operation of transformers connected in parallel. The impedance of the transformers must be such that the voltage drop from no load to full load is the same in all transformer units in both magnitude and phase. In most transformers, the voltage drop due to reactance is large compared to the voltage drop due to resistance. For our purposes, load sharing of similar transformers can be calculated using total impedance instead of reactance. If the full-load impedances of two similar transformers are the same, the load will be divided equally between them.

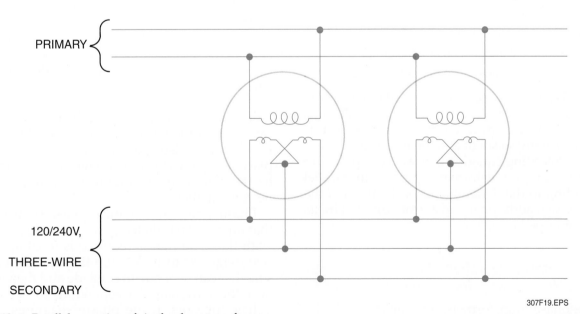

Figure 19 ◆ Parallel operation of single-phase transformers.

Transformers

These photos show single-phase and three-phase transformers. Note the orange marking on the high leg of the three-phase transformer in accordance with *NEC Section 110.15.*

100kVA SINGLE-PHASE TRANSFORMER
480V PRIMARY, 120V/240V SECONDARY

300kVA THREE-PHASE CENTER-TAPPED
DELTA TRANSFORMER
480V PRIMARY, 120V/240V SECONDARY

307SA01.EPS

The following equation may be used to obtain the division of loads between two transformer banks operating in parallel on single-phase systems:

$$\text{Power} = \frac{kVA_1 \div Z_1}{(kVA_1 \div Z_1) + (kVA_2 \div Z_2)} \times \text{total kVA load}$$

Where:

kVA_1 = kVA rating of transformer 1

kVA_2 = kVA rating of transformer 2

Z_1 = percent impedance of transformer 1

Z_2 = percent impedance of transformer 2

In this equation, it can be assumed that the ratio of resistance to reactance is the same in all units since the error introduced by differences in this ratio is usually so small as to be negligible.

The preceding equation may also be applied to more than two transformers operated in parallel by adding to the denominator of the fraction the kVA of each additional transformer divided by its percent impedance.

5.3.2 Parallel Operation of Three-Phase Transformers

Three-phase transformers, or banks of single-phase transformers, may be connected in parallel,

provided each of the three primary leads in one three-phase transformer is connected in parallel with a corresponding primary lead of the other transformer. The secondaries are then connected in the same way. The corresponding leads are the leads that have the same potential at all times and the same polarity. Furthermore, the transformers must have the same voltage ratio and the same impedance voltage drop.

When three-phase transformer banks operate in parallel and the three units in each bank are similar, the division of the load can be determined by the same method previously described for single-phase transformers connected in parallel on a single-phase system.

In addition to the requirements of polarity, ratio, and impedance, paralleling of three-phase transformers also requires that the angular displacement between the voltages in the windings be taken into consideration when they are connected together.

Phasor diagrams of three-phase transformers that are to be paralleled greatly simplify matters. With these, all that is required is to compare the two diagrams to make sure they consist of phasors that can be made to coincide and then to connect the terminals corresponding to coinciding voltage phasors. If the phasor diagrams can be made to coincide, leads that are connected together will have the same potential at all times. This is one of the fundamental requirements for paralleling. Phasor diagrams are covered in more detail later in this module.

6.0.0 ◆ AUTOTRANSFORMERS

An **autotransformer** is a transformer whose primary and secondary circuits have part of a winding in common; therefore, the two circuits are not isolated from each other. See *Figure 20*. The application of an autotransformer is a good choice where a 480Y/277V or 208Y/120V, three-phase, four-wire distribution system is used. Some of the advantages are:

- Lower purchase price
- Lower operating cost due to lower losses
- Smaller size, easier to install
- Better voltage regulation
- Lower sound levels

For example, when the ratio of transformation from the primary to the secondary voltage is small, the most economical way of stepping down the voltage is by using autotransformers, as shown in *Figure 21*. For this application, it is necessary that the neutral of the autotransformer bank be connected to the system neutral, similar to a wye connection.

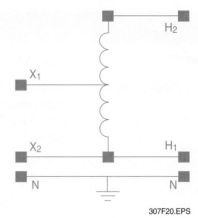

Figure 20 ◆ Step-down autotransformer.

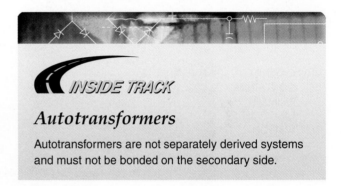

Autotransformers

Autotransformers are not separately derived systems and must not be bonded on the secondary side.

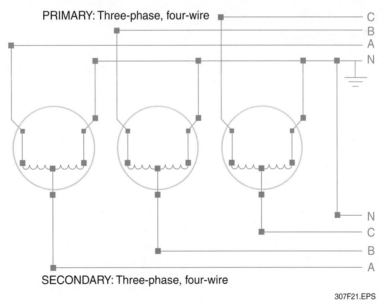

Figure 21 ◆ Autotransformers supplying power from a three-phase, four-wire system.

Liquid-Filled Transformers

This 2,500kVA liquid-filled transformer has cable bus secondary conductors and a single protection and control interface cabinet. Larger transformers may have two or more interface cabinets, depending on the number of heaters, fans, CTs, and other devices.

TRANSITION CONTAINING BUSBAR CONNECTION BETWEEN TRANSFORMER AND DISCONNECT

GAUGES MEASURE WINDING AND LIQUID TEMPERATURES, PRESSURE, AND LIQUID LEVEL

CABLE BUS DUCT

CABINET CONTAINING PRIMARY 15kV DISCONNECT SWITCH

PROTECTION AND CONTROL INTERFACE CABINET

DRAIN VALVE FOR FLUID SAMPLING

EXTERNAL EQUIPMENT GROUNDING CONDUCTOR BONDS TRANSFORMER CASE TO FACILITY GROUNDING GRID

307SA02.EPS

7.0.0 ◆ TRANSFORMER DATA

NEC Section 450.11 requires that each transformer must be provided with a nameplate giving the manufacturer, rated kVA, frequency, primary and secondary voltage, impedance of transformers 25kVA and larger, required clearances for transformers with ventilating openings, and the amount and type of insulating liquid (where used). In addition, the nameplate of each transformer must include the temperature class for the insulation system.

In addition, most manufacturers include a wiring diagram and a connection chart, as shown in *Figure 22* for a 480V delta primary to 208Y/120V wye secondary. It is recommended that all transformers be connected as shown on the manufacturer's nameplate.

In general, this wiring diagram and accompanying table indicate that the 480V, three-phase, three-wire primary conductors are connected to terminals H_1, H_2, and H_3, respectively, regardless of the desired voltage on the primary. A neutral conductor, if required, is carried from the primary, through the transformer, to the secondary. Two variations are possible on the secondary side of this transformer: 208V, three-phase, three-wire or four-wire or 120V, single-phase, two-wire. To connect the secondary side of the transformer as a 208V, three-phase, three-wire system, the secondary conductors are connected to terminals X_1, X_2, and X_3; the neutral is carried through with conductors usually terminating at a solid neutral bus in the transformer.

Another popular transformer connection is the 480V primary to 240V delta/120V secondary. This configuration is shown in *Figure 23*. Again, the primary conductors are connected to transformer terminals H_1, H_2, and H_3. The secondary connections for the desired voltages are made as indicated in the table.

8.0.0 ◆ CONTROL TRANSFORMERS

Control transformers are available in numerous types, but most are dry-type, step-down units with the secondary control circuit isolated from the primary line circuit to ensure maximum safety. See *Figure 24*. Industrial control transformers are designed to accommodate the momentary current inrush caused when electromagnetic components are energized without sacrificing secondary voltage stability beyond practical limits.

Other types of control transformers, sometimes referred to as control and signal transformers, are constant-potential, air-cooled transformers. Their purpose is to supply the proper reduced voltage for control circuits of electrically operated switches, signal circuits, or other equipment. These transformers do not normally require the industrial regulation characteristics found in other transformers. Some are of the open type with no protective casing over the windings; others are enclosed within a metal casing.

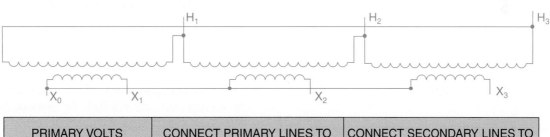

PRIMARY VOLTS	CONNECT PRIMARY LINES TO	CONNECT SECONDARY LINES TO
480V	H_1, H_2, H_3	———
SECONDARY VOLTS		
208V	———	X_1, X_2, X_3
120V SINGLE-PHASE	———	X_1 to X_0 X_2 to X_0 X_3 to X_0

307F22.EPS

Figure 22 ◆ Typical manufacturer's wiring diagram for a delta-wye transformer.

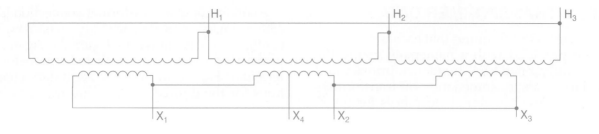

PRIMARY VOLTS	CONNECT PRIMARY LINES TO	CONNECT SECONDARY LINES TO
480V	H_1, H_2, H_3	———
SECONDARY VOLTS		
240V	———	X_1, X_2, X_3
120V	———	X_1, X_4 or X_2, X_4

307F23.EPS

Figure 23 ◆ 480V delta to 240V delta transformer connections.

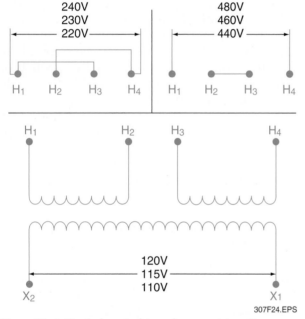

307F24.EPS

Figure 24 ◆ Typical control transformer wiring diagram.

When choosing control transformers for any application, the loads must be calculated and completely analyzed before the proper transformer selection can be made. This analysis must consider every electrically energized component in the control circuit. To select an appropriate control transformer, first determine the voltage and frequency of the supply circuit. Next, determine the total inrush volt-amperes (watts) of the control circuit. In doing so, do not neglect the current requirements of indicating lights and timing devices that do not have inrush volt-amperes but are energized at the same time as the other components in the circuit. Their total volt-amperes should be added to the total inrush volt-amperes. Control transformers will be covered in more detail in Level Four.

9.0.0 ◆ *NEC*® REQUIREMENTS

The *NEC*® requirements for transformer installations are based on the type of insulating fluid (if any), the transformer rating, and whether they are installed indoors or outdoors. There are five basic classifications:

- *Dry-type transformers* – These transformers can be ventilated, air-cooled, or sealed. See *NEC Sections 450.21 and 450.22*. These transformers are not required to be installed in a vault unless rated greater than 35kV.
- *Less-flammable liquid-insulated transformers* – These transformers contain a liquid with a fire point not less than 300°C. See *NEC Section 450.23*. These transformers are not required to be installed in a vault unless rated greater than 35kV and additional conditions are satisfied. Otherwise indoor or outdoor installation requirements are the same as for oil-insulated transformers.

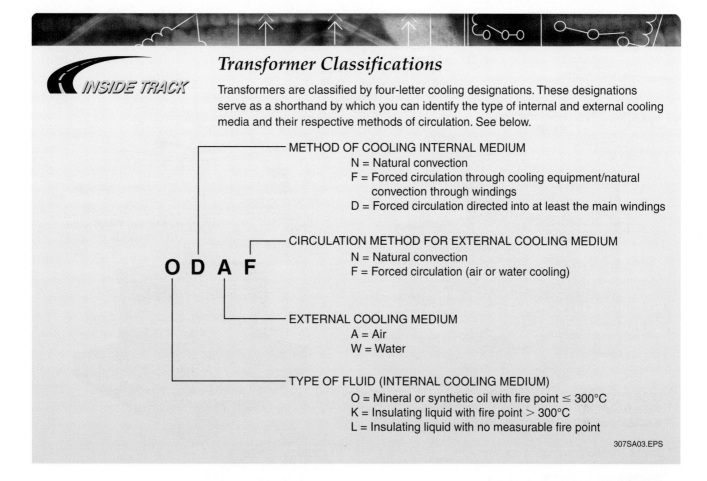

- *Nonflammable fluid-insulated transformers* – These transformers contain a nonflammable dielectric fluid that does not have a flash or fire point and is not flammable in air. See *NEC Section 450.24*. These transformers are not required to be installed in a vault unless rated greater than 35kV.

- *Askarel-insulated transformers* – These transformers contain Askarel fluid, which is a generic name for a type of synthetic, nonflammable liquid that may contain polychlorinated biphenyls (PCBs). Because of the possible presence of PCBs, these transformers have special safety requirements. See *NEC Section 450.25*. These transformers are not required to be installed in a vault unless rated greater than 35kV.

- *Oil-insulated transformers* – These transformers contain flammable oil and are usually installed in vaults. Exceptions to this requirement can be found in *NEC Section 450.26*. The requirements for oil-insulated transformers installed outdoors can be found in *NEC Section 450.27*.

Transformers must normally be accessible for inspection, except for dry-type transformers under certain specified conditions. Certain types of transformers with a high voltage or kVA rating are required to be enclosed in transformer rooms or vaults when installed indoors. The construction of these vaults is covered in *NEC Article 450* and described in *Figure 25* and *Table 1*.

In general, the *NEC*® specifies that the walls and roofs of vaults must be constructed of materials that have adequate structural strength for the conditions with a minimum fire resistance of three hours. However, where transformers are protected with an automatic sprinkler system, water spray, carbon dioxide, or halon, the fire resistance construction may be lowered to only one hour. The floors of vaults in contact with the earth must be of concrete and not less than 4" thick. If the vault is built with a vacant space or other floors (stories) below it, the floor must have adequate structural strength for the load imposed thereon and a minimum fire resistance of three hours. Again, if the fire extinguishing facilities are provided, as outlined above, the fire resistance construction need only be one hour. The *NEC*® does not permit the use of studs and wallboard construction for transformer vaults.

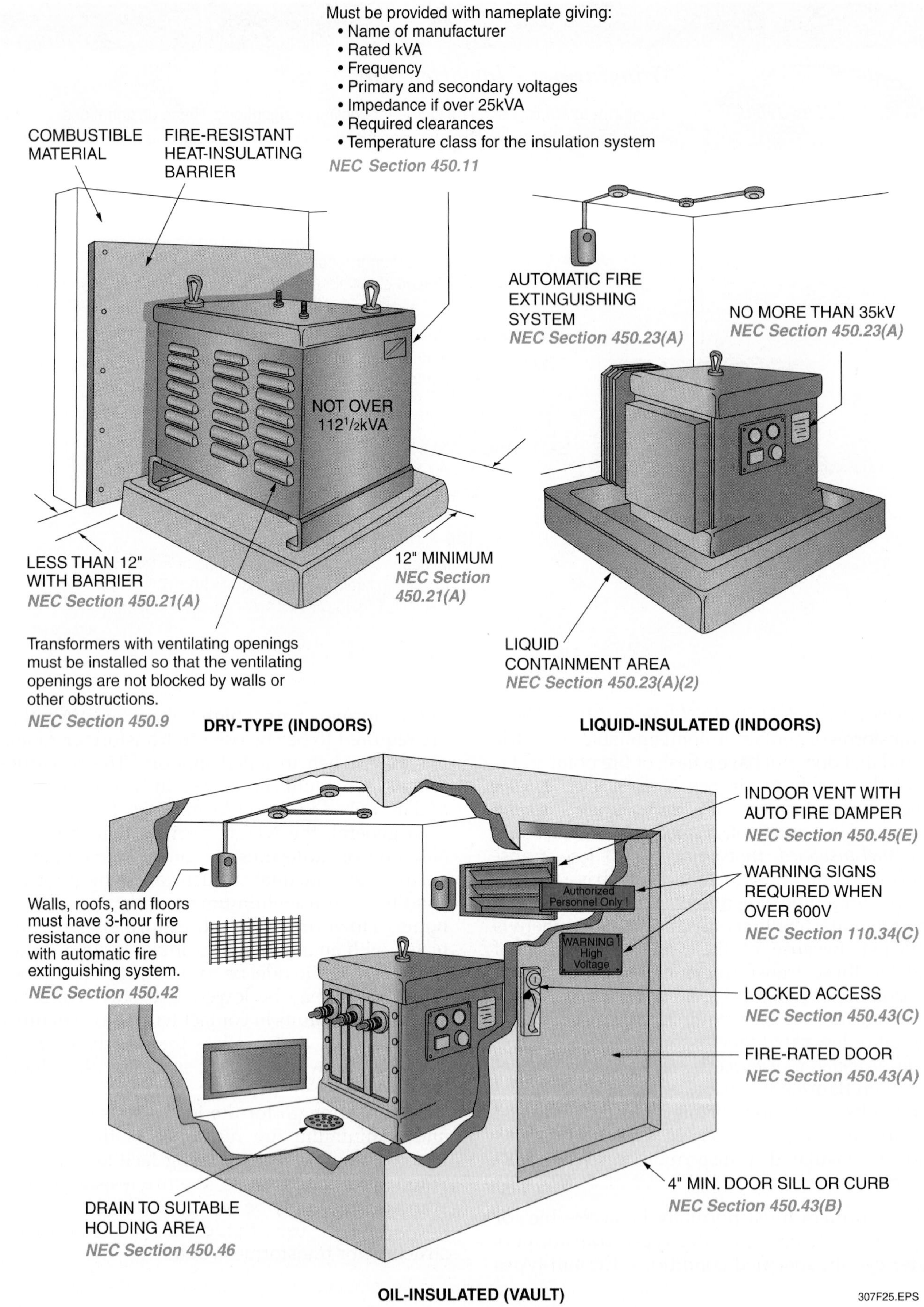

Must be provided with nameplate giving:
- Name of manufacturer
- Rated kVA
- Frequency
- Primary and secondary voltages
- Impedance if over 25kVA
- Required clearances
- Temperature class for the insulation system

NEC Section 450.11

COMBUSTIBLE
MATERIAL

FIRE-RESISTANT
HEAT-INSULATING
BARRIER

AUTOMATIC FIRE
EXTINGUISHING
SYSTEM
NEC Section 450.23(A)

NO MORE THAN 35kV
NEC Section 450.23(A)

NOT OVER
112¹/₂kVA

LESS THAN 12"
WITH BARRIER
NEC Section 450.21(A)

12" MINIMUM
*NEC Section
450.21(A)*

Transformers with ventilating openings
must be installed so that the ventilating
openings are not blocked by walls or
other obstructions.
NEC Section 450.9

LIQUID
CONTAINMENT AREA
NEC Section 450.23(A)(2)

DRY-TYPE (INDOORS)

LIQUID-INSULATED (INDOORS)

INDOOR VENT WITH
AUTO FIRE DAMPER
NEC Section 450.45(E)

Walls, roofs, and floors
must have 3-hour fire
resistance or one hour
with automatic fire
extinguishing system.
NEC Section 450.42

Authorized
Personnel Only !

WARNING SIGNS
REQUIRED WHEN
OVER 600V
NEC Section 110.34(C)

WARNING !
High
Voltage

LOCKED ACCESS
NEC Section 450.43(C)

FIRE-RATED DOOR
NEC Section 450.43(A)

4" MIN. DOOR SILL OR CURB
NEC Section 450.43(B)

DRAIN TO SUITABLE
HOLDING AREA
NEC Section 450.46

OIL-INSULATED (VAULT)

307F25.EPS

Figure 25 ◆ Transformer installation requirements.

Table 1 Summary of *NEC*® Transformer Installation and Overcurrent Protection Requirements

Application	*NEC*® Regulation	*NEC*® Reference
Location	Transformers must be readily accessible to qualified personnel for maintenance and inspection.	*NEC Section 450.13*
	Dry-type transformers rated at 112½kVA or less may be located out in the open provided they are separated from combustible material by 12" or a suitable fire/heat barrier.	*NEC Section 450.21(A)*
	Dry-type transformers rated at more than 112½kVA must be installed in a transformer room of fire-resistant construction.	*NEC Section 450.21(B)*
	Dry-type transformers not exceeding 600V and 50kVA are not required to be readily accessible and are permitted in fire-resistant hollow spaces of a building under the conditions specified in the *NEC*®.	*NEC Sections 450.13(A) and (B)*
	Dry-type transformers installed outdoors must have a weatherproof enclosure.	*NEC Section 450.22*
	Liquid-filled transformers must be installed as specified in the *NEC*® and usually in vaults when installed indoors.	*NEC Section 450.23*
Overcurrent protection	The primary protection must be rated or set as follows: • 9A or more—125% • Less than 9A—167% • Less than 2A—300% If the primary current (line side) is 9A or more, the next higher standard size overcurrent protective device greater than 125% of the primary current is used. For example, if the primary current is 15A, 125% of 15A = 18.75A. The next standard size circuit breaker is 20A. Therefore, this size, 20A, may be used. Conductors on the secondary side of a single-phase transformer with a two-wire secondary may be protected by the primary overcurrent device under certain *NEC*® conditions.	*NEC Table 450.3(B)*
Transformers used in motor control circuits	Special rules apply to these circuits for the various types of transformers.	*NEC Section 430.72(C)*
Over 600V	Special *NEC*® rules apply to transformers operating at over 600V.	*NEC Table 450.3(A)*

9.1.0 Overcurrent Protection for Transformers (600V or Less)

The overcurrent protection for transformers is based on their rated current, not on the load to be served. The primary circuit may be protected by a device rated or set at not more than 125% of the rated primary current of the transformer for transformers with a rated primary current of 9A or more.

Instead of individual protection on the primary side, the transformer may be protected only on the secondary side if all of the following conditions are met:

• The overcurrent device on the secondary side is rated or set at not more than 125% of the rated secondary current.

• The primary feeder overcurrent device is rated or set at not more than 250% of the rated primary current.

For example, if a 12kVA transformer has a primary voltage rating of 480V, calculate the amperage as follows:

$$\frac{12{,}000\text{VA}}{480\text{V}} = 25\text{A}$$

With a secondary voltage rated at 120V, the amperage becomes:

$$\frac{12{,}000\text{VA}}{120\text{V}} = 100\text{A}$$

The individual primary protection must be set at:

$$1.25 \times 25\text{A} = 31.25\text{A}$$

In this case, a standard 30A cartridge fuse rated at 600V could be used, as could a circuit breaker approved for use on 480V. However, if certain conditions are met, individual primary protection for the transformer is not necessary if the feeder overcurrent protective device is rated at not more than:

$$2.5 \times 25A = 62.5A$$

In addition, the protection of the secondary side must be set at not more than:

$$1.25 \times 100A = 125A$$

In this case, a standard 125A circuit breaker could be used.

NOTE

The example cited is for the transformer only, not the secondary conductors. The secondary conductors must be provided with overcurrent protection as outlined in *NEC Section 210.20(B)*.

The requirements of *NEC Section 450.3* cover only transformer protection; in practice, other components must be considered when applying circuit overcurrent protection. Circuits with transformers must meet the requirements for conductor protection in *NEC Articles 240 and 310*. Panelboards must meet the requirements of *NEC Article 408*.

- *Primary fuse protection only* – If secondary fuse protection is not provided, then the primary fuses must not be sized larger than 125% of the transformer primary full-load amperes (FLA), except if the transformer primary FLA is that shown in *NEC Table 450.3(B)*. See *Figure 26*. Individual transformer primary fuses are not necessary where the primary circuit fuse provides this protection.
- *Primary and secondary protection* – According to *NEC Table 450.3(A)*, a transformer with a primary voltage over 600V, located in unsupervised areas, is permitted to have the primary fuse sized at a maximum of 300%. If the secondary is also over 600V, the secondary fuses can be sized at a maximum of 250% for transformers with impedances not greater than 6% and 225% for transformers with impedances greater than 6% and not more than 10%. If the secondary is 600V or below, the secondary fuses can be sized at a maximum of 125%. Where these settings do not correspond to a standard fuse size, the next higher standard size is permitted.

In supervised locations, the maximum settings are as shown in *Figure 27*, except for secondary voltages of 600V or below, where the secondary fuses can be sized at a maximum of 250%.

- *Primary protection only* – In supervised locations, the primary fuses can be sized at a maximum of 250% or the next larger standard size if 250% does not correspond to a standard fuse size.

NOTE

The use of primary protection *only* does not remove the requirements for overcurrent compliance found in *NEC Articles 240 and 408*. See FPN No. 1 in *NEC Section 450.3*, which references *NEC Sections 240.4, 240.21, 240.100, and 240.101*, for proper protection of secondary conductors.

9.1.1 Overcurrent Protection for Small Power Transformers

Low-amperage, E-rated, medium-voltage fuses are general-purpose, current-limiting fuses. The E rating defines the melting time current characteristic of the fuse and permits electrical interchangeability of fuses with the same E rating. For a general-purpose fuse to have an E rating, the current responsive element shall melt in 300 seconds at an rms current within the range of 200% to 240% of the continuous current rating of the fuse, fuse refill, or link (*ANSI C37.46*).

FUSE MUST NOT BE LARGER THAN 125% OF TRANSFORMER PRIMARY FLA WHEN NO TRANSFORMER SECONDARY PROTECTION IS PROVIDED

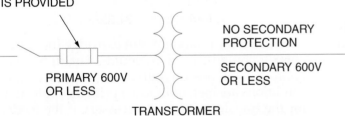

Figure 26 ◆ Transformer circuit with primary fuse only.

PRIMARY CURRENT	PRIMARY FUSE RATING
9A or more	125% or next higher standard rating if 125% does not correspond to a standard fuse size
2A to 9A	167% maximum
Less than 2A	300% maximum

307F26.EPS

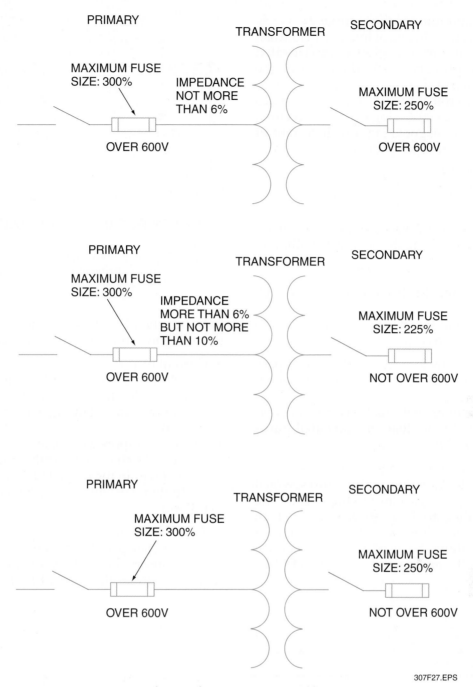

Figure 27 ◆ Minimum overcurrent protection for transformers in supervised locations.

Low-amperage, E-rated fuses are designed to provide primary protection for potential, small service, and control transformers. These fuses offer a high level of fault current interruption in a self-contained, non-venting package that can be mounted indoors or in an enclosure.

As for all current-limiting fuses, the basic application rules found in the *NEC*® and the manufacturer's literature should be adhered to. In addition, potential transformer fuses must have sufficient inrush capacity to successfully pass through the magnetizing inrush current of the transformer. If the fuse is not sized properly, it will open before the load is energized. The maximum magnetizing inrush currents to the transformer at system voltage and the duration of this inrush current vary with the transformer design. Magnetizing inrush currents are usually denoted as a percentage of the transformer full-load current (10X, 12X, 15X, etc.). The inrush current duration is usually given in seconds. Where this information is available, an easy check can be made on the appropriate minimum melting curve to verify proper fuse selection. In lieu of transformer inrush data, the rule of thumb is to select a fuse size rated at 300% of the primary full-load current or the next larger standard size.

For example, a transformer manufacturer states that an 800VA, 2,400V, single-phase potential transformer has a magnetizing inrush current of 12X lasting for 0.1 second. Therefore:

$$I = \frac{800VA}{2,400V} = 0.333A$$
$$Current = 12A \times 0.333A = 4A$$

Since the voltage is 2,400V, we can use either a JCW or a JCD fuse. Using the 300% rule of thumb:

$$300\% \text{ of } 0.333A = 0.999A$$

Therefore, we would choose a JCW-1E or JCD-1E fuse.

Typical potential transformer connections can be grouped into two categories:

- Those connections that require the fuse to pass only the magnetizing inrush of one potential transformer (*Figure 28*)
- Those connections that must pass the magnetizing inrush of more than one potential transformer (*Figure 29*)

Fuses for medium-voltage transformers and feeders are E-rated, medium-voltage fuses, which are general-purpose, current-limiting fuses. The fuses carry either an E or an X rating, which defines the melting time current characteristic of the fuse. The ratings are used to allow electrical interchangeability among different manufacturers' fuses.

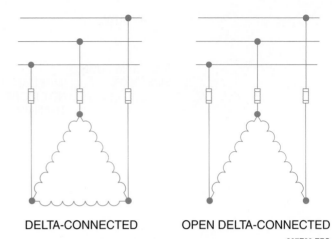

DELTA-CONNECTED   OPEN DELTA-CONNECTED

307F29.EPS

Figure 29 ◆ Connections requiring fuses to pass the magnetizing inrush of more than one transformer.

For a general-purpose fuse to have an E rating, the following conditions must be met:

- Current responsive elements with ratings 100A or below shall melt in 300 seconds at an rms current within the range of 200% to 240% of the continuous current rating of the fuse unit (*ANSI C37.46*).
- Current responsive elements with ratings above 100A shall melt in 600 seconds at an rms current within the range of 220% to 264% of the continuous current rating of the fuse unit (*ANSI C37.46*).

A fuse with an X rating does not meet the electrical interchangeability for an E-rated fuse, but it offers the user other ratings that may provide better protection for the particular application.

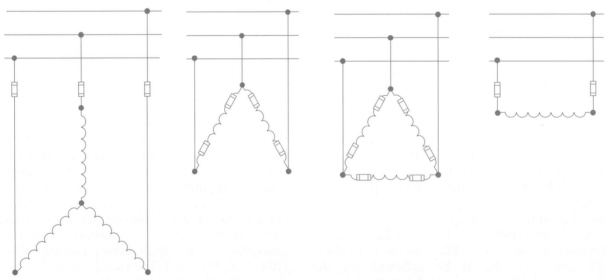

307F28.EPS

Figure 28 ◆ Connections requiring fuses to pass only the magnetizing inrush of one transformer.

Transformer protection is the most popular application of E-rated fuses. The fuse is applied to the primary of the transformer and is solely used to prevent rupture of the transformer due to short circuits. It is important, therefore, to size the fuse so that it does not clear on system inrush or permissible overload currents. Magnetizing inrush must also be considered when sizing a fuse. In general, power transformers have a magnetizing inrush current of 12X, which is the full-load rating for a duration of $\frac{1}{10}$ second.

9.2.0 Transformer Grounding

Grounding is necessary to remove static electricity and also as a precautionary measure in case the transformer windings accidentally come in contact with the core or enclosure. All transformers should be grounded and bonded to meet *NEC*® requirements and also local codes where applicable.

The case of every power transformer should be grounded to eliminate the possibility of obtaining static shocks from it or being injured by accidental grounding of the winding to the case. A grounding lug is provided on the base of most transformers for the purpose of grounding the case and fittings.

The *NEC*® specifically states the requirements for grounding and should be followed in every respect. Furthermore, certain advisory rules recommended by manufacturers provide additional protection beyond that of the *NEC*®. In general, the code requires that separately derived alternating current systems be grounded as stated in *NEC Section 250.30*.

10.0.0 ◆ POWER FACTOR

Power factor was covered in the Level Two module entitled *Alternating Current*, so a brief review of the subject should suffice here. The equation for power factor is:

$$\text{Power factor (pf)} = \frac{kW}{kVA}$$

Where:

kW = kilowatts

kVA = kilovolt-amperes

Calculating the power factor of an electrical system requires that the true power, inductive reactance, and capacitive reactance of the system be determined. An analogy should enhance your understanding of these terms.

Imagine a farm wagon to which three horses are hitched, as shown in *Figure 30*. The horse in the middle (#1) is pulling straight ahead; we will call this horse true power because all of the effort is in the direction that the work should be done. Horse

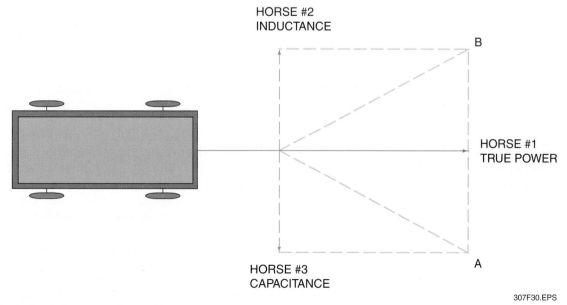

Figure 30 ◆ Depicting power factor.

Increasing Power Factors

Industrial facilities with large inductive loads that must rely on commercial power to operate pay a rate depending on power factor and consumption. The rate increases when the power factor is low because the utility must supply additional kVARs not metered by the watt-hour meter. In order to decrease the cost of electricity caused by these inductive loads, capacitors are often installed to increase the power factor.

#2 wants to nibble at the grass growing along the side of the road. This horse does not contribute an ounce of pull in the desired direction and causes a problem by pulling the wagon toward the ditch. We will call horse #2 inductance. The third horse has about the same strength as horse #2 and enjoys the grass on the opposite side of the road, which causes this horse to pull in the exact opposite direction of horse #2. Horse #3 also contributes nothing to the forward motion. This horse is called **capacitance**.

We will forget about horse #3 for the moment. If only horse #1 and horse #2 were pulling, the wagon would go in the direction of dotted line A. Notice that the length of that line is greater than the line to #1, so the horse named inductance has an effect on the final result. The direction and length of A might well be called apparent power and happens to be the hypotenuse (or diagonal) of a right angle triangle.

If the #2 horse were unhitched, then horses #1 and #3 would cause the wagon to move in the direction of line B, and the length of that line would also be apparent power. If all three horses were pulling, horses #2 and #3 would cancel one another out, and the only useful animal would be reliable horse #1, true power.

In an AC circuit, there are always three forces working in varying lengths. Inductance (horse #2) is present in every magnetic circuit and always works at a 90-degree angle with true power. Therefore, we have a power factor of less than 100% because the wagon does not move straight ahead, but travels toward the right due to the pull of horse #2. However, we can improve the power factor by adding horse #3, capacitance, which tends to cancel out the pull of inductance, enabling the wagon to travel straight ahead.

10.1.0 Capacitors

NEC Article 460 states specific rules for the installation and protection of capacitors other than surge capacitors or capacitors that are part of another apparatus. The chief use of capacitors is to improve the power factor of an electrical installation or an individual piece of electrically operated equipment. In general, this efficiency lowers the cost of power. The *NEC®* requirements for capacitors operating under 600V are summarized in *Table 2*.

Since capacitors may store an electrical charge and hold a voltage that is present even when a capacitor is disconnected from a circuit, capacitors must be enclosed, guarded, or located so that persons cannot accidentally contact the terminals. In most installations, capacitors are installed out of reach or are placed in an enclosure accessible only to qualified persons. The stored charge of a capacitor must be drained by a discharge circuit either permanently connected to the capacitor or automatically connected when the line voltage of the capacitor circuit is removed. The windings of a motor or a circuit consisting of resistors and reactors will serve to drain the capacitor charge.

Capacitor circuit conductors must have an ampacity of not less than 135% of the rated current of the capacitor. This current is determined from the VA rating of the capacitor as for any other load. For example, a 100kVA (100,000VA), three-phase capacitor operating at 480V has a rated current of:

$$I = \frac{100{,}000VA}{1.732 \times 480V} = 120.3A$$

The minimum conductor ampacity is then:

$$I = 1.35 \times 120.3A = 162.4A$$

When a capacitor is switched into a circuit, a large inrush current results to charge the capacitor to the circuit voltage. Therefore, an overcurrent protective device for the capacitor must be rated or set high enough to allow the capacitor to charge. Although the exact setting is not specified in the *NEC®*, typical settings vary between 150% and 250% of the rated capacitor current.

In addition to overcurrent protection, a capacitor must have a disconnecting means rated at not less than 135% of the rated current of the capacitor unless the capacitor is connected to the

Table 2 NEC® Capacitor Installation Requirements

Application	NEC® Regulation	NEC® Reference
Enclosing and guarding	Capacitors must be enclosed, located, or guarded so that persons cannot come into accidental contact or bring conducting materials into accidental contact with exposed energized parts, terminals, or buses associated with them. However, no additional guarding is required for enclosures accessible only to authorized and qualified persons.	NEC Section 460.2(B)
Stored charge	Capacitors must be provided with a means of draining the stored charge. The discharge circuit must be either permanently connected to the terminals of the capacitor or capacitor bank or provided with automatic means of connecting it to the terminals of the capacitor bank on removal of voltage from the line. Manual means of switching or connecting the discharge circuit shall not be used.	NEC Section 460.6
Capacitors on circuits over 600V	Special NEC® regulations apply to capacitors operating at over 600V.	NEC Articles 460, Part II and 490
Conductor ampacity	The ampacity of capacitor circuit conductors must not be less than 135% of the rated current of the capacitor.	NEC Section 460.8(A)
Capacitors on motor circuits	The ampacity of conductors that connect a capacitor to the terminals of a motor or to motor circuit conductors shall not be less than one-third the ampacity of the motor circuit conductors and in no case less than 135% of the rated current of the capacitor.	NEC Section 460.8(A)
Overcurrent protection	Overcurrent protection is required in each ungrounded conductor unless the capacitor is connected on the load side of a motor running overcurrent device. The setting must be as low as practicable.	NEC Section 460.8(B)
Disconnecting means	A disconnecting means is required for a capacitor unless it is connected to the load side of a motor controller. The rating must be not less than 135% of the rated current of the capacitor.	NEC Section 460.8(C)
Overcurrent protection for improved power factor	If the power factor is improved, the motor running overcurrent device must be selected based on the reduced current draw, not the full-load current of the motor.	NEC Section 460.9
Grounding	Capacitor cases must be grounded except when the system is designed to operate at other than ground potential.	NEC Section 460.10

load side of the motor running overcurrent device. In this case, the motor disconnecting means would serve to disconnect the capacitor and the motor.

A capacitor connected to a motor circuit serves to increase the power factor and reduce the total kVA required by the motor capacitor circuit. As stated earlier, the power factor (pf) is defined as the true power in kilowatts divided by the total kVA, or:

$$pf = \frac{kW}{kVA}$$

A power factor of less than one represents a lagging current for motors and inductive devices. The capacitor introduces a leading current that reduces the total kVA and raises the power factor to a value closer to unity (one). If the inductive load of the motor is completely balanced by the capacitor, a maximum power factor of unity results, and all of the input energy serves to perform useful work.

The capacitor circuit conductors for a power factor correction capacitor must have an ampacity of not less than 135% of the rated current of the capacitor. In addition, the ampacity must not be less than one-third the ampacity of the motor circuit conductors.

The connection of a capacitor reduces the current in the feeder up to the point of connection. If the capacitor is connected on the load side of the motor running overcurrent device, the current through this device is reduced, and its rating must be based on the actual current, not on the full-load current of the motor.

10.2.0 Resistors and Reactors

NEC Article 470 covers the installation of separate resistors and reactors on electric circuits. However, this article does not cover such devices that are component parts of other machines and equipment.

In general, *NEC Section 470.2* requires resistors and reactors to be installed where they will not be exposed to physical damage. Therefore, such devices are normally installed in a protective enclosure, such as a controller housing or other type of cabinet. When these enclosures are constructed of metal, they must be grounded as specified in *NEC Article 250*. Furthermore, a thermal barrier must be provided between resistors and/or reactors and any combustible material that is less than 12" away. A space of 12" or more between the devices and combustible material is considered a sufficient distance so as not to require a thermal barrier.

Insulated conductors used for connections between resistors and motor controllers must be rated at not less than 90°C (194°F) except for motor starting service. In this case, other conductor insulation is permitted, provided other sections of the *NEC®* are not violated.

10.3.0 Diodes and Rectifiers

The diode and the rectifier are the simplest form of electronic components. The major difference between the two is their current rating. A component that is rated less than 1A is called a diode; a similar component rated above 1A is called a rectifier. The main purpose of either device is to convert or rectify alternating current to direct current.

INSIDE TRACK

Types N and P Diodes or Rectifiers

Type N semiconductor diodes or rectifiers are called donor diodes because they have an abundance of free electrons and will give away electrons. Type P semiconductor diodes or rectifiers are referred to as acceptor diodes because they have a shortage of free electrons and will receive electrons. This helps you remember which is which; the word donor has the letter *n* in it, while the word acceptor has the letter *p* in it.

Diodes and rectifiers are composed of two types of semiconductor materials and are classified as either P or N types. One has free electrons; the other has a shortage of electrons. When the two types of material are bonded together, a solid-state component is produced that will allow electrons to flow in one direction and act as an insulator when the voltage is reversed.

Diodes and rectifiers are used extensively in control circuits. For example, *Figure 31* shows an AC voltage supplying a control transformer that must supply a DC electronic controller. Consequently, two rectifiers are installed on the secondary side of the transformer to change AC to DC. The rectifiers allow current to flow in one direction but will not allow the normal alternating current reversal, simulating direct current.

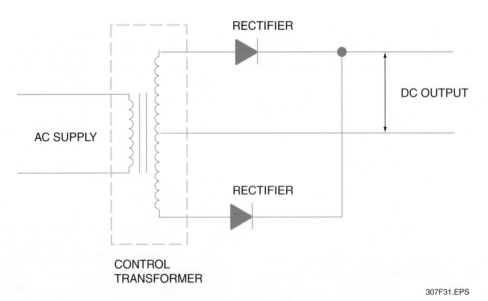

307F31.EPS

Figure 31 ◆ Rectifiers used in a control circuit to change AC to DC.

11.0.0 ◆ VECTORS

The theoretical study of transformers includes the use of phasor diagrams, or vectors, that graphically represent voltages and currents in transformer windings.

A vector or phasor diagram is a line with direction and length. Reading a vector diagram is like reading a road map and is not much more difficult, just a bit more refined. For example, road directions that instruct you to go east 40 miles and then south 30 miles to get to your destination are simple to understand. In other words, you had to drive 70 miles to get there. However, had you been able to drive as the crow flies, the distance would have been shorter—only 50 miles. See *Figure 32(A)*. If the miles were converted into electrical terms, this same triangle would be the classic 3/4/5 triangle or the 80% power factor relationship, as shown in *Figure 32(B)*.

Referring to *Figure 32(B)*, if the working current (line A-B) is 4A and the inductive current (line B-C) is 3A, then the diagonal line A-C, or hypotenuse, will be 5A. This may be obtained mathematically using the Pythagorean theorem, which states that the hypotenuse is equal to the square root of the sum of the squares of the other two sides:

$$AC = \sqrt{AB^2 + BC^2}$$
$$AC = \sqrt{4^2 + 3^2} = \sqrt{16 + 9} = \sqrt{25}$$
$$AC = 5$$

In this example, the working current is related to kilowatts (kW), and the total current is related to kVA. The ratio between the two is $4 \div 5 = 0.80$. Since the power factor equals $kW \div kVA$, the power factor of this circuit is 80%.

11.1.0 Practical Applications of Phasor Diagrams

When three single-phase transformers are used as a three-phase bank, the direction of the voltage in each of the six phase windings may be represented by a voltage phasor. A voltage phasor diagram of the six voltages involved provides a convenient way to study the relative direction and amounts of the primary and secondary voltages.

The same is true for one three-phase transformer, which also has six voltages to be considered, because its phase windings on the high-voltage and low-voltage sides are connected together in the same way as the phase windings of three single-phase transformers.

To show how a phasor diagram is drawn, consider the three-phase transformer shown in *Figure 33*. Here we have a wye/delta-connected, three-phase transformer with three legs, with each leg carrying a high-voltage and a low-voltage winding. The high-voltage windings are connected in wye (with a common neutral point at N) with the leads to the high-voltage terminals designated H_1, H_2, and H_3. The three low-voltage windings are connected in delta. The junction points of the three windings serve as low-voltage terminals X_1, X_2, and X_3.

The voltages in the low-voltage windings are assumed to be equal to each other in amount but are displaced from each other by 120 degrees.

When drawing the phasor diagram for the low-voltage windings, phasor X_1X_2 is drawn first in any selected direction and to any convenient scale. The arrowhead indicates the instantaneous direction of the alternating voltage in winding X_1X_2, and the length of the phasor represents the amount of voltage in the winding. The broken lines that extend past the arrowheads represent reference lines for phase angles.

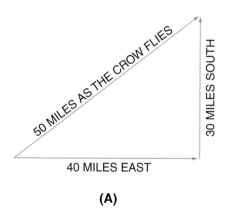

(A)

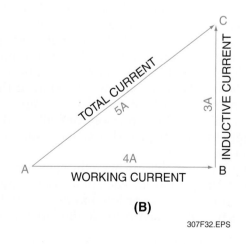

(B)

307F32.EPS

Figure 32 ◆ Typical vectors.

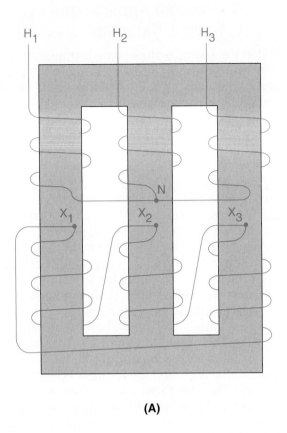

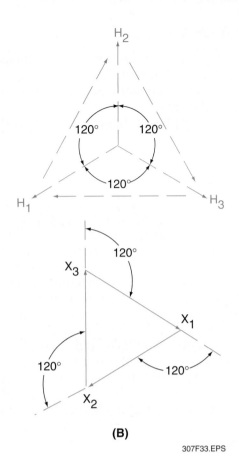

(A)

(B)

307F33.EPS

Figure 33 ◆ Windings and phasor diagrams of a three-phase transformer.

Since winding X_2X_3 is physically connected to the end of X_2 in the X_1X_2 winding, phasor X_2X_3 will be started at point X_2, 120 degrees out of phase with phasor X_1X_2 in the clockwise direction. The length of phasor X_2X_3 is equal to the length of phasor X_1X_2. Winding X_3X_1 is drawn in a similar manner.

The high-voltage phasor comes next. Since the low-voltage winding X_1X_2 is wound on the same leg as the high-voltage winding NH_1, the voltages in these two windings are in phase and are represented by parallel voltage phasors. Therefore, the phasor NH_1 is drawn from a selected point N parallel to X_1X_2. Note that all three high-voltage windings are physically connected to a common point N. Therefore, the phasors representing the voltages NH_1, NH_2, and NH_3 will all start at the common point (point N in the phasor diagram). Again, the high-voltage phasors are all of the same length but are displaced from each other by 120 degrees as shown.

The high-voltage phasors have the same length as the low-voltage phasors. Each phase of the high voltage is, however, proportional to the low voltage in the same phase according to the turns ratio,

making the line voltage 1.732 times higher than the phase voltage in any one winding. It can be geometrically proven, for example, that $H_1H_2 \times NH_1 = 1.732 \times NH_2$ and so forth for each line voltage.

When comparing three-phase transformers for possible operation in parallel, draw the voltage phasor diagram for transformer bank A, and mark the terminals as shown in *Figure 34(A)*. Next, draw the phasor diagram for bank B on drafting or other transparent paper; then draw a reference line m-n as shown in *Figure 34(B)*. Cut the transparent diagram into two parts as indicated by the dashed lines in the drawing. The two parts of the reference line are now marked m and n. Place diagram m on the high-voltage diagram in *Figure 34(A)* so that the terminals that are desired to be connected together coincide. Place diagram n on the low-voltage diagram in *Figure 34(A)* so that the reference line of n is parallel to the reference line of m. If the terminals of n can be made to coincide with the low-voltage terminals, the terminals that coincide can be connected together for parallel operation. If the low-voltage terminals cannot be made to coincide, parallel operation is not possible with the assumed high-voltage connection.

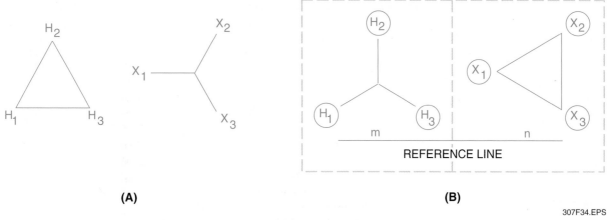

(A)   **(B)**

307F34.EPS

Figure 34 ◆ Phasor diagrams for three-phase transformer banks in parallel.

11.2.0 Voltage Drop

Figure 35 shows a simple 100% power factor circuit in which a 1Ω resistance appears between the source of power and the load. The current in this series circuit is 10A, and the voltage at the source is 120V. Because Ohm's law states that E = IR, we must have a voltage drop across the resistance equal to 10 times 1, or 10V. With the voltage and current in phase, only 110V will be available at the load because we must subtract the resistance drop from the source voltage. Voltage drop was covered more thoroughly earlier in your training.

Voltage drop, however, becomes a little more complicated when inductance is introduced into the circuit, as shown in *Figure 36*. In this case, the load voltage will be equal to the source voltage minus the voltage drops through R and X, but they cannot be added arithmetically. The vector

diagram in *Figure 37* shows that the dotted line A-C is the combination of the two voltages across R and X and represents the voltage drop in the line only.

Figure 37 shows that line G-A, which is the load voltage E_L, is less than the source voltage G-C due to the voltage drop in line A-C. Because of the effect of the reactance in the line, this voltage drop cannot be subtracted arithmetically from the source voltage to obtain the load voltage; vectors must be used.

Calculations of impedance can be simplified by using equivalent circuits. The reactance voltage drop is governed by the leakage flux, and the voltage regulation depends on the power factor of the load. To determine transformer efficiency at various loads, it is necessary to first calculate the core loss, hysteresis loss, eddy current loss, and load loss.

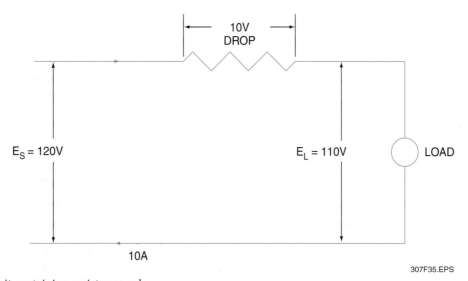

307F35.EPS

Figure 35 ◆ Circuit containing resistance only.

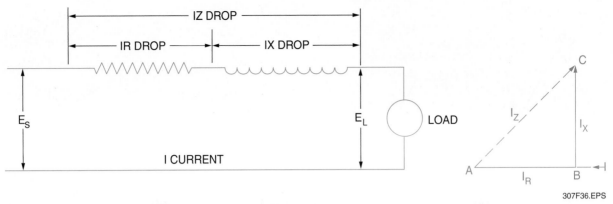

Figure 36 ◆ Electrical circuit with both resistance and inductance.

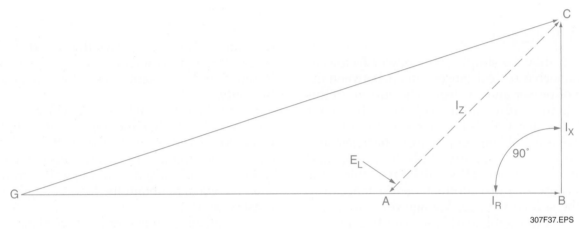

Figure 37 ◆ Effect of voltage drop in an AC circuit.

However, for all practical purposes in electrical construction applications, a transformer's efficiency may be considered to be 100%. Therefore, for our purposes, a transformer may be defined as a device that transfers power from its primary circuit to the secondary circuit without any significant loss.

Since apparent power (VA) equals voltage (E) times current (I), if E_PI_P represents the primary apparent power and E_SI_S represents the secondary apparent power, then $E_PI_P = E_SI_S$ (the subscript

P = primary, and the subscript S = secondary). See *Figure 38*. If the primary and secondary voltages are equal, the primary and secondary currents must also be equal. Assume that E_P is twice as large as E_S. For E_PI_P to equal E_SI_S, I_P must be one-half of I_S. Therefore, a transformer that steps voltage down always steps current up. Conversely, a transformer that steps voltage up always steps current down. However, transformers are classified as step-up or step-down only in relation to their effect on voltage.

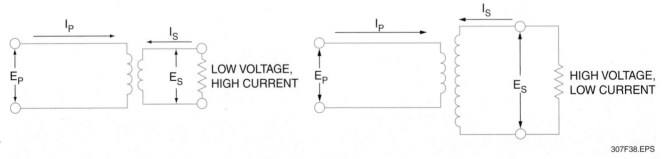

Figure 38 ◆ Voltage-current relationship in a transformer.

12.0.0 ◆ TROUBLESHOOTING

Since transformers are an essential part of every electrical installation, electricians must know how to test and locate problems that develop in transformers, especially in the smaller power supply or control transformers. The procedure for accomplishing this is commonly known as troubleshooting.

The term troubleshooting, as used in this module, covers the investigation, analysis, and corrective action required to eliminate faults in electrical systems, including circuits, components, and equipment. Most problems are simple and easily corrected, such as an open circuit, a ground fault or short circuit, or a change in resistance.

There are many useful troubleshooting charts available. Most charts list the complaint on the left side of the chart, then the possible cause, followed by the proper corrective action.

When troubleshooting, think before acting, study the problem thoroughly, then ask yourself these questions:

- What were the warning signs preceding the trouble?
- What previous repair and maintenance work has been done?
- Has similar trouble occurred before?
- If the circuit, component, or piece of equipment still operates, is it safe to continue operation before further testing?

The answers to these questions can usually be obtained by:

- Questioning the owner of the equipment
- Taking time to think the problem through
- Looking for additional symptoms
- Consulting troubleshooting charts
- Checking the simplest things first
- Referring to repair and maintenance records
- Checking with calibrated instruments
- Double-checking all conclusions before beginning any repair

NOTE

Always check the easiest and most obvious things first; following this simple rule will save time and trouble.

12.1.0 Double-Check Before Beginning

The source of many problems can be traced not to one part alone but to the relationship of one part with another. For instance, a tripped circuit breaker may be reset to restart a piece of equipment, but what caused the breaker to trip in the first place? It could have been caused by a vibrating hot conductor momentarily coming into contact with a ground, a loose connection, or any number of other causes.

Too often, electrically operated equipment is completely disassembled in search of the cause of a certain complaint and all evidence is destroyed during disassembly. Check again to be certain an easy solution to the problem has not been overlooked.

12.2.0 Find and Correct the Basic Cause of the Trouble

After an electrical failure has been corrected in any type of electrical circuit or piece of equipment, be sure to locate and correct the cause so the same failure will not be repeated. Further investigation may reveal other faulty components.

Also be aware that although troubleshooting charts and procedures greatly help in diagnosing malfunctions, they can never be complete. There are too many variations and solutions for any given problem.

To solve electrical problems consistently, you must first understand the basic parts of electrical circuits, how they function, and for what purpose. If you know that a particular part is not performing its job, then the cause of the malfunction must be within this part or series of parts.

12.3.0 Common Transformer Problems

Common transformer problems include open circuits, shorted turns, complete shorts, and grounded windings.

12.3.1 Open Circuit

Should one of the windings in a transformer develop a break or open condition, no current can flow, and therefore, the transformer will not deliver any output. The main symptom of an open circuit in a transformer is that the circuits that derive power from the transformer are de-energized or dead. Use an AC voltmeter or a volt-ohm-milliammeter (VOM) to check across the transformer output terminals, as shown in *Figure 39*. A reading of zero volts indicates an open circuit.

Next, take a voltage reading across the input terminals. If a voltage reading is present, then the conclusion is that one of the windings in the transformer is open. However, if no voltage reading is on the input terminals either, then the conclusion is that the open is elsewhere on the line side of the circuit; perhaps a disconnect switch is open.

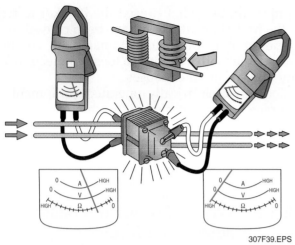

Figure 39 ◆ Checking a transformer for an open winding using a VOM to measure voltage.

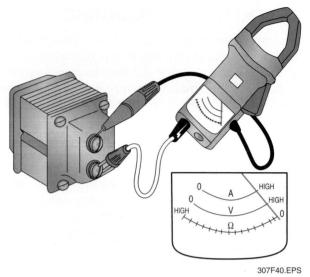

Figure 40 ◆ Checking for an open winding with a continuity test using a VOM.

((⦿)) **WARNING!**

Make absolutely certain that your testing instruments are designed for the job and are calibrated for the correct voltage. Never test the primary side of any transformer over 600V unless you are qualified and have the correct high-voltage testing instruments and the test is made under the proper supervision. Always set the range selector switch on the test equipment to the highest setting, and decrease it in increments until the most accurate reading is indicated.

If voltage is present on the line or primary side and is not present on the secondary or load side, open the switch to de-energize the circuit, and place a warning tag (tagout and lock) on this switch so that it is not inadvertently closed again while someone is working on the circuit. Disconnect all of the transformer primary and secondary leads; then check each winding in the transformer for continuity, as indicated by a resistance reading taken with an ohmmeter (*Figure 40*).

Continuity is indicated by a relatively low resistance reading on control transformers; an open winding will be indicated by an infinite resistance reading on the ohmmeter. In most cases, small transformers will have to be replaced unless the break is accessible and can be repaired.

12.3.2 Shorted Turns

Sometimes a few turns in the secondary winding of a transformer will acquire a partial short, which in turn will cause a voltage drop across the secondary. The symptom of this condition is usually

overheating of the transformer caused by large circulating currents flowing in the shorted windings. The most accurate way to check for this condition is with a transformer turns ratio tester (TTR). However, another way to check for this condition is with a VOM set at the proper voltage scale (*Figure 41*). Take a reading on the line or primary side of the transformer first to make certain normal voltage is present; then take a reading on the secondary side. If the transformer has a partial short or ground fault, the voltage reading should be lower than normal.

Replace the faulty transformer with a new one, and again take a reading on the secondary. If the voltage reading is now normal and the circuit

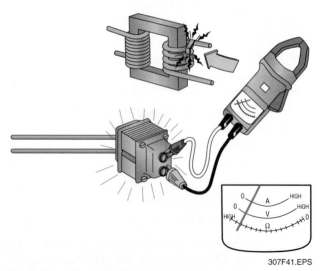

Figure 41 ◆ Transformers that overheat usually have a partial short in the windings.

Open and Unbonded Neutrals

THINK ABOUT IT

What is the difference between an open neutral and an unbonded neutral?

operates satisfactorily, leave the replacement transformer in the circuit, and either discard or repair the original transformer.

A highly sensitive ohmmeter may also be used to test for this condition when the system is de-energized and the leads are disconnected; a lower than normal reading on the ohmmeter indicates this condition. However, the difference will usually be so slight that the average ohmmeter is not sensitive enough to detect it. Therefore, the recommended method is to use the voltmeter test.

12.3.3 Complete Short

Occasionally, a transformer winding will become completely shorted. In most cases, this will activate the overload protective device and de-energize the circuit, but in other instances, the transformer may continue trying to operate with excessive overheating due to the very large circulating current. This heat will often melt the wax or insulation inside the transformer, which is easily detected by the odor. Also, there will be no voltage output across the shorted winding, and the circuit across the winding will be dead.

The short may be in the external secondary circuit, or it may be in the transformer's winding. To determine its location, disconnect the external secondary circuit from the winding, and take a reading with a voltmeter. If the voltage is normal with the external circuit disconnected, then the problem lies within the external circuit. However, if the voltage reading is still zero across the secondary leads, the transformer is shorted and will have to be replaced.

12.3.4 Grounded Windings

Insulation breakdown is quite common in older transformers, especially those that have been overloaded. At some point, the insulation breaks or deteriorates, and the wire becomes exposed. The exposed wire often comes into contact with the transformer housing and grounds the winding.

If a winding develops a ground and a point in the external circuit connected to this winding is also grounded, part of the winding will be shorted out. The symptoms will be overheating, which is usually detected by heat or a burning smell and a low resistance reading from the secondary winding to the core (case), as indicated on the VOM in *Figure 42*. In most cases, transformers with this condition will have to be replaced.

A megohmmeter (megger) is the best test instrument to check for this condition. Disconnect the leads from both the primary and secondary windings. Tests can then be performed on either winding by connecting the megger negative test lead to an associated ground and the positive test lead to the winding to be measured.

 WARNING!
Do not use a megger unless you are qualified and properly supervised.

The insulation resistance should then be measured between the windings themselves. This is accomplished by connecting one test lead to the primary and the second test lead to the secondary

Testing External Circuits

INSIDE TRACK

Always retest the external circuits supplied by the transformer before energizing a replacement transformer. This is done to avoid damaging or destroying the replacement due to a shorted external circuit, which may have caused the original transformer to burn out.

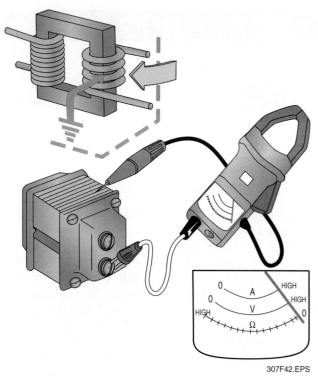

Figure 42 ◆ Testing a transformer for a ground fault by measuring resistance using a VOM.

and ground. After that, connect the first lead to the secondary and the second test lead to the primary and ground. All such tests should be noted on a record card under the proper identifying labels.

13.0.0 ◆ TRANSFORMER MAINTENANCE

Regular transformer maintenance can help to reduce energy use and equipment failure. The following sections identify general transformer maintenance tasks. Always refer to the manufacturer's instructions and policies in your facility for specific service and maintenance procedures.

 WARNING!

Transformer maintenance may only be performed by qualified individuals operating under the appropriate safe work plan or permit. De-energize and lock out the transformer circuits and conductors before performing any cleaning procedures. Never use solvents or detergents as they may damage components or leave a coating that attracts dust or other contaminants.

Dry-Type Transformers

• The buildup of dust on cooling surfaces will cause increased heating under load. Regularly remove dust using a vacuum cleaner or other manufacturer-approved method.
• When access panels are removed for cleaning, check the internal components for signs of discoloration or heat damage.
• Check for loose or damaged connections and bracing. Inspect for physical damage, broken insulation, tightness of connections, defective wiring, and general mechanical and electrical conditions.
• For dry-type transformers above 600V, check for cracked insulators.
• Check and clean cooling surfaces.
• Check for proper clearances to ensure adequate ventilation. If the transformer is in a vault, check for proper operation of the ventilation system.
• Dry-type sealed transformers use an inert gas blanket under positive pressure to prevent condensation. They are equipped with pressure gauges that should be monitored and the results plotted over time.

 CAUTION

If a transformer is to remain de-energized for an extended period in a humid atmosphere, it may be necessary to use electric space heaters to prevent condensation and moisture buildup.

Liquid-Filled Transformers

• Inspect for physical damage, cracked bushings, leaks, loose connections, and general mechanical and electrical conditions. Dust or grime may cake on exposed high-voltage bushings, causing surface tracking and reduced insulation values.
• Verify the proper operation of auxiliary devices such as fans and indicators. Liquid-filled transformers have indicating gauges to monitor temperature, pressure, and liquid level. These gauges must be checked regularly and the results plotted over time. Pressures and levels will vary according to ambient conditions and transformer loading. Manufacturer guidelines and user experience provide acceptable ranges according to conditions.

- Samples of insulating oil or liquid are drawn for visual inspection and to test the dielectric strength of the fluid. Many facilities also do detailed analysis of oil samples over time. Gases are generated from electrical corona, thermal decomposition of insulating material and oil, and electrical arcing. Trend analysis of dissolved gas in oil and in the gas blanket above the oil can indicate problems and a reduced life cycle.

CAUTION

In order to prevent contamination and get a representative sample for testing, the sample must be taken using strict procedures and an approved collection vessel. Testing laboratories can supply specific instructions.

INSIDE TRACK

Generator Step-Up (GSU) Transformers

This GSU steps up the generating system voltage to a transmission voltage of 230kV. It has a rated load capacity of 185MVA, which equates to 185,000,000 watts. This installation includes a deluge system for fire protection, a huge containment basin to protect the surrounding area from oil leaks, and a reinforced blast wall to shield adjacent equipment in the event of an explosion.

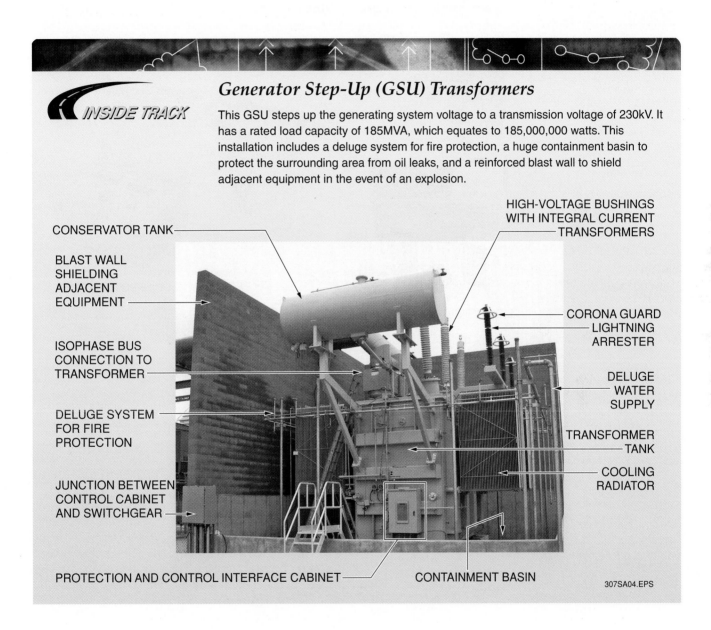

CONSERVATOR TANK

BLAST WALL SHIELDING ADJACENT EQUIPMENT

ISOPHASE BUS CONNECTION TO TRANSFORMER

DELUGE SYSTEM FOR FIRE PROTECTION

JUNCTION BETWEEN CONTROL CABINET AND SWITCHGEAR

PROTECTION AND CONTROL INTERFACE CABINET

HIGH-VOLTAGE BUSHINGS WITH INTEGRAL CURRENT TRANSFORMERS

CORONA GUARD LIGHTNING ARRESTER

DELUGE WATER SUPPLY

TRANSFORMER TANK

COOLING RADIATOR

CONTAINMENT BASIN

307SA04.EPS

Maintenance Testing of Transformers

In commercial and industrial applications involving a number of large liquid-filled or dry transformers that are not automatically monitored, manual periodic preventive maintenance testing and performance trending is normally instituted in a manner similar to that for large motors. These tests include liquid dielectric tests to measure the breakdown voltage of samples of insulating fluids from transformers and DC hi-pot tests of transformer winding insulation. Some DC hi-pot testers are also available with a test pot for high-voltage components. The test sets shown here are capable of performing tests ranging up to 160kV.

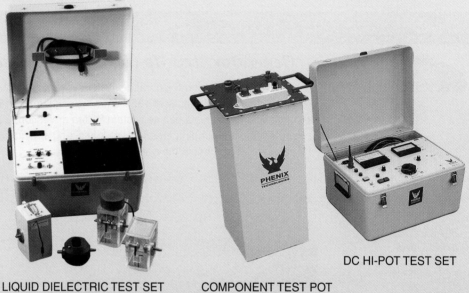

LIQUID DIELECTRIC TEST SET   COMPONENT TEST POT   DC HI-POT TEST SET

307SA05.EPS

1. The main purpose of a transformer is to _____.
   a. change the current
   b. improve the power factor
   c. change the output voltage
   d. enter impedance into the circuit

2. The three parts of a basic transformer are the _____.
   a. housing, lifting hooks, and base
   b. dry, oil-filled, and gas-filled chambers
   c. shell, open winding, and closed winding
   d. primary winding, secondary winding, and core

3. When AC flows through a transformer coil, a _____ field is generated around the coils.
   a. magnetic
   b. non-magnetic
   c. high-impedance
   d. rotating

4. When the field from one coil cuts through the turns of a second coil, _____.
   a. the second coil will not be affected
   b. the coils will rotate
   c. voltage will be generated
   d. no current will flow in the circuit

5. What causes voltage to be induced in a transformer?
   a. Transformer taps
   b. Mutual induction
   c. Reluctance
   d. Capacitance

6. In a transformer with a turns ratio of 5:1 (the primary has five times the number of turns as the secondary), what will be the voltage on the secondary if the primary voltage is 120V?
   a. 12V
   b. 24V
   c. 48V
   d. 60V

7. The three basic types of iron core transformers are _____.
   a. dry, oil-filled, and auto
   b. closed core, open core, and shell
   c. control, power, and lighting
   d. metal, nonmetallic, and high-temperature

8. One effect caused by magnetic leakage in transformers is a _____.
   a. reactance voltage drop
   b. low impedance
   c. higher secondary voltage
   d. lower secondary voltage

9. If one of the transformers is damaged in a delta system, a(n) _____ configuration may be used to temporarily carry the load.
   a. open delta
   b. center-tapped delta
   c. delta-wye
   d. delta-delta

10. The windings in an autotransformer are not isolated from one another.
    a. True
    b. False

11. Most control transformers are _____.
    a. dry-type, step-up
    b. dry-type, step-down
    c. oil-filled, step-up
    d. oil-filled, step-down

12. A dry-type transformer must be housed in a vault if rated at more than _____.
    a. 20kV
    b. 25kV
    c. 30kV
    d. 35kV

13. The power factor of a circuit can be improved by increasing the _____.
    a. voltage
    b. current
    c. inductance
    d. capacitance

14. A symptom of a transformer with an open circuit is _____.
    a. high voltage on the secondary
    b. excessive overheating
    c. no output on the secondary
    d. voltage drop across the secondary

15. A(n) _____ should be used to measure insulation resistance.
    a. ammeter
    b. megger
    c. voltmeter
    d. TTR

Summary

This module covered the basic components and applications of distribution system transformers. When the AC voltage needed for an application is lower or higher than the voltage available from the source, a transformer is used. The essential components of a transformer are the primary winding, which is connected to the source, and the secondary winding, which is connected to the load. Both are wound on an iron core. The two windings are not physically connected, with the exception of autotransformers. The alternating voltage in the primary winding induces an alternating voltage in the secondary winding. The ratio of the primary and secondary voltages is equal to the ratio of the number of turns in the primary and secondary windings. Transformers may step up the voltage applied to the primary winding and have a higher voltage at the secondary terminals, or they may step down the voltage applied to the primary winding and have a lower voltage available at the secondary terminals.

Notes

Trade Terms
Introduced in This Module

Ampere turn: The product of amperes times the number of turns in a coil.

Autotransformer: Any transformer in which primary and secondary connections are made to a single winding. The application of an auto-transformer is a good choice where a 480Y/277V or 208Y/120V, three-phase, four-wire distribution system is used.

Capacitance: The storage of electricity in a capacitor or the opposition to voltage change. Capacitance is measured in farads (F) or microfarads (μF).

Flux: The rate of energy flow across or through a surface. Also a substance used to promote or facilitate soldering or welding by removing surface oxides.

Induction: The production of magnetization or electrification in a body by the mere proximity of magnetized or electrified bodies, or the production of an electric current in a conductor by the variation of the magnetic field in its vicinity.

Kilovolt-amperes (kVA): 1,000 volt-amperes (VA).

Loss: The power expended without doing useful work.

Magnetic field: The area around a magnet in which the effect of the magnet can be felt.

Magnetic induction: The number of magnetic lines or the magnetic flux per unit of cross-sectional area perpendicular to the direction of the flux.

Mutual induction: The condition of voltage in a second conductor because of current in another conductor.

Power transformer: A transformer that is designed to transfer electrical power from the primary circuit to the secondary circuit(s) to step up the secondary voltage at less current or step down the secondary voltage at more current, with the voltage-current product being constant for either the primary or secondary.

Reactance: The opposition to AC due to capacitance and/or inductance.

Rectifiers: Devices used to change alternating current to direct current.

Transformer: A static device consisting of one or more windings with a magnetic core. Transformers are used for introducing mutual coupling by induction between circuits.

Turn: The basic coil element that forms a single conducting loop comprised of one insulated conductor.

Turns ratio: The ratio between the number of turns between windings in a transformer; normally the primary to the secondary, except for current transformers, in which it is the ratio of the secondary to the primary.

This module is intended to present thorough resources for task training. The following reference work is suggested for further study. This is optional material for continued education rather than for task training.

National Electrical Code® Handbook, Latest Edition. Quincy, MA: National Fire Protection Association.

NCCER makes every effort to keep these textbooks up-to-date and free of technical errors. We appreciate your help in this process. If you have an idea for improving this textbook, or if you find an error, a typographical mistake, or an inaccuracy in NCCER's Contren® textbooks, please write us, using this form or a photocopy. Be sure to include the exact module number, page number, a detailed description, and the correction, if applicable. Your input will be brought to the attention of the Technical Review Committee. Thank you for your assistance.

Instructors – If you found that additional materials were necessary in order to teach this module effectively, please let us know so that we may include them in the Equipment/Materials list in the Annotated Instructor's Guide.

Write: Product Development and Revision
National Center for Construction Education and Research
3600 NW 43rd St., Bldg. G, Gainesville, FL 32606

Fax: 352-334-0932

E-mail: curriculum@nccer.org

Craft _____ Module Name _____

Copyright Date _____ Module Number _____ Page Number(s) _____

Description _____

_____

_____

_____

(Optional) Correction _____

_____

_____

(Optional) Your Name and Address _____

_____

_____

Commercial Electrical Services

Statue of Liberty goes green

The Statue of Liberty celebrated her 120th birthday in 2006 by going 100 percent green. The Statue is "green powered" by windmill generators in West Virginia and Pennsylvania.

26308-08

26308-08
Commercial Electrical Services

Topics to be presented in this module include:

Overview

Electrical services can range in size from a small single circuit service of 15 amps, which is the minimum allowed by *NEC Section 230.79(A)*, to large distribution systems rated at thousands of amps, such as those found in large switchyards and substations. Regardless of the size of the electrical service, its components serve the same purpose. The service provides the connection point from the utility to the premises wiring system.

Objectives

When you have completed this module, you will be able to do the following:

1. Describe various types of electric services for commercial and industrial installations.
2. Read electrical diagrams describing service installations.
3. Select service-entrance equipment for various applications.
4. Explain the role of the *National Electrical Code*® in service installations.
5. Install main disconnect switches, panelboards, and overcurrent protection devices.
6. Identify the *National Electrical Code*® requirements and purposes of service grounding.
7. Describe single-phase service connections.
8. Describe both wye- and delta-connected three-phase services.

Trade Terms

Cold sequence metering
Delta-connected
Service
Service conductors

Service drop
Service entrance
Service equipment
Service lateral

Required Trainee Materials

1. Pencil and paper
2. Appropriate personal protective equipment
3. Copy of the latest edition of the *National Electrical Code*®

Prerequisites

Before you begin this module, it is recommended that you successfully complete *Core Curriculum; Electrical Level One; Electrical Level Two;* and *Electrical Level Three,* Modules 26301-08 through 26307-08.

This course map shows all of the modules in *Electrical Level Three.* The suggested training order begins at the bottom and proceeds up. Skill levels increase as you advance on the course map. The local Training Program Sponsor may adjust the training order.

26311-08
Motor Controls

26310-08
Voice, Data, and Video

26309-08
Motor Calculations

26308-08
Commercial Electrical Services

26307-08
Transformers

26306-08
Distribution Equipment

26305-08
Overcurrent Protection

26304-08
Hazardous Locations

26303-08 Practical
Applications of Lighting

26302-08 Conductor
Selection and Calculations

26301-08 Load Calculations –
Branch and Feeder Circuits

ELECTRICAL LEVEL TWO

ELECTRICAL LEVEL ONE

CORE CURRICULUM:
Introductory Craft Skills

ELECTRICAL LEVEL THREE

308CMAP.EPS

1.0.0 ◆ INTRODUCTION

The electrical contractor is typically responsible for the installation of all electrical systems including the electric service. Service equipment consists of conduit and service conductors, the panelboard, and the overcurrent protective devices. *NEC Article 230* lists the requirements for calculating the size of the service, service-entrance conductors, and raceways, along with numerous additional requirements for service installations.

2.0.0 ◆ DRAWINGS AND SPECIFICATIONS

Electric services are usually designed by qualified electrical engineers who produce drawings and specifications showing the design and installation requirements for the service. The drawings include the floor plan, one-line diagram, site plan, panelboard schedules, and specifications. A service one-line or single-line drawing shows service characteristics including the size and voltage rating, fault current rating, service-entrance conduit and conductor sizes, grounding requirements, transformers, and panelboards. The one-line drawing also shows whether the service is overhead or underground. If the power connects to the electric service overhead, the incoming service supply is referred to as a service drop. If the service is supplied underground, the supply is defined as a service lateral. The one-line for a small installation typically shows the conduit and conductor sizes for the service. Larger service one-line diagrams often include a schedule or table indicating the conduit and conductor sizes for service entrance, distribution, and panelboard feeders.

NOTE

Electricians install the electric service but connections to the utility transformer are typically made by the utility company.

A building layout shows the location of the electric service and the location of distribution and branch circuit panelboards within the structure. It is important to review all of the project drawings and specifications to obtain all of the information needed to install the electric service. For example, the building elevation drawing will show the amount of exterior wall space available at the location where the service equipment might be mounted. It is also important to know the type of construction so that the proper anchoring systems can be ordered for the installation.

3.0.0 ◆ GENERAL INSTALLATION CONSIDERATIONS

A qualified electrician should have the ability to work with the supervisor, the project manager, the utility company, and the electrical engineer to install the service components in a workmanlike, professional, and efficient manner. Safety considerations come first in every installation; this means that the required personal protective equipment must be used to allow the installation to proceed in a safe manner. All equipment must be installed level and plumb and in a manner that meets all *NEC*® requirements and all of the requirements presented in the project electrical drawings and specifications. All equipment and materials must be new and installed per the manufacturer's guidelines. The electrician may be required to install the conduit, service-entrance conductors, wireways, bussed gutter, service disconnects, current transformer (CT) can, meter can, and other fittings as needed to comply with *NEC*® requirements, local building department requirements and codes, project specifications, and even the local fire marshal's requirements.

The location of the bus and ground bars and the service-side entrance compartment are important considerations when installing the service-entrance conductors. The service grounding conductors are usually routed to and bonded in the first service disconnecting means. When service-entrance conductors terminate in a wireway or bussed gutter to serve multiple service disconnecting means, the service grounding conductors may be terminated to and bonded to the grounded service conductor in the wireway or bussed gutter enclosure. In many instances, the local Authority Having Jurisdiction (AHJ) will make the determination regarding specific grounding decisions to ensure that these connections meet *NEC*® requirements. Another important consideration is whether housekeeping pads are required to mount or raise the switchgear or transformers above the finished floor.

In most cases, electricians are not required to design electrical systems except for design-build projects that require a stamped set of drawings and the oversight of an electrical engineer. Usually the electrician is only required to interpret the engineer's drawings and apply them to the installation of the project. Other important factors relating to the installation of service equipment are the inspection and acceptance of service equipment from the delivery service and the megging of the

Receiving Equipment

Freight claims for damage caused during shipping must be made by written notation on the freight bill when the equipment is received at the job site. If freight damage is not noted on the freight bill, a claim for damages may not be possible at a later time. All boxes should be examined for hidden damage prior to uncrating.

conductors and equipment before energizing the equipment.

The following considerations apply to the selection and installation of electric services:

- *Codes and standards* – All services must be designed and installed in accordance with the provisions of the *NEC®* and all local codes. Suitability of service equipment for the intended use is evidenced by listing or labeling.
- *Mechanical protection* – Mechanical strength and durability, including the adequacy of the protection provided, must be considered.
- *Wiring space* – Wire bending and connection space in distribution equipment is provided according to UL standards. When unusual conditions are encountered requiring additional wiring space, larger wireways and/or termination cabinets must be considered.
- *Classification* – Classification according to type, size, voltage rating, current capacity, interrupting rating, and specific use must be considered when selecting service equipment. Loads may be continuous or noncontinuous, and load calculations per *NEC®* requirements must be completed in order to determine the proper sizing of service equipment. All distribution equipment carries labels showing the maximum voltage and current ratings for the equipment. A typical example is shown in *Figure 1*. The supply voltage and current ratings must not be exceeded. In addition, always observe the minimum and maximum values listed for conductor terminations.
- *Heating effects* – All installations must consider heating effects both under normal operating conditions and under any extreme conditions of use that might reasonably be expected. Ambient heat conditions must be considered when specifying conductor insulation for the specific application.
- *Arcing effects* – The normal arcing effects of overcurrent protective devices must be considered when the application is in or near combustible materials or vapors (classified/hazardous locations). Enclosures and wiring methods must be selected based on *NEC®* requirements.

308F01.EPS

Figure 1 ◆ Typical maximum voltage and current rating label.

- *Personal protection* – Every service installation should begin with a review of safety policies and a site visit to determine any special safety requirements relating to the specific site and type of construction. Always wear all personal protective equipment required for the installation. Safe work practices are extremely important when installing service equipment that may be heavy and awkward to lift and place. Be aware of potential pinch points as well as all other work site conditions that may present specific safety hazards. Safety should always be a top priority when installing electric services.

4.0.0 ◆ SERVICE COMPONENTS

All service components must be carefully selected and coordinated to ensure proper system operation. Common service components include the service disconnect, meter, transformers, wireways, gutters, weatherhead and service mast, and panelboards.

4.1.0 Service Disconnecting Means

Whether a service is single-phase or three-phase, overhead or underground, *NEC Article 230,*

Dual Services

Some large installations have two or more services. Dual services are provided with separate disconnects, as shown here.

308SA01.EPS

Part VI requires that a means be provided to disconnect all conductors in a building from the service-entrance conductors. The service disconnecting means must be in a readily accessible location, marked as a service disconnect, and listed as suitable for use as a service disconnecting means. The service disconnect can be a fusible switch or a circuit breaker and is typically located at the point where the service-entrance conductors come into the building or structure. In some cases, the utility company requires cold sequence metering, in which the service disconnect is installed ahead of the electric meter. Cold sequence metering provides the utility company the opportunity to work on the metering equipment with the power disconnected.

4.2.0 Metering

The metering assembly, single meter enclosure, multiple meter enclosures, or CT metering is installed at the point of service entrance. A self-contained meter installation refers to a single meter enclosure with a single watt-hour or demand meter. Typically, the electrician is responsible for the meter enclosure installation and the utility company will provide and install the electric meter. A CT meter installation refers to the installation of a CT enclosure or cabinet. The CT cabinet is a listed piece of manufactured equipment designed to accept the installation of current transformers used to measure electric usage. A CT meter enclosure is installed alongside or within a short distance of the CT cabinet. The electrician is usually responsible for the installation of the CT cabinet and the CT meter base. The utility company will then install the current transformers, the CT meter, and the wiring between the CT meter enclosure and the CT cabinet.

Most watt-hour meters use five dials (see *Figure 2*). The dial farthest to the right on the meter counts the kilowatt-hours singly. The second dial from the right counts by tens, the third dial by hundreds, the fourth dial by thousands, and the left-hand dial by ten thousands. The dials may seem a little strange at first, but they are actually

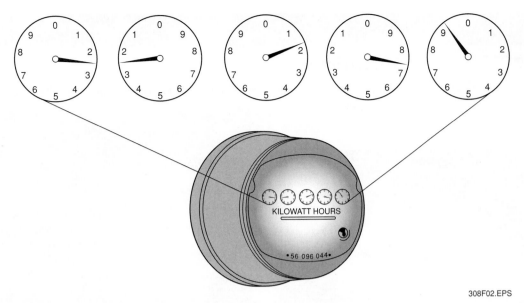

Figure 2 ◆ Typical watt-hour meter.

308F02.EPS

Reading a Watt-Hour Meter

What is the reading on this meter?

308SA02.EPS

very simple to read. The number that the dial has passed is the reading. For example, look at the dial on the far left. Note that the pointer is about halfway between the number 2 and the number 3. Since it has passed the number 2, but has not yet reached 3, the dial reading is 2. The same is true of the second dial from the left; that is, the pointer is between 2 and 3. Consequently, the reading of this dial is also 2. Following this procedure, the complete reading for the meter in *Figure 2* is 22,179 kilowatt-hours.

Although knowing how to read an electric meter is useful, most electricians will be involved only with installing the meter base and making the connections therein. Once these connections are made and inspected, the local power company will install and seal the meter.

4.3.0 Current Transformers

Meters used to record electric usage normally respond to a current level that varies from 0A to 5A. To respond to the actual current, each meter is provided with current transformers. If the peak demand is 100A, a 100:5 current transformer is used. If the peak demand is expected to be 200A, a 200:5 current transformer is used.

Services rated 400A and above usually utilize a group of current transformers, one for each ungrounded conductor or set of conductors. There are two basic types of current transformers: the bus-bar type and the doughnut type. The doughnut-type CT encircles the ungrounded conductors in the system to measure the current flow, much the same as a clamp-on ammeter. See *Figure 3*. The bus-bar type CT has each transformer connected in

Meter Types

Watt-hour meters are available in both analog and digital forms, as shown here.

ANALOG

308SA03A.EPS

DIGITAL

308SA03B.EPS

308F03.EPS

Figure 3 ◆ Clamp-on ammeters operate on the same principle as doughnut-type current transformers.

series with a busbar and does not encircle the conductor.

Current transformers are normally enclosed in a CT cabinet. *Figure 4* shows a typical CT cabinet with current transformers and their related

wiring. In some cases, the current transformers may be mounted exposed on overhead conductors, but this arrangement is more of an exception than the rule.

Utility companies have specifications for the location and wiring of CTs and CT cabinets, depending upon the location. The following are requirements for one utility company. Always check requirements for the local utility.

- The meter base and meter may be located on either the side or top of the current transformer cabinet, or they may be located at a distance, if approved by the utility company and as long as the conduit containing the instrument wiring is run exposed.
- In no case shall more than one set of conductors terminate in the instrument transformer cabinet. Subfeeders and branch circuits are to terminate at the customer's distribution panel. The instrument transformer cabinet shall not be used as a junction box.
- When service-entrance conductors enter or leave the back of the cabinet, the size of the CT cabinet must be increased to provide additional working space.

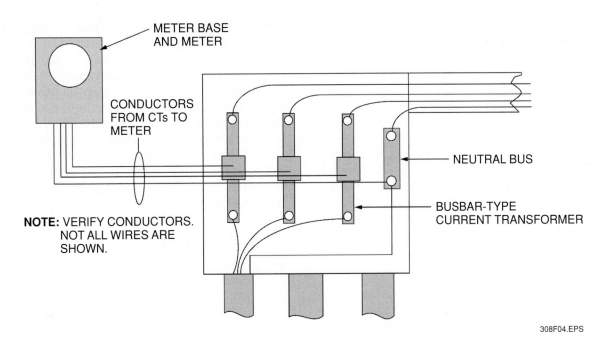

Figure 4 ◆ Typical CT cabinet arrangement.

- For services at higher voltages, additional space must be provided in the transformer cabinet for mounting potential transformers. Consult the local utility company for dimensions.
- If recording demand instruments are required, increase the height of the meter mounting by 12" to accommodate the recording instrument.

4.4.0 Metal Wireways

Metal wireways are defined in *NEC Article 376* as sheet metal troughs with hinged or removable covers for housing electrical wires. As part of a service installation, wireways are used to provide space for splicing service-entrance conductors to conductors supplying individual metering equipment or service disconnects. Follow *NEC Table 376.22* to determine the appropriate size wireway to use in a particular service installation. Wireways installed outdoors must be listed for outdoor use.

4.5.0 Bussed Gutters

NEC Article 368 defines a bussed gutter as a listed piece of equipment consisting of an enclosure and busbars mounted inside the enclosure (*Figure 5*). Bussed gutters are used to provide a means of tapping the service-entrance conductors or distribution feeders to supply the service disconnecting means, self-contained metering equipment, CT metering equipment, or for other purposes such as disconnects for equipment feeders in an industrial plant or manufacturing facility. Bussed gutters are manufactured to meet the project specifications for their intended use. These specifications include

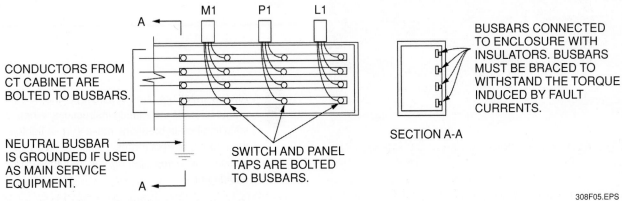

Figure 5 ◆ Three-phase bussed gutter.

the voltage, ampacity, and fault current ratings, along with the NEMA enclosure classification (e.g., NEMA 3R for outdoor use) and the gutter length. Bussed gutters can be manufactured with copper bus or plated aluminum bus. It may include factory-installed mechanical lugs of specific sizes to accommodate line and load feeders. Bussed gutters may be bottom fed or top fed.

Bussed gutters are one of the most common types of wiring methods for use with multi-switch service installations. One advantage of a bussed gutter is the ease of installation and modification. Adding disconnect switches or changing switches is relatively easy.

4.5.1 Bus Bracing

One characteristic of fault currents is an induced torque in conductors carrying the fault. Because of this torque, the busbars must be mounted in the enclosure in such a manner as to allow them to withstand the fault current that may be imposed. This is accomplished by the use of bracing. *Figure 6* shows typical busbar bracing. Larger available fault current values require more substantial bracing and braces that are placed closer together. For example, bussed gutters may be manufactured with an AIC rating of 20,000 AIC, 30,000 AIC, or 200,000 AIC. Manufacturers typically have standard ratings to accommodate project requirements. The listing for bussed gutters must include the fault current rating.

4.6.0 Weatherhead and Service Mast

Overhead services include a weatherhead and a service mast. Weatherheads may be cast aluminum,

galvanized steel, or PVC. Service masts are typically constructed of galvanized steel and supported as required by the *NEC®* and the utility company. The service mast terminates in the hub of the meter housing, CT cabinet, or other service equipment and must use watertight fittings when the components are installed outdoors. The service-entrance riser conductors terminate in the meter housing or the service disconnecting means if the service disconnect is required before the meter (cold sequence metering) or after the meter (hot sequence metering).

4.7.0 Panelboards

The final component in a commercial electric service is the panelboard. Panelboards are designed to distribute electric energy through branch circuit breakers. In a small commercial service, a panelboard may include the service disconnecting means together with all of the branch circuit breakers needed for the building. Panelboards may also be designated as distribution panelboards. Distribution panels are used to provide overcurrent protection for individual panel feeders when the service is large enough to require numerous branch circuit panelboards. The selection of the components used in a panelboard is based on the following criteria:

- Feeder voltage and amperage
- Phase (single or three-phase) and number of conductors
- Type of busbar material (copper or aluminum)
- Number of required spaces or blanks
- Whether the panelboard is configured as a main breaker type of panel or main-lugs only (MLO) type of panel

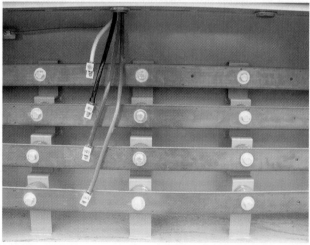

308F06.EPS

Figure 6 ◆ Typical busbar bracing.

NOTE
MLO panels do not have a MAIN circuit breaker.

CAUTION
Always install breakers according to the manufacturer's installation instructions. When installing bolt-in breakers, make sure to tighten them to the correct torque. Failure to do so could result in improper operation of the service equipment.

Panelboards

After the panelboard is assembled, the appropriate number and type of circuit breakers must be ordered separately and installed in the panel.

308SA04.EPS

5.0.0 ◆ *NEC*® REQUIREMENTS

NEC Section 230.2 states that a building shall be supplied by only one service except when special conditions listed under *NEC Section 230.2(A) through (D)* are present. Per *NEC Section 230.2(A)*, an additional service is permitted to supply a fire pump, emergency systems, legally required or optional standby systems, parallel power production systems, or systems connected to multiple supply sources. Additional services are permitted for sufficiently large buildings or certain multiple occupancy buildings per *NEC Section 230.2(B)*, capacity requirements as outlined in *NEC Section 230.2(C)*, and for different characteristics as identified in *NEC Section 230.2(D)*. *NEC Section 230.66* requires service equipment to be marked as suitable for its intended use. Distribution panelboards and switchgear assemblies are generally intended to carry and control electric current, but are not intended to dissipate or use energy.

In accordance with *NEC Section 230.82(3)*, a service disconnect switch connected to the supply side of the meter (cold sequence) must be rated at no more than 600V. In addition, it must have a short-circuit rating equal to or greater than the available short-circuit current, provided the housings and service enclosures are grounded and bonded in accordance with *NEC Article 250, Parts V and VII*. The switch must be capable of interrupting the load served.

5.1.0 Services Passing Through Buildings

Per *NEC Section 230.3*, service conductors supplying one building cannot pass through the interior of another building or structure. *NEC Section 230.6* states that conductors are considered to be outside of a building if any of the following conditions are met:

- Where installed beneath the building under at least 2" of concrete
- Where buried under at least 18" of earth in conduit
- Located inside the building if encased in at least 2" of concrete or brick
- Installed in a vault that meets the requirements of *NEC Article 450, Part III*

5.2.0 Grounding and Bonding

NEC Article 250 covers the requirements for grounding and bonding electric services. In general, the *NEC®* requires that a premises AC service be grounded with a grounding electrode conductor connected to a grounding electrode. It follows that all of the requirements for grounding and bonding identified in *NEC Figure 250.1 and Table 250.3* must be considered and applied to every electric service.

The grounding electrode conductor must be grounded to the grounded service conductor (neutral) at any accessible point from the load end of the service drop or service lateral up to and including the terminal bus to which the grounded service conductor is connected at the service disconnecting means. A grounding connection must not be made to any grounded circuit conductor on the load side of the service disconnecting means.

NEC Section 250.50 requires all grounding electrodes to be bonded. *NEC Section 250.52* lists the grounding electrodes as metal underground water pipe, the metal frame of the building or

structure, concrete-encased electrode, ground ring, rod and pipe electrodes, other listed electrodes, and plate electrodes. These requirements apply to all electric services, regardless of the voltage rating of the service or the size (ampacity rating) of the service. *NEC Section 250.53* guides the installation of the grounding electrode system. *Table 1* gives the required sizes of the grounding electrode conductor for electric services.

5.3.0 High-Leg Marking

The conductor or busbar with a higher voltage to ground in a **delta-connected** four-wire system, also known as the high leg, is required to be durably and permanently marked with an outer finish that is orange in color per *NEC Section 110.15*. If the grounded conductor is also present, this marking must appear at each point where a connection is made. In panelboards, the phase having the higher voltage to ground is the B phase.

5.4.0 Clearances

NEC Section 230.24 lists the minimum service drop clearances for various applications. Service drop conductors require a vertical clearance of no less than 8' above roof surfaces per *NEC Section 230.24(A)* and no less than 10' at the electrical service entrance to buildings, at the lowest point of the drip loop, and over sidewalks accessible to pedestrians only per *NEC Section 230.24(B)(1)*. Note that this is the minimum for the lowest point of the drip loop and the weatherhead or tie point for the service drop should be a minimum of 13'-6" above grade. This accounts for the height of the weatherhead itself, the insulator that attaches

Table 1   Grounding Electrode Conductors for AC Systems

Size of Largest Service-Entrance Conductor or Equivalent for Parallel Conductors		Size of Grounding Electrode Conductor	
Copper	**Aluminum or Copper-Clad Aluminum**	**Copper**	**Aluminum or Copper-Clad Aluminum**
2 or smaller	1/0 or smaller	8	6
1 or 1/0	2/0 or 3/0	6	4
2/0 or 3/0	4/0 or 250 kcmil	4	2
Over 3/0 through 350 kcmil	Over 250 kcmil through 500 kcmil	2	1/0
Over 350 kcmil through 600 kcmil	Over 500 kcmil through 900 kcmil	1/0	3/0
Over 600 kcmil through 1,100 kcmil	Over 900 kcmil through 1,750 kcmil	2/0	4/0
Over 1,100 kcmil	Over 1,750 kcmil	3/0	250 kcmil

to the mast below the weatherhead, the drip loop, and the conductor sag (lowest point) in a service drop, which varies by conductor size and length of drop. The insulator, whether attached to the mast or to the building, is the point of attachment for the service drop. If the service mast is to support the service drop, it must be well secured per *NEC Section 230.28*. Meter bases should be installed with anchors to hold the weight of the meter as well as the raceway system resting on the meter base.

Service conductor clearances on buildings, porches, and platforms are covered in *NEC Section 230.9*. They must be located at least 3' from windows, doors, porches, balconies, ladders, stairs, fire escapes, and similar locations. These clearances do not apply to windows that are not designed to be opened, or to locations where the conductor is run above the top level of the window [*NEC Section 230.9(A) Exception*].

6.0.0 ◆ TYPICAL INSTALLATIONS

The most common service size for small commercial and other small loads is a 120/240V, single-phase, three-wire service. Common services for larger commercial installations are 120/240V, three-phase, four-wire; 120/208V, three-phase, four-wire; and 277/480V, three-phase, four-wire installations. To understand the function of each component of an electrical service, we will examine five specific installations of commercial electric services. These include a single-phase service, a small service utilizing a wireway to connect multiple meters and service disconnects, a bussed gutter installation, a service entrance using switchgear, and a multi-family service.

6.1.0 208Y/120V Overhead Service

Figure 7 shows an overhead 208Y/120V service. The service is rated at 400A and the grounding requirements are established based on a 400A service. The two service-entrance conductors consist of No. 3/0 THWN copper in 2" conduit. Because these are parallel service-entrance conductors, the GEC is sized at No. 2 copper (or 1/0 aluminum) using *NEC Table 250.66*. Note that there are drip loops formed at the weatherheads as required by *NEC Section 230.24(B)(1)*. The parallel service-entrance conductors terminate in a wireway. The service disconnects are connected to the wireway with rigid galvanized steel nipples and the conductors supplying each meter are terminated in the wireway. The terminations in the wireway may be mechanical (split-bolt) type connections or they may be mechanical connections made using a

Figure 7 ◆ Overhead service.

308F07.EPS

multi-tap insulated connector. In this example, the GEC is bonded to the grounded service conductor (neutral) in the wireway. There are two 200A, 240V three-phase meters and service disconnects, and one 100A, 240V single-phase meter and disconnect. Because the building has a 208Y/120V three-phase service, the voltage to the single-phase meters is also 208V. Note the use of weatherheads, drip loops, conduit supports, and the arrangement of the service equipment. This service illustrates hot sequence metering.

Because this equipment is installed on the exterior of the building, it is rated for outdoor use. In this installation, the point of attachment is bolted through the masonry wall to make it secure enough to support the service drop (also taking into account the wind load and ice weight) and includes a porcelain insulator. Porcelain insulator attachments are preferable and may be required by the local utility. Always check with the local utility before proceeding with any installation. Porcelain insulator attachments provide a separation between the service drop conductors (which may become worn and short over time) and other metal components.

Figure 8 is the one-line diagram for the service just described. Note that the one-line diagram includes sizes for service-entrance conduit and conductors, sizes and fuse requirements for the service disconnects, the size and length of the wireway, and sizes for the grounding electrode conductors.

6.2.0 208Y/120V Three-Phase Underground Service – Wireway

Figure 9 shows an underground 208Y/120V three-phase service. The service is rated at 800A, and the

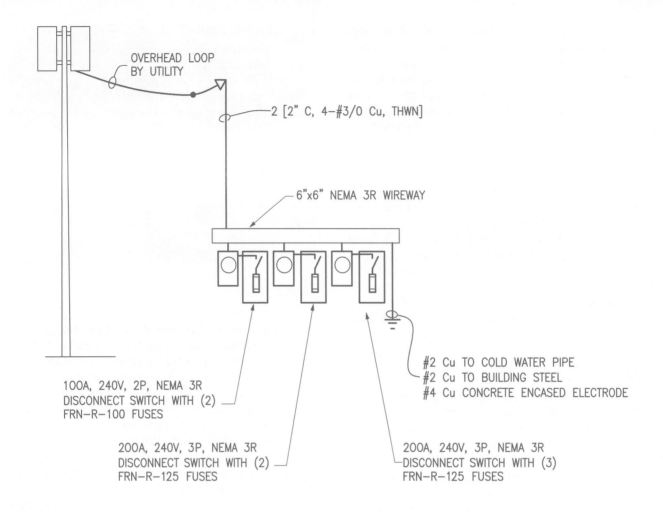

OVERHEAD LOOP
BY UTILITY

2 [2" C, 4-#3/0 Cu, THWN]

6"x6" NEMA 3R WIREWAY

100A, 240V, 2P, NEMA 3R
DISCONNECT SWITCH WITH (2)
FRN-R-100 FUSES

200A, 240V, 3P, NEMA 3R
DISCONNECT SWITCH WITH (2)
FRN-R-125 FUSES

#2 Cu TO COLD WATER PIPE
#2 Cu TO BUILDING STEEL
#4 Cu CONCRETE ENCASED ELECTRODE

200A, 240V, 3P, NEMA 3R
DISCONNECT SWITCH WITH (3)
FRN-R-125 FUSES

ONE-LINE DIAGRAM

208Y/120V 3Ø, 4W SERVICE USING A
WIREWAY AND FUSED DISCONNECT
SWITCHES. NOTE THAT SINGLE-PHASE
AND THREE-PHASE TENANTS ARE
CONNECTED TO THE SAME SERVICE
EQUIPMENT.

308F08.EPS

Figure 8 ◆ One-line diagram for overhead service.

grounding requirements are established based on an 800A service. The service-entrance conductors consist of three cables, with each cable being four-conductor, 400 kcmil THWN aluminum. Each cable is in an underground 3" conduit (the fourth conduit is a spare). Because these are parallel service-entrance conductors, the GEC is sized at 2/0 copper (or 4/0 aluminum) using *NEC Table 250.66*. The parallel service-entrance conductors terminate in a main service disconnect. The tenant service disconnects are connected to the wireway with EMT conduit with Myers-type hubs used where the conduit terminates into the wireway, and the conductors supplying each tenant disconnect are terminated in the wireway. The terminations in the wireway are mechanical connections made using a multi-tap insulated connector. Per *NEC Section 310.4(B)*, parallel conductors must be terminated in the same manner. In this example, the GEC is bonded to the grounded service conductor (neutral) in the wireway. In this installation, there is one 100A fusible disconnect ahead of

Figure 9 ◆ Underground wireway service.

308F09.EPS

(cold sequence) the meter serving panel HP (the house panel) which in turn supplies branch circuit power to the lighting controller for exterior building lighting and parking lot lighting. There is one 200A fusible tenant disconnect ahead of the meter serving a tenant space and one 400A CT cabinet ahead of a 400A fusible disconnect serving a larger tenant space. The vertical line originating at the top of the 200A tenant meter indicates the feeder to the tenant panel located in the tenant space. Note that the feeder to the 400A tenant panel is not shown because it is routed inside the building and terminates into the back of the 400A tenant disconnect.

Note that the *NEC*® would not require a main service disconnect for this installation because there are fewer than six switches to disconnect power to the building and the ampacity of the feeders serving panel HP and the tenant services does not exceed the ampacity of the service-entrance conductors. However, the service for this retail building was designed before any tenants were identified to occupy the building, and provided flexibility to allow a tenant mix that could have required more than six disconnects for tenant services. This design also provides flexibility for future changes in the tenant mix, making it unnecessary to rework the entire service to

accommodate changes. Because this equipment is installed on the exterior of the building, all of the equipment is rated for outdoor use.

Figure 10 shows the one-line diagram for the service just described. Note that the one-line diagram includes sizes for service-entrance conduit and conductors, sizes and fuse requirements for the service disconnects, the size and length of the wireway, the sizes of panel feeder conduit and conductors, and sizes for the grounding electrode conductors. The branch circuit panelboard for the lighting contactors is also shown as part of the one-line diagram.

6.3.0 208Y/120V Three-Phase Service – Bussed Gutter with Multiple Service Disconnects

Figure 11 shows an underground 208Y/120V three-phase service. The service is rated at 800A and the grounding requirements are established based on an 800A service. The service-entrance conductors consist of two cables, with each cable being a three-conductor, 500 kcmil THWN copper in 3" conduit. Because these are parallel service-entrance conductors, the GEC is sized at 2/0 copper (or 4/0 aluminum) using *NEC Table 250.66*. The parallel service-entrance conductors terminate in a bussed gutter. The service disconnects are connected to the bussed gutter with RGS nipples and the conductors supplying each meter are terminated in the bussed gutter using mechanical lugs. It is important to follow the manufacturer's requirements for torque when terminating conductors using mechanical lugs.

Because this equipment is installed on the exterior of the building, all of the equipment is rated for outdoor use.

Figure 12 is the one-line diagram for the service just described. Note that the one-line diagram includes sizes for service-entrance conduit and conductors, sizes and fuse requirements for the service disconnects, the size and length of the wireway, the sizes of panel feeder conduit and conductors and sizes for the grounding electrode conductors. The branch circuit panelboards are also shown as part of the one-line diagram.

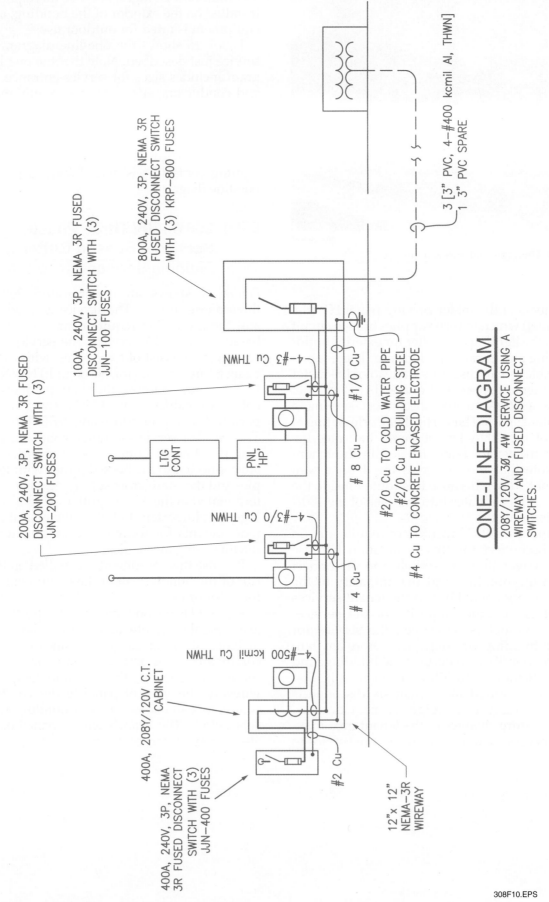

ONE-LINE DIAGRAM

208Y/120V 3Ø, 4W SERVICE USING A WIREWAY AND FUSED DISCONNECT SWITCHES.

800A, 240V, 3P, NEMA 3R FUSED DISCONNECT SWITCH WITH (3) KRP–800 FUSES

100A, 240V, 3P, NEMA 3R FUSED DISCONNECT SWITCH WITH (3) JJN–100 FUSES

200A, 240V, 3P, NEMA 3R FUSED DISCONNECT SWITCH WITH (3) JJN–200 FUSES

400A, 208Y/120V C.T. CABINET

400A, 240V, 3P, NEMA 3R FUSED DISCONNECT SWITCH WITH (3) JJN–400 FUSES

4–#3 Cu THWN

4–#3/0 Cu THWN

4–#500 kcmil Cu THWN

LTG CONT

PNL 'HP'

#1/0 Cu

8 Cu

4 Cu

#2 Cu

#2/0 Cu TO COLD WATER PIPE
#2/0 Cu TO BUILDING STEEL
#4 Cu TO CONCRETE ENCASED ELECTRODE

12"x 12" NEMA–3R WIREWAY

3 [3" PVC, 4–#400 kcmil Al, THWN]
1 3" PVC SPARE

308F10.EPS

Figure 10 ◆ One-line diagram for an underground wireway service.

308F11A.EPS   308F11B.EPS

Figure 11 ◆ Underground bussed gutter service.

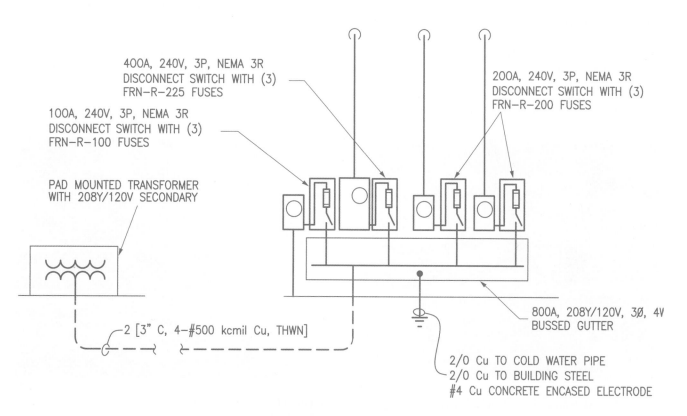

400A, 240V, 3P, NEMA 3R
DISCONNECT SWITCH WITH (3)
FRN–R–225 FUSES

200A, 240V, 3P, NEMA 3R
DISCONNECT SWITCH WITH (3)
FRN–R–200 FUSES

100A, 240V, 3P, NEMA 3R
DISCONNECT SWITCH WITH (3)
FRN–R–100 FUSES

PAD MOUNTED TRANSFORMER
WITH 208Y/120V SECONDARY

800A, 208Y/120V, 3Ø, 4V
BUSSED GUTTER

2 [3" C, 4–#500 kcmil Cu, THWN]

2/0 Cu TO COLD WATER PIPE
2/0 Cu TO BUILDING STEEL
#4 Cu CONCRETE ENCASED ELECTRODE

ONE-LINE DIAGRAM
208Y/120V 3Ø, 4W SERVICE USING A
BUSSED GUTTER AND FUSED
DISCONNECT SWITCHES

308F12.EPS

Figure 12 ◆ One-line diagram for an underground bussed gutter service.

6.4.0 480Y/277V Three-Phase Service – Switchgear

Figure 13 shows a switchgear 480Y/277V three-phase service. The service is rated at 2,000A and the grounding requirements are established based upon a 2,000A service. The fault current rating for this equipment is 50,000 AIC. The service-entrance conductors consist of seven cables, with each cable being a four-conductor, 500 kcmil THWN copper in 3½" conduit. Because these are parallel service-entrance conductors, the GEC is sized at 3/0 copper (or 250 kcmil aluminum) using *NEC Table 250.66*. The switchgear consists of a network of busbars installed horizontally. The busbars extend through each metal switchgear cabinet to supply power to individual disconnects. If a service is of sufficient size to require more than six disconnects, a main disconnecting means must be included. The service shown in this example has four service disconnects and spaces for two additional future service disconnects. A main disconnecting means is not required in this configuration.

The service-entrance conductors enter underground and are terminated in a CT metering section that is part of the panel main switchboard (MSB). The metering section includes provisions for terminating the service-entrance conductors to the switchgear bus. The bus then carries current through the switchgear to fusible disconnects supplying power to panelboards (*Figure 14*) serving mechanical equipment loads, a lighting panelboard, a transformer, and a shunt-trip circuit breaker supplying an elevator. Because this equipment is installed on the interior of the building, all of the equipment is rated for indoor use.

Figure 15 is the one-line diagram for the service just described. Note that the 600A switch with 500A fuses feeds a 300kVA transformer to the panel SSB—a 1,200A switchgear supplying power for all 208V and 120V loads. This one-line diagram includes sizes for service-entrance conduit and conductors, the size and ratings for the switchgear, transformers, and panelboard feeders. The branch circuit panelboards are also shown as part of the one-line diagram. *Figure 16* represents panel schedules for two of the panelboards in this installation.

308F13.EPS

Figure 13 ◆ MSB and SSB switchgear service.

308F14.EPS

Figure 14 ◆ Panelboards.

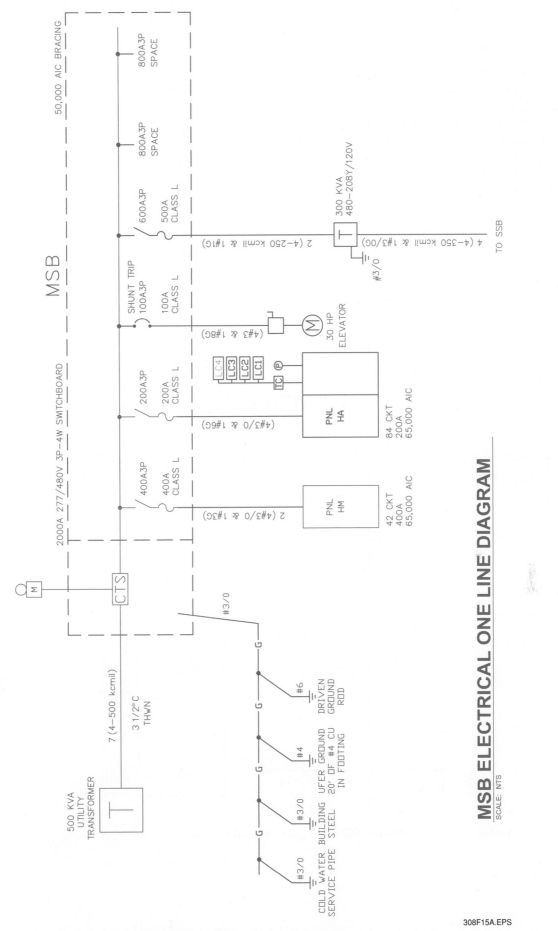

Figure 15 ◆ One-line diagram for MSB and SSB switchgear service (2 sheets).

308F15A.EPS

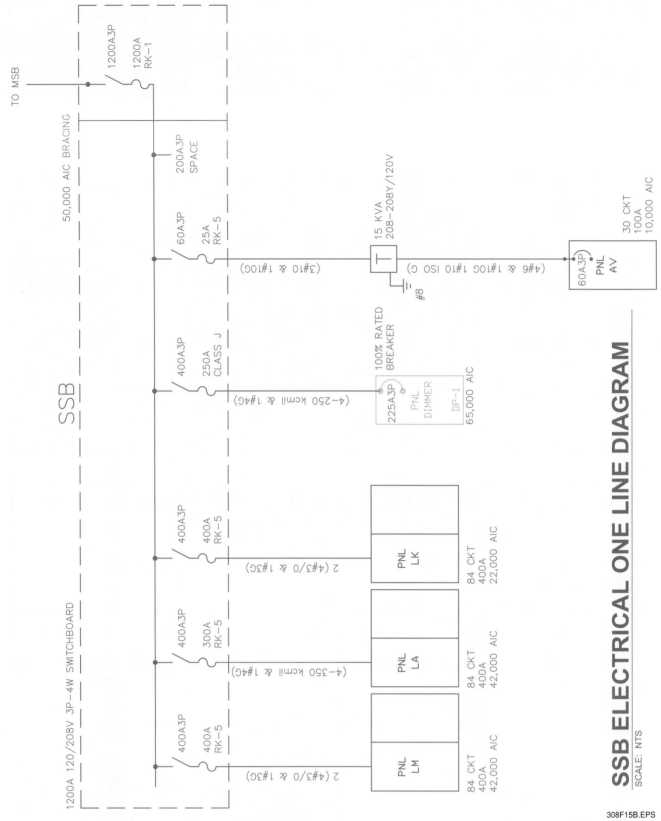

Figure 15 ◆ One-line diagram for MSB and SSB switchgear service (2 sheets).

308F15B.EPS

PANEL SCHEDULE

PANEL NAME   **HA**

PROJECT:	CROSSROADS CHURCH	
JOB NO.:	M-1904	
LOCATION:	ELECTRICAL ROOM	
MINIMUM BUS CAPACITY (A):	200	
MAIN O.C. DEVICE (A):	MLO	
DESIGN CAPACITY (A):	200	

VOLTAGE L-L (V):	480
VOLTAGE L-N (V):	277
TYPE:	3P 4W
SHORT CIRCUIT RATING (A):	65000
MOUNTING:	SURFACE
COMMENTS:	

DEVICE AMPS	POLE	LTNG (VA)	RCPT (VA)	M/LM/E/A/S (VA)	DESCRIPTION	CKT NO.	PHASE	CKT NO.	DESCRIPTION	M/LM/E/A/S (VA)	RCPT (VA)	LTNG (VA)	POLE	DEVICE AMPS
20	1	1200			SE MONUMENT SIGN	1	A	2	N DRIVE SITE LTG			2530	2	20
20	1	1200			E MONUMENT SIGN	3	B	4	/			2530	/	/
20	1	1200			E MONUMENT SIGN	5	C	6	SE DRIVE SITE LTG			1610	2	20
20	1	1032			BLDG ACCENT LTG	7	A	8				1610	/	/
20	1	100			TREE ACCENT LTG	9	B	10	W PARKING LOT LTG			3220	2	20
20	1	683			ARCADE DOWN LTG	11	C	12	/			3220	/	/
20	1	394			ARCADE ACCENT LTG	13	A	14	SPARE				2	20
20	1	200			ENTRY LTG	15	B	16	/				/	/
20	1	3795			CLASSRM #139-#147 LTG	17	C	18	SPARE				2	20
20	1	3566			CORR/STOR/RR/CSRM LTG	19	A	20	/				/	/
20	1	4094			1ST FLR NE LTG	21	B	22	EXTER WALLMNT LTG			72	1	20
20	1	2800			NARTHEX LTG	23	C	24	EXTER WALLMNT LTG			459	1	20
20	1	3108			1ST FLR NW LTG	25	A	26	SPARE				1	20
20	1	3670			1ST FLR N LTG	27	B	28	SPARE				1	20
20	1				SPARE	29	C	30	MULTI RM LTG STEP1			1680	1	20
20	1				SPARE	31	A	32	MULTI RM LTG STEP2			1680	1	20
20	1				SPARE	33	B	34	MULTI RM LTG STEP3			1680	1	20
20	1				SPARE	35	C	36	MULTI RM LTG STEP 4			1680	1	20
20	1	2975			2ND FLR BEV/STAR/CORR LT	37	A	38	MULTI RM LTG STEP1			1680	1	20
20	1	2607			2ND FLR S LTG	39	B	40	MULTI RM LTG STEP2			1680	1	20
20	1	2935			2ND FLR S LTG	41	C	42	MULTI RM LTG STEP3			1680	1	20
20	1	2614			2ND FLR NE LTG	43	A	44	MULTI RM LTG STEP 4			1680	1	20
20	1				SPARE	45	B	46	SPARE				1	20
20	1				SPARE	47	C	48	SPARE				1	20
20	1				SPARE	49	A	50	SPARE				1	20
20	1				SPARE	51	B	52	SPARE				1	20
20	1				SPARE	53	C	54	SPARE				1	20
20	1				SPARE	55	A	56	SPARE				1	20
20	1				SPARE	57	B	58	SPARE				1	20
20	1				SPARE	59	C	60	SPARE				1	20

CONNECTED VA PHASE A:	24069		DEMANDED VA PHASE A:	30086	
CONNECTED VA PHASE B:	21053		DEMANDED VA PHASE B:	26316	
CONNECTED VA PHASE C:	21742		DEMANDED VA PHASE C:	27178	

	CONNECTED	D.F.	DEMAND		
LIGHTING LOAD:	66864	1.25	83580	DEMAND LOAD (A) =	101
RECEPTACLE (FIRST 10 KVA)	0	1.00	0	SPARE CAPACITY (A) =	99
RECEPTACLE (REMAINDER)	0	0.50	0		
LARGEST MOTOR:	0	1.25	0		
REMAINING MOTORS:	0	1.00	0		
APPLIANCES:	0	0.65	0		
EQUIPMENT:	0	1.00	0		
SUB FED PANEL:	0	1.00	0		
TOTAL:	66864		83580		
LOAD (AMPS):	80.4		100.5		

M = MOTOR   E = EQUIPMENT   S = SUB FEED PANEL
LM = LARGEST MOTOR   A = APPLIANCE

NOTES:
     1 CONTROL VIA CONTACTOR

308F16A.EPS

Figure 16 ◆ Panel schedules for a 480Y/277V panelboard and a 208Y/120V panelboard (2 sheets).

PANEL SCHEDULE

PANEL NAME   LA

PROJECT:	CROSSROADS CHURCH	VOLTAGE L-L (V):	208
JOB NO.:	M-1904	VOLTAGE L-N (V):	120
LOCATION:	ELECTRICAL ROOM	TYPE:	3P 4W
MINIMUM BUS CAPACITY (A):	400	SHORT CIRCUIT RATING (A):	42000
MAIN O.C. DEVICE (A):	MLO	MOUNTING:	SURFACE
DESIGN CAPACITY (A):	300	COMMENTS:	

DEVICE AMPS	POLE	LTNG (VA)	RCPT (VA)	M/LM/E/A/S (VA)	DESCRIPTION	CKT NO.	PHASE	CKT NO.	DESCRIPTION	M/LM/E/A/S (VA)	RCPT (VA)	LTNG (VA)	POLE	DEVICE AMPS
20	1		720		CLASSRM #140		A	2	TOILET #145		180		1	20
20	1		900		CLASSRM #140	3	B	4	EWC-1B	M 244			1	20
20	1		1080		CLASSRM #143	5	C	6	TOILET #133		180		1	20
20	1		1260		CLASSRM #146		A	8	RESTROOMS #149/150		360		1	20
20	1		1260		CLASSRM #147	9	B	10	STORAGE #151		540		1	20
20	1		1260		TODDLERS #135	11	C	12	CORRIDOR #125/129		1440		1	20
20	1		1080		NURSERY #131	13	A	14	MAINT WORKSHP		720		1	20
20	1			M 1582	WASHER	15	B	16	MAINT WORKSHP		900		1	20
50	2			E 4800	DRYER	17	C	18	STOR/ELEC RM		720		1	20
/	/			E 4800	/	19	A	20	YOUTH RM #124, #121		1440		1	20
20	1		1080		EXTERIOR RCPT	21	B	22	SPARE				1	20
20	1		720		RESTROOM #119/120	23	C	24	WELCOME NARTHEX		1080		1	20
20	1			M 244	EWC-1A	25	A	26	MULTI PURPOSE RM		900		1	20
20	1			M 244	EWC-1C	27	B	28	MULTI PURPOSE RM		540		1	20
20	1		1260		LOCKER/STOR/JAN	29	C	30	STORAGE		900		1	20
20	1	555			COFFEE BAR LTG	31	A	32	PA SYSTEM	E 200			1	20
20	1	740			ENTRY LTG	33	B	34	PA SYSTEM	E 200			1	20
20	1			M 1000	ENTRY DOORS	35	C	36	FIRE SMOKE DAMPERS	E 400			1	20
20	1			LM 1656	MOTORIZED HOOP	37	A	38	SPARE				1	20
20	1		1080		YOUTH #224 / STORAGE	39	B	40	SPARE				1	20
20	1			M 1656	MOTORIZED HOOP	41	C	42	SPARE				1	20
20	1		1260		CLASSRM #206	43	A	44	CONF RM #213		540		1	20
20	1		1080		CLASSRM #207	45	B	46	RECEP #214		720		1	20
20	1		1260		CLASSRM #209	47	C	48	KITCHENETTE #211	E 1000	180		1	20
20	1		900		OFFICE #210A	49	A	50	KITCHENETTE #211		180		1	20
20	1				SPARE	51	B	52	WORK ROOM #212		360		1	20
20	1		1080		OFFICE #215/216	53	C	54	COPIER	E 1000			1	20
20	1		900		CLASSRM #204	55	A	56	TELE/COMP RM #205		1080		1	20
20	1		900		CLASSRM #203	57	B	58	STAIR/CORRIDOR		360		1	20
20	1		900		STORAGE/MEDIA RM	59	C	60	RESTROOMS		360		1	20
20	1		1080		YOUTH #225	61	A	62	BEV AREA		360		1	20
20	1		1080		YOUTH #224	63	B	64	BEV AREA		540		1	20
20	1	72			BEV AREA LTG	65	C	66	CORRIDOR /BEV AREA		900		1	20
20	1			M 244	EWC-1D	67	A	68	SPARE		360		1	20
20	1		180		PA SYSTEM	69	B	70	SECURITY GRILL	M 1175			1	20
20	1		180		PA SYSTEM	71	C	72	YOUTH TRACK			1600	1	20
20	1			E 900	BEV AREA REFRIG	73	A	74	YOUTH TRACK			1600	1	20
20	1				SPARE	75	B	76	YOUTH TRACK			1600	1	20
20	1				SPARE	77	C	78	YOUTH TRACK			1600	1	20
20	1				SPARE	79	A	80	YOUTH STAGE			900	1	20
20	1			E 600	CONTACTOR COILS	81	B	82	ROOFTOP RCPT		720		1	20
20	1		540		EXTERIOR RCPT	83	C	84	ROOFTOP RCPT		540		1	20

CONNECTED VA PHASE A:	24419	DEMANDED VA PHASE A:	20573
CONNECTED VA PHASE B:	18625	DEMANDED VA PHASE B:	14187
CONNECTED VA PHASE C:	27708	DEMANDED VA PHASE C:	23503

	CONNECTED	D.F.	DEMAND	
LIGHTING LOAD:	8667	1.25	10834	DEMAND LOAD (A) = 162
RECEPTACLE (FIRST 10 KVA):	10000	1.00	10000	SPARE CAPACITY (A) = 138
RECEPTACLE (REMAINDER):	30140	0.50	15070	
LARGEST MOTOR:	1656	1.25	2070	
REMAINING MOTORS:	6389	1.00	6389	
APPLIANCES:	0	0.65	0	
EQUIPMENT:	13900	1.00	13900	
SUB FED PANEL:	0	1.00	0	
TOTAL:	70752		58263	
LOAD (AMPS):	196.4		161.7	

M = MOTOR   E = EQUIPMENT   S = SUB FEED PANEL
LM = LARGEST MOTOR   A = APPLIANCE

NOTES:

Figure 16 ◆ Panel schedules for a 480Y/277V panelboard and a 208Y/120V panelboard (2 sheets).

6.5.0 240/120V Multi-Family Service

Figure 17 shows a 240/120V single-phase service for a multi-family apartment building. This is a 24-unit apartment building with individual tenant meters located on each end of the building. Grounding for this service is sized for parallel (3 sets) of 350 kcmil aluminum service-entrance conductors. The service equipment is designed for a fault current rating of 100,000 AIC. The service-entrance conductors consist of three cables with each cable being a three-conductor, 350 kcmil THWN aluminum in 3" PVC conduit. Because these are parallel service-entrance conductors, the GEC is sized at 2/0 copper (or 4/0 aluminum) using *NEC Table 250.66* (3 × 350 kcmil = 1,050 kcmil). The service consists of a main fused disconnect switch and meter stacks connected with horizontal busbars to the main disconnect. The busbars extend horizontally from the main disconnect through each meter stack and then vertical bus is connected to supply power to each

meter. Individual meters in the meter stack consist of the meter base and a two-pole, 100A circuit breaker that serves as the overcurrent and short-circuit protection for the feeder supplying the apartment. This circuit breaker is sized for the load to be served. The meter breakers may not be equal size—they are sized for the load to be served and the load for each apartment must be calculated to determine the feeder size. The red sign on the main service disconnect reads: A DISCONNECT ON OTHER END OF BUILDING. This plaque informs the fire department and other authorities that more than one disconnect controls power to the building.

This service complies with *NEC Section 230.72(C)*, which requires that each occupant have access to his or her own service disconnecting means. This exception allows the tenant meters and service disconnecting means to be placed inside the building under the conditions listed. *Figure 18* is a one-line diagram for the multi-family service just described.

308F17.EPS

Figure 17 ◆ Multi-family service.

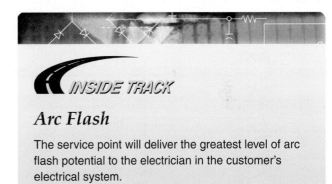

INSIDE TRACK

Arc Flash

The service point will deliver the greatest level of arc flash potential to the electrician in the customer's electrical system.

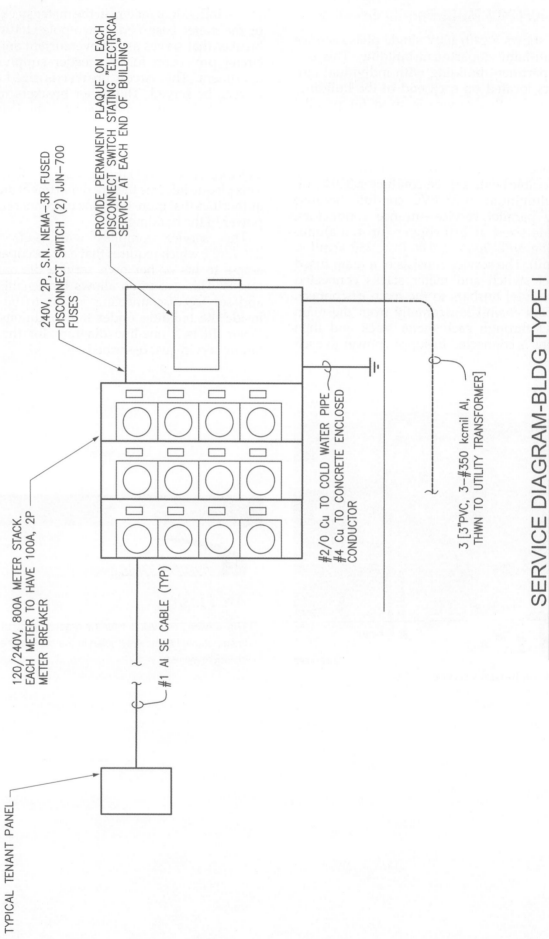

Figure 18 ◆ One-line diagram for a multi-family service.

TYPICAL TENANT PANEL

120/240V, 800A METER STACK. EACH METER TO HAVE 100A, 2P METER BREAKER

#1 Al SE CABLE (TYP)

240V, 2P, S.N. NEMA–3R FUSED DISCONNECT SWITCH (2) JJN–700 FUSES

PROVIDE PERMANENT PLAQUE ON EACH DISCONNECT SWITCH STATING "ELECTRICAL SERVICE AT EACH END OF BUILDING"

#2/0 Cu TO COLD WATER PIPE
#4 Cu TO CONCRETE ENCLOSED CONDUCTOR

3 [3"PVC, 3–#350 kcmil Al, THWN TO UTILITY TRANSFORMER]

SERVICE DIAGRAM-BLDG TYPE I

308F18.EPS

Call First Before Digging

When installing any underground electrical, a call to the local utility should be the first order of business. This allows the utility company's third-party locating service to come out and mark where the existing underground services are present. These may include existing gas, telephone, fiber-optic lines, or other electrical services.

Putting It All Together

Examine the service at your school or workplace. Is the service entrance properly installed and protected? Is the service-entrance cable cracked or frayed? What about the electrical panel? Are the breakers properly labeled? Is there a main service disconnect switch?

1. The NEC requirements for services can be found in _____.
   a. *NEC Article 225*
   b. *NEC Article 230*
   c. *NEC Article 240*
   d. *NEC Article 250*

2. All electrical service distribution equipment is labeled with its maximum voltage and current ratings.
   a. True
   b. False

3. Cold sequence metering is accomplished by _____.
   a. unplugging the meter from the meter base
   b. installing the service disconnect after the meter
   c. installing the service disconnect before the meter
   d. placing a switch in CT leads to a meter

4. Most analog watt-hour meters have _____ dials.
   a. four
   b. five
   c. six
   d. seven

5. CT metering is typically used in services rated _____ and above.
   a. 200A
   b. 300A
   c. 400A
   d. 500A

6. The term bussed gutter is defined in _____.
   a. *NEC Article 368*
   b. *NEC Article 376*
   c. *NEC Article 378*
   d. *NEC Article 392*

7. Bracing of bussed gutter busbars is required primarily to _____.
   a. support the weight of the busbars
   b. allow easier connection of taps
   c. resist the torque of fault current
   d. remove heat from the busbar

8. Service conductors passing through a building are permitted if the conductors are _____.
   a. insulated and fireproofed
   b. in heavy insulated conduit
   c. in conduit encased by 2" of concrete
   d. out of reach

9. The weatherhead for a service drop should be a minimum of _____ above grade.
   a. 10'
   b. 13'-6"
   c. 18'
   d. 21'-6"

10. Grounding electrode conductors are sized using _____.
    a. *NEC Table 250.66*
    b. *NEC Table 250.122*
    c. *NEC Table 310.16*
    d. *NEC Table 310.19*

Summary

This module covered typical commercial and small industrial electric service applications, including the installation of service-rated components and their related electrical equipment. Regardless of the size or complexity, all electric services serve the same purpose: to deliver electrical energy safely from the supply system to the wiring system on the premises served. The basic components of a commercial electric service usually include a utility transformer, a service drop or lateral, a service entrance, and service equipment. *NEC Article 230* lists the requirements for electric services.

Notes

Trade Terms
Introduced in This Module

Cold sequence metering: A service installation where the service disconnect is installed ahead of the metering in order for power to be disconnected to allow for safe maintenance of the metering.

Delta-connected: A three-phase transformer connection in which the terminals are connected in a triangular shape like the Greek letter delta (Δ).

Service: The electric power delivered to the premises.

Service conductors: The conductors between the point of termination of the overhead service drop or underground service lateral and the main disconnecting device in the building or on the premises.

Service drop: The overhead conductors through which electrical service is supplied between the last power company pole and the point of their connection to the service-entrance conductors located at the building or other support used for the purpose.

Service entrance: All components between the point of termination of the overhead service drop or underground service lateral and the building's main disconnecting device, except for metering equipment.

Service equipment: The necessary equipment, usually consisting of a circuit breaker or switch and fuses and their accessories, located near the point of entrance of supply conductors to a building and intended to constitute the main control and cutoff means for the electric supply to the building.

Service lateral: The underground service conductors between the street main, including any risers at a pole or other structure or from transformers, and the first point of connection to the service-entrance conductors in a terminal box, meter, or other enclosure with adequate space, inside or outside the building wall.

This module is intended to present thorough resources for task training. The following reference work is suggested for further study. This is optional material for continued education rather than for task training.

National Electrical Code® Handbook, Latest Edition. Quincy, MA: National Fire Protection Association.

CONTREN® LEARNING SERIES – USER UPDATE

NCCER makes every effort to keep these textbooks up-to-date and free of technical errors. We appreciate your help in this process. If you have an idea for improving this textbook, or if you find an error, a typographical mistake, or an inaccuracy in NCCER's Contren® textbooks, please write us, using this form or a photocopy. Be sure to include the exact module number, page number, a detailed description, and the correction, if applicable. Your input will be brought to the attention of the Technical Review Committee. Thank you for your assistance.

Instructors – If you found that additional materials were necessary in order to teach this module effectively, please let us know so that we may include them in the Equipment/Materials list in the Annotated Instructor's Guide.

Write: Product Development and Revision
National Center for Construction Education and Research
3600 NW 43rd St., Bldg. G, Gainesville, FL 32606

Fax: 352-334-0932

E-mail: curriculum@nccer.org

Craft _____ Module Name _____

Copyright Date _____ Module Number _____ Page Number(s) _____

Description _____

_____

_____

_____

(Optional) Correction _____

_____

_____

(Optional) Your Name and Address _____

_____

_____

Motor Calculations

Inn at the Ballpark

Houston, Texas, is home to this award-winning project built by HOAR Construction, LLC. It won in the renovation $10–99 Million category in the Award of Excellence competition sponsored by Associated Builders and Contractors, Inc.

26309-08

26309-08
Motor Calculations

Topics to be presented in this module include:

Overview

There are three basic types of motors: squirrel cage induction, wound-rotor induction, and synchronous. Motors may be single-phase or three-phase. There are two basic parts to an alternating current motor, the rotor and the stator. The rotor is the part that turns or spins. The stator is the stationary winding assembly in which the rotor turns.

In order to calculate conductor sizes and overcurrent protection for motor circuits, you must know the full load amperage (FLA) of the motor. FLAs for motors may be found on the nameplate of the motor or approximated by referring to the full load amperage tables located in *NEC Article 430*. Multiple motors connected to a single feeder circuit require the application of specific formulas in order to determine circuit ratings. Conductor sizing and branch circuit ratings for all motors and motor controllers are also regulated by *NEC Article 430*. Motor overload protection should be calculated based on the FLA of the motor when running in a normal state. Fuses protecting motor circuits must be selected based on the starting amperage of the motor. Some motors draw as much as five or six times normal FLA during startup. This is called locked rotor amperes or LRA. Time delay fuses are typically used in these motor circuits to allow the motor to reach full running speed without opening the circuit.

Objectives

When you have completed this module, you will be able to do the following:

1. Size branch circuits and feeders for electric motors.
2. Size and select overcurrent protective devices for motors.
3. Size and select overload relays for electric motors.
4. Size and select devices to improve the power factor at motor locations.
5. Size motor short circuit protectors.
6. Size multi-motor branch circuits.
7. Size motor disconnects.

Trade Terms

Circuit interrupter
Rating
Service factor
Terminal
Torque

Required Trainee Materials

1. Pencil and paper
2. Appropriate personal protective equipment
3. Copy of the latest edition of the *National Electrical Code*®

Prerequisites

Before you begin this module, it is recommended that you successfully complete *Core Curriculum, Electrical Level One; Electrical Level Two; Electrical Level Three*, Modules 26301-08 through 26308-08.

This course map shows all of the modules in *Electrical Level Three*. The suggested training order begins at the bottom and proceeds up. Skill levels increase as you advance on the course map. The local Training Program Sponsor may adjust the training order.

26311-08
Motor Controls

26310-08
Voice, Data, and Video

26309-08
Motor Calculations

26308-08
Commercial Electrical Services

26307-08
Transformers

26306-08
Distribution Equipment

26305-08
Overcurrent Protection

26304-08
Hazardous Locations

26303-08 Practical Applications of Lighting

26302-08 Conductor Selection and Calculations

26301-08 Load Calculations – Branch and Feeder Circuits

ELECTRICAL LEVEL THREE

ELECTRICAL LEVEL TWO

ELECTRICAL LEVEL ONE

CORE CURRICULUM:
Introductory Craft Skills

309CMAP.EPS

1.0.0 ◆ INTRODUCTION

Electric motors are used in almost every type of installation imaginable, from residential appliances to heavy industrial machines. Many types of motors are available, from small shaded-pole motors (used mostly in household fans) to huge synchronous motors for use in large industrial installations. There are numerous types in between to fill every conceivable niche. None, however, have the wide application possibilities of the three-phase motor. This is the type of motor that electricians encounter most frequently. Therefore, the majority of the material in this module will deal with three-phase motors.

There are three basic types of three-phase motors:

- Squirrel cage induction motor
- Wound-rotor induction motor
- Synchronous motor

The type of three-phase motor is determined by the rotor or rotating member (*Figure 1*). The stator winding is basically the same for all three motor types.

The principle of operation for all three-phase motors is the rotating magnetic field. There are three factors that cause the magnetic field to rotate:

- The voltages of a three-phase electrical system are 120° out of phase with each other.
- The three voltages change polarity at regular intervals.
- The stator windings around the inside of the motor are arranged in a specific manner to induce rotation.

The *National Electrical Code®* (*NEC®*) plays an important role in the installation of electric motors. *NEC Article 430* covers the application and installation of motor circuits and motor control connections, including conductors, short-circuit and ground-fault protection, controllers, disconnects, and overload protection.

NEC Article 440 contains provisions for motor-driven air conditioning and refrigerating equipment, including the branch circuits and controllers for the equipment. It also takes into account the special considerations involved with sealed (hermetic) motor compressors, in which the motor operates under the cooling effect of the refrigeration. When referring to *NEC Article 440*, be aware that the rules in this article are in addition to, or are amendments to, the rules given in *NEC Article 430*. Motors are also covered to some degree in *NEC Articles 422 and 424*.

2.0.0 ◆ MOTOR BASICS

The rotor of an AC squirrel cage induction motor (*Figure 2*) consists of a structure of steel laminations mounted on a shaft. Embedded in the rotor

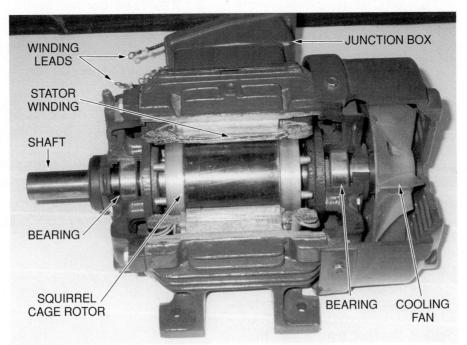

Figure 1 ◆ Basic parts of a three-phase motor.

309F01.EPS

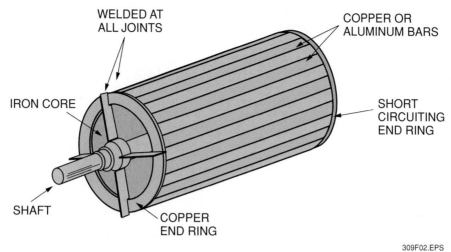

Figure 2 ◆ Squirrel cage rotor.

309F02.EPS

is the rotor winding, which is a series of copper or aluminum bars, short circuited at each end by a metallic end ring. The stator consists of steel laminations mounted in a frame. Slots in the stator hold stator windings that can be either copper or aluminum coils or bars. These are connected to form a circuit.

Energizing the stator coils with an AC supply voltage causes current to flow in the coils. The current produces an electromagnetic field that causes magnetic poles to be created in the stator iron. The strength and polarity of these poles vary as the AC current flows in one direction, then the other. This change causes the poles around the stator to alternate between being south and north poles, thus producing a rotating magnetic field.

The rotating magnetic field cuts through the rotor, inducing a current in the rotor bars. This induced current only circulates in the rotor, which in turn causes a rotor magnetic field. As with two conventional bar magnets, the north pole of the rotor field attempts to line up with the south pole of the stator magnetic field, and the south pole attempts to line up with the north pole. However, because the stator magnetic field is rotating, the rotor chases the stator field. The rotor field never quite catches up due to the need to furnish torque to the mechanical load.

2.1.0 Synchronous Speed

The speed at which the magnetic field rotates is known as the synchronous speed. The synchronous speed of a three-phase motor is determined by two factors:

- Number of stator poles
- Frequency of the AC line in hertz (Hz)

The synchronous speeds for various 60Hz motors are as follows:

- Two poles–3,600 rpm
- Four poles–1,800 rpm
- Six poles–1,200 rpm
- Eight poles–900 rpm

These speeds illustrate that the rpm of a three-phase motor decreases as the number of poles increases. The formula for synchronous speed is:

$$\text{Synchronous speed} = \frac{120F}{P}$$

Where:
   F = frequency
   P = number of poles

2.2.0 Stator Windings

The stator windings of three-phase motors are connected in either a wye or a delta configuration (*Figure 3*). Some motor stators are designed to operate both ways; that is, they are started as a wye-connected motor to help reduce starting current, and then changed to a delta configuration for running.

Many three-phase motors have dual-voltage stators. These stators are designed to be connected to either 240V or 480V. The leads of a dual-voltage stator use a standard numbering system. *Figure 4* shows a dual-voltage, wye-connected stator.

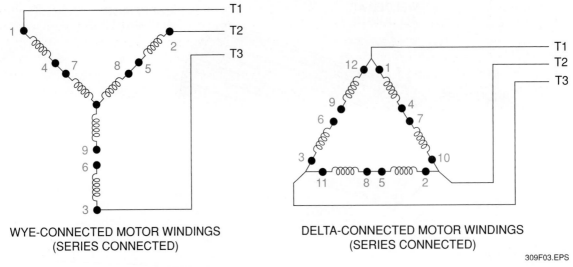

WYE-CONNECTED MOTOR WINDINGS
(SERIES CONNECTED)

DELTA-CONNECTED MOTOR WINDINGS
(SERIES CONNECTED)

309F03.EPS

Figure 3 ◆ Types of windings found in three-phase motors.

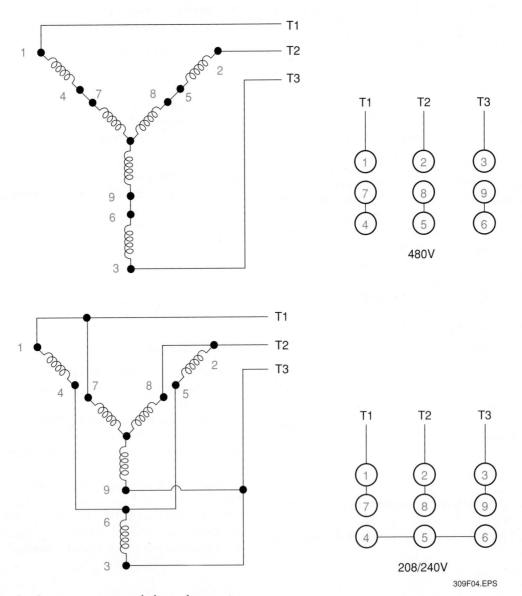

480V

208/240V

309F04.EPS

Figure 4 ◆ Dual-voltage, wye-connected, three-phase motors.

Note that the nine motor leads are numbered in a spiral. The leads are connected in series for use on the higher voltage and in parallel for use on the lower voltage. Therefore, for the higher voltage, leads 4 and 7, 5 and 8, and 6 and 9 are connected together. For the lower voltage, leads 4, 5, and 6 are connected together; further connections are 1 and 7, 2 and 8, and 3 and 9, which are then connected to the three-phase power source. *Figure 5*

shows the equivalent parallel circuit when the motor is connected for use on the lower voltage.

The same standard numbering system is used for delta-connected motors, and many delta-wound motors also have nine leads, as shown in *Figure 6*. However, there are only three circuits of three leads each. The high-voltage and low-voltage connections for a three-phase, delta-wound, dual-voltage motor are shown in *Figure 7*.

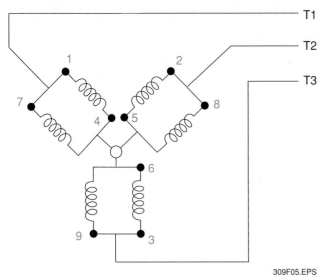

309F05.EPS

Figure 5 ◆ Equivalent parallel circuit.

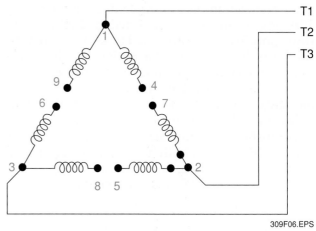

309F06.EPS

Figure 6 ◆ Arrangement of leads in a nine-lead, delta-wound, dual-voltage motor.

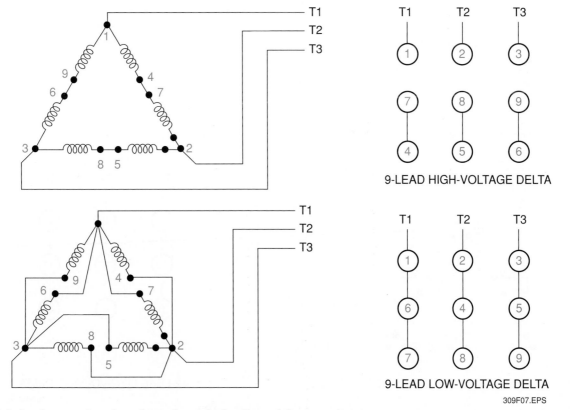

309F07.EPS

Figure 7 ◆ Lead connections for a three-phase, dual-voltage, delta-wound motor.

In some instances, a dual-voltage motor connected in a delta configuration will have 12 leads instead of nine. *Figure 8* shows the high-voltage and low-voltage connections for a dual-voltage, 12-lead, delta-wound motor.

2.2.1 Principles of Dual-Voltage Connections

When a motor is operated at 240V, the current draw of the motor is double the current draw of a 480V connection. For example, if a motor draws 10A of current when connected to 240V, it will draw only 5A when connected to 480V. The reason for this is the difference of impedance in the windings between a 240V connection and a 480V connection. Remember that the low-voltage windings are always connected in parallel, while the high-voltage windings are connected in series.

For instance, assume that the stator windings of a motor (R1 and R2) both have a resistance of 48Ω. If the stator windings are connected in parallel, the total resistance (R_t) may be found as follows:

$$R_t = \frac{R1 \times R2}{R1 + R2}$$

$$R_t = \frac{48\Omega \times 48\Omega}{48\Omega + 48\Omega}$$

$$R_t = \frac{2,304\Omega}{96\Omega}$$

$$R_t = 24\Omega$$

Therefore, the total resistance (R) of the motor winding connected in parallel is 24Ω, and if a voltage (E) of 240V is applied to this connection, the following current (I) will flow:

$$I = \frac{E}{R}$$

$$I = \frac{240V}{24\Omega}$$

$$I = 10A$$

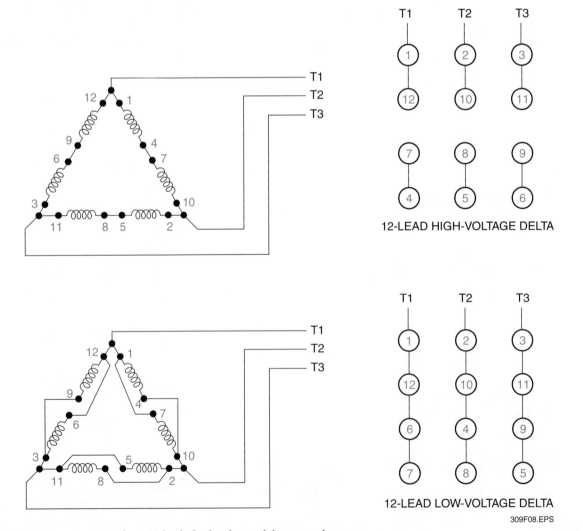

Figure 8 ◆ Lead connections for a 12-lead, dual-voltage, delta-wound motor.

If the windings are connected in series for operation on 480V, the total resistance of the winding is:

$$R_t = R1 + R2$$

$$R_t = 48\Omega + 48\Omega$$

$$R_t = 96\Omega$$

Consequently, if 480V is applied to this winding, the following current will flow:

$$I = \frac{E}{R}$$

$$I = \frac{480V}{96\Omega}$$

$$I = 5A$$

It is obvious that twice the voltage means half the current flow, or vice versa.

2.3.0 Special Connections

Some three-phase motors designed for operation on voltages higher than 600V may have more than 12 leads. Motors with 15 or 18 leads are common in high-voltage installations. A 15-lead motor has three coils per phase, as shown in *Figure 9*. Notice that the leads are numbered in the same spiral sequence as a nine-lead, wye-wound motor.

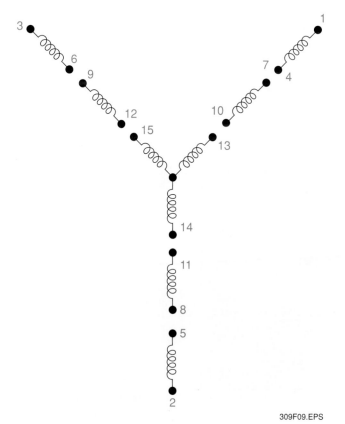

309F09.EPS

Figure 9 ◆ Fifteen-lead motor.

3.0.0 ◆ CALCULATING MOTOR CIRCUIT CONDUCTORS

The basic elements of a motor circuit are shown in *Figure 10*. Although these elements are shown separately in this illustration, there are certain cases in which the *NEC®* permits a single device to serve more than one function. For example, in some cases, one switch can serve as both the disconnecting means and the controller. In other cases, short circuit protection and overload protection can be combined in a single circuit breaker or set of fuses.

NOTE

NEC Section 430.6(A)(1) states that, with the exception of low-speed (less than 1,200 rpm), high-torque, and multi-speed motors, the full-load current (FLC) values given in *NEC Table 430.250* for 3Ø AC standard squirrel cage induction, wound rotor, and synchronous motors shall be used to determine the ampacity of feeder and branch circuit conductors, ratings of switches, controllers, branch circuit short-circuit values, and ground fault protection. The FLC values shall be used instead of any actual current rating marked on the motor nameplate (except for some non-continuous duty motor applications). Where a motor is marked in amperes, but not horsepower, the horsepower rating shall be assumed to be that corresponding to the values given in *NEC Table 430.250*, interpolated if necessary. In addition, *NEC Section 430.22(A)* states that for a single continuous duty motor, the conductors shall have an ampacity on both sides of the controller of not less than 125% of the motor's FLC as determined from the tables. Considerations for other types of motors, including wye-start/delta-run, multi-speed, high-torque, part-winding, and dual-voltage motors as well as wound rotor motors with external resistors or motors operating non-continuously, are not covered in the example motor calculations reflected in this module. *NEC Annex D* provides motor calculation examples.

WARNING!

The motor circuit calculations reflected in this module and described in *NEC Article 430, Part II, Motor Circuit Conductors* apply only to motor circuits rated at a nominal 600V or less as stated in *NEC Section 430.21*. Calculations for motor circuits over 600V are not covered in this module. *NEC Article 430, Part IX* amends or adds to the other provisions in *NEC Article 430* for motor circuits over 600V.

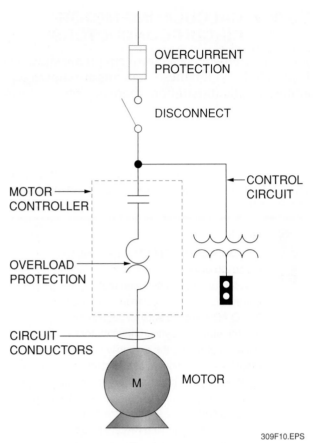

309F10.EPS

Figure 10 ◆ Basic elements of any motor circuit.

A typical motor control center and branch circuits feeding four different motors are shown in *Figure 11*. We will see how the feeder and branch circuit conductors are sized for these motors.

Step 1 Refer to *NEC Table 430.250* for the full-load current of each motor.

Step 2 Determine the full-load current of the largest motor in the group.

Step 3 Calculate the sum of the full-load current ratings for the remaining motors in the group.

Step 4 Multiply the full-load current of the largest motor by 1.25 (125%) and then add the sum of the remaining motors to the result *(NEC Section 430.24)*. The combined total will give the minimum feeder size.

When sizing feeder or branch circuit conductors for motors, be aware that the previous and following procedures will give the minimum conductor ampacity rating based on the *NEC*® minimum only. Consequently, it is often necessary to increase the size of conductors to compensate for voltage drop and power loss in the circuit.

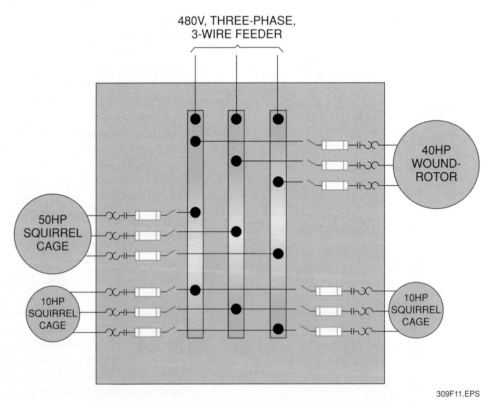

309F11.EPS

Figure 11 ◆ Typical motor control center.

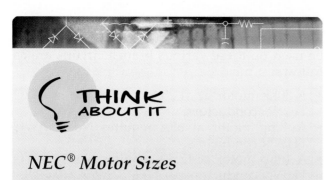

THINK
ABOUT IT

NEC® Motor Sizes

How is the largest motor in a group determined by the *NEC*®? Is it by frame size, horsepower, weight, or the motor's full-load amps (FLA)?

Another factor that can affect conductor sizing for standard squirrel cage induction, wound rotor, and synchronous motors is the motor circuit component rating. *NEC Section 110.14(C)* states that conductor sizes shall be selected so that the lowest temperature rating of any terminations, connected conductors, or connected devices (excluding the internal motor windings) are not exceeded. This means that regardless of a higher temperature rating for a conductor, the allowable ampacity must be selected from *NEC Tables 430.247 through 430.250* that does not exceed the lowest temperature rating of any connected terminations, devices, or other conductors. If the temperature rating of any of the connected terminations, devices, or other conductors is unknown, the conductor sizes must be selected from the 60°C ampacity column of the tables. In the case of termination provisions for equipment rated at 100A or less or marked for No. 14 AWG through No. 1 AWG, conductor sizes shall be selected from the 60°C ampacity columns of the tables.

INSIDE TRACK

Voltage Ratings

The motor voltage ratings used in the *NEC*® can be somewhat confusing. Motors are built and tested at various voltages. For example, the 460V referenced in *NEC Tables 430.248 and 430.250* is based on rated motor voltages; however, in order to standardize calculations and maintain minimum safety standards, the *NEC*® uses a higher value of 480V for motor calculations. See *NEC Section 220.5(A)*.

However, for motors marked with NEMA design code B, C, or D, conductors rated at 75°C or higher may be used if the ampacity required is 100A or less. In the case of termination provisions of equipment rated over 100A or marked for No. 1 AWG or larger, conductor sizes shall be selected from the 75°C column of the tables.

NOTE

The motor circuit calculation worksheet shown in the *Appendix* can be used for the types of motor circuits covered in this module. It is designed for use in motor circuit calculations for one or more standard single-speed, 3Ø, 60Hz continuous duty squirrel cage induction, synchronous, and wound rotor motors (without external resistors), all rated at 600V or less and with separate overcurrent devices and controllers.

Now we will complete the conductor calculations for the motor circuits in *Figure 11*.

Referring to *NEC Table 430.250*, the motor horsepower is shown in the far left-hand column. Follow across the appropriate row until you come to the column titled *460V*, which is the center voltage (440V to 480V) of the motor circuits in *Figure 11*. We find that the ampere ratings for the motors in question are as follows:

- 50hp = 65A
- 40hp = 52A
- 10hp = 14A

The largest motor in this group is the 50hp squirrel cage motor, which has a full-load current of 65A.

The sum of the remaining motors is:

$$52A + 14A + 14A = 80A$$

Now, multiply the full-load current of the largest motor by 125% (1.25) and then add the total amperage of the remaining motors:

$$(1.25 \times 65A) + 80A = 161.25A$$

Therefore, the feeders for the 460V, three-phase, three-wire motor control center will have a minimum ampacity of 161.25A. Referring to *NEC Table 310.16* under the column headed 75°C, the closest conductor size is 2/0 copper (rated at 175A) or 4/0 aluminum (rated at 180A).

The branch circuit conductors feeding the individual motors are calculated somewhat differently. *NEC Section 430.22(A)* requires that the ampacity of branch circuit conductors supplying a single continuous-duty motor must not be less than 125% of the motor full-load current rating.

INSIDE TRACK

Squirrel Cage Motor Fuse Selection

Most squirrel cage motors manufactured prior to 1996 are not marked with a design code B, C, or D on their nameplates. However, unless otherwise marked, most common squirrel cage motors fall into the design code B classification and are considered to be design code B motors for the purpose of selecting overcurrent protection.

Therefore, the current-carrying capacity of the branch circuit conductors feeding the four motors in question are calculated as follows:

$$50\text{hp motor} = 65\text{A} \times 1.25 = 81.25\text{A}$$
$$40\text{hp motor} = 52\text{A} \times 1.25 = 65\text{A}$$
$$10\text{hp motor} = 14\text{A} \times 1.25 = 17.5\text{A}$$

Referring to *NEC Table 310.16*, the closest size 75°C THWN copper conductors permitted to be used on these various branch circuits are as follows:

- A 50hp motor at 81.25A requires No. 4 AWG THWN conductors.
- A 40hp motor at 65A requires No. 6 AWG THWN conductors.
- A 10hp motor at 17.5A requires No. 14 AWG THWN conductors per *NEC Section 240.4(D)*.

Refer to *Figure 12* for a summary of the conductors used to feed our example motor control center, along with the branch circuits supplying the individual motors.

If voltage drop and/or power loss must be taken into consideration, please refer to the Level Three module, *Conductor Selection and Calculations*.

For motors with other voltages (up to 2,300V) or for synchronous motors, refer to *NEC Table 430.250*.

In accordance with *NEC Section 430.22(E)*, branch circuit conductors serving motors used for short-time, intermittent, or other varying duty must have an ampacity not less than the percentage of

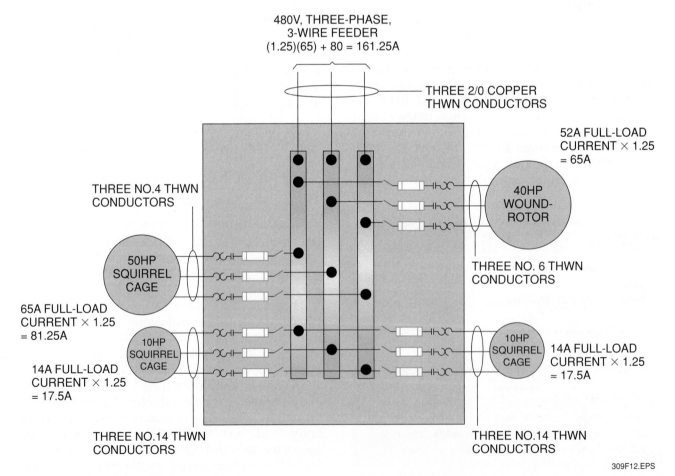

Figure 12 ◆ Sizing motor branch circuits.

the motor nameplate current rating shown in *NEC Table 430.22(E)*. However, to qualify as a short-time, intermittent motor, the nature of the apparatus that the motor drives must be arranged so that the motor cannot operate continuously with a load under any condition of use. Otherwise, the motor must be considered continuous duty. Consequently, the majority of motors encountered in the electrical trade must be rated for continuous duty, and the branch circuit conductors sized accordingly.

3.1.0 Wound-Rotor Motors

The primary full-load current ratings for wound-rotor motors are listed in *NEC Table 430.250* and are the same as those for squirrel cage motors. Conductors connecting the secondary leads of wound-rotor induction motors to their controllers must have a current-carrying capacity at least equal to 125% of the motor full-load secondary current if the motor is used for continuous duty. If the motor is used for less than continuous duty, the conductors must have a current-carrying capacity of not less than the percentage of the full-load secondary nameplate current given in *NEC*

Table 430.22(E). Conductors from the controller of a wound-rotor induction motor to its starting resistors must have an ampacity in accordance with *NEC Table 430.23(C)*.

> **NOTE**
>
> *NEC Section 430.6(A)(1)* specifies that for general motor applications (excluding applications of torque motors and sealed hermetic-type refrigeration compressor motors), the values given in *NEC Tables 430.247, 430.248, 430.249, and 430.250* should be used instead of the actual current rating marked on the motor nameplate when sizing conductors, switches, and overcurrent protection. Overload protection, however, is based on the marked motor nameplate.

3.2.0 Conductors for DC Motors

NEC Sections 430.22(A), Exception 1 and 430.29 cover the rules governing the sizing of conductors from a power source to a DC motor controller and from the controller to separate resistors for power accelerating and dynamic braking. *NEC Section 430.29*, with its table of conductor ampacity percentages, assures proper application of DC constant-potential motor controls and power resistors. However, when selecting overload protection, the actual motor nameplate current rating must be used.

3.3.0 Conductors for Miscellaneous Motor Applications

NEC Section 430.6 should be referred to for torque motors, shaded-pole motors, permanent split capacitor motors, and AC adjustable-voltage motors.

Feeder Size

If the 40hp wound-rotor motor in *Figure 12* was replaced with another 50hp squirrel cage motor, the #6 THWN conductors feeding that motor must be replaced with #4 THWN conductors. What size THWN, 75°C feeders are needed to accommodate the change?

Duty Ratings

Not all motors are rated for continuous duty. Some are rated for 5-, 15-, 30-, or 60-minute operation. The *NEC®* makes special allowances for these duty ratings. Refer to *NEC Section 430.22(E)*.

Breakdown Torque

Certain listed appliances are now rated by the breakdown torque (that is, the force it takes to stop the motor from rotating at full speed). This allows the use of the FLA directly from the nameplate of the appliance rather than having to refer to *NEC Tables 430.248 and 430.250*.

NEC Section 430.6(B) specifically states that the motor's nameplate full-load current rating is used to size ground-fault protection for a torque motor. However, both the branch circuit conductors and the overcurrent protection are sized by the provisions listed in *NEC Article 430, Part II* and *NEC Section 430.52*.

For sealed (hermetic) refrigeration compressor motors, the actual nameplate full-load running current of the motor must be used in determining the current rating of the disconnecting means, controller, branch circuit conductor, overcurrent protective devices, and motor overload protection.

4.0.0 ◆ MOTOR PROTECTIVE DEVICES

NEC Sections 430.51 through 430.58 require that the branch circuit protection for motor controls protect the circuit conductors, control apparatus, and the motor itself against overcurrent due to short circuits or ground faults.

Motors and motor circuits have unique operating characteristics and circuit components. Therefore, these circuits must be dealt with differently from other types of loads. Generally, two levels of overcurrent protection are required for motor branch circuits:

- *Overload protection* – Motor running overload protection is intended to protect the system components and motor from damaging overload currents.
- *Short circuit protection (includes ground fault protection)* – Short-circuit protection is intended to protect the motor circuit components such as the conductors, switches, controllers, overload relays, motor, etc., against short circuit currents or grounds. This level of protection is commonly referred to as motor branch circuit protection. Dual-element fuses are designed to give this protection, provided they are sized correctly.

There are a variety of ways to protect a motor circuit, depending upon the application. The ampere rating of a fuse selected for motor protection depends on whether the fuse is of the dual-element, time-delay type or the nontime-delay type.

In general, *NEC Table 430.52* specifies that short-circuit/ground-fault protection nontime-delay fuses can be sized at 300% of the motor full-load current for ordinary motors, while those for wound-rotor or direct current motors may be sized at 150% of the motor full-load current. The sizes of nontime-delay fuses for the four motors previously mentioned are listed in *Figure 13*.

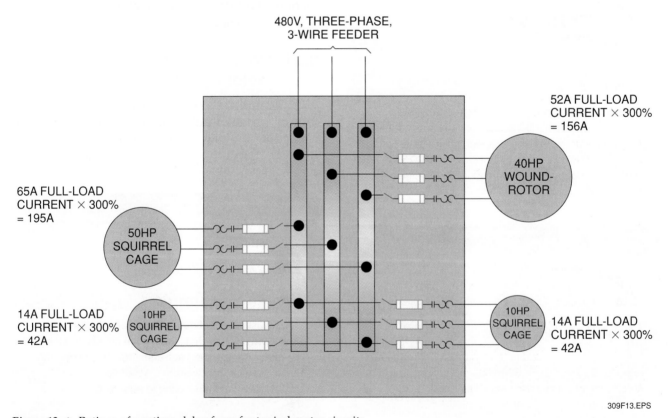

480V, THREE-PHASE, 3-WIRE FEEDER

52A FULL-LOAD CURRENT × 300% = 156A

40HP WOUND-ROTOR

65A FULL-LOAD CURRENT × 300% = 195A

50HP SQUIRREL CAGE

14A FULL-LOAD CURRENT × 300% = 42A

10HP SQUIRREL CAGE

10HP SQUIRREL CAGE

14A FULL-LOAD CURRENT × 300% = 42A

309F13.EPS

Figure 13 ◆ Ratings of nontime-delay fuses for typical motor circuits.

Because none of these sizes are standard, *NEC Section 430.52(C)(1), Exception No. 1* permits the size of the fuses to be increased to a standard size. Also, where absolutely necessary to permit motor starting, the size of the overcurrent device may be further increased, but must never be more than 400% of the full-load current *[NEC Section 430.52(C)(1), Exception No. 2(a)]*. In actual practice, most electricians would use a 200A nontime-delay fuse for the 50hp motor, a 175A fuse for the 40hp motor, and 45A fuses for the 10hp motors. If any of these fuses do not allow the motor to start without blowing, the fuses for the 50hp motor may be increased to a maximum of 400% of the full-load currents, which are 360A for the 50hp motor, 208A for the 40hp motor, and 56A for the 10hp motors.

Per *NEC Table 430.52*, dual-element, time-delay fuses are able to withstand normal motor starting current and can be sized closer to the actual motor rating than nontime-delay fuses. If necessary for proper motor operation, dual-element, time-delay fuses may be sized up to 175% of the motor's full-load current for all standard motors with the exception of wound-rotor and direct current motors. These motors must not have fuses sized for more than 150% of the motor's full-load current rating. Where absolutely necessary for proper operation, the rating of dual-element, time-delay fuses may be increased, but must never be more than 225% of the motor full-load current rating *[NEC Section 430.52(C)(1), Exception No. 2(b)]*. To size dual-element fuses at 175% for the four motors in *Figure 13*, proceed as follows:

50hp motor = 65A × 175 = 113.75A

40hp motor = 52A × 175 = 91A

10hp motors = 14A × 175 = 24.5A

Figure 14 gives general fuse application guidelines for motor branch circuits (*NEC Article 430, Part IV*). Bear in mind that in many cases, the maximum fuse size depends on the motor design letter, motor type, and starting method.

4.1.0 Practical Applications

For various reasons, motors are often oversized. For instance, a 5hp motor may be installed when the load demand is only 3hp. In these cases, a much higher degree of overload protection can be obtained by sizing the overload relay elements and/or dual-element, time-delay fuses based on the actual full-load current draw. In existing installations, the procedure for providing the maximum overcurrent protection for oversized motors is as follows:

Step 1 With a clamp-on ammeter, determine the running rms current when the motor is at normal full-load, as shown in *Figure 15*. Be sure this current does not exceed the nameplate current rating. The advantage of this method is realized when a lightly loaded motor (especially those over 50hp) experiences a single-phase condition. Even though the relays and fuses may be sized correctly based on the motor nameplate, circulating currents within the motor may cause damage. If unable to meter the motor current, take the current rating off the motor nameplate.

Step 2 Size the overload relay elements and/or overcurrent protection based on this current.

Step 3 Use a labeling system to mark the type and ampere rating of the fuse that should be in the fuse clips. This simple system makes it easy to run spot checks for proper fuse replacements.

NOTE

When installing the proper fuses in the switch to give the desired level of protection, it is often advisable to leave spare fuses on top of the disconnect or starter enclosure, or in a cabinet adjacent to the motor control center. This way, if the fuses open, the proper fuses can be readily reinstalled.

Individual motor disconnect switches must have an ampere rating of at least 115% of the motor full-load ampere rating *[NEC Section 430.110(A)]* or sized as specified in *NEC Section 430.109*. The next larger size switches with fuse reducers may sometimes be required.

Some installations may require larger dual-element fuses when:

• The motor uses dual-element fuses in high ambient temperature environments.
• The motor is started frequently or rapidly reversed.
• The motor is directly connected to a machine that cannot be brought up to full speed quickly (e.g., centrifugal machines such as extractors and pulverizers, machines having large fly wheels such as large punch presses, etc.).
• This is a Design B energy-efficient motor with full-voltage start.

Type of Motor	Dual-Element, Time-Delay Fuses			Nontime-Delay Fuses
	Motor Overload and Short Circuit	Backup Overload and Short Circuit	Short Circuit Only (Based on *NEC Tables 430.247 through 430.250* current ratings)	Short Circuit Only (Based on *NEC Tables 430.247 through 430.250* current ratings)
Service Factor 1.15 or Greater or 40°C Temp. Rise or Less	125% or less of motor nameplate current	125% or next standard size (not to exceed 140% of motor nameplate current)	150% to 175%	150% to 300%
Service Factor Less Than 1.15 or Greater Than 40°C Temp. Rise	115% or less of motor nameplate current	115% or next standard size (not to exceed 130% of motor nameplate current)	150% to 175%	150% to 300%

Fuses give overload and short circuit protection.

Overload relay gives overload protection and fuses provide backup overload protection.

Overload relay provides overload protection and fuses provide only short circuit protection.

Overload relay provides overload protection and fuses provide only short circuit protection.

309F14.EPS

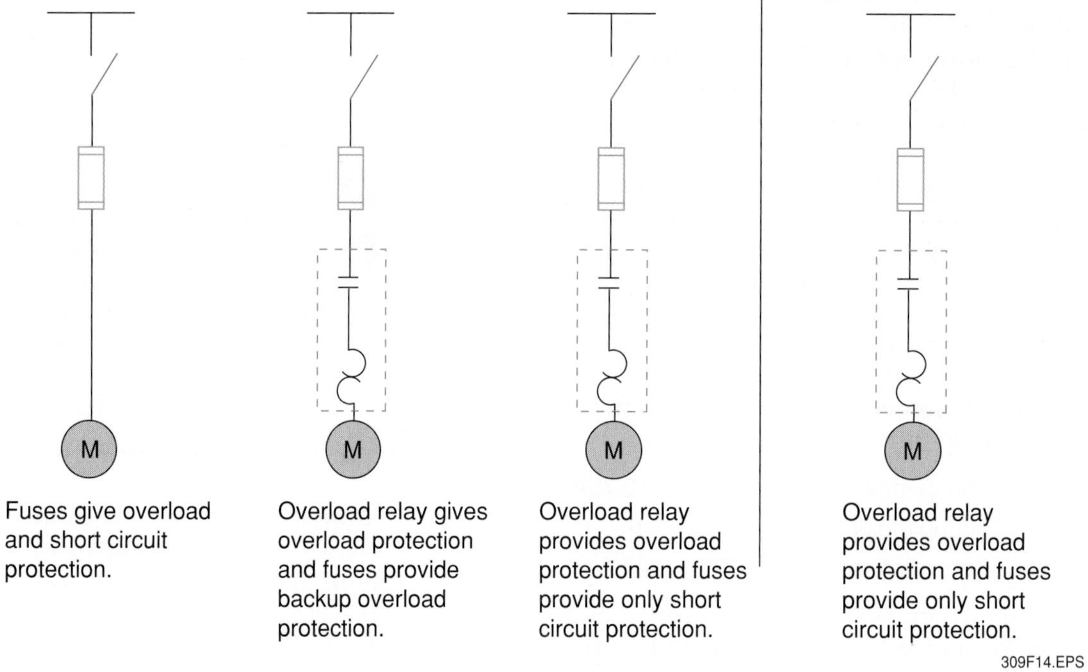

Figure 14 ◆ Fuse application guidelines for motor branch circuits.

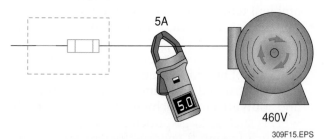

309F15.EPS

Figure 15 ◆ Determining running current with an ammeter.

4.2.0 Motor Overload Protection

A high-quality electric motor that is properly cooled and protected against overloads can be expected to have a long life. The goal of proper motor protection is to prolong motor life and postpone the failure that ultimately takes place. Good electrical protection consists of providing both proper overload protection and current-limiting short-circuit protection. AC motors and other types of high inrush loads require protective devices with special characteristics. Normal, full-

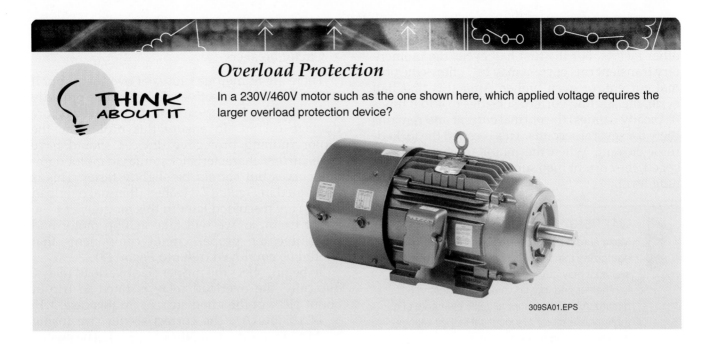

Overload Protection

THINK ABOUT IT

In a 230V/460V motor such as the one shown here, which applied voltage requires the larger overload protection device?

309SA01.EPS

load running currents of motors are substantially less than the currents that result when motors start or are subjected to temporary mechanical overloads. This is illustrated by the typical motor starting current curve shown in *Figure 16*.

At the moment an AC motor circuit is energized, the starting current rapidly rises to many times the normal running current and the rotor begins to rotate. As the rotor accelerates and reaches running speed, the current declines to the normal running current. Thus, for a period of time, the overcurrent protective devices in the motor circuit must be able to tolerate the rather substantial temporary overload. Motor starting

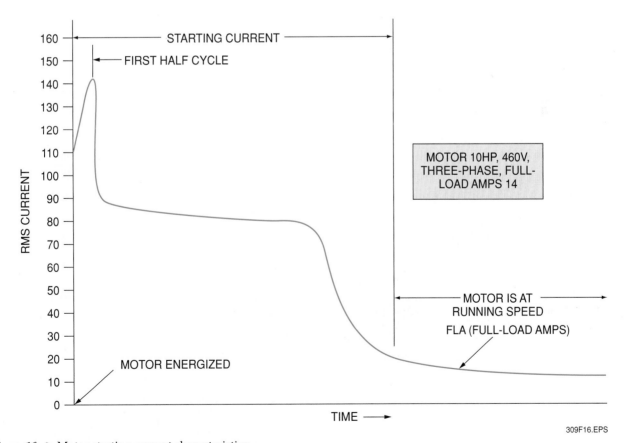

309F16.EPS

Figure 16 ◆ Motor starting current characteristics.

currents can vary substantially, depending on the motor type, load type, starting methods, and other factors. For the first half cycle, the momentary transient rms current may be 11 times the normal current, or even higher. After this first half cycle, the starting current subsides to 4 to 8 times (typically 6 times) the normal current and remains there for several seconds. This is called the locked-rotor current. When the motor reaches running speed, the current then subsides to its normal running level.

WARNING!

Make sure fuses are fully seated before energizing the device. If a fuse is not fully seated it may arc, causing damage to the fuseholder and creating a potential safety hazard. Also, make sure all relay elements of a three-phase motor are identical in their ratings before installing them.

Motor overload protective devices must withstand the temporary overload caused by motor starting currents, and, at the same time, protect the motor from continuous or damaging overloads. The main types of devices used to provide motor overload protection include:

- Overload relays
- Fuses
- Circuit breakers

There are numerous causes of overloads, but if the overload protective devices are properly responsive, such overloads can be removed before damage occurs. To ensure this protection, the motor running protective devices should have time-current characteristics similar to motor damage curves but should be slightly faster. This is illustrated in *Figure 17*. Heaters (thermal overload relays) are discussed later in this module.

For example, we will take a 10hp motor and determine the proper circuit components that should be employed (refer to *Figure 18*).

To begin, select the proper size overload relays. Typically, the overload relay is rated to trip at about 115% of the rated current (in this case, 1.15 × 14A = 16.1A). The correct starter size (using NEMA standards) is a NEMA Type 1. The switch size that should be used is 30A. Switch sizes are based on *NEC*® requirements; dual-element, time-delay fuses allow the use of smaller switches.

For short-circuit protection on large motors with currents in excess of 600A, low-peak time-delay fuses are recommended. Most motors of this size will have reduced voltage starters, and the inrush currents are not as rigorous. Low-peak

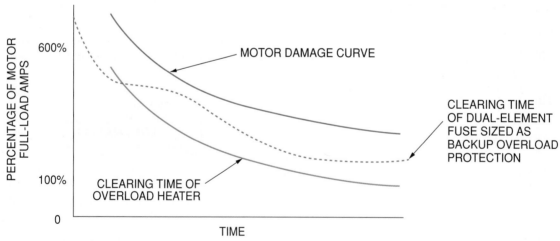

309F17.EPS

Figure 17 ◆ Time-current characteristics of dual-element fuses and overload heaters.

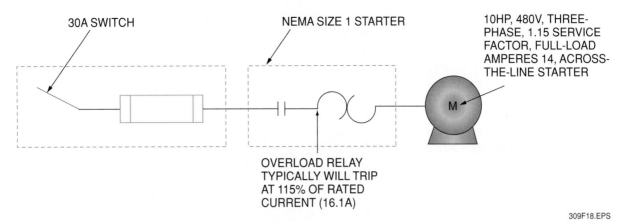

30A SWITCH

NEMA SIZE 1 STARTER

10HP, 480V, THREE-PHASE, 1.15 SERVICE FACTOR, FULL-LOAD AMPERES 14, ACROSS-THE-LINE STARTER

OVERLOAD RELAY TYPICALLY WILL TRIP AT 115% OF RATED CURRENT (16.1A)

309F18.EPS

Figure 18 ◆ Circuit components of a typical 10-horsepower motor.

fuses should be sized at approximately 150% to 175% of the motor full-load current.

Motor controllers with overload relays commonly used on motor circuits provide motor running overload protection. The overload relay setting or selection must comply with *NEC Section 430.32*. On overload conditions, the overload relays should operate to protect the motor. For motor backup protection, size dual-element fuses at the next ampere rating greater than the overload relay trip setting. This can typically be achieved by sizing dual-element fuses at 125% for 1.15 **service factor** motors and 115% for 1.0 service factor motors. The service factor is the number by which the horsepower rating is multiplied to determine the maximum safe load that a motor may be expected to carry continuously at its rated voltage and frequency.

5.0.0 ◆ CIRCUIT BREAKERS

The *NEC®* recognizes the use of instantaneous trip circuit breakers (without time delay) for short circuit protection of motor branch circuits. Such breakers are acceptable only if they are adjustable and are used in combination motor starters. Such starters must have coordinated overload, short-circuit, and ground-fault protection for each conductor and must be approved for the purpose in accordance with *NEC Section 430.52(C)(3)*. This permits the use of smaller circuit breakers than would be allowed if a standard thermal-magnetic circuit breaker was used. In this case, smaller circuit breakers can offer faster operation for greater protection against grounds and short circuits. *Figure 19* shows a schematic diagram of motor circuit protector (MCP) used in a combination motor starter.

MCPs are used only in combination with motor starters having integral overload protection. However, heaters in the motor starter protect the entire circuit and all equipment against overloads up to and including locked-rotor current. Heaters (thermal overload relays) are commonly set at 115% to 125% of the motor full-load current.

In dealing with such circuits, an adjustable circuit breaker can be set to take over the interrupting task at currents above locked-rotor current and up to the short circuit duty of the supply system at the point of the installation. The magnetic trip in such breakers can typically be adjusted from 3 to 13 times the breaker current rating. For example, a 100A circuit breaker can be adjusted to trip anywhere between 300A and 1,300A. Consequently, the circuit breaker may serve as motor short circuit protection. However, instantaneous trip circuit breakers used in these installations cannot be adjusted to more than the value specified in *NEC Table 430.52*.

5.1.0 Application of MCPs

We will compare the use of both thermal-magnetic and magnetic-only circuit breakers in the motor circuit shown in *Figure 20*. In doing so, our job is to select a circuit breaker that will provide short-circuit protection and also qualify as the motor circuit disconnecting means.

Step 1 Determine the motor full-load current from *NEC Table 430.250*. This is found to be 80A.

Step 2 A circuit breaker suitable for use as a motor disconnecting means must have a current rating of at least 115% of the motor full-load current. Therefore:

$$1.15 \times 80A = 92A$$

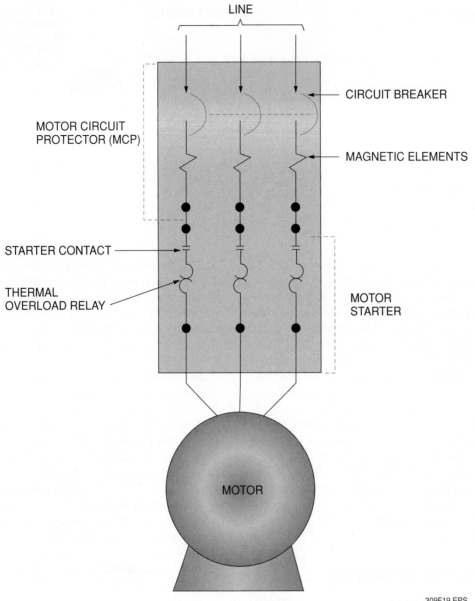

LINE

CIRCUIT BREAKER

MOTOR CIRCUIT
PROTECTOR (MCP)

MAGNETIC ELEMENTS

STARTER CONTACT

THERMAL
OVERLOAD RELAY

MOTOR
STARTER

MOTOR

309F19.EPS

Figure 19 ◆ Combination motor starter with motor circuit protection (MCP).

NOTE

NEC Table 430.52 permits the use of an inverse-time circuit breaker rated at not more than 250% of the motor full-load current. However, a circuit breaker could be rated as high as 400% of the motor full-load current if necessary to hold the motor starting current without opening, according to *NEC Section 430.52(C)(1), Exception 2(c)*.

Step 3 Assuming that a circuit breaker rated at 250% of the motor full-load current will be used, perform the following calculation:

$$2.5 \times 80A = 200A$$

Step 4 Select a regular thermal-magnetic circuit breaker with a 225A frame that is set to trip at 200A.

Step 5 Refer to *Figure 20* and note that this is a NEMA Design B energy-efficient motor. Refer to *NEC Table 430.52* and note that an instantaneous breaker for a Design B energy-efficient motor should be no more than 1,100% of the motor full-load current.

Step 6 Determine the circuit breaker rating by multiplying the full-load current by 1,100%:

$$80A \times 1,100\% \ (11.00) = 880A$$

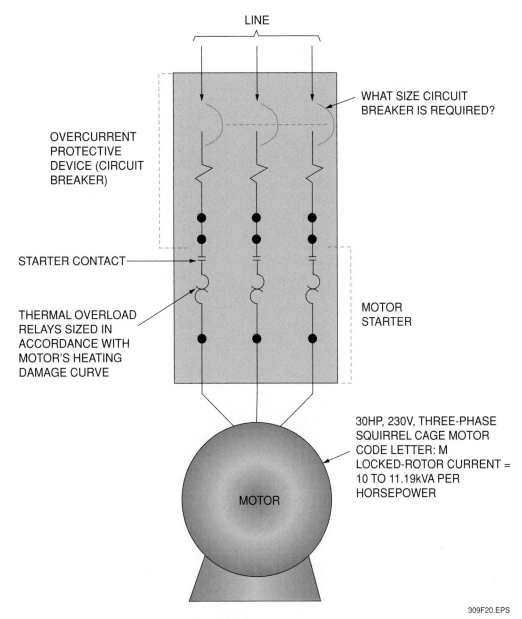

LINE

WHAT SIZE CIRCUIT BREAKER IS REQUIRED?

OVERCURRENT PROTECTIVE DEVICE (CIRCUIT BREAKER)

STARTER CONTACT

THERMAL OVERLOAD RELAYS SIZED IN ACCORDANCE WITH MOTOR'S HEATING DAMAGE CURVE

MOTOR STARTER

MOTOR

30HP, 230V, THREE-PHASE SQUIRREL CAGE MOTOR CODE LETTER: M LOCKED-ROTOR CURRENT = 10 TO 11.19kVA PER HORSEPOWER

309F20.EPS

Figure 20 ◆ Typical 30-horsepower Design B energy-efficient motor circuit.

The thermal-magnetic circuit breaker selected in Step 4 will provide protection for grounds and short circuits without interfering with motor overload protection. Note, however, that the instantaneous trip setting of a 200A circuit breaker will be about 10 times the current rating, or:

$$200A \times 10 = 2,000A$$

Now consider the use of a 100A circuit breaker with thermal and adjustable magnetic trips. The instantaneous trip setting at 10 times the normal current rating would be:

$$100A \times 10 = 1,000A$$

Although this 1,000A instantaneous trip setting is above the 880A locked-rotor current of the 30hp motor in question, the starting current would probably trip the thermal element and open the circuit breaker.

This problem can be solved by removing the circuit breaker's thermal element and leaving only the magnetic element in the circuit breaker. Then the conditions of overload can be cleared by the overload devices (heaters) in the motor starter. If the setting of the instantaneous trip circuit breaker will not hold under the starting load in *NEC Section 430.52(C)(3)*, then *Exception No. 1* will, under engineering evaluation, permit increasing the trip setting up to, but not exceeding, 1,300% (1,700% for NEMA Design B energy-efficient motors).

Therefore, since it has been determined that the 30hp motor in question has a full-load ampere rating of 80A, the maximum trip must not be set higher than:

$$80A \times 17 = 1{,}360A \text{ or about } 1{,}300A$$

This circuit breaker would qualify as the circuit disconnect because it has a rating higher than 115% of the motor full-load current (80A × 1.15 = 92A). However, the use of a magnetic-only circuit breaker does not protect against low-level grounds and short circuits in the branch circuit conductors on the line side of the motor starter overload relays—such an application must be made only where the circuit breaker and motor starter are installed as a combination motor starter in a single enclosure.

5.2.0 Motor Short Circuit Protectors

Motor short circuit protectors (MSCPs) are fuse-like devices designed for use only in a special type of fusible-switch combination motor starter. The combination offers short-circuit protection, overload protection, disconnecting means, and motor control, all with assured coordination between the short circuit interrupter and the overload devices.

The *NEC*® recognizes MSCPs in *NEC Section 430.52(C)(7)*, provided the combination is identified for the purpose (i.e., a combination motor starter equipped with an MSCP and listed by Underwriters Laboratories or another nationally recognized third-party testing lab as a package called an MSCP starter).

6.0.0 ◆ MULTI-MOTOR BRANCH CIRCUITS

NEC Sections 430.53(A),(B), and (C) permit the use of more than one motor on a branch circuit, provided the following conditions are met:

- Two or more motors, each rated at not more than 1hp, and each drawing a full-load current not exceeding 6A, may be used on a branch circuit protected at not more than 20A at 120V or less, or 15A at 600V or less. The rating of the branch circuit protective device marked on any of the controllers must not be exceeded. Individual overload protection is necessary in such circuits unless the motor is not permanently installed, or is manually started and is within

sight of the controller location, or has sufficient winding impedance to prevent overheating due to locked-rotor current, or is part of an approved assembly which does not subject the motor to overloads and which incorporates protection for the motor against locked-rotor current, or the motor cannot operate continuously under load.

- Two or more motors of any rating, each having individual overload protection, may be connected to a single branch circuit that is protected by a short circuit protective device (MSCP). The protective device must be selected in accordance with the maximum rating or setting that could protect an individual circuit to the motor of the smallest rating. This may be done only where it can be determined that the branch circuit device so selected will not open under the most severe normal conditions of service that might be encountered. This *NEC*® section offers wide application of more than one motor on a single circuit, particularly in the use of small integral-horsepower motors installed on 208V, 240V, and 480V, three-phase industrial and commercial systems. Only such three-phase motors have full-load operating currents low enough to permit more than one motor on circuits fed from 15A protective devices.

Using these *NEC*® rules, we will take a typical branch circuit (*Figure 21*) with more than one motor connected and see how the calculations are made.

The full-load current of each motor is taken from *NEC Table 430.250* as required by *NEC Section 430.6(A)(1)*. A circuit breaker must be chosen that does not exceed the maximum value of short-circuit protection (250%) required by *NEC Section 430.52* and *NEC Table 430.52* for the smallest motor in the group (in this case, 1.5hp). Since the listed full-load current for the smallest motor (1.5hp) is 3A, the calculation is made as follows:

$$3A \times 2.5 \text{ (250\%)} = 7.5A$$

NOTE

NEC Section 430.52, Exception No. 1 allows the next higher size rating or setting for a standard circuit breaker. Since a 15A circuit breaker is the smallest standard rating recognized by *NEC Section 240.6*, a 15A, three-pole circuit breaker may be used.

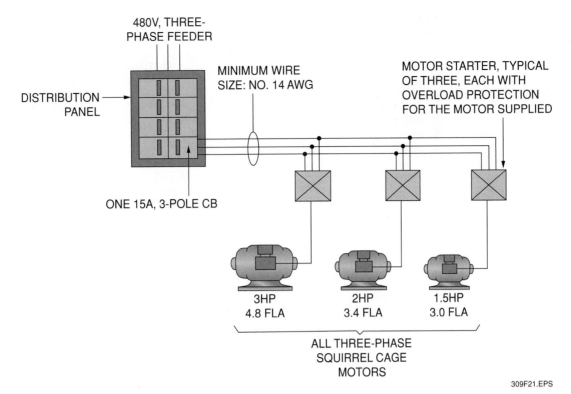

Figure 21 ◆ Several motors on one branch circuit.

The total load of the motor currents must be calculated as follows:

$$4.8A + 3.4A + 3.0A = 11.2A$$

The total full-load current for the three motors (11.2A) is well within the 15A circuit breaker rating, which has a sufficient time delay in its operation to permit starting of any one of these motors with the other two already operating. The torque characteristics of the loads on starting are not high. Therefore, the circuit breaker will not open under the most severe normal service.

Make certain that each motor is provided with the properly rated individual overload protection in the motor starter.

Branch circuit conductors are sized in accordance with *NEC Section 430.24*. In this case:

$$4.8A + 3.4A + 3.0A + [25\% \text{ of the largest motor}$$
$$(4.8A \times 0.25 = 1.2A)] = 12.4A$$

No. 14 AWG conductors rated at 75°C will fully satisfy this application, as long as the overcurrent protection device does not exceed 15A, according to *NEC Section 240.4(D)*.

Another multi-motor situation is shown in *Figure 22*. In this case, smaller motors are used. In general, *NEC Section 430.53(B)* requires branch circuit protection to be no greater than the maximum amperes permitted by *NEC Section 430.52* for the lowest rated motor of the group, which in this case is 1.1A for the ½hp motors. With this information in mind, we will size the circuit components for this application.

From *NEC Section 430.52* and *NEC Table 430.52*, the maximum protection rating for a circuit breaker is 250% of the lowest rated motor. Since this rating is 1.1A, the calculation is performed as follows:

$$2.5 \times 1.1A = 2.75A$$

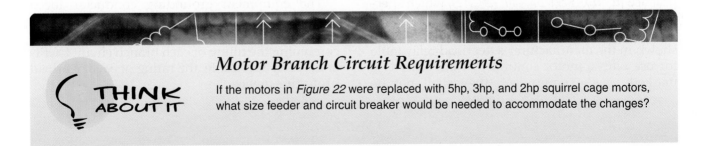

Motor Branch Circuit Requirements

THINK ABOUT IT

If the motors in *Figure 22* were replaced with 5hp, 3hp, and 2hp squirrel cage motors, what size feeder and circuit breaker would be needed to accommodate the changes?

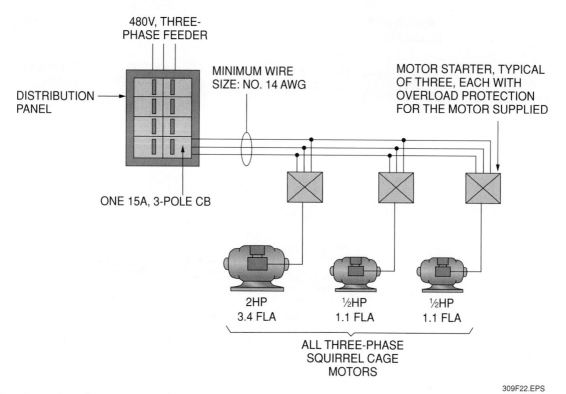

Figure 22 ◆ Several smaller motors supplied by one branch circuit.

309F22.EPS

> **NOTE**
>
> Since 2.75A is not a standard rating for a circuit breaker *(NEC Section 240.6)*, *NEC Section 430.52(C)(1), Exception No. 1* permits the use of the next higher rating. Because 15A is the lowest standard rating of circuit breakers, it is the next higher device rating above 2.75A and satisfies *NEC®* rules governing the rating of the branch circuit protection.

These two previous applications permit the use of several motors up to the circuit capacity, based on *NEC Sections 430.24 and 430.53(B)* and on starting torque characteristics, operating duty cycles of the motors and their loads, and the time delay of the circuit breaker. Such applications greatly reduce the number of circuit breakers and panels and the amount of wire used in the total system. One limitation, however, is placed on this practice in *NEC Section 430.52(C)(2)*, which specifies that where maximum branch circuit short circuit and ground fault protective device ratings are shown in the manufacturer's overload relay table for use with a motor controller or are otherwise marked on the equipment, they shall not be exceeded even if higher values are allowed, as shown in the preceding examples.

7.0.0 ◆ EQUIPMENT GROUNDING CONDUCTORS FOR MOTOR FEEDER AND BRANCH CIRCUITS

NEC Section 250.122 states that equipment grounding conductors of the wire type shall not be smaller than shown in *NEC Table 250.122*. For branch motor circuits of continuous duty motors, the sizing is based on the overcurrent protective device (OCPD) but is not required to be larger than the circuit conductors supplying the equipment. Additionally, where the OCPD is an instantaneous-trip device, the equipment grounding conductor is permitted to be based on the rating of the overload protective device, but shall not be less than the size shown in *NEC Table 250.122* for that rating.

The equipment grounding conductor for a motor feeder circuit furnishing the branch circuits for continuous-duty motors is determined by combining the rating of each branch circuit OCPD and, when applicable, the rating of each overload protective device.

8.0.0 ◆ POWER FACTOR CORRECTION AT MOTOR TERMINALS

Generally, the most effective method of power factor correction is the installation of capacitors at the cause of the poor power factor—the induction motor. This not only increases the power factor, but also releases system capacity, improves voltage stability, and reduces power losses.

When power factor correction capacitors are used, the total corrective kVAR on the load side of the motor controller should not exceed the value required to raise the no-load power factor to unity. Corrective kVAR in excess of this value may cause over-excitation that results in high transient voltages, currents, and torques that can increase safety hazards to personnel and possibly damage the motor or driven equipment.

Do not connect power factor correction capacitors at motor **terminals** on elevator motors; multi-speed motors; plugging or jogging applications; or open transition, wye-delta, autotransformer starting, and some part-winding start motors.

If possible, capacitors should be located at position No. 2, as shown in *Figure 23*. This does not change the current flowing through the motor overload protectors.

The connection of capacitors at position No. 3 requires a change of overload protectors. Capacitors should be located at position No. 1 for any of the following applications:

- Elevator motors
- Multi-speed motors
- Plugging or jogging applications
- Open transition, wye-delta, autotransformer starting motors
- Some part-winding motors

NOTE

To avoid over-excitation, make sure that the bus power factor is not increased above 95% under all loading conditions.

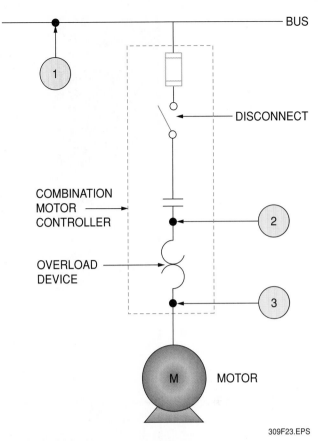

Figure 23 ◆ Placement of capacitors in motor circuit.

Table 1 allows the determination of corrective kVAR required where capacitors are individually connected at motor leads. These values should be considered the maximum capacitor rating when the motor and capacitor are switched as a unit. The figures given are for three-phase, 60Hz, NEMA Class B motors to raise the full-load power factor to 95%.

Table 1 Motor Power Factor Correction Table

Induction Motor Horsepower Rating	Nominal Motor Speed in RPM											
	3600		1800		1200		900		720		600	
	Capacitor Rating kVAR	Line Current Reduction %	Capacitor Rating kVAR	Line Current Reduction %	Capacitor Rating kVAR	Line Current Reduction %	Capacitor Rating kVAR	Line Current Reduction %	Capacitor Rating kVAR	Line Current Reduction %	Capacitor Rating kVAR	Line Current Reduction %
3	1.5	14	1.5	15	1.5	20	2	27	2.5	35	3.5	41
5	2	12	2	13	2	17	3	25	4	32	4.5	37
7½	2.5	11	2.5	12	3	15	4	22	5.5	30	6	34
10	3	10	3	11	3.5	14	5	21	6.5	27	7.5	31
15	4	9	4	10	5	13	6.5	18	8	23	9.5	27
20	5	9	5	10	6.5	12	7.5	16	9	21	12	25
25	6	9	6	10	7.5	11	9	15	11	20	14	23
30	7	8	7	9	9	11	10	14	12	18	16	22
40	9	8	9	9	11	10	12	13	15	16	20	20
50	12	8	11	9	13	10	15	12	19	15	24	19
60	14	8	14	8	15	10	18	11	22	15	27	19
75	17	8	16	8	18	10	21	10	26	14	32.5	18
100	22	8	21	8	25	9	27	10	32.5	13	40	17
125	27	8	26	8	30	9	32.5	10	40	13	47.5	16
150	32.5	8	30	8	35	9	37.5	10	47.5	12	52.5	15
200	40	8	37.5	8	42.5	9	47.5	10	60	12	65	14
250	50	8	45	7	52.5	8	57.5	9	70	11	77.5	13

1. Stator windings can best be described as _____.
   a. a structure of copper or aluminum wire coils
   b. insulating fibers mounted on a shaft
   c. copper bars mounted on a spindle
   d. steel wires mounted on a shaft

2. The synchronous speed of a 60Hz, three-phase induction motor with two poles is _____.
   a. 3,600 rpm
   b. 1,800 rpm
   c. 1,200 rpm
   d. 900 rpm

3. The two most common three-phase motor configurations are _____.
   a. synchronous and rms
   b. box and star
   c. wye and delta
   d. star and wye

4. The most common number of motor leads found on a three-phase, wye-wound motor is _____.
   a. 3
   b. 6
   c. 9
   d. 12

5. Doubling the voltage on a dual-voltage motor _____.
   a. doubles the synchronous speed
   b. doubles the full-load current
   c. halves the full-load current
   d. halves the synchronous speed

6. The total resistance of the stator windings in a three-phase motor, if the windings are connected in parallel and the resistance of each winding is 96Ω, is _____.
   a. 48Ω
   b. 96Ω
   c. 192Ω
   d. 220Ω

7. If a squirrel cage induction motor draws 2A of current at 240V, the amperage will be _____ if connected for use on 480V.
   a. 1A
   b. 2A
   c. 3A
   d. 4A

8. Of the following motors, the one most likely to have 15 or 18 motor leads is the _____.
   a. single-phase capacitor-start motor
   b. 120V shaded-pole motor
   c. 480V three-phase squirrel cage motor
   d. 2,400V three-phase motor

9. The main purpose of motor overload protection is to protect the motor _____.
   a. against short circuits
   b. against ground faults
   c. from damaging overload currents
   d. against locked-rotor current

10. A circuit breaker suitable for use as a motor disconnecting means must have a current rating of at least _____% of the motor full-load current.
   a. 250
   b. 175
   c. 300
   d. 115

Summary

The *NEC*® plays an important role in the selection and application of motors, including branch circuit conductors, disconnects, controllers, overcurrent protection, and overload protection. For example, *NEC Article 430* covers the application and installation of motor circuits and motor control connections, including conductors, short-circuit and ground-fault protection, controllers, disconnects, and overload protection.

NEC Article 440 contains provisions for motor-driven air conditioning and refrigerating equipment, including the branch circuits and controllers for the equipment. It also takes into account the special considerations involved with sealed (hermetic) motor compressors, in which the motor operates under the cooling effect of the refrigeration. When referring to *NEC Article 440*, be aware that the rules in this article are in addition to, or are amendments to, the rules given in *NEC Article 430*.

Notes

Trade Terms
Introduced in This Module

Circuit interrupter: A non-automatic, manually operated device designed to open a current-carrying circuit without injury to itself.

Rating: A designated limit of operating characteristics based on definite conditions. Such operating characteristics as load, voltage, frequency, etc., may be given in the rating.

Service factor: The number by which the horsepower rating is multiplied to determine the maximum safe load that a motor may be expected to carry continuously at its rated voltage and frequency.

Terminal: A point at which an electrical component may be connected to another electrical component.

Torque: A force that produces or tends to produce rotation. Common units of measurement of torque are foot-pounds and inch-pounds.

Motor Circuit Calculation Worksheet

MOTOR CIRCUIT CALCULATION WORKSHEET

For standard single-speed, 3Ø, 60Hz continuous-duty squirrel cage induction, synchronous, and wound-rotor motors (without external resistors) rated 600V or less with separate overcurrent devices and controllers

MOTOR NAMEPLATE DATA

	Motor 1	Motor 2	Motor 3	Motor 4	Motor 5
TYPE OF MOTOR					
Rated volts					
Rated full-load amps (FLA)[1]					
Rated horsepower (hp)[2]					
Design code letter (B, C, or D)					
Thermally protected (yes or no)					
Service factor					
Temperature rise					

[1] If rated in kVA, convert kVA to VA, then use VA divided by rated voltage × 1.73 to determine current.

[2] If not rated, use the rated FLA to find the horsepower that corresponds with the current in *NEC Table 430.250*, interpolating as necessary.

BRANCH CIRCUIT CALCULATIONS

(Controller for each motor is marked at no less than the hp of the motor and with integral selectable overload relays)

Calculations	*NEC*® Reference	Motor 1	Motor 2	Motor 3	Motor 4	Motor 5
1. Lowest temperature rating of termination provisions or devices in circuit [1]	*NEC Section 110.14(C)*					
2. Select FLA from table for motor type, voltage, and hp	*NEC Table 430.250*					
3. Select conductor size from table using FLA × 125%[2]	*NEC Table 310.16*					
4. Select overcurrent protective device (OCPD) using FLA × percentage from *NEC Table 430.52* [3]	*NEC Table 430.52* *NEC Section 240.6* *NEC Section 430.52(C)(1) & (2)*					
5. Select grounding conductor size from *NEC Table 250.122* using OCPD value	*NEC Table 250.122*					
6. Select overload protection size for motors not thermally protected[4]	*NEC Section 430.32 (A)(1) & (B)(1)*					

[1] If marked for 100A or less, for use of No. 14 through 1 AWG conductors, or if unknown, use 60°C. For Design Code B, C, or D motors use 75°C. For termination provisions or devices rated for more than 100A or AWG conductor sizes larger than 1 AWG, use 75°C.

[2] Use the ampacity column for the temperature entered on line 1 to select the conductor size from the table. Higher temperature rated conductors for the same conductor size may be used if desired.

[3] If the result is not a standard size, use the next standard size specified in *NEC Section 240.6*. Values can be increased as specified in *NEC Section 430.52(C)(1) Exception 2*, but cannot exceed the manufacturer's maximum overcurrent device rating per *NEC Section 430.52(C)(2)*.

[4] For motor service factors of 1.15 or greater or marked with a temperature rise of 40°C or less, multiply the nameplate FLA by 125%. For all other motors, multiply the nameplate FLA by 115%.

FEEDER CIRCUIT CALCULATIONS (MORE THAN ONE MOTOR)

Determine	Using Data from Branch Circuit Calculations	Result
Conductor size (*NEC Table 310.16*)[1]	From line 2 above, select the largest motor FLA and multiply by 125%, then add the remaining FLAs. Use the total to determine conductor size from *NEC Table 310.16*.	
Overcurrent protective device (OCPD) [2]	From line 4 above, select the largest OCPD and add the remaining motor FLAs from line 2 to determine the total for the feeder circuit OPCD per *NEC Section 240.6*.	
Equipment grounding conductor from *NEC Table 250.122*	Add all OCPDs from line 4 above to determine the total current rating required for the conductor. Use the total to determine conductor size from *NEC Table 250.122*.	

[1] If termination provisions or devices in a feeder circuit are marked for 100A or less, for use of No. 14 through 1 AWG conductors, or if unknown, use 60°C column. For termination provisions or devices rated for more than 100A or AWG conductor sizes larger than 1 AWG, use 75°C.

[2] If the total is not a standard OCPD, use the next smaller size listed in *NEC Section 240.6*.

309A01.EPS

This module is intended to present thorough resources for task training. The following reference work is suggested for further study. This is optional material for continued education rather than for task training.

National Electrical Code® Handbook, Latest Edition. Quincy, MA: National Fire Protection Association.

CONTREN® LEARNING SERIES – USER UPDATE

NCCER makes every effort to keep these textbooks up-to-date and free of technical errors. We appreciate your help in this process. If you have an idea for improving this textbook, or if you find an error, a typographical mistake, or an inaccuracy in NCCER's Contren® textbooks, please write us, using this form or a photocopy. Be sure to include the exact module number, page number, a detailed description, and the correction, if applicable. Your input will be brought to the attention of the Technical Review Committee. Thank you for your assistance.

Instructors – If you found that additional materials were necessary in order to teach this module effectively, please let us know so that we may include them in the Equipment/Materials list in the Annotated Instructor's Guide.

Write: Product Development and Revision
National Center for Construction Education and Research
3600 NW 43rd St., Bldg. G, Gainesville, FL 32606

Fax: 352-334-0932

E-mail: curriculum@nccer.org

Craft _____ Module Name _____

Copyright Date _____ Module Number _____ Page Number(s) _____

Description _____

_____

_____

_____

(Optional) Correction _____

_____

_____

(Optional) Your Name and Address _____

_____

_____

Voice, Data, and Video

St. Vincent's North Tower

Birmingham, Alabama, is home to this $25–99 Million award winner in the Health Care category in the Award of Excellence competition sponsored by Associated Builders and Contractors, Inc. Its builder was Brasfield & Gorrie, LLC.

26310-08

26310-08
Voice, Data, and Video

Topics to be presented in this module include:

Overview

A variety of cable systems are used to supply power to voice, data, and video systems. These include the unshielded twisted pair cable used in standard telephone systems, the coaxial cable associated with video systems, and the fiber-optic cable that is increasingly used due to its ability to transfer very large volumes of information over long distances. Regardless of the cable type, each must be carefully installed and terminated in order to provide a reliable signal. After the cable has been installed, the system must be tested using various meters and test equipment.

Objectives

When you have completed this module, you will be able to do the following:

1. Define the different categories for voice-data-video (VDV) cabling systems.
2. Install raceways, boxes, and enclosures for VDV systems.
3. Interpret and apply *NEC*® requirements for installing and grounding VDV systems.
4. Explain the requirements for firestopping.

Trade Terms

Attenuation
Backbone
Bandwidth
Coaxial cable
Cross connect
Delay skew
ELFEXT

Ethernet
Firestopping
Gigabit
Innerduct
Megabit
NEXT
Patch cord

Required Trainee Materials

1. Pencil and paper
2. Appropriate personal protective equipment

Prerequisites

Before you begin this module, it is recommended that you successfully complete *Core Curriculum; Electrical Level One; Electrical Level Two;* and *Electrical Level Three*, Modules 26301-08 through 26309-08.

This course map shows all of the modules in *Electrical Level Three*. The suggested training order begins at the bottom and proceeds up. Skill levels increase as you advance on the course map. The local Training Program Sponsor may adjust the training order.

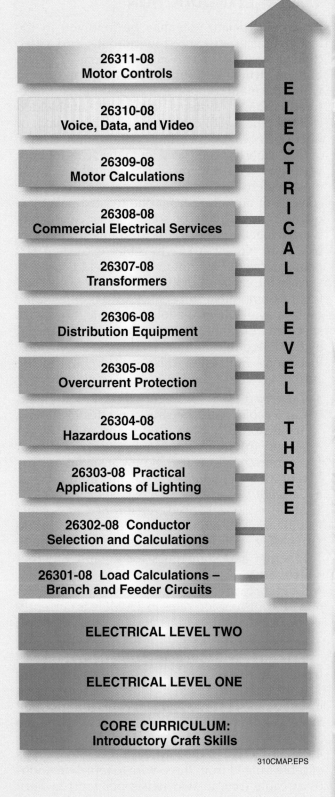

26311-08
Motor Controls

26310-08
Voice, Data, and Video

26309-08
Motor Calculations

26308-08
Commercial Electrical Services

26307-08
Transformers

26306-08
Distribution Equipment

26305-08
Overcurrent Protection

26304-08
Hazardous Locations

26303-08 Practical
Applications of Lighting

26302-08 Conductor
Selection and Calculations

26301-08 Load Calculations –
Branch and Feeder Circuits

ELECTRICAL LEVEL THREE

ELECTRICAL LEVEL TWO

ELECTRICAL LEVEL ONE

CORE CURRICULUM:
Introductory Craft Skills

310CMAP.EPS

1.0.0 ◆ INTRODUCTION

As an electrician, you may be involved in the installation of structured voice, data, and video cabling systems in various commercial applications. A structured cabling system is a system in which the main components do not change. Structured cabling systems may use unshielded twisted pair (UTP) cable, coaxial cable (coax), or fiber-optic cable, depending on the system design.

UTP cable has been used for many years for both voice and data transmission. Currently, it is the most widely used cable in these applications. It consists of anywhere from one twisted pair to as many as 1,800 pairs of solid copper conductors twisted together. The conductors range in size from No. 24 to No. 22 AWG. In cables exceeding 600 pairs, an overall aluminum-steel shield is used to enclose the cable conductors. Each pair of conductors has a nominal impedance of 100 ohms. UTP cable is available in the following categories with specified transmission rates that are recognized by the Telecommunications Industry Association/Electronics Industries Alliance (TIA/EIA) in the current version of the *TIA/EIA 568* standard:

- *Category 3 cable* – Category 3 cable, also called Cat 3 cable, with transmission rates to 16MHz has historically been used for 10 megabit per second (Mbit/s) Ethernet networks. It is not recommended for use in new installations except for analog data or voice only applications.
- *Category 5e cable* – Category 5e cable (Cat 5e cable) is an enhanced version of Cat 5 cable with transmission rates up to and exceeding 100MHz. It is frequently used for 100Mbit/s Ethernet and sometimes for gigabit Ethernet networks.
- *Category 6 cable* – Category 6 cable (Cat 6 cable) with transmission rates to 250MHz is currently the fastest copper transmission cable. It is recommended for use in new installations.

Category 1 and Category 2 cable (also known as Level 1 and Level 2 cable) are not recognized by TIA/EIA and are never used for new commercial installations; however, they may exist in old Bell System commercial private branch exchange (PBX) installations or in residential installations for analog voice or data applications. Older types of cable ranged from three-wire to four-wire long-twist cable (one to two twists per foot) to very large multi-pair cables using the old Bell System PBX conductor color codes. Cat 4 cable was used for 16Mbit/s token ring networks, but is not recognized by TIA/EIA. Cat 5 cable is no longer recognized by TIA/EIA for use in new installations, but exists in present residential and commercial installations. Currently, the TIA/EIA is in the process of considering higher performance Cat 6a and Cat 7 cables for addition to a future version of the TIA/EIA standard.

The video cables used in new installations are normally RG6 coax types. RG means radio guide. Today, RG6 tri-shield and quad-shield are the most common, but dual-shield is also used. Dual-shield cable consists of a layer of foil that is usually bonded to the dielectric core surrounding the center conductor. The foil shield is, in turn, covered with a braid shield followed by an insulation cover over the outside of the cable (foil/braid construction). Tri-shield cable includes a layer of foil shield over the braid shield of a dual-shield cable followed by the cable cover (foil/braid/foil construction). Quad-shield includes a braid shield over the foil of a tri-shield cable followed by the cable cover (foil/braid/foil/braid construction).

2.0.0 ◆ STRUCTURED CABLING SYSTEMS

Structured cabling systems are divided into five subsystems, as shown in *Figure 1*. The subsystems are the campus backbone, equipment room, riser, horizontal cabling, and work area.

These subsystems are defined as follows:

- *Campus backbone subsystem* – Outside plant (OSP) connectivity between buildings in a multi-building campus environment. This includes the building entrances for each building.
- *Equipment room subsystem* – This is the area where all subsystems for an individual building (or the entire campus) connect. This can be an equipment room (ER), telecom room (TR), or campus or building main cross connect (MC).
- *Riser subsystem* – This consists of the blocks, cables, pathways, and hardware that provide connectivity between ERs, TRs, and MCs.
- *Horizontal cabling subsystem* – This is made up of the patch panels, cables, support structures, and hardware that deliver voice/data connectivity to the end user outlet location.
- *Work area subsystem* – Consists of the telecommunications (telecom) outlets at the end user location and associated patch cables providing final connectivity.

> **NOTE**
>
> Any common pathway used for both telecom and power distribution, such as utility columns, must be equipped with a barrier and must comply with applicable electrical codes. When a metallic barrier is used, it must be bonded to ground.

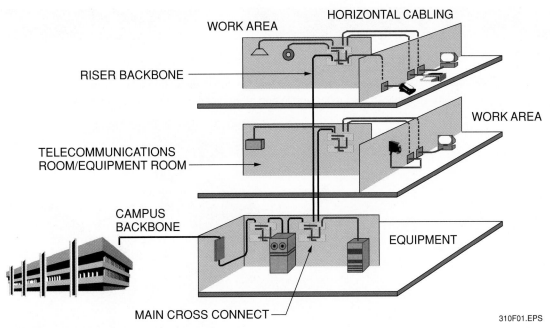

Figure 1 ◆ Diagram of a campus structured cabling system.

For each subsystem in a structured cabling system, there are specific requirements for how the cable is run, what environmental protection must be in place, standard support structures, expectations for workmanship, and standards for labeling. The following sections describe these areas in each of the five subsystems.

2.1.0 Campus Backbone Subsystem

The design of the backbone subsystem routes must consider the most economical and flexible solution. Any hazards (roads, railroad tracks, ponds, seasonal waterways, etc.) that have the potential to cause an inconvenience when adding cables should be avoided. Any pathways that have a greater potential for disturbance such as those near sprinkler system manifolds or utility entrances (gas, water, electric, or CATV) should be avoided. Minimize any opportunity for others to disturb system pathways.

When designing campus backbone conduit systems, install pull boxes (sized appropriate to the installed conduit) no more than 140' apart on straight runs and after every 180 degrees of bends (cumulative). Do not use 90-degree conduit bodies (condulets), also known as Type L. To minimize issues with water, maintain a negative slope away from the buildings. In addition, observe the following minimum distances for installation:

- Keep telecom conduit 3" away from power cables when encased in concrete, 4" away from power cables when surrounded by masonry, and 12" away from power cables when surrounded by well-tamped earth.
- Keep telecom conduit 6" away from pipes (gas, oil, water, etc.) when perpendicular and 12" away when parallel.
- Keep telecom conduit buried 36" below the top of street car rails and 50" below the top of railroad rails.

Aerial distribution systems must also be designed with safe clearances in mind. When designing aerial distribution systems, observe the following minimum attachment clearances:

- Keep telecom cable 40" away from all power cable.
- Keep telecom cable 40" away (+0.5" per kV over 8.7kV) from open supply conductors.
- Keep telecom cable 12" away from traffic signal and luminary drip loops.
- Keep telecom cable 30" away from grounded supply equipment (transformers, etc.).

NOTE

Mid-span clearances must be 75% of the attachment clearances.

Innerduct

This innerduct raceway system is used in direct burial and concrete encasement applications for fiber-optic and other types of communications cable.

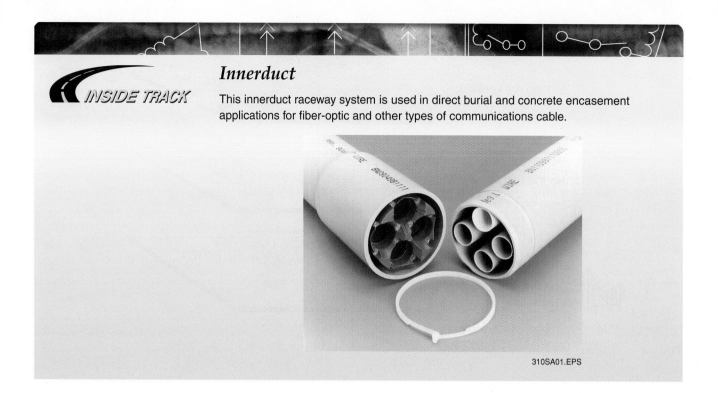

310SA01.EPS

2.1.1 Environmental Protection

When installing nonmetallic (non-armored fiber) cabling underground, either direct buried or in nonmetallic pipe, add a metallic tape strip along the length of the cable to aid in locating the cable for future repair or maintenance. If metallic tape is not available, use copper wire (No. 18 AWG or larger) or two-pair or four-pair cable (No. 24 AWG).

All incoming conduit and innerduct should be plugged with water blocking compounds or mechanical devices (jack plugs) at the last break (manhole, handhole, junction box, or weatherhead) before the building entrance and at the first break (conduit stub, junction box, etc.) after the building entrance.

Cables with a UV-resistant OSP sheath must be terminated and/or spliced to indoor-rated (riser or plenum) cables within 50' of the building entrance or must access and run in conduit to its final termination point.

At the building entrance or the final termination point (whichever method was used), ground any metallic members in the sheath of the incoming cable to the building telecommunications main grounding busbar (TMGB). Terminate individual metallic conductors on grounded lightning protection panels using solid-state protection fuses.

2.1.2 Support Structure

OSP conduit must be buried a minimum of 24" below grade (or frost line depending on local climate). Live and dead load ratings must be calculated prior to installation. Soil type and backfill composition must be taken into account to ensure there is no conduit deformation. A soil bed may be required to provide a stable base to eliminate deformation.

When installing plastic or metallic OSP buried conduit (NEMA TC-6 and TC-8), all 90-degree bends (5' sweep minimum) and areas under vehicular traffic (roadways or railways) must be encased in concrete capable of withstanding 2,500 psi.

When placing self supporting cables aerially, use standardized tables to calculate span lengths according to the cable type (BHBS, BHAS, etc.), weight, and number of pairs. When supporting cables with metallic strand, never use 2.2M strand.

2.1.3 Workmanship

OSP pathways and cables for a campus backbone subsystem must be run in a parallel and perpendicular manner as much as possible. Avoid diagonal runs.

2.1.4 Labeling

All campus backbone subsystem cable, protection blocks, conduit, innerduct, handholes, and pedestals must be labeled. Labels must be machine-printed as large as reasonable, protected with a clear shield, and affixed to the following system components:

- Cable within 12" of the termination point
- Protection blocks at the top center of the front cover
- Conduit within 6" of the termination point
- Innerduct within 12" of the termination point
- Handhole cover(s)
- Pedestal at the top of the removable face or nearest the street-facing side

In addition, all OSP conduit must be labeled on each end and in each handhole, manhole, and pedestal. All innerduct within each conduit must be labeled on each end as well as any point where it exits the conduit, handhole, manhole, or pedestal.

2.2.0 Equipment/Telecom Room Subsystem

The design of the administrative subsystem routes requires a layout for all systems (fire alarm, A/V, security, access controls, etc.) to be housed in the equipment room subsystem area or main cross connect (MC). The first concern when laying out the MC is maintaining appropriate clearances from sources of electromagnetic interference (EMI). The following are general safe distance clearances:

- Keep telecom cable 5" away from fluorescent lighting.
- Keep telecom cable 12" away from conduit and cables used for electrical distribution.
- Keep telecom cable 48" away from motors and transformers.

Lay out the MC backboard to minimize the number of times cross connect (interconnect) cables must cross fields. Start left to right (preferable) or right to left, and dedicate 16" vertically on

Labeling

One of the most important aspects of cable installation is proper labeling. In fact, some jobs prohibit the use of handwritten labels. Labels are easy to produce using a handheld labeling machine, such as the one shown.

310SA02.EPS

ENT

ENT is a flexible, lightweight conduit suitable for indoor installations. It is often used to carry communications and signaling circuits.

310SA03.EPS

the backboard for each of the following fields: carrier, switch in (if a wall-mounted switch is used, allocate 32" minimum between switch-in and switch-out fields), switch out, all risers in order, and the MC field.

Mount backboards (¾" 4' × 8' A/C plywood, painted on all six sides with two coats of fire-resistant paint) at 7'-4" minimum above the finished floor (AFF). Minimize any opportunity for others to disturb system pathways. When designing administrative subsystem conduit systems, enter MC near the corners and keep the conduit at least 4" away from the nearest wall.

2.2.1 Environmental Protection

Equipment/telecom rooms contain sensitive electronic equipment that must be protected from temperature and humidity extremes. The room should be positively pressurized, and the heating, ventilation, and air conditioning (HVAC) system must be in service 24 hours a day and 365 days per year. Maintain a temperature of 64°F to 75°F, 30% to 55% relative humidity (non-condensing). Proper lighting is also essential. Paint the walls and ceiling white. Provide 50 footcandles (fc) of light when measured 3' from the floor.

All cables entering a main cross connect from outdoors must be properly grounded with lightning protection installed. All incoming conduit and innerduct (non OSP) should include **firestopping**.

2.2.2 Support Structure

Whenever possible, do not use shared cable tray to distribute both telecom and power cables. If trays or wireways are shared, the power and telecom cables must be separated by a grounded metallic barrier.

Ladder rack should be mounted at 7' AFF and allow for a minimum of 12" of clear vertical space above the tray. Ladder rack support systems include wall triangles or supports, pipe stands, auxiliary bar structures, equipment racks, independent threaded rod, or any combination of these. Support centers must be on 5' centers (maximum span), unless they are designed for greater spans. A support must also be placed within 2.5' on each side of any connection to a fitting.

CAUTION

When installing supports, ensure that all fasteners seat securely in wall studs or solid anchor points. Failure to do so could result in collapse of the ladder rack system.

All metallic cable trays must be grounded, but may also be used as a ground conductor. Clearly mark any cable tray that is used as an equipment grounding conductor.

2.2.3 Workmanship

The workmanship requirements for an equipment room/telecom room subsystem are described in this section and shown in *Figures 2* through *5*. When using standard cross connect wire (No. 24 AWG twisted pair), provide a two-finger slack at each block.

Install all cables with a 15' minimum service loop. The loop should be neatly coiled, labeled, and secured overhead. Cables on a ladder rack in an ER, TR, or MC should be neatly bundled using tie wraps, Velcro® ties, or approved cord. Velcro® is preferable because it is easy to undo if necessary. Cables must be dressed from the point of entry in the room to the termination point. Dressing means that all cables are lying straight next to each other with no knots, crossovers, or divers. A diver is a cable in the bundle that crosses and disappears into the center of the bundle. An example of proper cable trunk dressing is shown in *Figure 3*.

Cables must be run on a support structure or secured with D-rings and button ties at all times in the ER, TR, and MC. All routing must be parallel and perpendicular to horizontal, with no bee-lining or diagonal routing. Cables should be laid in sequentially on the back of the patch panels. Field locations should also be laid out in a sequential manner whenever possible.

2.2.4 Labeling

All equipment/telecom room subsystem cable, conduit, innerduct, racks, and patch panels must be labeled. Labels must be machine-printed as

310F02.EPS

Figure 2 ◆ Cables dressed in an underfloor environment with tie wrapping and bundle breakouts for each patch panel.

310F04.EPS

Figure 4 ◆ Another view of bend radius, cable dressing, and tie wrap tensioning.

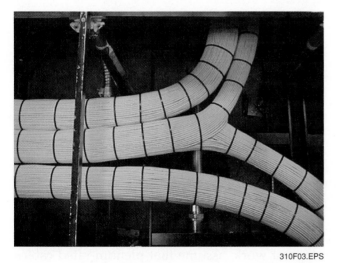

310F03.EPS

Figure 3 ◆ Proper cable trunk dressing.

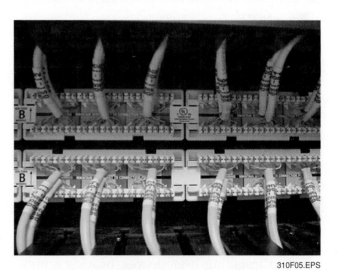

310F05.EPS

Figure 5 ◆ Cables labeled on the closet end with the preferred method of termination.

large as reasonable, protected with a clear shield, and affixed to the following system components:

- Cable within 12" of the termination point
- Conduit within 6" of the termination point
- Innerduct within 12" of the termination point
- Rack top front crossmember
- Patch panel at the top center of the front of the panel

In addition, all OSP conduit must be labeled on each end and in each handhole, manhole, and pedestal. All innerduct within each conduit must be labeled on each end as well as any point where it exits the conduit, handhole, manhole, pedestal, etc. Manholes must be labeled via engraving on the lid.

2.3.0 Riser Subsystem

The routes for a riser subsystem should consider what low-voltage services are needed at each location (ER/TR/MC). The connectivity requirements, including single-mode (SM) fiber, multimode (MM) fiber, coax, 100-pair UTP, and so on must also be determined. Backbone pathway viability should also be determined based on capacity requirements for all systems (fire alarm, A/V, security, access controls, etc.). Design routes around the following maximum distances:

- *Analog/digital voice* – 2,600' on UTP
- *Digital data (10/100/1000)/voice over internet protocol (VOIP)* – 295' on UTP
- *Digital data (10/100/1000)/VOIP* – 810' on multimode fiber
- *Closed circuit television (CCTV)* – Distribution distance based on cable used

---

NOTE

These distances are rules of thumb and may vary by application.

Riser pathways should cross perpendicular to fluorescent lighting and run no closer than 5" away. Keep telecom cables 12" away from electrical power cables or conduit and 48" away from motors and transformers. Run riser pathways in the most direct route possible (through secure space vertically and secured in public space horizontally) and maintain the same vertical and/or horizontal plane as much as possible.

When installing riser sleeves, provide a minimum of three 4" sleeves between floors vertically and TRs horizontally. Allow a 4" offset from the

nearest wall. Terminate conduit that protrudes through the structural floor 1" to 3" above the surface. Firestop all sleeves (whether installed or existing). Fill riser sleeves according to the firestop system specifications. Ream all sleeve ends and fit them with an insulated bushing.

When installing telecom cabling in a riser where the vertical run spans more than 30' continuously, all cables must be secured in 30' intervals (maximum span) using a split Kellum's grip or similar mechanical device. This may require installing vertical ladder racking to support high pair count copper riser cables. Use devices designed for this purpose as improper riser support can degrade copper, fiber, and coax cable performance.

Riser conduit should be 4' from motors or transformers, 1' from conduit and cables used for electrical power distribution, and 5" from fluorescent lighting. All conduit should be reamed and bushed on each end and have no more than two 90-degree bends between pull points or pull boxes. Conduit should contain no 90-degree conduit bodies (also known as Type L) as they will make the pull too difficult. For the same reason, they should contain no continuous sections longer than 100' and must be clean, dry, and unobstructed. Conduit must be bonded to common ground on one or both ends and terminated near the TR/ER corner of the room nearest the backboard. It should be capped/firestopped for protection upon installation, labeled for identification, and equipped with a plastic or nylon line (also called a fish tape or pull cord) with a minimum test rating of 200 pounds.

See *Figure 6* for conduit and sleeve capacity information. This chart provides guidelines for horizontal conduit with no more than two 90-degree bends (180 degrees total) and no longer than 100'.

NOTE

The number of cables that can be installed horizontally is limited by the maximum allowable pulling tension on the cable. The number of cables that can be installed vertically is determined by the firestopping system requirements.

2.3.1 Environmental Protection

All cables run in the riser subsystem should be rated for indoor use (CMR, CMP, etc.). If PVC or riser-rated cable is not specifically called for in the scope of work, assume that plenum-rated cables are required in the riser.

CABLE TYPE	CABLE O.D.	CABLE AREA	¾"	1"	1¼"	1½"	2"	2½"	3"	3½"	4"
		CONDUIT AREA =	0.533	0.864	1.495	2.035	3.354	4.785	7.389	9.882	12.724
4 PAIR CAT 3	0.180	0.025	8	13	23	32	52	75	116	155	200
4 PAIR CAT 5	0.190	0.028	7	12	21	28	47	67	104	139	179
4 PAIR CAT 5	0.220	0.038	5	9	15	21	35	50	77	104	133
25 PAIR CAT 5	0.570	0.255	0	1	2	3	5	7	11	15	19

CABLE TYPE	CABLE O.D.	CABLE AREA	¾"	1"	1¼"	1½"	2"	2½"	3"	3½"	4"
		CONDUIT AREA =	0.533	0.864	1.495	2.035	3.354	4.785	7.389	9.882	12.724
RG-59U (NP)[1]	0.265	0.055	3	6	10	14	24	34	53	71	92
RG-62A/U (P)[2]	0.208	0.034	6	10	17	23	39	56	87	116	149
RG-62U (NP)	0.260	0.053	4	6	11	15	25	36	55	74	95

CABLE TYPE	CABLE O.D.	CABLE AREA	¾"	1"	1¼"	1½"	2"	2½"	3"	3½"	4"
		CONDUIT AREA =	0.533	0.864	1.495	2.035	3.354	4.785	7.389	9.882	12.724
25 PAIR NCC[3]	0.340	0.091	2	3	6	8	14	21	32	43	56
50 PAIR NCC	0.470	0.173	1	1	3	4	7	11	17	22	29
100 PAIR NCC	0.640	0.322	0	1	1	2	4	5	9	12	15
200 PAIR NCC	0.970	0.739	0	0	0	1	1	2	4	5	6
300 PAIR NCC	1.070	0.899	0	0	0	0	1	2	3	4	5
400 PAIR NCC	1.300	1.327	0	0	0	0	1	1	2	2	3
600 PAIR NCC	1.500	1.766	0	0	0	0	0	1	1	2	2

CABLE TYPE	CABLE O.D.	CABLE AREA	¾"	1"	1¼"	1½"	2"	2½"	3"	3½"	4"
		CONDUIT AREA =	0.533	0.864	1.495	2.035	3.354	4.785	7.389	9.882	12.724
100 PAIR ARMM[4]	0.890	0.622	0	0	0	1	2	3	4	6	8
200 PAIR ARMM	1.190	1.112	0	0	0	0	1	1	2	3	4
300 PAIR ARMM	1.410	1.561	0	0	0	0	0	1	1	2	3
400 PAIR ARMM	1.480	1.719	0	0	0	0	0	1	1	2	2
600 PAIR ARMM	1.900	2.834	0	0	0	0	0	0	1	1	1
900 PAIR ARMM	2.280	4.081	0	0	0	0	0	0	0	0	1
1200 PAIR ARMM	2.600	5.307	0	0	0	0	0	0	0	0	0

[1] Nonplenum cable
[2] Plenum cable
[3] Network communication cable
[4] Armored riser multipurpose multi-pair

310F06.EPS

Figure 6 ◆ Horizontal conduit cable capacity chart.

Any penetrations through walls or floors that are required for the riser pathways must be neatly sleeved and firestopped immediately upon completing the penetration. Size floor slots adequately but not extravagantly—remember that they will need to be firestopped. If a previously installed core hole/sleeve does not contain firestop material, install firestopping when the work is completed.

CAUTION

Do not run riser pathways over or adjacent to boilers, incinerators, hot water lines, or steam lines.

2.3.2 Support Structure

Ladder rack or tray used in riser subsystems should be mounted at a minimum of 4" above the ceiling grid with 12" vertical clearance above the side rails of the tray. Whenever possible, avoid using shared cable trays to distribute both telecom and power cables.

NOTE

If trays or wireways must be shared, the power and telecom cables must be separated by a grounded metallic barrier.

Riser cable support systems are not required to be continuous, but they must support the live loads exerted when placing cable. When installing telecom cabling in a riser where the vertical run spans more than 30' continuously, all cables must be secured in 30' intervals (maximum span) using a split Kellum's grip or similar mechanical device. This may require installing vertical ladder racking to support high pair count copper riser cables. Use devices designed for this purpose as improper riser support can degrade copper, fiber, and coax cable performance.

When installing riser cables horizontally, the live loads exerted during cable placement require attention to lateral sway support. Field conditions, such as threaded rod diameter and length, will determine when and where this is required. Tray support centers must be on 5' centers (maximum span), unless they are designed for greater spans. A support must also be placed within 2.5' on each side of any connection to a fitting. When using wall triangles or supports, ensure that fasteners anchor to wall studs.

All metallic cable trays must be grounded (*Figure 7*), but may also be used as a ground conductor. Clearly mark any cable tray that is used as an equipment grounding conductor.

2.3.3 Workmanship

The riser support structure and cable must be straight, level, and plumb. Use sweeping 90-degree transitions with strict adherence to the cable manufacturer's bend radius. Cables should be dressed cable for cable and end to end as much as possible. Cables must be neatly bundled using tie wraps, Velcro® ties (*Figure 8*), or approved cord.

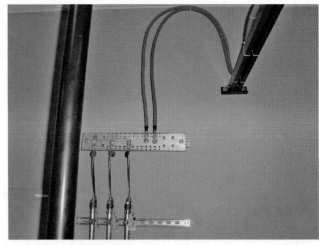

310F07.EPS

Figure 7 ◆ Telecommunications main ground bus (TMGB) with conductors direct from the building ground ring.

310F08.EPS

Figure 8 ◆ Cable bundle secured with Velcro® ties.

2.3.4 Labeling

All riser subsystem cable, conduit, and innerduct must be labeled. Labels must be machine-printed as large as reasonable, protected with a clear shield, and affixed to both ends of the following system components:

• Cable within 12" of the termination point (label must show pair count, cable start point, and cable termination point)
• Conduit within 6" of the termination point (label must note start point and conduit termination point)
• Innerduct within 12" of the termination point (label must list duct number, start point, and duct termination point)

2.4.0 Horizontal Subsystem

Routes for horizontal subsystems should be selected based on the best long-term route for the customer. Preferred routes will minimize interruption to the customer's daily operations. Typically, this means above hallways and common areas as opposed to directly overhead in offices or cubicle areas. Whenever possible, do not use shared cable trays to distribute telecom and power cables. If trays or wireways are shared, the power and telecom cables must be separated by a grounded metallic barrier.

Horizontal trunks should cross perpendicular to fluorescent lighting and run no closer than 5" away. Keep telecom cables 12" away from electrical power cables or conduit and 48" away from motors or transformers. Run in the most direct route possible (over hallways as much as possible). Run in conduit through non-secure hallways and elsewhere when required by local code. Trunks must be a minimum of 4" above the ceiling grid and have a minimum of 12" clear space directly above it. Remain on the same vertical or horizontal plane as much as possible. Minimize any opportunity for others to disturb system pathways. Provide segregation from other sources of electromagnetic interference (EMI), such as low-voltage cables or systems.

When installing horizontal (ER, TR, or MC) sleeves, provide a minimum of 3" between the ER, TR, or MC and the horizontal subsystem. Allow a minimum of 4" offset above any office space ceiling grid. Also allow a 4" offset from the nearest wall. Terminate sleeves that penetrate walls no less than 2" to 3" from the wall surface on each side. Adequately secure sleeves to prevent movement when cabling is installed. Firestop all sleeves on both sides (both installed and existing). Fill horizontal sleeves according to firestop system specifications. Ream all sleeve ends and fit them with an insulated bushing.

Horizontal (field) conduit/stubs should be run in the most direct route possible (parallel and perpendicular to building lines). They should be located a minimum of 4' away from motors or transformers, 1' from conduit and cable used for electrical power distribution, and 5" from fluorescent lighting. In addition, they must be reamed and bushed on each end and have no more than two 90-degree bends between pull points or pull boxes. To ensure a smooth pull, they must be clean, dry, and unobstructed. They should contain no 90-degree conduit bodies and no continuous sections longer than 100'. They must be bonded to common ground on one or both ends (home-run conduit) and terminated near the tray/trunk (no closer than 6" on either side and vertically within 4" to 16" of the tray top), but not directly over it. The conduit must be bushed, capped, and firestopped upon finishing installation (*Figure 9*), labeled for identification, and equipped with a plastic or nylon pull cord with a minimum test rating of 200 pounds.

When penetrating fire walls, always use firestop products that allow for re-entry and reuse. Allow a minimum of 4" offset above any office space ceiling grid. Cut conduit that protrudes through the wall to leave 1" to 3" on both

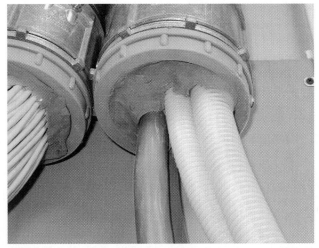

310F09.EPS

Figure 9 ◆ Fire-sealed conduit.

sides. Adequately secure the sleeve to prevent movement when cabling is installed. Use fire caulk to seal the rough opening around the conduit. Firestop all sleeves (both installed and existing) according to firestopping system and wall rating specifications. Ream all conduit ends and fit them with an insulated bushing to eliminate sharp edges that can damage cables during installation or service. Where ladder rack or cable racking penetrates fire-rated walls, make the opening no bigger than necessary, and use fire bricks or other approved material to properly stop the opening according to the chosen firestop manufacturer's specifications.

When penetrating nonfire-rated walls, install sleeves but use standard caulk as opposed to fire caulk. When penetrating a ceiling (drop tile, gypsum, etc.), always secure conduit sleeves adequately to minimize movement when pulling cable through the sleeve.

2.4.1 Environmental Protection

All cables entering a plenum-rated horizontal subsystem must also be plenum-rated. Plenum subsystems require plenum-rated cable ties, etc.

CAUTION

Do not run trunks over or adjacent to boilers, incinerators, hot water lines, or steam lines.

2.4.2 Support Structure

The support structure for a horizontal subsystem can consist of J-hooks, bags, bridal rings, and ladder racks/trays.

- *J-hooks* – When using J-hooks, locate them on 48" to 60" centers to adequately support and distribute the cable's weight. These types of supports may typically hold up to fifty 0.25" diameter cables. Suspended cables must be installed with at least 3" of clear vertical space above the ceiling tiles and support channels (T-bars). For large quantities of cables (50 to 75) that convene at the telecom closet and other areas, provide cable trays or other special supports that are specifically designed to support the required cable weight and volume without adversely affecting cable performance.
- *Bags* – When using bags, locate them on 48" to 60" centers to adequately support and distribute the cable weight. These types of supports may typically hold up to fifty 0.25" diameter cables. Suspended cables must be installed

with at least 3" of clear vertical space above the ceiling tiles and support channels (T-bars). For large quantities of cables (50 to 75) that convene at the telecom closet and other areas, provide cable trays or other special supports that are specifically designed to support the required cable weight and volume without adversely affecting cable performance.
- *Bridal rings* – When using bridal rings, the weight-bearing surface area is relatively small, therefore use these only for short span or single user location bundles. Locate them on 36" to 48" centers to adequately support and distribute the cable weight. Suspended cables must be installed with at least 3" of clear vertical space above the ceiling tiles and support channels (T-bars). For large quantities of cables use J-hooks, bags, or other supports that are specifically designed to support the required cable weight and volume without adversely affecting cable performance.
- *Ladder racks/trays* – Ladder rack/tray should be mounted 4" above the top of ceiling tile grid. Allow for a minimum of 12" of clear vertical space above the tray. Ladder rack/tray support systems include independent threaded rod, wall triangles or supports, pipe stands, auxiliary bar structures, equipment racks, or any combination of these components.

CAUTION

When installing supports, ensure that all fasteners seat securely in wall studs or solid anchor points. Failure to do so could result in collapse of the ladder rack system.

NOTE

Support stringers for horizontal cable support must be on 5' centers (maximum span). A support must also be placed within 2.5' on each side of any connection to a ladder rack/tray splice.

All metallic cable trays must be grounded, but may also be used as a ground conductor. Clearly mark any cable tray that is used as an equipment grounding conductor. *Figure 10* shows a large-scale ladder rack/equipment rack grounding system. The trunk and lateral cables are bonded using H-taps with covers, with approved cord securing the covers.

Figure 10 ◆ Ladder rack/equipment rack grounding system.

310F10.EPS

2.4.3 Workmanship

The horizontal support structure and cable must be straight, level, and plumb. Use sweeping 90-degree transitions with strict adherence to the manufacturer's bend radius. Cables should be dressed cleanly with no visible kinks, knots, divers, holes, or wrapping. All cables must be installed with a service loop of 10' minimum on the ER, TR, and MC end, and 18" on the field end. The loop must be neatly coiled, labeled, and secured overhead. Cables in any horizontal support structure must be neatly bundled using tie wraps, Velcro® ties, or approved cord at 18" to 30" intervals. All routing must be parallel and perpendicular to horizontal, with no diagonal routing.

2.4.4 Labeling

All horizontal subsystem cable, conduit, innerduct, racks, and patch panels must be labeled.

Labels must be machine-printed as large as reasonable, protected with a clear shield, and affixed to the following system components:

- Cable within 6" of the termination point
- Conduit within 6" of the termination point
- Innerduct within 12" of the termination point
- Rack top front crossmember
- Patch panel at the top center of the front of the panel

NOTE

All innerduct within each conduit must be labeled on each end as well as at any point where it exits the conduit (pull boxes, junction boxes, etc.).

2.5.0 Work Area Subsystem

Determine the furniture layout before designing the work area subsystem. Locate outlet boxes within 3' of the electrical outlet for the intended device. Patch cables should be long enough to serve the intended device but no longer than necessary to minimize cable clutter.

The center line of outlets should be situated at 12" AFF when mounted vertically or horizontally. The preferred configuration for a single wall outlet is ¾" EMT into a double-gang box with a single-gang adapter over the front. Wall phones should be mounted at 48" AFF using a stainless steel 8-pin modular faceplate.

When routing cables in furniture pathways, maintain the following standard separations:

- 5" away from fluorescent lighting
- 12" away from conduit and cables used for electrical distribution
- 48" away from motors and transformers

Wireless

More than a hundred years ago, telegraphs used a binary code (Morse code) to wirelessly transfer information over long distances. Sending a message via telegraph was known as sending a wireless. Today's wireless technology works in basically the same way. It transfers binary computer data over various high radio frequencies between antennas. Wireless computers contain a combination receiver/transmitter that enables the computer to both transmit and receive data. Because wireless technology transfers information over very high frequencies, it produces a very short wavelength and therefore only works well at short distances. Some systems rely on direct connections for long distances but use a wireless connection as a local area network (LAN).

2.5.1 Environmental Protection

When installing boxes, leave no more than 6" to 12" of excess cable coiled within the box (double gang), with the service coil (where required) tied off above the conduit stub. When installing surface-mounted boxes under desktops, place split wire loom over the exposed cables. When installing surface-mounted boxes, never position jacks such that the plug enters from directly above. In the case of floor outlets, make sure a spring-loaded cover is installed.

2.5.2 Support Structure

Wall boxes must be secured to the nearest structural member. If a new location must be cut in, box eliminators should be used. Stamped sheet metal with folding ears is not acceptable. Surface-mounted boxes and faceplates must be secured by a mechanical means, such as screws, clips, snap-ins, etc. Double-sided tape is not acceptable. Utility columns that are used for both telecom and power distribution must be equipped with a barrier and must comply with applicable electrical codes. When a metallic barrier is used, it must be bonded to ground.

2.5.3 Workmanship

All faceplates, surface-mounted boxes, etc. must be mounted flush, plumb, and level. Patch cables must be routed in such a manner as to limit cable clutter. All cables must be installed with a service loop of 18". The loop must be neatly coiled, labeled, and secured overhead.

2.5.4 Labeling

All work area subsystem cables, faceplates, and jacks must be labeled. Labels must be machine-printed as large as reasonable, protected with a clear shield, and affixed to the cable within 12" of the termination point and in the windows of all faceplates and boxes (*Figure 11*).

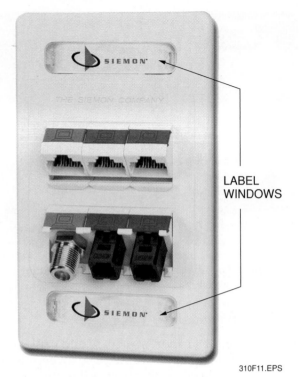

Figure 11 ♦ Label windows.

310F11.EPS

3.0.0 ♦ UTP AND COAX CABLE TERMINATIONS

This section covers UTP and coax cable terminations using jacks and plugs. It also covers UTP **patch cord** construction.

3.1.0 UTP Jack and Plug Terminations

UTP jacks and plugs for new structured cabling networks wired with Cat 3, Cat 5e, or Cat 6 UTP 4-pair cables must be wired in accordance with *TIA/EIA 568A* and *568B* wiring schemes using keystone Cat 3-, Cat 5e-, or Cat 6-rated RJ45 modular jacks and mating RJ45 or Cat 3 RJ11 plugs as required. The term RJ means registered jack. The jacks are connected to horizontal homerun 4-pair UTP cables that originate from a patch panel and terminate in the work areas. The two wiring schemes were originally included in the *568A* standard to accommodate installations that had been or were being installed to the B wiring scheme {formerly known as Western Electric Company [WECO] or *AT&T Standard 258A*}. The preferred scheme for new installations was supposed to be the A wiring scheme; however, the most popular method used today is the B scheme because of the continued interface with AT&T equipment and systems that use the B scheme. This has caused much confusion. In the past, most patch panels and modular jacks were made using either the A or B scheme, but were not labeled as A or B, and most suppliers carried only the B version. Today, almost all patch panels and modular jacks show diagrams for both versions (see *Figure 12*). The only difference between the two is the interchanging of the pins used for the second and third pairs.

For Cat 3, Cat 5e, and Cat 6 structured cable systems with 4-pair cables, typical keystone 8-pin, 8-conductor (8P8C) RJ45 modular jacks are used, as shown in *Figure 13*. Keystone jacks are shaped so that they can accommodate 8-pin, 6-pin, or 4-pin plugs including RJ11 plugs. Jacks may be crimped, punchdown types, or tool-less (self punching). For the tool-less versions, a hinged cover is opened and the wire pairs are inserted at the proper pins. When the cover is closed and latched, the wires are automatically punched down. To prevent system performance degradation, both types of jacks must be rated to match or exceed the category of UTP used for the system.

For 5e and Cat 6 patch cord construction, appropriately rated 8P8C RJ45 plugs are normally used. For Cat 3 and certain other analog applications in higher category systems, RJ11 plugs are sometimes used. *Figure 14* shows typical instructions for constructing patch cords for Cat 6 cable. RJ11 patch cords for analog voice or data are constructed in a similar manner using two-pair ribbon cable and two-pin or four-pin RJ11 plugs. The ribbon cable is a flat cable (untwisted) with the conductors identified by the old style Bell System colors of black, red, green, and yellow. For two-pin plugs, pair 1 is the two inner pins (red and green of the flat cable). The other pin slots contain no pins for the black and yellow conductors. For four-pin jacks, pair 1 is the two inner pins (red and green), and pair 2 is the two outer pins (black and yellow).

For straight-through patch cords, it doesn't make any difference whether both ends of the cord are wired to any color scheme so long as the same scheme is used on both ends. Any scheme will pass the signal, unchanged, from one jack to another jack if both are wired to the same scheme. Today, most straight-through patch cords for Cat 3, 5e, and 6 cable are wired to the B scheme, including commercial patch cords, because most systems are wired using the B scheme. However, when making a crossover patch cord, one end is

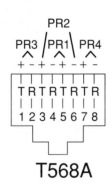

T568A

568A WIRING		
Pair #	Wire	Pin #
1 – White/Blue	White/Blue	5
	Blue/White	4
2 – White/Orange	White/Orange	3
	Orange/White	6
3 – White/Green	White/Green	1
	Green/White	2
4 – White/Brown	White/Brown	7
	Brown/White	8

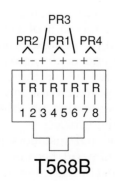

T568B

568B WIRING		
Pair #	Wire	Pin #
1 – White/Blue	White/Blue	5
	Blue/White	4
2 – White/Orange	White/Orange	1
	Orange/White	2
3 – White/Green	White/Green	3
	Green/White	6
4 – White/Brown	White/Brown	7
	Brown/White	8

310F12.EPS

Figure 12 ◆ *TIA/EIA 568A and 568B wiring schemes.*

(A) MODULAR CONNECTOR CRIMPING TOOLS

(B) PUNCHDOWN WITH 110/66 BLADE

CAT 6 JACKS

CAT 3 OR CAT 5e JACKS   CAT 6 TOOL-LESS JACK   CAT 5e TOOL-LESS JACKS

(C) JACKS

310F13.EPS

Figure 13 ◆ Typical punchdown and tool-less RJ45 modular jacks and tools.

wired to the A scheme and the other is wired to the B scheme. This allows signal transition from a modular jack wired to one scheme to another modular jack wired to the other scheme.

3.2.0 RG6 Coax F-Type Terminations

A variety of coax F-type connector terminations are available, including crimp-on, push-on, twist-on, etc. Before a connection can be made, the cable must first be stripped using a special coax stripping tool (*Figure 15*). Select the proper tool based on the cable diameter and stripping requirements. In all cases, the first blade of a coax stripper is set to cut almost to the conductor at the proper distance from the end of the coax. A second blade, spaced behind the first at the proper distance, is adjusted to cut almost to the lowest layer of braid in the cable, depending on whether it is dual-, tri-, or quad-shield coax.

Most professional installers use compression connectors (*Figure 16*) because they are fast to install, the connection is more secure, and there is less chance of damaging the coax.

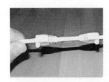
1. If you are planning to use boots, slide them onto the cable as shown. If you prefer not to use boots, start at Step 2.

Note: Making cables can be very labor intensive. Factory-made cables typically have better tolerances and consequently better quality than field-made cables.

2. Remove approximately 1½" of the cable jacket using a cable stripper.

3. Partially untwist the pairs, leaving one twist remaining at the bottom. Be careful not to untwist into the cable jacket. Straighten and organize the conductors.

Note: Choose 568B (most common) or 568A wiring.

4. (Optional) Cut the ends of the conductors at an angle while holding them in the proper order. This will make it easier to install the load bar in the next step.

5. Slide the conductors into the load bar in their proper order with the hollow portion of the load bar facing the jacket. The holes in the load bar alternate up and down. For that reason, you may find it easier to insert the conductors one at a time. Double-check the color order.

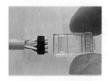
6. Make sure the pairs are twisted at least ¼" beyond the cable jacket. Push the load bar as far down as it will go. Then cut the conductors straight across approximately ⅛" from the front of the load bar. It is very important to get a straight and even cut. The use of a pair of electrician's scissors is highly recommended.

7. Pull the load bar back up near to the cut end of the conductors. Then slide the wires and load bar into the connector body, holding it with the pins facing you. A very slight amount of jiggling may be needed to make the wires find their slots in the connector body.

8. Once all of the wires have entered their slots, firmly push the connector body toward the cable. You will need to be sure that a) the wires have reached the end of the connector body, and b) the cable jacket is about halfway into the connector and past the first crimp point (the jacket crimp).

9. Crimp the connector using a high-quality crimp tool such as a ratchet-type RJ45 tool.

310F14.EPS

Figure 14 ◆ Typical Cat 6 patch cable construction.

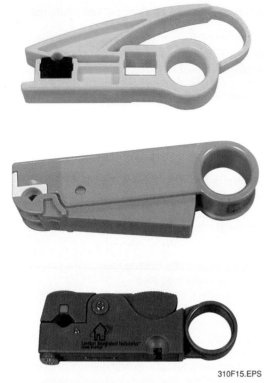

Figure 15 ◆ Coax stripping tools.

310F15.EPS

310F16.EPS

Figure 16 ◆ Typical coax compression tool and connector.

When using compression tools for F-type compression connectors, be aware that some tools are designed to fit only 19.6mm connector barrels or 20.3mm connector barrels. Other compression tools are designed to handle both sizes. Proper tool selection and termination is essential. Many problems in cable systems can be traced to poor connections. *Figure 17* provides typical instructions for installing an F-type compression connector on RG6 coax.

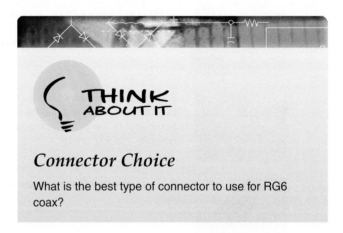

THINK ABOUT IT

Connector Choice

What is the best type of connector to use for RG6 coax?

1. Cut the end of the cable off square with a set of sharp cable cutters.

2. Place the coax in the stripper part of the combination tool, with the end flush against the side of the tool or use a separate stripper.

3. Twirl the strip tool around the cable until the crunching sound stops (5 to 10 turns).

4. Remove the combination or strip tool.

5. Pull off the stripped material.

6. Fold back the remaining braid over the cable cover so that there is only one layer of foil left against the center white dielectric.

Note: If the braid is not pushed back far enough, it will be very difficult to push on the connector.

7. Untwist the ring from the connector and slide it onto the cable, large end first.

8. Insert the cable into the connector.

9. When inserted properly, the white insulator of the cable should be flush with the metal flange. If the coax does not go in all the way, pull it out and push it in again. Sometimes the cable catches on the inner ring. If the cable jacket is loose, kink the cable slightly in your palm while pressing it into the connector (no more than 45 degrees).

Note: Various tools are available to hold the connector and/or cable for this operation.

10. Slide the ring into the connector and lay the assembly into the crimp tool.

11. Squeeze the handle until the ring seats all the way into the connector. You should hear or feel a click as it pops into position.

12. Remove the coax and connector from the crimper.

310F17.EPS

Figure 17 ◆ Typical F-type compression connector installation.

4.0.0 ◆ FIBER-OPTIC INSTALLATION CONSIDERATIONS

Fiber optics is the transmission of light through glass fibers. Fiber-optic transmission offers the benefits of high speed and low signal loss over long distances. It is also lightweight, immune to electromagnetic interference and corrosion, and capable of carrying a much greater amount of information than copper or coaxial wire due to extremely wide system bandwidths. In addition, it provides a very high level of safety and security because fiber-optic cable presents no spark hazard, and it cannot be tapped without breaking the integrity of the system. A fiber-optic system is ideal for making connections between many fixed points, as in a telephone network, or between two fixed points, as between a computer and a remote terminal.

Fiber-optic cable can consist of a single cable (simplex), two cables (duplex), or many cables (multi-fiber). As with any other type of cable, fiber-optic cable is not 100% efficient. The signal loss is known as attenuation, and is the result of several factors, the most crucial of which are proper installation and termination. The tools and techniques used with fiber-optic cable differ greatly from those used for copper wire. Another area of difference is in the safety precautions you must take. Although there is no worry of being shocked by fibers, there is the possibility of eye damage if you are exposed to sufficient light energy.

WARNING!

To avoid both retinal damage from light energy and physical damage from glass fiber entering the eyes, eye protection must be worn when repairing or troubleshooting an operating fiber-optic system and when cutting fiber ends. In addition, the ends of stripped glass fibers are very hazardous to eyes and skin. The cleaved ends are very sharp and can easily penetrate the skin. Take care to avoid this hazard by keeping the work area cleaned of bits of fiber. A piece of double-stick tape on the bench can be used to temporarily store the fiber pieces. In addition, never eat or drink in the area of fiber handling.

Due to their light weight and flexibility, fiber-optic cables are often more easily installed than their copper counterparts. They can be more easily handled and pulled over greater distances. The minimum bend radius and maximum tensile rating are the critical specifications for any fiber-optic cable, both during the installation process and over the installed life of the cable. Careful planning of the installation layout will ensure that the specifications are not exceeded. In addition, the installation process itself must be carefully planned.

The minimum bend radius and maximum tensile loading allowed on a cable differ during and after installation. An increase in tensile load first causes a reversible attenuation increase, then an irreversible attenuation increase, and finally, cracking of the fiber. The tensile loading allowed during installation is higher than that allowed after installation.

The minimum bend radius allowed during installation is larger than the bend radius allowed after installation. One reason is that the allowable bend radius increases with tensile loading. Because the fiber is under load during installation, the minimum bend radius must be larger. The allowable bend radius after installation depends on the tensile load.

Most indoor cables must be placed in conduit or trays. Because standard fiber-optic cables are electrically nonconductive, they may be placed in the same ducts as high-voltage cables without the special insulation required by copper wire. Many cables cannot, however, be placed inside air conditioning or ventilation ducts for the same reason that PVC-insulated wire should not be placed in these areas: A fire inside these ducts could cause the outer jacket to burn and produce toxic gases that could be carried throughout the structure. Plenum cables, however, can be placed in any plenum area within a building without special restrictions. The material used in these cables does not produce toxic fumes when it burns.

4.1.0 Tray and Duct Installation

The primary consideration in selecting a route for fiber-optic cable through trays and ducts is to avoid potential cutting edges and sharp bends. Areas where particular caution must be taken are corners and exit slots in sides of trays. If a fiber-optic cable is in the same tray or duct with very large, heavy electrical cables, avoid placing excessive crushing forces on the fiber-optic cable, particularly where the heavy cables cross over the fiber-optic cable. In general, cables in trays and ducts are not subjected to tensile forces; however, keep in mind that in long vertical runs, the weight of the cable itself will create a tensile load of approximately 0.16 pound per foot for simplex cable and 0.27 pound per foot for duplex cable.

This tensile load must be considered when determining the minimum bend radius at the top of the vertical run.

Long vertical runs should be clamped at intermediate points (preferably every 3' to 6') to prevent excessive tensile loading on the cable. The absolute maximum distance between clamping points is 330' for duplex cable and 690' for simplex cable. Clamping forces should be no more than is necessary to prevent the possibility of slippage. It is best determined experimentally because it is highly dependent on the type of clamping material used and the presence of surface contaminants, both on the clamp and on the jacket of the optical cable. Even so, the clamping force must not exceed 57 pounds per inch and must be applied uniformly across the full width of the cable. The clamping force should be applied over as long a length of the cable as practical, and the clamping surfaces should be made of a soft material such as rubber or plastic. The tensile load during a vertical installation is reduced by beginning at the top and running the cable downward.

4.2.0 Conduit Installation

Fibers are pulled through conduit by a wire or synthetic rope attached to the cable. Any pulling forces must be applied to the cable members and not to the fiber or jacket. For cables without connectors, pull wire can be tied to Kevlar® strength members, or a pulling grip can be taped to the cable jacket or sheath (if approved by the cable manufacturer). Whenever possible, use an automated puller with tension control and a swivel pulling eye. Use extra caution when pulling cables with connectors installed.

The first factor that must be considered when determining the suitability of a conduit for a fiber-optic cable is the clearance between the walls of the conduit and other cables that may be present. Sufficient clearance must be available to allow the fiber-optic cable to be pulled through without excess friction or binding, because the maximum pulling force that can be used is 90 pounds. Because the minimum bend radius increases with increased pulling force, bends in the conduit itself and any fittings through which the cable must be pulled should not require the cable to make a bend with a radius of less than 5.9". Fittings in particular should be checked carefully to ensure that they will not cause the cable to make sharp bends or press against corners. If the conduit must make a 90-degree turn, a turn fitting must be used to allow the cable to be pulled in a straight line and to avoid sharp bends in the cable.

Pull boxes should be used on straight runs at intervals of 250' to 300' to reduce the length of cable that must be pulled at any one time, which reduces the pulling force. Also, pull boxes should be located in any area where the conduit makes several bends that total more than 180 degrees. To guarantee that the cable will not be bent too tightly while pulling the slack into the pull box, use a pull box with an opening length equal to at least four times the minimum bend radius. The tensile loading effect of vertical runs discussed previously is also applicable to conduit installations. Since it is more difficult to properly clamp fiber-optic cables in a conduit than in a duct or tray, long vertical runs should be avoided if possible. If a clamp is required, the best type is a fitting that grips the cable in a rubber ring. Also, the tensile load caused by the weight of the cable must be considered along with the pulling force to determine the maximum total tensile load being applied to the cable.

4.3.0 Splice Closures/Organizers

A splice closure is a standard piece of hardware used in the telephone industry to protect cable splices. Splices are protected mechanically and environmentally within the sealed closure. The body of the closure serves to join the outer sheaths of the two joined cables.

A standard universal telephone closure can also be used to house spliced fiber-optic cable. An organizer panel holds the splices. Most organizers contain provisions for securing cable strength members and for routing and securing fibers. Metal strength members may be grounded through the closure.

Organizers are specific to the splice they are designed to hold. In a typical application, the cable is removed to expose the fibers at the point at which it enters the closure. The length of fiber exposed is sufficient to loop it one or more times around the organizer (*Figure 18*). Such routing provides extra fiber for re-splicing or rearrangement of splices. Closures can hold one or more organizers to accommodate from 12 to 144 splices.

Closures have two main applications. The first is a straight splicing of two cables when an installation span is longer than can be accommodated by a single cable. The second is to switch between cable types for various reasons. For example, a 48-fiber cable can be brought into one end of the closure. Three 12-fiber cables, all going to different locations, can be spliced to the first cable and brought out the other end of the closure.

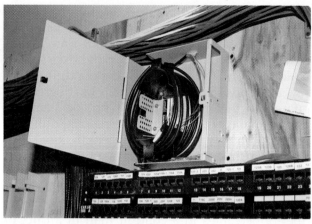

Figure 18 ◆ Splice organizer.

310F18.EPS

4.4.0 Distribution Hardware

Within a large fiber-optic system, signals must be routed to their final destination. Such routing involves the distribution of signals and the fibers that carry them. There are various types of hardware that allow transitions from one fiber type to another or the distribution of fibers to different points.

In indoor applications, a wiring center is used as a central distribution point. From this point, fibers can be routed to their destinations. A typical example is an outdoor multi-fiber cable brought into a control room to the distribution point. At the wiring center, each fiber in the outdoor cable is spliced to a simplex fiber that is routed to a different location, such as a different control panel.

A rack box containing an organizer similar to the organizer used in the splice closure serves such distribution needs. The boxes are mounted in 19" and 23" equipment racks, which are standard sizes in the telephone and electronics industries (*Figure 19*). Besides the organizer, rack boxes contain provisions for securing fibers and strain-relieving cables and provide storage room for service loops.

4.5.0 Patch Panels

Patch panels provide a convenient way to rearrange fiber connections and circuits. A simple patch panel is a metal frame containing bushings on either side into which fiber-optic connectors may be plugged. One side of the panel is usually fixed, which means the fibers are not intended to be disconnected. On the other side of the panel, fibers can be connected and disconnected to arrange the circuits as required (*Figure 20*). Like rack boxes, patch panels are compatible with 19" and 23" equipment racks.

Patch panels are widely used in the telephone industry to connect circuits to transmission equipment. They can also be used in the wiring center of a building to rearrange connections. The splice organizer and the patch panel serve the similar

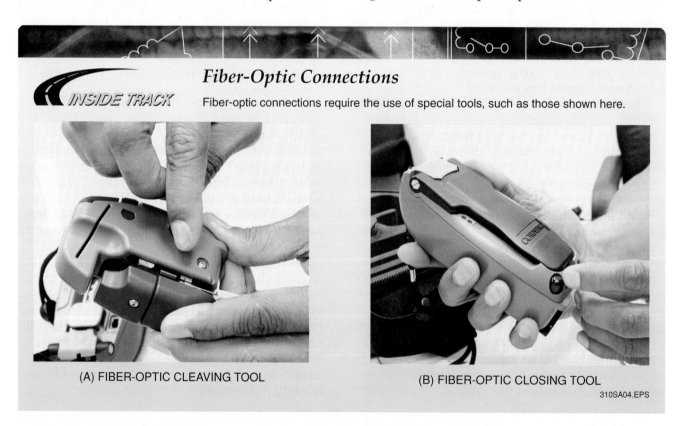

INSIDE TRACK

Fiber-Optic Connections

Fiber-optic connections require the use of special tools, such as those shown here.

(A) FIBER-OPTIC CLEAVING TOOL

(B) FIBER-OPTIC CLOSING TOOL

310SA04.EPS

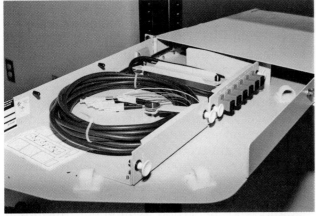

Figure 19 ◆ Open rack-mounted fiber termination box.

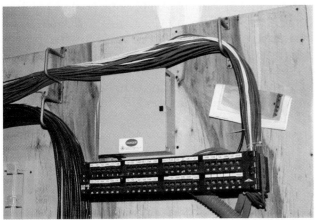

Figure 20 ◆ Rack-mounted fiber patch panel.

function of distribution. The difference is that the organizer is intended for fixed, unchanged connections, whereas a patch panel is used for flexible, changeable connections.

Some applications combine a splice organizer with a patch panel. A large fiber is brought into the splice organizer and then spliced to other cables running directly into the fixed side of a patch panel. From the panel, cables can then be routed as necessary.

4.6.0 Outlet Boxes

Fiber-optic outlet boxes are similar to electrical outlet boxes except that they allow connection to cables carrying optical signals rather than current. *Figure 21* shows typical fiber-optic outlet boxes, which mount in a wall in a similar manner as electrical boxes.

(A) MULTI-USER OUTLET ASSEMBLY

(B) FIBER OUTLET BOX

310F21.EPS

Figure 21 ◆ Typical fiber-optic outlet boxes.

In a control room wired with fiber optics, the box serves as a transition point between the cabling and the instruments. A short simplex or duplex cable (called a jumper or drop cable) runs from the junction box to the equipment being served.

5.0.0 ◆ GROUNDING AND BONDING

All conductors for grounding must run in straight lines and be protected from physical damage. The minimum size grounding conductor for fiber-optic cables is No. 14 AWG per

NEC Section 770.100(A)(3). The grounding means in buildings or structures shall be connected to the nearest accessible location on the:

- Building grounding electrode conductor
- Grounded interior metal water pipe within 5' of the entrance to the building
- Power service accessible means external to enclosure
- Metallic power raceway
- Service equipment enclosure
- Grounding electrode conductor

The bonding conductor size shall be no smaller than No. 6 AWG when connected to the power grounding electrode per *NEC Section 770.100(D)*. The conductors shall be copper, stranded or solid.

The minimum size grounding conductor for communication circuits is No. 14 AWG per *NEC Section 800.100(A)(3)*. The length shall be as short as possible, not to exceed 20' in one and two family dwellings per *NEC Section 800.100(A)(4)*.

The minimum size grounding conductor for network-powered broadband circuits is No. 14 AWG and is not required to be larger than No. 6 AWG per *NEC Section 830.100(A)(3)*. It must be able to carry the capacity approximately equal to that of the grounded metallic members and protected conductors. The length shall be as short as possible, not to exceed 20' in single-family and multi-family dwellings as stated in *NEC Section 830.100(A)(4)*.

Community antenna television and radio systems have similar requirements for grounding. The minimum size grounding conductor is No. 14 AWG and is not required to be larger than No. 6 AWG per *NEC Section 820.100(A)(3)*. It must be able to carry the capacity approximately equal to the outer conductor of the coax cable.

Radio and television equipment (including satellite dishes) have slightly different requirements for grounding. The minimum size grounding conductor is No. 10 AWG copper, No. 8 AWG aluminum, or No. 17 AWG copper-clad steel or bronze as listed in *NEC Section 810.21(H)*. The bonding jumper used to connect to the power grounding system shall not be smaller than No. 6 AWG per *NEC Section 810.21(J)*. Other requirements for these systems can be found in the appropriate articles in the *NEC®*, some of which are listed in *Table 1*.

Table 1 *NEC®* Requirements for VDV Systems

Description	Fiber Optics *NEC Article 770*	Communications Circuits *NEC Article 800*	Network-Powered Broadband Systems *NEC Article 830*
Definitions	NEC Section 770.2	NEC Section 800.2	NEC Section 830.2
Other Articles	NEC Section 770.3	NEC Section 800.3	NEC Section 830.3
Mechanical Execution of Work	NEC Section 770.24	NEC Section 800.24	NEC Section 830.24
Abandoned Cables	NEC Section 770.25	NEC Section 800.25	NEC Section 830.25
Spread of Fire	NEC Section 770.26	NEC Section 800.26	NEC Section 830.26
Overhead Wires	n/a	NEC Section 800.44	NEC Section 830.44
Protection	NEC Section 770.93	NEC Section 800.90	NEC Section 830.90
Cable Grounding	NEC Section 770.100	NEC Section 800.100	NEC Section 830.100
Cable Installation	NEC Section 770.133	NEC Section 800.133	NEC Section 800.133
Cable Substitution Hierarchy	NEC Table 770.154(E)	NEC Table 800.154(E)	NEC Table 830.154
Cable Marking	NEC Table 770.179	NEC Table 800.179	NEC Section 800.179(A)

6.0.0 ◆ TESTING

Installation is not complete without testing. After testing, double-check the quality of the job before leaving the work site. All testing can be done with scanners such as those shown in *Figure 22*. The test set shown in *Figure 22(A)* is a basic certification tester for UTP and ScTP cable. The test set shown in *Figure 22(B)* can also be used to certify fiber-optic cable.

The following is a list of the tests that must be performed:

- *Cat 3* – Tests include length, wire map, attenuation, and near end crosstalk (**NEXT**) up to 16MHz.
- *Cat 5e* – Tests include length, wire map, delay, **delay skew**, attenuation, NEXT, return loss, equal level far end crosstalk (**ELFEXT**), and power sum NEXT up to 100MHz.
- *Cat 6* – Tests include length, wire map, delay, delay skew, attenuation, NEXT, return loss, ELFEXT, and power sum NEXT up to 250MHz.
- *Multi-mode fiber* – Tests include length, delay, loss/km @850, and loss/km @1300.
- *Single-mode fiber* – Tests include length, delay, and loss/km @1310.
- *Coax* – Tests include length, open, shorts, and loss.

(A) AGILENT CABLE CERTIFICATION TESTER

(B) OMNI CABLE CERTIFICATION TESTER

310F22.EPS

Figure 22 ◆ Cable certification test sets.

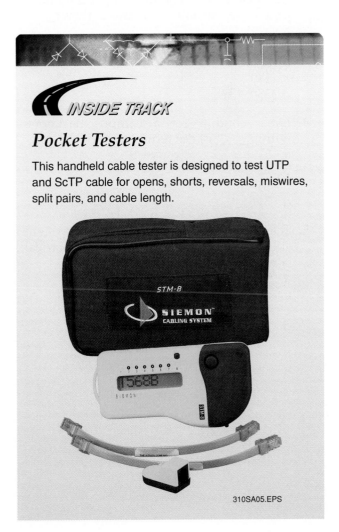

INSIDE TRACK

Pocket Testers

This handheld cable tester is designed to test UTP and ScTP cable for opens, shorts, reversals, miswires, split pairs, and cable length.

310SA05.EPS

1. The video cable used in new installations is normally _____.
   a. Cat 5
   b. Cat 6
   c. RG5
   d. RG6

2. A structured cabling subsystem that connects buildings together in a multi-building environment is called a(n) _____ subsystem.
   a. riser
   b. equipment room
   c. campus backbone
   d. work area

3. Campus straight conduit runs should have pull boxes installed no more than _____ apart.
   a. 20'
   b. 80'
   c. 140'
   d. 350'

4. Label conduit for a campus backbone within _____ of the termination point.
   a. 6"
   b. 12"
   c. 18"
   d. 24"

5. Telecom cables in an ER or TR subsystem should be kept at least _____ from electrical conduit or cable.
   a. 5"
   b. 8"
   c. 10"
   d. 12"

6. Riser pathways should run perpendicular to and be spaced at least _____ from fluorescent lighting.
   a. 5"
   b. 8"
   c. 10"
   d. 12"

7. Vertical cabling spans exceeding 30' in riser support subsystems must be supported in _____ increments.
   a. 15'
   b. 20'
   c. 30'
   d. 40'

8. When installing horizontal ER/TR/MC sleeves, a minimum of _____ must be provided between ER/TR/MCs and the horizontal subsystem.
   a. 3"
   b. 4"
   c. 5"
   d. 6"

9. RJ45 connectors used for 4-pair UTP are identified as _____ connectors.
   a. 2P2C
   b. 3P6C
   c. 6P8C
   d. 8P8C

10. The minimum size grounding conductor for fiber-optic cable is _____ AWG.
    a. No. 10
    b. No. 12
    c. No. 14
    d. No. 16

Summary

This module covered generic installation guidelines for structured voice, data, and video system cabling. A structured cabling system is a system in which the main components do not change. Structured cabling systems are divided into five subsystems: the campus backbone, equipment room, riser, horizontal cabling, and work area. Each subsystem has specific installation requirements. Structured cabling systems may use unshielded twisted pair (UTP) cable, coaxial cable, or fiber-optic cable. UTP is the most widely used cable for voice and data applications, while coaxial cable is commonly used for video. However, fiber-optic cable offers many advantages and its use is increasing rapidly. Regardless of the cable type, proper terminations are essential, as is proper system grounding and acceptance testing.

Notes

Trade Terms
Introduced in This Module

Attenuation: The reduction of a signal due to system losses.

Backbone: A high-capacity network link between two sections of a network.

Bandwidth: The range of frequencies that will pass through a particular portion of a system.

Coaxial cable: A type of cable that has a center conductor separated from the outer conductor by a dielectric. Coaxial cable is used to transmit very small currents or high-frequency signals.

Cross connect: A facility for connecting cable runs, subsystems, and equipment using patch cords or jumper cable.

Delay skew: A change of timing or phase in a transmission signal.

ELFEXT: Equal level far end crosstalk; a value derived by subtracting the attenuation of the interfering pair from the far end crosstalk (FEXT) that it has caused in the interfered pair.

Ethernet: A local area network (LAN) developed by Xerox Corporation that uses a bus or star topology. Ethernet is currently the most popular type of LAN.

Firestopping: A material or mechanical device used to block openings in walls, ceilings, or floors to prevent the passage of fire or smoke.

Gigabit: An extremely fast rate of data transfer equal to 1,000,000,000 bits per second.

Innerduct: A type of high-density polyethylene (HDPE) conduit that is available in both corrugated and smooth types, and is often used to run communications cable.

Megabit: A rate of data transfer equal to 1,000,000 bits per second.

NEXT: Near end crosstalk; a measurement of the interference (crosstalk) between two wire pairs.

Patch cord: A connecting cable used with cross-connect equipment and work area cables. It has three conductive portions separated by insulators.

This module is intended to present thorough resources for task training. The following reference works are suggested for further study. These are optional materials for continued education rather than for task training.

Cisco Home Technology Integration Fundamentals and Certification, 2004. Englewood, CO: Upper Saddle River, NJ: Pearson Education, Inc.

National Electrical Code® Handbook, Latest Edition. Quincy, MA: National Fire Protection Association.

The Cabling Handbook, 2nd Edition. 2000. Upper Saddle River, NJ: Pearson Education, Inc.

TIA/EIA Telecommunication Building Wiring Standards, Latest Edition. Englewood, CO: Global Engineering Documents.

NCCER makes every effort to keep these textbooks up-to-date and free of technical errors. We appreciate your help in this process. If you have an idea for improving this textbook, or if you find an error, a typographical mistake, or an inaccuracy in NCCER's Contren® textbooks, please write us, using this form or a photocopy. Be sure to include the exact module number, page number, a detailed description, and the correction, if applicable. Your input will be brought to the attention of the Technical Review Committee. Thank you for your assistance.

Instructors – If you found that additional materials were necessary in order to teach this module effectively, please let us know so that we may include them in the Equipment/Materials list in the Annotated Instructor's Guide.

Write: Product Development and Revision
National Center for Construction Education and Research
3600 NW 43rd St., Bldg. G, Gainesville, FL 32606

Fax: 352-334-0932

E-mail: curriculum@nccer.org

Craft _____ Module Name _____

Copyright Date _____ Module Number _____ Page Number(s) _____

Description _____

_____

_____

_____

(Optional) Correction _____

_____

_____

(Optional) Your Name and Address _____

_____

_____

Motor Controls

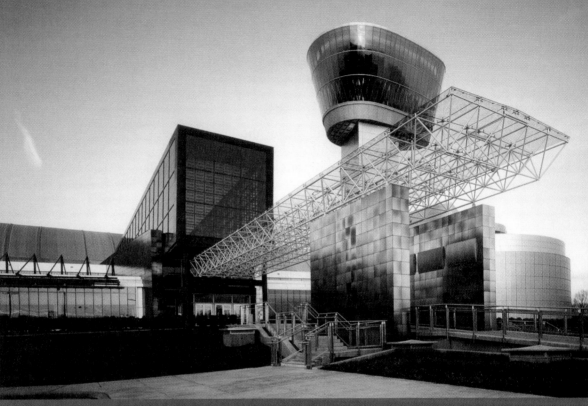

Steven F. Udvar-Hazy Center – National Air and Space Museum

Chantilly, Virginia is home to this winner in the Mega-Projects over $100 Million category in the Award of Excellence competition sponsored by Associated Builders and Contractors, Inc. Hensel Phelps Construction Co. was the builder.

26311-08

26311-08
Motor Controls

Topics to be presented in this module include:

Overview

Small motors that require very little current to operate are often controlled through a simple OFF-ON switch. As the size of motors increases, the way in which they are controlled becomes more complex.

Devices such as relays, contactors, and magnetic motor starters typically use an electromagnetic coil assembly and rod to move one or more sets of contacts. These devices operate by causing a conductive rod or bar to be pulled into the center of an energized coil by magnetic force. Contact arrangements range from a single set of contacts that open and close with one coil, or multiple sets of contacts operated by a single coil. Contactors are nothing more than large relays with multiple sets of contacts that can handle a high current. Magnetic motor starters are contactors equipped with overload protection devices for the motor.

Motor overload protective devices operate by monitoring the current through each motor lead and opening the current path to the control coil if an overcurrent condition occurs. Various technologies are applied in motor overload devices, including melting-alloy thermal overloads, bimetallic overload relays, and magnetic overload relays. Motor controls are regulated by *NEC Article 430*.

Objectives

When you have completed this module, you will be able to do the following:

1. Identify contactors and relays both physically and schematically and describe their operating principles.
2. Identify pilot devices both physically and schematically and describe their operating principles.
3. Interpret motor control wiring, connection, and ladder diagrams.
4. Select and size contactors and relays for use in specific electrical motor control systems.
5. Select and size pilot devices for use in specific electrical motor control systems.
6. Connect motor controllers for specific applications according to *National Electrical Code®* (*NEC®*) requirements.

Trade Terms

Actuator
Auxiliary contact
Cam
Dashpot
Deadband
Dielectric constant
Dropout voltage
Eddy current
Electromechanical relay (EMR)
Infrared (IR)
International Electrotechnical Commission (IEC)
Jogging (inching)

Line-powered sensor
Load-powered sensor
Maintained contact switch
Momentary contact switch
Motor control circuit
Operator
Pickup voltage
Pilot devices
Plugging
Proximity sensors (switches)
Seal-in voltage
Setpoint

Required Trainee Materials

1. Pencil and paper
2. Appropriate personal protective equipment
3. Copy of the latest edition of the *National Electrical Code®*

Prerequisites

Before you begin this module, it is recommended that you successfully complete *Core Curriculum; Electrical Level One; Electrical Level Two; Electrical Level Three*, Modules 26301-08 through 26310-08.

This course map shows all of the modules in *Electrical Level Three*. The suggested training order begins at the bottom and proceeds up. Skill levels increase as you advance on the course map. The local Training Program Sponsor may adjust the training order.

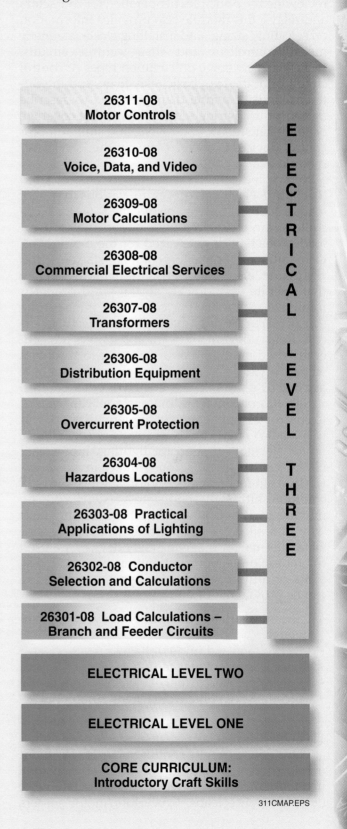

26311-08
Motor Controls

26310-08
Voice, Data, and Video

26309-08
Motor Calculations

26308-08
Commercial Electrical Services

26307-08
Transformers

26306-08
Distribution Equipment

26305-08
Overcurrent Protection

26304-08
Hazardous Locations

26303-08 Practical Applications of Lighting

26302-08 Conductor Selection and Calculations

26301-08 Load Calculations – Branch and Feeder Circuits

ELECTRICAL LEVEL TWO

ELECTRICAL LEVEL ONE

CORE CURRICULUM: Introductory Craft Skills

ELECTRICAL LEVEL THREE

311CMAP.EPS

1.0.0 ◆ INTRODUCTION

This module is the second of three related modules that cover electrical motor controls. It expands upon the basic material presented in the Level Two module, *Contactors and Relays*. In addition to reviewing the principles of operation for contactors and relays, this module also provides information on selecting, sizing, and installing electromagnetic motor controllers and their control circuits. Included is coverage on common types of control circuit pilot devices, basic relay logic, and the different kinds of wiring diagrams used to describe motor control wiring schemes. Note that coverage of solid-state relays, controllers, and similar motor control devices is covered later in the Level Four module, *Advanced Controls*.

Relays and contactors/motor starters are used to control electrical power to various types of loads. Motor starters, also called motor controllers, are contactors with added motor overload protection devices sized for the motor load. Conventional contactors/motor starters and relays are electromagnetic, mechanical devices that operate very quickly and, if sized properly, minimize contact arcing caused by the opening (breaking) or closing (making) of an electrical circuit. Relays or contactors/motor starters are generally used in motor control circuits to amplify a signal. In this application, a low-current signal applied by a pilot device (switch or other control) to the relay or contactor/motor starter can be used to control the application of higher-current power to a load. A large variety of relays or contactors/motor starters are used for lighting, motor, and HVAC control circuits. Electromechanical relays (EMRs) and contactors/motor starters are still common, although solid-state versions with no moving parts or contacts are rapidly finding their way into all types of applications. Special solid-state, thermal, or magnetic relay devices are used for motor protection. Depending on the device, some or all of the following types of protection can be provided:

- Current overload (time delay and/or instantaneous)
- Overvoltage and undervoltage
- Phase loss or unbalance
- Directional overcurrent
- Percentage of voltage or current differential

Contactors/motor starters are designed to handle higher current loads than most relays using relatively low-current control signals. Without overload protection, contactors are used to handle high-current, noninductive loads such as lighting or other resistance loads. Contactors/motor starters

and their enclosures are available with National Electrical Manufacturers Association (NEMA) or International Electrotechnical Commission (IEC) ratings. For identical load ratings, IEC-rated contactors are less expensive and of lighter-duty construction than the NEMA-rated contactors.

2.0.0 ◆ ELECTROMECHANICAL RELAYS

Electromechanical relays (EMRs) are normally used in control circuits to operate low-current loads. They range in size from subminiature versions rated for milliamp loads to power relays or mercury-displacement relays with average load ratings of 30A maximum at 277VAC or 15A maximum at 600VAC. Most relays, except for subminiature, reed, power, industrial, or mercury relays, are configured as plug-in devices. Subminiature relays are very small, sealed, electromechanical or solid-state relays that are typically used on printed circuit boards. Industrial relays are similar to the contactors covered later in this module. Reed relays are sensitive electromechanical relays with low-current contacts sealed in a glass envelope.

2.1.0 General-Purpose Relay Configurations and Designation

General-purpose EMRs are mechanical switches operated by a magnetic solenoid coil. They are available in various AC and DC voltage designs and current ratings. Coils can be specified in DC voltage ranges of 5V to 24V or in AC ranges of 12V to 240V, single phase. DC coils can be activated with as little as 4mA at 5VDC, making them compatible with certain integrated circuit logic gates. Relays can have up to 12 poles (separate, insulated, switching circuits) per device with various combinations of normally open and normally closed contacts. A normally closed (N.C.) contact or normally open (N.O.) contact is defined as the contact state that occurs when the coil is de-energized. *Figure 1* is an example of a single phase, 25A, 240V, open-frame AC power relay with two poles (double pole or DP). Each pole has an N.C. and N.O. contact with a common single break (SB) moving contact. When a pole is equipped with both N.C. and N.O. contacts, it is designated as a double-throw (DT) pole. The power relay shown is thus defined as a DPDT-SB relay. *Figure 2* is a miniature, single phase, 5A, 240VAC plug-in relay with four poles (4P). Each pole has an N.C. and N.O. contact (DT) with a common SB moving contact. This relay is defined as a 4PDT-SB relay.

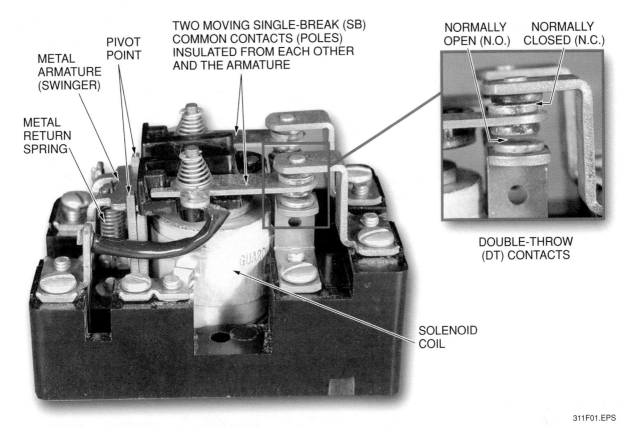

SOLENOID COIL

DOUBLE-THROW (DT) CONTACTS

Figure 1 ◆ Typical double-pole, double-throw, single-break (DPDT-SB), open-frame power relay.

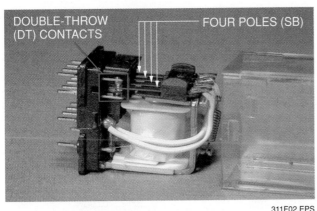

DOUBLE-THROW (DT) CONTACTS

FOUR POLES (SB)

311F02.EPS

Figure 2 ◆ Miniature four-pole, double-throw, single-break (4PDT-SB), plug-in relay.

Table 1 Relay Contact and Pole Designations

Designation	Meaning
ST	Single Throw
DT	Double Throw
N.O.	Normally Open
N.C.	Normally Closed
SB	Single Break
DB	Double Break (industrial relays)
SP	Single Pole
DP	Double Pole
3P	Three Pole
4P	Four Pole
5P	Five Pole
6P	Six Pole
(N)P	N = numeric number of poles

As just explained, relays are designated by their number of poles, throws, and breaks. Relay pole and contact designations are shown in *Table 1*, and various relay contact configurations are shown in *Figure 3*. Some of these pole and contact designations are also used for contactors.

Manufacturers use a common code to simplify the contact sequence identification for relays. The code uses form letters to indicate the type of contact sequencing applicable to each pole of a relay, as shown in *Figure 4*.

2.2.0 Typical Operation

In the single-pole, single-break version of an electromechanical power relay shown in *Figure 5*, operation occurs when a control signal voltage is applied to the relay solenoid coil. The magnetic field created through the core of the solenoid coil instantly attracts the metal armature and draws it into contact with the core. The moving contact, attached to the armature by the insulating block

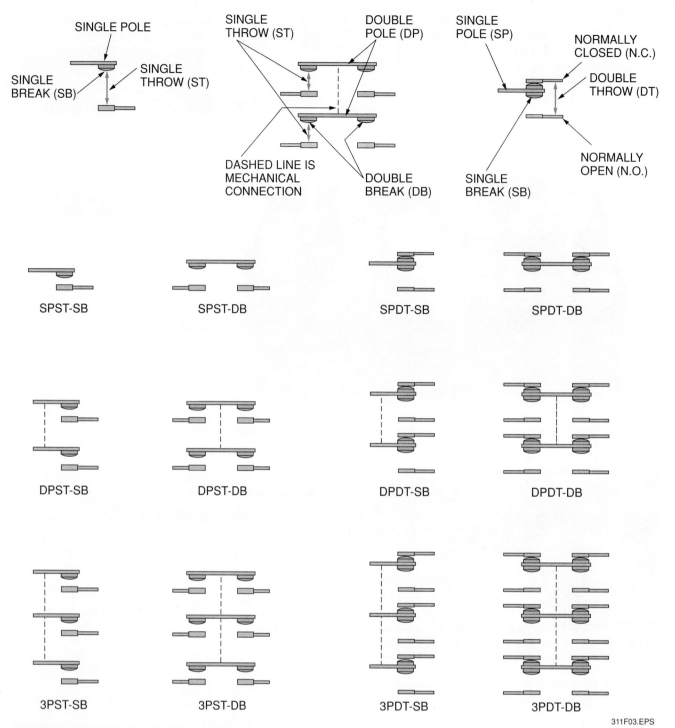

Figure 3 ◆ Examples of relay contact configurations.

311F03.EPS

and contact pressure spring (see *Figures 5* and *6*), breaks from the N.C. contact and makes with the N.O. contact. The input power at the common (COM) terminal is routed through the moving contact to the circuits connected to the N.O. terminal. When the control signal is removed from the solenoid coil, a return spring pulls the armature back to a de-energized position, thus causing the moving contact to break with the N.O. contact and make with the N.C. contact. Input power from the COM terminal is then switched back from the N.O. contact circuits to the N.C. contact circuits.

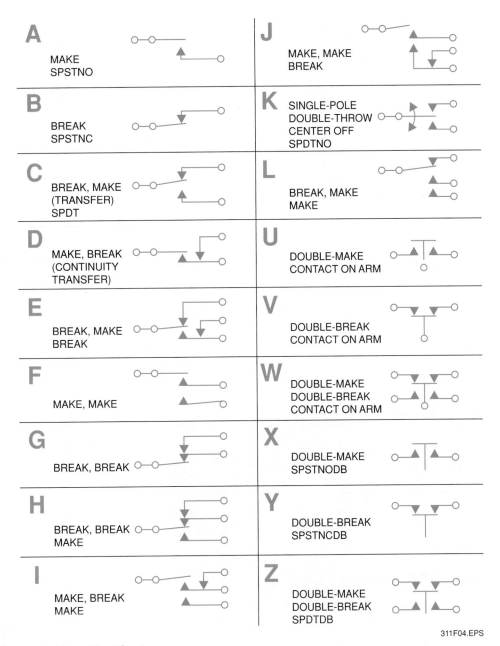

Figure 4 ◆ Relay contact form identification.

When the solenoid coil is specified for use with an AC control signal, a shading coil (aluminum or copper) at the top of the core (see *Figures 5* and *6*) is used to create a weak, out of phase, auxiliary magnetic field. As the main field collapses when the AC current periodically drops to zero, the weak field generated by the shading coil is strong enough to keep the armature in contact with the core and prevent the relay from chattering at a 120Hz rate. If the shading coil is loose or missing, the relay will produce excessive noise and be subject to abnormal wear and coil heat buildup. In a power relay of this size, the moving contact arm must be made of relatively thick metal to accommodate the current

rating of the relay. Lighter-duty relays use a flexible, thin-gauge copper spring stock for the moving contact arm. Because the armature is designed to swing over a slightly greater distance than necessary in both directions (armature over travel), the flexible moving arm bends slightly in the normally open or closed positions. This eliminates any contact-pressure mating or bounce problems. In relays (or contactors) that use an inflexible moving contact arm, a contact-pressure spring is used to allow flexible positioning of the moving contact arm so that contact-pressure mating and bounce problems are eliminated. Like light-duty relays, this is accomplished by armature overtravel.

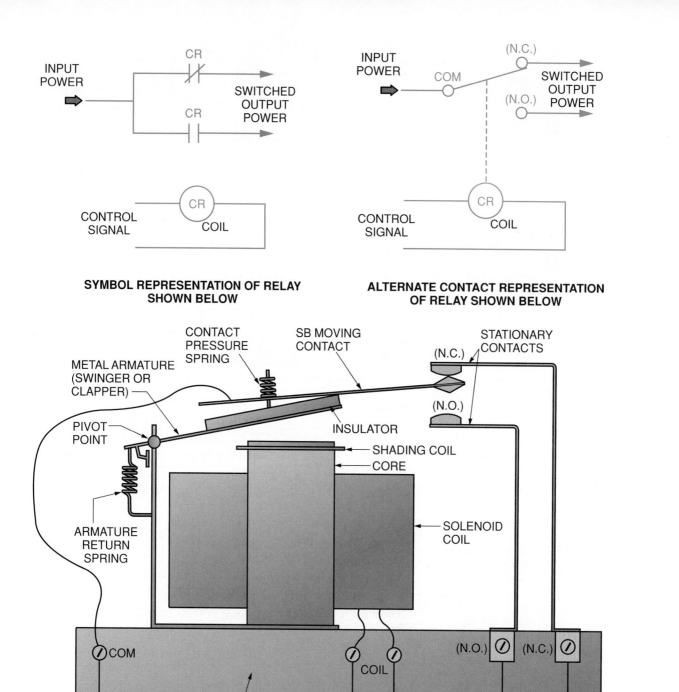

SYMBOL REPRESENTATION OF RELAY SHOWN BELOW

ALTERNATE CONTACT REPRESENTATION OF RELAY SHOWN BELOW

Figure 5 ◆ Typical SPDT-SB power relay and symbol representation.

311F05.EPS

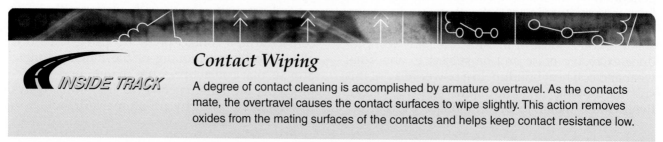

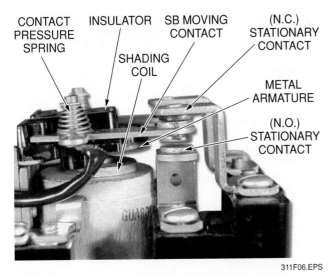

CONTACT PRESSURE SPRING   INSULATOR   SB MOVING CONTACT   (N.C.) STATIONARY CONTACT

SHADING COIL

METAL ARMATURE

(N.O.) STATIONARY CONTACT

311F06.EPS

Figure 6 ◆ Detail view of a power relay.

2.3.0 Relay Selection Criteria

This section covers relay selection criteria, including coil voltage characteristics, contact ratings, and contact materials.

2.3.1 Coil Voltage Characteristics

The **pickup voltage** is the minimum allowable coil control voltage that will cause an electromechanical device to energize. Once energized, the **seal-in voltage** is the minimum allowable coil control voltage that will keep the device energized. It is usually less than the pickup voltage. The **dropout voltage** is defined as the coil voltage that is reached when the armature return spring overpowers the magnetic field of the coil and the contacts of the device change position. The dropout voltage is less than the seal-in voltage. Most coils on electromechanical devices such as relays and contactors must be designed so that the dropout does not occur until the voltage is reduced to 85% of the nominal coil voltage. They must also be designed so that they pick up when the voltage rises to more than 85% of the nominal coil voltage. These voltage levels are set by NEMA and are conservative. Most electromechanical devices manufactured today drop out and pick up at lower voltages.

2.3.2 Contact Ratings

The most important consideration in the selection of a relay (or contactor) for a particular application is the current rating of the contacts. However, the rating of contacts can be confusing. Sometimes only one rating is given, and it is important to know the definition of the rating. Many relays and other devices have three published ratings:

- Inrush current (make contact) capacity
- Normal or continuous-carrying capacity
- Current break (opening) capacity

For example, a typical industrial relay may have the following contact ratings at a particular AC voltage:

- 15A noninductive (resistive) continuous load
- 8A inductive continuous load
- 75A inrush, 50A break (inductive or resistive)

Two examples of resistive loads are heating elements and incandescent lights. Inductive loads are coils such as those used in solenoids, relay coils, and motor starter coils. The type of load and its inrush current, along with switching frequency, can cause contact welding. Therefore, for loads with inrush current, the steady-state and inrush currents should be measured to determine the selection of the proper contactor or relay contact rating. Some typical loads and their approximate inrush currents are summarized in *Table 2*.

2.3.3 Contact Materials

The contacts used in relays or contactors are available in a number of materials. These materials have certain advantages and disadvantages for various applications, as shown in *Table 3*.

2.4.0 Contact Arc Suppression

The typical contact life of a relay ranges between 100,000 and 500,000 operations. The contact rating of a relay is based on the contact's full-rated power. When contacts switch loads that are less than their full-rated power, contact life is increased. If loads exceed the contact rating, or if arcing occurs, the life of the contact is shortened due to burning and overheating.

Table 2 Typical Loads and Their Inrush Currents

Type of Load	Approximate Inrush Current
Resistive heating	Steady-state current
Sodium vapor lamps	1 to 3 times steady-state current
Mercury lamps	About 3 times steady-state current
Motors	5 to 10 times steady-state current
Transformers	5 to 15 times steady-state current
Incandescent lamps	10 to 15 times steady-state current
Solenoids	10 to 20 times steady-state current
Capacitive loads	20 to 40 times steady-state current

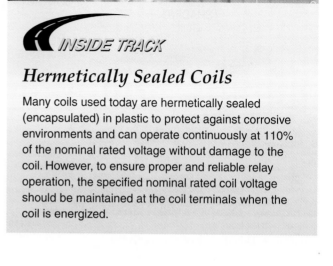

In some applications involving control relays, arc protection circuits can be added to the relay-controlled circuit to reduce the arcing caused by inductive loads. The circuit provides a nondestructive path for the voltage generated by the collapsing field of the inductive load when the relay contacts open. *Figure 7* shows three differ-ent protection circuits that can be used for arc suppression.

In the DC diode protection circuit, a diode can be placed in parallel with the load so that it will oppose any current flow when the load is energized by the relay contacts. When the relay contacts open, the reverse polarity voltage generated by the collapsing field is shorted out by conduction through the diode. By shorting out the

Table 3 Contact Material Characteristics

Material	Advantages	Disadvantages
Silver	This metal has the highest conductivity and thermal properties of any contact material.	This metal is subject to welding and sticking under arcing conditions. Rapid sulfidation (tarnishing) in many applications creates a film that increases contact resistance. Normal contact wiping usually removes the film. It is not used in intermittent or arcing applications.
Silver-cadmium alloy or silver-cadmium oxide alloy	These alloys have good conductivity and thermal properties. Cadmium oxide alloy conducts even when oxide forms on surface of contacts.	These alloys resist arcing damage but are subject to some sulfidation. Normal contact wiping usually removes the film. When used in circuits drawing several amperes at more than 12V, any sulfidation is burned off.
Gold-flashed silver	This metal has the same advantages as silver. Gold flashing protects against sulfidation. It is used in intermittent applications and is good for switching current of 1A or less.	This metal is not used in applications where arcing occurs because gold burns off quickly.
Tungsten or tungsten-carbide alloy	These alloys experience minimal damage from arcing due to their high melting temperature. They are good for high-voltage and repetitive switching applications.	These alloys offer higher contact resistance than other materials.
Silver tungsten-carbide alloy	This alloy has the same advantages as tungsten or tungsten-carbide but lower contact resistance.	This alloy is subject to minor sulfidation; however, wiping action or any arcing removes the film.

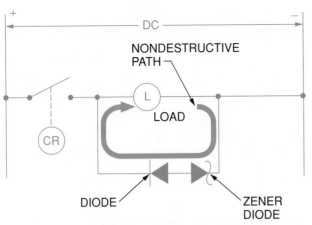

DC DIODE PROTECTION CIRCUIT

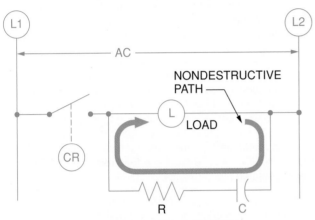

RC PROTECTION CIRCUIT

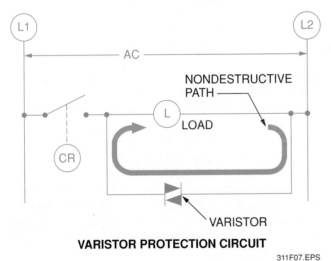

VARISTOR PROTECTION CIRCUIT

311F07.EPS

Figure 7 ◆ Contact protection circuits.

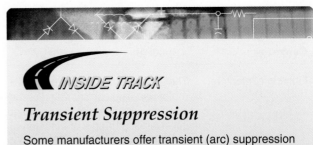

In an AC circuit, arc suppression is less of a problem because the arc is extinguished wherever the current passes through zero. This is why contacts of relays rated for AC use can tolerate a much smaller separation at break than DC contacts. Arcing can still occur, however, and a resistor/capacitor (RC) circuit or a varistor can be used for contact protection to extend contact life. In the RC protection circuit, the resistor/capacitor combination is selected so that the circuit is effectively resistive and its time constant is approximately equal to the time constant of the load. This provides a path through the load to dissipate the generated voltage and current when the relay contact is opened. In the varistor protection circuit, the varistor is a solid-state device whose resistance is inversely proportional to the voltage applied across it. When the relay contacts are closed and voltage appears across the load and the varistor, the varistor resistance is high. When the relay contacts open, a low-resistance path is provided through the varistor to dissipate the generated voltage and current across the load. The disadvantages of these circuits are that they slow down coil reaction time and they cannot be used in electronic timer-controlled circuits.

Other circuits that are placed across the control relay contacts can be used for arc suppression, but they are not fail-safe. If a component shorts in a suppression circuit across the relay contacts, the load may remain energized even though the relay contact opens.

3.0.0 ◆ MAGNETIC CONTACTORS

Magnetic contactors, like relays, are control devices that energize or de-energize loads by the use of a control signal. They are used instead of relays to repeatedly establish and interrupt higher current loads. They are usually available from most manufacturers in four categories for the applications summarized in *Table 4*.

generated voltage, the resulting current is dissipated across the load instead of the relay contacts. The diode on-time and resulting dissipation time can be controlled by the use of a zener diode. The cutoff voltage characteristic of the zener can be used to halt the dissipation across the load very shortly after the contacts have opened.

Table 4 Contactor Categories and Applications

Category	Application
Lighting contactors	These contactors are current-rated only for resistive loads and do not have horsepower ratings for motor use. The current rating of these contactors is for the maximum continuous current required by a resistive load. At the maximum current rating, the contactors are designed to withstand the large initial inrush currents of tungsten and ballast lamp loads, as well as nonmotor (resistive) loads, without contact welding. They are generally available in versions that are locally or remotely controlled via AC or DC control circuits and are magnetically held, mechanically held, or magnetically latched. The mechanically held or latched contactors are quiet (no AC hum) and remain closed during power interruptions so that their loads will come back on when power is restored.
Definite-purpose contactors and motor starters	Low-duty cycle, definite-purpose contactors and motor starters are intended for use in applications where the control requirements are well defined. These contactors carry dual ratings (ampere and horsepower) for either resistive or inductive loads found in applications such as refrigeration, air conditioning, resistance heating, and other Standard Industrial Classification (SIC) applications. The motor starters are usually equipped with bimetallic motor overcurrent protective devices (overload relays).
NEMA-rated contactors and motor starters	These are contactors for general use and for use as motor starters when equipped with overcurrent protective devices (overload relays). They are identified in eleven overlapping ranges (sizes) and rated in horsepower (for motor starters) and/or continuous current capacity. They are primarily designed for use with inductive motor loads and to withstand the interrupt current of a locked rotor. They are designed with reserve capacity to perform over a broad range of applications without the need for an assessment of life requirements. The motor starters are either manual or magnetically actuated and are equipped with melting-alloy, bimetallic, or solid-state overload relays. Most of these contactors have replaceable contacts and encapsulated (sealed) coils.
IEC-rated contactors and motor starters	These contactors and motor starters perform the same functions as NEMA-rated devices but are smaller than NEMA devices for the same horsepower or current ratings. As a result, they are very application sensitive and may require rating and life assessment matches to the load. U.S. manufacturers normally provide size tables similar to NEMA tables to facilitate selection. The motor starters are usually equipped only with bimetallic overload relays. Most of these contactors do not have sealed coils and, except for the larger sizes, have nonreplaceable contacts.

3.1.0 Magnetic Contactor Construction and Operation

Like a relay, a magnetic contactor is actuated by a solenoid. *Figure 8* shows three types of mechanical actions used for solenoid closure of contactors. In all cases, energizing the solenoid coil assembly, made up of a frame, coil, and moving armature, causes a moving contact assembly with spring-loaded, double break (DB) bridging contacts (one DB set of contacts for each pole) to mate with corresponding pairs of stationary contacts to close a circuit. Contactors typically have between one and six poles. The bridging contacts and the materials used for the contacts themselves allow the higher current ratings for contactors. The angled contacts allow a slight, random wiping (cleaning) action as the contacts open and close. Coil voltage characteristics and contact rating information is the same as for relays.

Figure 9 is a typical NEMA-rated magnetic contactor. It is a three-pole (3Ø) power device that is closed by a mechanical bell crank when

the solenoid coil assembly is energized. The energized indicator shows if the contactor is de-energized (flush) or energized (depressed). It can also be manually depressed to check if the contactor mechanism is free to move and the contacts can be closed. Except for certain IEC or small contactors, most contactors can be disassembled and individual components replaced. For the contactor shown, the coil and contacts can be individually replaced. Other types of contactors, including definite-purpose and lighting contactors, are similarly constructed.

For coil replacement, the two cover screws (see *Figure 9*) can be loosened or removed and the cover removed (*Figure 10*). This exposes the solenoid coil assembly and its positioning clips that are released when the cover is removed. For this contactor, lugs on the cover are forced down between the clips and the contact mechanism housing to clamp the solenoid assembly in place horizontally when the cover is replaced. After the wires to the solenoid coil assembly are removed, the assembly can be lifted out of the contact

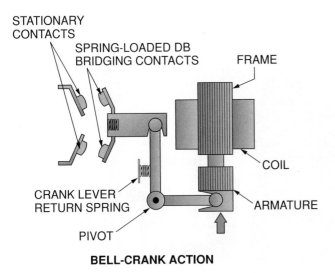

STATIONARY
CONTACTS

SPRING-LOADED DB
BRIDGING CONTACTS

FRAME

COIL

CRANK LEVER
RETURN SPRING

ARMATURE

PIVOT

BELL-CRANK ACTION

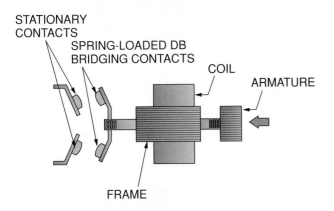

STATIONARY
CONTACTS

SPRING-LOADED DB
BRIDGING CONTACTS

COIL

ARMATURE

FRAME

HORIZONTAL ACTION

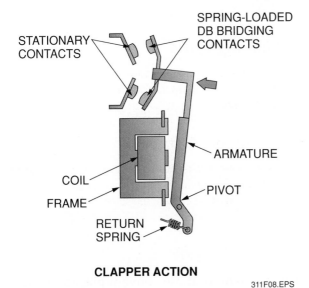

SPRING-LOADED
DB BRIDGING
CONTACTS

STATIONARY
CONTACTS

ARMATURE

COIL

PIVOT

FRAME

RETURN
SPRING

CLAPPER ACTION

311F08.EPS

Figure 8 ◆ Mechanical actions used for magnetic contactors.

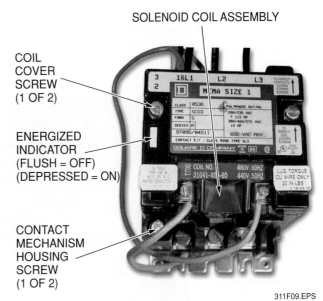

SOLENOID COIL ASSEMBLY

COIL
COVER
SCREW
(1 OF 2)

ENERGIZED
INDICATOR
(FLUSH = OFF)
(DEPRESSED = ON)

CONTACT
MECHANISM
HOUSING
SCREW
(1 OF 2)

311F09.EPS

Figure 9 ◆ Typical bell-crank actuated three-pole magnetic contactor.

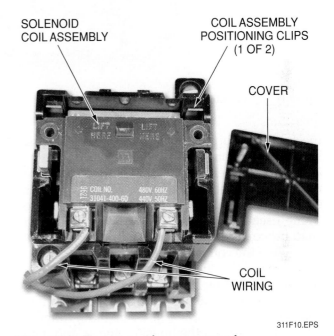

SOLENOID
COIL ASSEMBLY

COIL ASSEMBLY
POSITIONING CLIPS
(1 OF 2)

COVER

COIL
WIRING

311F10.EPS

Figure 10 ◆ Contactor with cover removed.

mechanism housing (*Figure 11*). When assembled, the bell-crank lever hook of the armature fits over the spring-returned bell-crank lever. Energizing the coil causes the armature to pull in and move the bell crank (against a return spring) so that the moving contact assembly is pushed down to mate with the stationary contacts under the contact mechanism housing. When the solenoid coil assembly is energized and the armature is pulled in, a slight air gap is maintained between the frame and armature for two reasons. The first is to prevent the laminated steel armature from slamming against the laminated steel frame, and

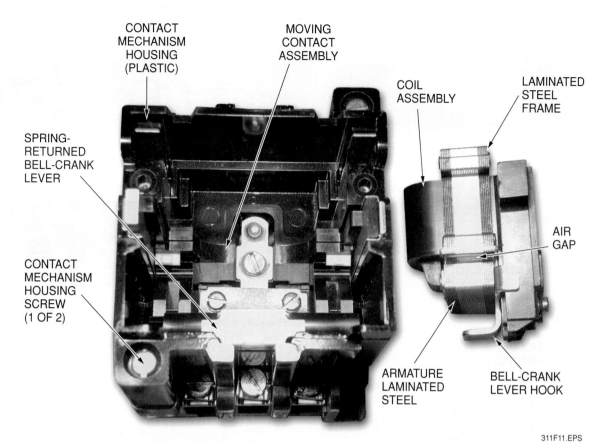

Figure 11 ◆ Contactor with solenoid coil assembly removed.

311F11.EPS

the second is to prevent any residual magnetism in the frame from holding the armature in when the coil is de-energized. *Figure 12* shows the laminated frame and armature separated from the encapsulated coil. Like a relay, the legs of the coil frame used with an AC coil are each fitted with shading coils to prevent chattering of the armature when the coil is energized.

To expose the contacts, the contact mechanism housing mounting screws can be loosened, and the contact mechanism housing removed from the contactor base (*Figure 13*). Even though shown disassembled from the housing, the solenoid coil assembly does not have to be removed to gain access to the contacts. A close-up of the moving contact assembly with replaceable spring-loaded bridging contacts and corresponding replaceable stationary contacts in the contactor base is shown in *Figure 14*. Each bridging contact is removed by sliding it out from under its spring-loaded clip in the moving contact assembly. Each pair of stationary contacts can be removed by removing the screws securing them to the contactor base.

3.2.0 Contactor Marking Conventions

Most contactors reflect various labeling/marking conventions that are in general use by manufac-

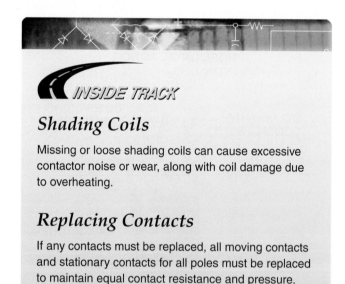

INSIDE TRACK

Shading Coils

Missing or loose shading coils can cause excessive contactor noise or wear, along with coil damage due to overheating.

Replacing Contacts

If any contacts must be replaced, all moving contacts and stationary contacts for all poles must be replaced to maintain equal contact resistance and pressure.

turers. Many of these conventions are because of Underwriters Laboratories, Inc. (UL) requirements or NEMA and IEC standards. Because many NEMA devices conform to IEC standards and vice versa, marking conventions used by manufacturers may include elements of both on the devices.

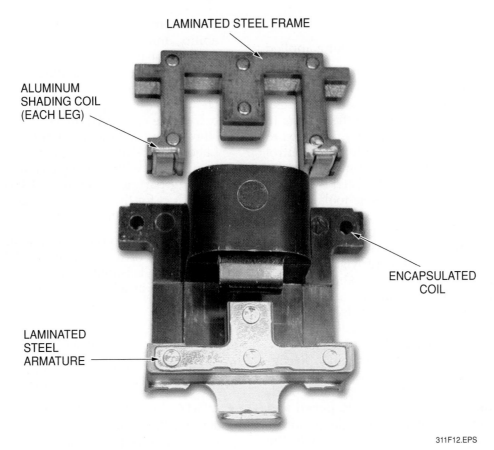

LAMINATED STEEL FRAME

ALUMINUM
SHADING COIL
(EACH LEG)

ENCAPSULATED
COIL

LAMINATED
STEEL
ARMATURE

311F12.EPS

Figure 12 ◆ Disassembled solenoid coil assembly.

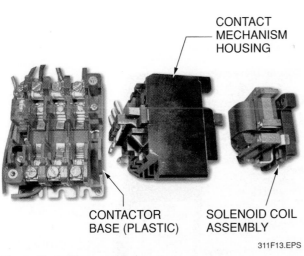

CONTACT
MECHANISM
HOUSING

CONTACTOR
BASE (PLASTIC)

SOLENOID COIL
ASSEMBLY

311F13.EPS

Figure 13 ◆ Disassembled contactor.

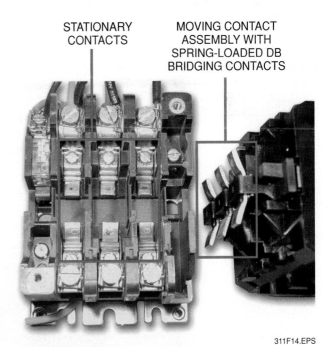

STATIONARY
CONTACTS

MOVING CONTACT
ASSEMBLY WITH
SPRING-LOADED DB
BRIDGING CONTACTS

311F14.EPS

Figure 14 ◆ Close-up view of stationary and moving
contacts.

3.2.1 UL Markings

There are three categories of UL marking applied to contactors and motor starters:

- *UL Listed* – Products that are UL Listed must have a UL Listing mark (UL enclosed in a circle) on their nameplates adjacent to the ratings to which the marking applies (*Figure 15*). Next to the mark, the words "Listed Industrial Control Equipment" or "Listed Ind. Cont. Eq." must appear. Equipment that is UL Listed is suitable for installation with general-use tools and for application at the ratings to which the listing mark is related. A UL Listed product complies with a governing UL Standard and qualifies for installation under specific provisions of the *NEC®*. Other ratings and certifications such as IEC horsepower or kW ratings or Canadian Certifications (CE or CSA) may also appear on the nameplate, but they are segregated from the ratings associated with the UL Listing mark.

- *UL Recognized* – A product that is UL Recognized carries the UL Recognized Component Mark (called a backward UR), a printed, bold-faced, back-slanted letter U and R, joined and reversed as they would appear in a mirror (ᴙU). This mark must appear adjacent to the manufacturer's identification and catalog number. UL Recognized equipment is not suitable for general use and must be combined with

other items, under stated condition of acceptability, into another product that can be UL Listed. The conditions of acceptability are published in a required manufacturer's UL Component Recognition Report.

- *UL Classified* – A product that is UL Classified carries the circled UL mark and the word classified along with a notation that Underwriters Laboratories has evaluated the product for compliance with a specific characteristic, standard, or part of a standard. UL Classification of a product for these items has no bearing on the product's ability to comply with the *NEC®*.

It is important to note that not all functions or ratings shown on a nameplate of a UL Listed, Recognized, or Classified product are qualified for use under all articles of the *NEC®* or meet other UL standards.

3.2.2 NEMA Nameplate Ratings

The following ratings are normally marked on NEMA-rated contactor or motor-starter nameplates:

- *NEMA size* – The NEMA size designation is a standardized rating system for contactors and motor-starters (see *Figure 15*). As defined by NEMA standards, horsepower, voltage, frequency, and/or current ratings are assigned for each size.

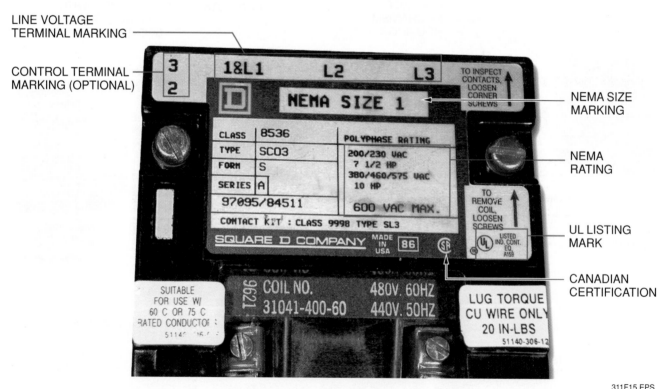

311F15.EPS

Figure 15 ◆ Typical NEMA contactor nameplate showing size, rating, and UL Listing mark.

- *Horsepower and voltage* – The maximum horsepower at various voltages that correspond with values assigned for the designated NEMA size.
- *Continuous current* – In addition to or instead of horsepower marking, some contactors are marked with the maximum current that an enclosed starter or contactor may carry continuously or switch without exceeding the temperature rise permitted by NEMA standards.

3.2.3 IEC Nameplate Ratings

The following ratings are normally marked on IEC contactor or motor-starter nameplates:

- *HP and kW* – The maximum rating for each operational voltage (U_e) and utilization category marked on the nameplate. The most common ratings given are for Utilization Category AC-3.
- *Utilization category* – Describes the types of service for which the controller is rated, such as AC-1, AC-2, AC-3.
- *Thermal current (I_{th})* – The maximum current a contactor or starter, outside an enclosure, can carry continuously without exceeding the temperature rise allowed by the IEC standard. This is not a load-switching rating.
- *Rated operational current (I_e)* – The maximum full-load current (FLC) at which a motor starter

or contactor may be used for a given combination of voltage, frequency, and utilization category. A device may have more than one operational current.

- *Rated insulation voltage (U_i)* – A parameter sometimes shown that defines the insulation properties of the controller. This parameter is not usually used for selection or application purposes.
- *Rated operational voltage (U_e)* – The voltage required for each listed hp or kW rating.
- *Standard designation* – The specific IEC standard to which the product has been tested.

3.2.4 NEMA and IEC Terminal Marking/Symbol Conventions

The symbols shown in *Figures 16* and *17* for NEMA are generally used for all North American and Canadian products including lighting, general-purpose, and IEC contactors sold in these countries. Many IEC contactors retain the IEC marking as additional information. However, imported equipment may only reflect the IEC marking.

The most commonly accepted wiring practice for NEMA-marked products is to connect all terminals together that have the same terminal marking. For example, in power circuits, the terminal

POWER TERMINALS

CONTROL TERMINALS

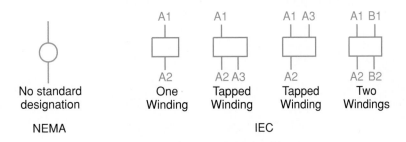

COIL TERMINAL MARKINGS

* Some manufacturers add a 1 before the L1 to designate a control-circuit voltage source connection (See *Figure 15*).
** Some manufacturers label a factory-installed control circuit (N.O.) with terminal numbers 2 and 3 (See *Figures 15* and *17*).

311F16.EPS

Figure 16 ◆ Conventional terminal markings.

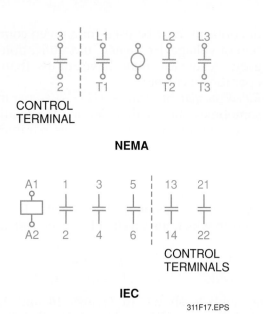

CONTROL
TERMINAL

NEMA

CONTROL
TERMINALS

IEC

311F17.EPS

Figure 17 ◆ Typical contactor/motor starter marking and symbols.

marked T1 on a motor starter is connected to the terminal marked T1 on a motor, and in control circuits, terminals marked 1, 2, and 3 on the starter are connected to terminals 1, 2, and 3 on the control station. The IEC devices use power terminals marked 1, 3, 5, and 2, 4, 6 that correspond to L1, L2, L3 and T1, T2, T3 on NEMA devices. IEC device terminals carrying the same numbers are not normally connected together. Control terminals on IEC contactors have two-digit markings; the first digit designates location sequence, and the second is the control-device function. Most NEMA coils are single voltage; however, IEC coils can be single or multi-voltage.

3.3.0 Two-Wire and Three-Wire Contactor Control

As shown in *Figure 18*, two-wire or three-wire control using pilot devices can be used to energize an electrically held magnetic contactor. Pilot devices are switches or sensors used to control a contactor.

Two-wire control of an electrically held magnetic contactor is accomplished with a simple on/off toggle switch, latching start/stop switch, or other toggling pilot device without the use of a holding contact. In this case, loss of voltage causes the contactor to de-energize, but when power is restored, the contactor will automatically re-energize if the pilot device is still closed.

Three-wire control is accomplished with a pair of momentary contact pushbutton switches (*Figure 19*) and is used when loss-of-voltage protection and/or personnel safety is involved. In the simplest configuration (see *Figures 18* and *19*), a three-phase contactor is initially energized when the START pushbutton switch is momentarily depressed to connect L1 (via wire 1) through the closed STOP switch and the contactor coil to L2. To keep the contactor energized, a holding (seal-in) auxiliary contact is paralleled across the START switch via wires 2 and 3. The holding contact is normally located on the left side of NEMA contactors and is mechanically closed by a plunger on the side of the moving contact assembly when the contactor is initially energized. If factory installed, the holding contact on NEMA contactors is sometimes marked as terminals 2 and 3.

When closed by the moving contactor assembly, the holding contact bypasses the START pushbutton switch contacts so that after the pushbutton is released, control voltage will remain applied to the coil. If the source of the control voltage is momentarily interrupted or the STOP pushbutton is momentarily depressed, the contactor de-energizes, the auxiliary contact opens, and the contactor remains de-energized until the START pushbutton is depressed again. In the pictured contactor, the auxiliary contact is a momentarily actuated snap-action switch assembly that can be replaced by lifting it out of its mounting slot in the contactor base.

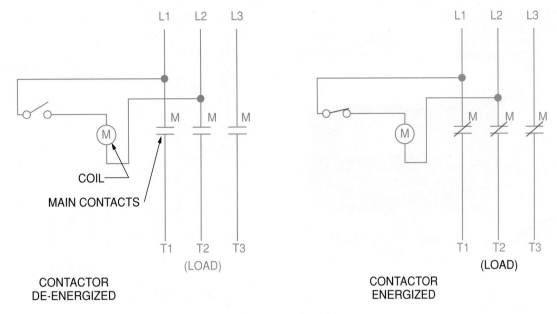

TWO-WIRE CONTROL

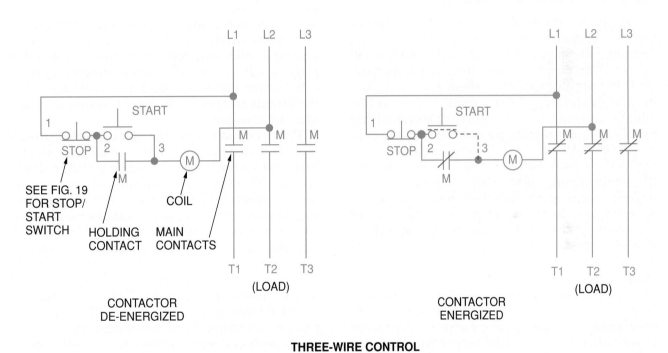

THREE-WIRE CONTROL

311F18.EPS

Figure 18 ◆ Two-wire and three-wire control schematics.

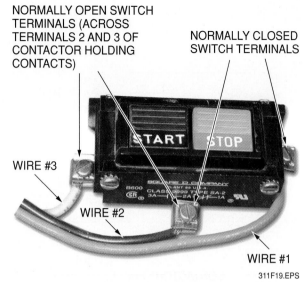

Figure 19 ◆ Momentary START/STOP switch assembly.

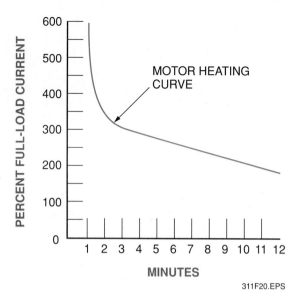

Figure 20 ◆ Typical motor heating curve.

4.0.0 ◆ OVERLOAD PROTECTION

Overload protection for an electric motor is necessary to prevent burnout and ensure maximum operating life. If permitted, an electric motor will operate at an output of more than its rated capacity. Conditions of motor overload may be caused by an overload from driven machinery, a low line voltage, or an open line in a polyphase system that results in single-phase operation. Under any condition of overload, a motor draws excessive current that causes overheating. Since motor winding insulation deteriorates when subjected to overheating, there are established limits on motor operating temperatures. To protect a motor from overheating, overload relays are employed on motor controls to limit the amount of current drawn. Overload protection is also known as running protection.

The ideal overload protection for a motor is an element with current-sensing properties that are very similar to the heating curve of the motor (*Figure 20*), which would act to open the motor circuit when the full-load current is exceeded. The operation of the protective device should allow the motor to carry short overloads but should also disconnect the motor from the line when an overload has persisted for too long.

Single-element, nontime-delay fuses are not designed to provide overload protection. Their basic function is to protect against short circuits (overcurrent protection). Motors draw a high inrush current when starting, and conventional single-element fuses have no way of distinguishing between this temporary and harmless inrush current and a damaging overload. Such fuses, if chosen based on motor full-load current, will blow every time the motor is started. On the other hand, if a fuse is chosen large enough to pass the starting or inrush current, it will not protect the motor against smaller long-term, harmful overloads that might occur later. Dual-element, time-delay fuses can provide motor overload protection but suffer the disadvantages of being nonrenewable and must be replaced.

The overload relay is the heart of motor protection. As stated previously, it normally has inverse trip-time characteristics, permitting short high-current draw during the motor accelerating period (when inrush current is drawn), yet providing protection on overloads above the full-load current when the motor is running. Unlike dual-element fuses, overload relays are renewable and can withstand repeated trip and reset cycles without requiring replacement. However, overload relays, unless also equipped with an instantaneous overcurrent trip device, cannot take the place of overcurrent protective equipment.

The overload relay consists of a current-sensing unit connected in line with the motor, plus a mechanism that is actuated by the sensing unit and serves to directly or indirectly break the circuit. In a manual starter, an overload trips a mechanical latch and causes the starter contacts to open and disconnect the motor from the line. In a magnetic starter, an overload opens a set of contacts within the overload relay itself. These contacts are wired in series with the starter coil in the control circuit of the magnetic starter. Breaking the coil circuit causes the starter main contacts to open, disconnecting the motor from the line.

Overload relays can be thermal, magnetic, or solid-state. Solid-state overload relays are covered in the Level Four module, *Advanced Motor Controls*. As the name implies, thermal overload relays rely on rising temperatures caused by an overload current to trip the overload mechanism. They are also known as inverse-time overload relays because their response time varies inversely to the amount of current flow. The ideal temperature curve of an overload relay corresponds to a motor overheating curve. Magnetic overload relays react only to current excesses and are not reflective of motor temperature.

Thermal overload relays can be further subdivided into melting-alloy and bimetallic types. Both melting-alloy and bimetallic types are usually available in three or four NEMA-rated classes of trip time. Class 10 devices must trip within 10 seconds, Class 15 within 15 seconds, Class 20 within 20 seconds, and Class 30 within 30 seconds. The three most common classes available for NEMA devices are Class 10, Class 20 (Standard), and Class 30. IEC starters usually use equivalent type Class 10 devices.

Thermal units (heaters) for both melting-alloy and bimetallic overload relays should always be selected based on the full-load current (FLC) rating given on the motor nameplate. The FLC is also known as the full-load ampere (FLA) rating of the motor. In the event that the plate is missing, the full-load current can be measured when the motor is operating under its maximum load conditions. As a last resort, manufacturers have charts that give approximate FLC (FLA) values for motors with various horsepower ratings operating at different voltages. However, any heaters that are chosen based on tables or measurements may have to be changed if nuisance tripping occurs.

4.1.0 Melting-Alloy Thermal Overload Relays

Figure 21 is a typical melting-alloy thermal overload relay used with a three-pole contactor. It comprises a housing with three changeable melting-alloy heater assemblies, an internal spring-loaded trip mechanism that operates an N.O. switch (overload contact), and a reset button. The alloy, called a eutectic alloy, is a mixture of low-melting point metals that always melts at a fixed temperature and rapidly changes from a solid to a liquid. The alloy is also unaffected by repeated melting and re-solidifying cycles. *Figure 22* shows a highly simplified version of a melting-alloy overload relay along with its symbol representation and a typical application. When the alloy in the solder pot is solid, the shaft of the ratchet wheel is locked in place and the trip lever holds the overload contacts closed. The closed

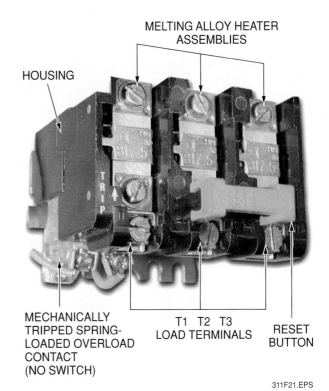

MELTING ALLOY HEATER ASSEMBLIES

HOUSING

MECHANICALLY TRIPPED SPRING-LOADED OVERLOAD CONTACT (NO SWITCH)

T1 T2 T3 LOAD TERMINALS

RESET BUTTON

311F21.EPS

Figure 21 ◆ Typical melting-alloy overload relay.

overload contact is in series with the coil of a contactor and allows the contactor to be energized. Motor current flows from the closed contactor contacts through the overload heaters and out to the motor load connected to T1, T2, and T3. Any excessive overload motor current, passing through any one of the heater elements, eventually raises the temperature enough to cause the alloy in the integral solder pot to melt. The ratchet wheel shaft then turns in the molten alloy because of the upward pressure of the spring-loaded trip lever. By mechanical linkage, this action causes the overload contact to open, which, in turn, de-energizes the contactor, stopping the motor.

Figure 23 shows one of the melting-alloy heater assemblies removed from the overload relay shown in *Figure 21* and turned over to show the ratchet wheel and solder pot containing the alloy. Also visible is one of the three trip levers that, through a common mechanism, opens the spring-loaded overload contact. A cooling-off period is required to allow the alloy in the solder pot to harden before the overload relay assembly can be manually reset and motor service restored.

Melting-alloy heater assemblies are interchangeable and of one-piece construction, which ensures a constant relationship between the heater element and solder pot and allows factory calibration, making them virtually tamperproof in the field. These important features are not possible with any other type of overload relay construction.

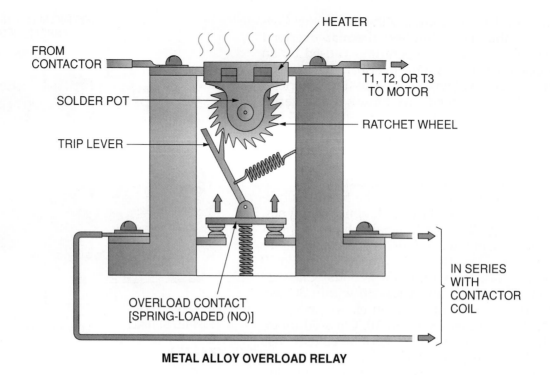

METAL ALLOY OVERLOAD RELAY

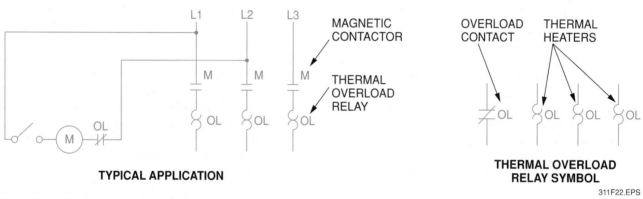

TYPICAL APPLICATION

THERMAL OVERLOAD RELAY SYMBOL

311F22.EPS

Figure 22 ◆ Mechanical and symbol representation of a thermal overload relay.

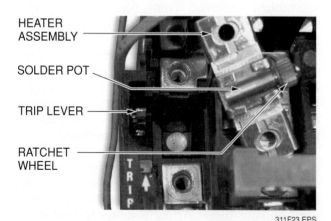

311F23.EPS

Figure 23 ◆ View of melting-alloy heater assembly and trip lever.

A wide selection of interchangeable heater assemblies is available to provide precise motor overload protection for any full-load current. This includes special heater assemblies used for partial correction of ambient temperature deviations at the overload relay location that are substantially different than at the motor location.

4.2.0 Bimetallic Overload Relays

Today a great many overload relays are of the bimetallic variety (*Figure 24*). These include the ones usually used in most IEC motor starters. Except for the IEC devices, which have nonchangeable fixed heaters, bimetallic overload relays use interchangeable heaters to accommodate the FLC of various

Overload Relays

Although seldom identified, overload relays are also categorized by NEMA for their ability to consider the cumulative heating effect of motor operation or overload. This characteristic is called operating memory and may be either volatile or nonvolatile, as shown below. All thermal overloads have a volatile operating memory (Category B) in that they have to cool off before they can be reset. This mimics the heat retention of the motor to some extent. Solid-state overload relays may have volatile and/or nonvolatile memory (Category B and/or A), depending on whether the memory is stored by charging a capacitor or as a value in a nonvolatile microprocessor memory. Magnetic overload relays normally have no memory (Category C) and can be instantly reset even though the motor may still be hot.

- *Category A*—Nonvolatile
- *Category B*—Volatile
- *Category C*—No memory

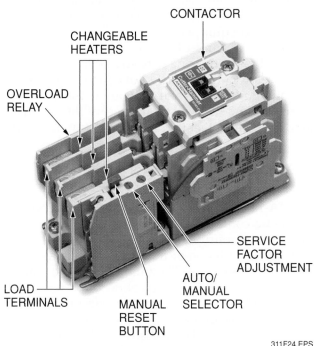

311F24.EPS

Figure 24 ◆ Typical three-pole bimetallic overload relay.

sizes of motors. If a different current range is required for an IEC fixed heater overload relay or its heater must be replaced, the entire overload relay must be changed.

A bimetallic overload relay has three advantages over a melting-alloy overload relay:

- The automatic reset feature is an advantage when the device is mounted in a location that is not easily accessible for manual operation or requires an automatic reset, such as in refrigeration units. Some bimetallic overload relays have a provision for selection of either automatic reset or manual reset.

- These relays usually have a motor service factor adjustment that is used to adjust to the trip range within 85% to 115% of the nominal trip rating of the heater unit. This feature is useful when the recommended heater size might result in unnecessary tripping, while the next larger size would not provide adequate protection.

- Most bimetallic overload relays employ automatic ambient temperature compensation so that the overload relay can be located in a different ambient temperature than the motor. The ambient compensation is accomplished by a second bimetal element that changes the trip point of the overload contact. If no automatic compensation is included, special heaters may sometimes be available to partially correct the difference.

Figure 25 is a single-pole, auto-reset, bimetallic overload relay with the side cover removed, showing the location of the contact, changeable heater, and heater bimetal element. As motor overcurrent increases, the increasing heat generated by the heater eventually causes the U-shaped heater bimetal element to straighten and press on the overload contact plunger. This action forces the snap-action opening of the ambient compensated overload contact. The contact is held open until the heater and bimetallic strip cool, allowing the snap action contact to toggle back to a closed position.

4.3.0 Magnetic Overload Relays

Magnetic overload relays are used for special applications in heavy industrial environments where the specific motor current draw must be

Automatic Reset

NEC Section 430.43 prohibits the use of automatic motor overload reset devices in applications where automatic restart can endanger personnel.

Instantaneous Overload Relays

Instantaneous overload relays, including magnetic or solid-state versions, are sometimes referred to as jam relays. In some applications, a magnetic overload relay may be combined with a thermal overload relay to provide both loadjamming and running overload protection.

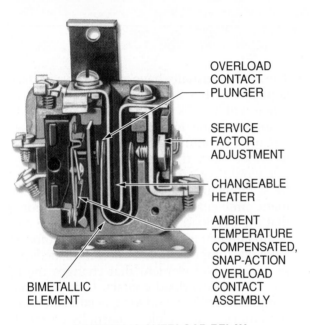

OVERLOAD
CONTACT
PLUNGER

SERVICE
FACTOR
ADJUSTMENT

CHANGEABLE
HEATER

AMBIENT
TEMPERATURE
COMPENSATED,
SNAP-ACTION
OVERLOAD
CONTACT
ASSEMBLY

BIMETALLIC
ELEMENT

BIMETALLIC OVERLOAD RELAY

311F25A.EPS

SYMBOL

311F25B.EPS

Figure 25 ◆ A single-pole, auto-reset, bimetallic overload relay and symbol.

detected or where prolonged inrush currents may be encountered when starting high inertia loads. Under either of these conditions, the time/current curves of thermal overload relays would be unsuitable. Magnetic overload relays may offer settings for either instantaneous or inverse-time trips along with adjustable time or current settings.

Magnetic overload relays operate by passing load current through a heavy coil that functions as a solenoid in each pole of the relay. The position of a moving core within the solenoid coil is adjustable so that the core is sensitive to current flow. When equipped or set for inverse-time trip purposes, a fluid-filled adjustable dashpot is attached to the moving core so that the time delay of the relay can be adjusted over a wide range for various currents. When activated, the instantaneous or time-delayed total travel of the core trips an overload contact that must be manually reset.

5.0.0 ◆ MAGNETIC AND MANUAL MOTOR STARTERS

NEMA-rated motor starters consist of NEMA magnetic contactors and overload relays assembled into a controller consisting of one or more contactors with one or more overload relays on a common base plate. They have complete control circuit wiring (except for connections to remote equipment) and complete power circuits from the line to load terminals. They are designed so that the factory wiring is not disturbed by the connections for line, load, or remote control circuit wiring. Typically, IEC-rated magnetic motor starters are not factory wired and must be assembled in the field after delivery. However, a number of North American manufacturers also offer IEC magnetic motor starters as factory-assembled, internally wired units, the same as NEMA units. Field assembly, when required, includes assembling and mounting the contactor(s) and overload relay(s), as well as furnishing and connecting all internal and external wiring. Both NEMA and IEC motor starters use contacts that are designed to withstand high inrush current (*Figure 26*) encountered when motors are started. *Table 5* summarizes the basic differences between IEC and NEMA devices.

Table 5 IEC and NEMA Product Comparison Summary

Subject	IEC	NEMA
Starter size	Physically smaller	Physically larger
Contactor performance	Electrical life = 1 million for AC-3 category operations with 30,000 for AC-4 category operations	Electrical life typically 2.5 to 4 times more than equivalently rated IEC device
Contactor application	Application sensitive—more knowledge and care in selection required	Application selection easier and less critical with fewer parameters to consider
Overload relay trip reset characteristics	Class 10 (fast) typical	Class 20 (Standard) typical
Overload relay adjustability	Fixed, noninterchangeable heaters; adjustable to suit different motors at the same horsepower	Field changeable heaters allow adjustment for motors of different horsepower
Overload relay reset mechanism	Manual/Auto typical; some use RESET/STOP dual function mechanism	Manual/Auto or Manual Only typical
Short circuit current rating	Typically designed for use with fast-acting, current-limiting fuses	Designed for use with common domestic current rating fuses and circuit breakers

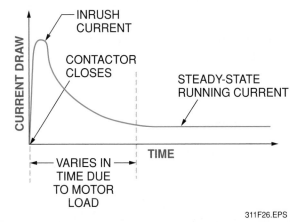

311F26.EPS

Figure 26 ◆ Typical motor inrush and running current.

5.1.0 Nonreversing and Reversing Magnetic Motor Starters

The most commonly encountered NEMA or IEC magnetic motor starters are nonreversing and reversing types for wye, delta, and reconfigurable wye/delta motors. *Figure 27* shows typical nonreversing and reversing NEMA motor starters along with their symbol representations. IEC motor starters are similar except for the terminal marking that has been described previously.

As shown in the symbol representations of the starters, normal convention for a nonreversing starter is to wire L1 to T1, L2 to T2, and L3 to T3 for forward motor rotation. In the case of the reversing motor starter, the same wiring convention is observed for the forward contactor (F). The wiring convention for the reverse contactor is to interchange L1 and L3 so that L1 is tied to T3, L2 is tied to T2, and L3 is tied to L1. Note that a mechanical interlock that is part of the contactor assembly is sometimes used to prevent both contactors from being closed at the same time. If pilot devices are used that are not mechanically interlocked, an electrical backup-interlock system can be used for protection against accidental energization of both contactors simultaneously. As shown, the electrical interlock uses a set of N.C. auxiliary contacts (F and R) that are wired in series with their opposite contactor coils. The N.O. auxiliary contacts can be used as holding contacts for start switches or other pilot devices if required.

5.2.0 NEMA Magnetic Contactors/Motor Starters

This section covers NEMA magnetic contactors and motor starters. Both rating and selection criteria are covered.

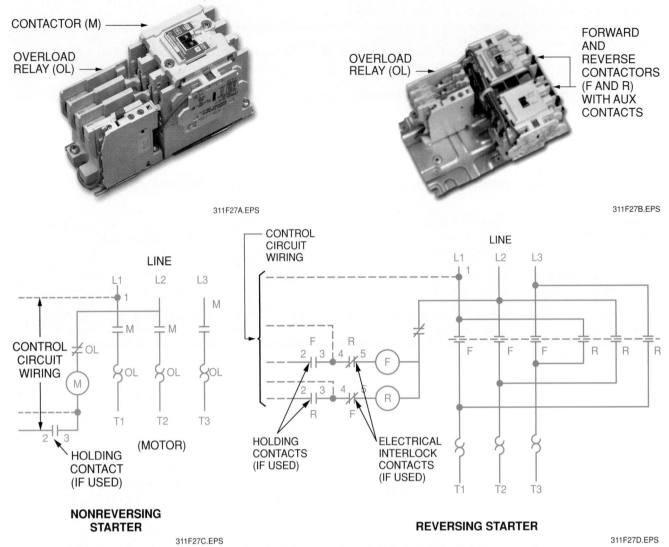

CONTACTOR (M)

OVERLOAD RELAY (OL)

311F27A.EPS

OVERLOAD RELAY (OL)

FORWARD AND REVERSE CONTACTORS (F AND R) WITH AUX CONTACTS

311F27B.EPS

LINE

CONTROL CIRCUIT WIRING

HOLDING CONTACT (IF USED)

(MOTOR)

NONREVERSING STARTER

311F27C.EPS

CONTROL CIRCUIT WIRING

HOLDING CONTACTS (IF USED)

ELECTRICAL INTERLOCK CONTACTS (IF USED)

LINE

REVERSING STARTER

311F27D.EPS

Figure 27 ◆ Typical NEMA nonreversing and reversing magnetic motor starters and symbol representations.

5.2.1 Ratings

A NEMA contactor/motor starter (*Figure 28*) is designed to meet the size rating specified in NEMA Standards. The standards are used to provide electrical interchangeability among manufacturers for a given NEMA size. *Table 6* lists the NEMA sizes for contactors/motor starters and gives the maximum allowable horsepower ratings at different line voltages for various applications of devices, including plugging and jogging (inching) operations. The table also lists the continuous current, kW, kVA, and kVAR ratings for other applications of the contactors/starters. Because NEMA contactors/motor starters must be able to safely interrupt the locked rotor current of a motor under nonplugging or nonjogging conditions, the maximum allowable horsepower ratings shown for a particular NEMA size are based on the

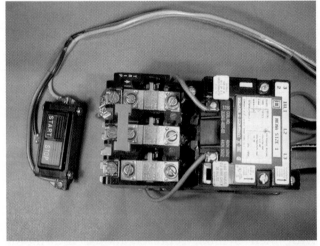

311F28.EPS

Figure 28 ◆ Typical NEMA Size 1 motor starter wired with a three-wire STOP/START switch.

Contactors Used for Plugging or Jogging Applications

Contactors/starters used regularly for plugging and jogging (inching) are derated. Plugging is the momentary stopping and/or reversing of a running motor. Jogging is the momentary starting of a stopped motor. Both plugging and jogging are used in industrial machine operations for drive-train positioning purposes that can subject contactors/starters to high inrush currents numerous times in a short period.

locked-rotor current for motors rated at the listed horsepower for that size. For plugging and jogging, the maximum allowable horsepower within any NEMA size is derated as reflected in *Table 6*.

5.2.2 Selection

The information found on a motor nameplate is essential in the selection of a contactor/motor starter and must be used instead of actual current measurements or manufacturers' tables if possible. On the motor nameplate shown in *Figure 29*, the following parameters must be noted and used as indicated in the selection process:

- *Voltage (volts) and frequency (Hz)* – 230V at 60Hz AC
- *FLC or FLA (amps)* – 13.0A
- *Phase* – Three-phase
- *NEMA Design Letter* – B
- *Horsepower (hp)* – 5hp
- *Service factor (SF)* – 1.0
- *Power factor* – 85.7 (Necessary only if capacitor correction is connected on the load side of overload relays. If so, the overload relay manufacturer must be consulted to determine current elements for the relays.)

Assuming the motor is not used for plugging or jogging, *Table 6* shows that a 230V, polyphase, 60Hz, 5hp motor can be controlled with a NEMA Size 1 starter. The overload relay would be selected and equipped for an FLC (FLA) of 13A in accordance with the manufacturer's motor-starter overload relay selection tables and adjusted (if allowed) for a service factor (SF) of 1.0. The service factor rating is the amount of extra power demand that can be placed on the motor for a short period without damage to the motor.

Common motor service factors range from 1.00 to 1.25, indicating that the motor can intermittently be required to produce 0% to 25% extra power over its normal rating. In the event that the SF is unknown, a value of 1.00 should be assumed.

The total excessive current required for intermittent power demands can be approximated by multiplying the motor FLC (FLA) by the motor SF. For the motor SF value of 1.0 shown in *Figure 29*, the allowed extra current is 0%. This means that any service factor adjustment would be equal to the value of the FLC (13A). If the SF had been 1.25, then the adjustment would be equal to 16.25A (13A × 1.25 = 16.25A). Any SF adjustment determined that is greater than the FLC permits the overload relay to allow any excessive intermittent power demands without nuisance tripping. For overload relays that are not ambient temperature compensated and where the starter is located in a different temperature environment than the motor, an appropriate overload relay/heating element must be selected as specified in special tables available from the motor-starter manufacturer.

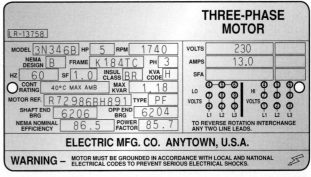

311F29.EPS

Figure 29 ◆ Typical motor nameplate.

Table 6 NEMA Sizes for Contactors and Motor Starters *(Reprinted by permission of National Electrical Manufacturers Association.)*

NEMA Size	Volts	Maximum Horsepower Rating—Nonplugging and Nonjogging Duty[1]		Maximum Horsepower Rating—Plugging and Jogging Duty[1]		Continuous Current Rating, Amperes 600V Max[2]
		Single Phase	Poly-Phase	Single Phase	Poly-Phase	
00	115	⅓	—	—	—	9
	200	—	1½	—	—	9
	230	1	1½	—	—	9
	380 (50Hz)	—	1½	—	—	9
	460	—	2	—	—	9
	575	—	2	—	—	9
0	115	1	—	½	—	18
	200	—	3	—	1½	18
	230	2	3	1	1½	18
	380 (50Hz)	—	5	—	1½	18
	460	—	5	—	2	18
	575	—	5	—	2	18
1	115	2	—	—	—	27
	200	—	7½	—	3	27
	230	3	7½	—	3	27
	380 (50Hz)	—	10	—	5	27
	460	—	10	—	5	27
	575	—	10	—	5	27
2	115	3	—	2	—	45
	200	—	10	—	7½	45
	230	7½	15	5	10	45
	380 (50Hz)	—	25	—	15	45
	460	—	25	—	15	45
	575	—	25	—	15	45
3	115	7½	—	—	—	90
	200	—	25	—	15	90
	230	15	30	—	20	90
	380 (50Hz)	—	50	—	30	90
	460	—	50	—	30	90
	575	—	50	—	30	90
4	200	—	40	—	25	135
	230	—	50	—	30	135
	380 (50Hz)	—	75	—	50	135
	460	—	100	—	60	135
	575	—	100	—	60	135
5	200	—	75	—	60	270
	230	—	100	—	75	270
	380 (50Hz)	—	150	—	125	270
	460	—	200	—	150	270
	575	—	200	—	150	270
6	200	—	150	—	125	540
	230	—	200	—	150	540
	380 (50Hz)	—	300	—	250	540
	460	—	400	—	300	540
	575	—	400	—	300	540
7	230	—	300	—	—	810
	460	—	600	—	—	810
	575	—	600	—	—	810
8	230	—	450	—	—	1215
	460	—	900	—	—	1215
	575	—	900	—	—	1215
9	230	—	800	—	—	2250
	460	—	1600	—	—	2250
	575	—	1600	—	—	2250

Notes: (1) These horsepower ratings are based on the locked-rotor current ratings given in *NEMA Standard ICS 2*. For motors having higher locked-rotor currents, a controller should be used so that its locked-rotor current rating is not exceeded. (2) The continuous current ratings represent the maximum rms current, in amperes, which the controller shall be permitted to carry continuously without

Service-Limit Current Rating, Amperes[3]	Tungsten and Infrared Lamp Load, Amperes 250V Max[2]	Resistance Heating Loads, kW other than Infrared Lamp Loads		KVA Rating for Switching Transformer Primaries at 50 or 60 Cycles		3-Phase Rating for Switching Capacitors Kvar
		Single Phase	Poly-Phase	Single Phase	Poly-Phase	
11	5	—	—	—	—	—
11	5	—	—	—	—	—
11	5	—	—	—	—	—
11	—	—	—	—	—	—
11	—	—	—	—	—	—
11	—	—	—	—	—	—
21	10	—	—	0.9	1.2	—
21	10	—	—	—	1.4	—
21	10	—	—	1.4	1.7	—
21	—	—	—	—	2.0	—
21	—	—	—	1.9	2.5	—
21	—	—	—	1.9	2.5	—
32	15	3	5	1.4	1.7	—
32	15	—	9.1	—	3.5	—
32	15	6	10	1.9	4.1	—
32	—	—	16.5	—	4.3	—
32	—	12	20	3	5.3	—
32	—	15	25	3	5.3	—
52	30	5	8.5	1.0	4.1	—
52	30	—	15.4	—	6.6	11.3
52	30	10	17	4.6	7.6	13
52	—	—	28	—	9.9	21
52	—	20	34	5.7	12	26
52	—	25	43	5.7	12	33
104	60	—	—	0.9	7.6	—
104	60	—	—	—	13	23.4
104	60	—	—	1.4	15	27
104	—	—	—	—	19	43.7
104	—	—	—	1.9	23	53
104	—	—	—	1.9	23	67
156	120	—	45	—	20	34
156	120	30	52	11	23	40
156	—	—	86.7	—	38	66
156	—	60	105	22	46	80
156	—	75	130	22	46	100
311	240	—	91	—	40	69
311	240	60	105	28	46	80
311	—	—	173	—	75	132
311	—	120	210	40	91	160
311	—	150	260	40	91	200
621	480	—	182	—	79	139
621	480	120	210	57	91	160
621	—	—	342	—	148	264
621	—	240	415	86	180	320
621	—	240	515	86	180	400
932	720	180	315	—	—	240
932	—	360	625	—	—	480
932	—	450	775	—	—	600
1400	1080	—	—	—	—	360
1400	—	—	—	—	—	720
1400	—	—	—	—	—	900
8590	—	—	—	—	—	—
2590	—	—	—	—	—	—
2590	—	—	—	—	—	—

exceeding the temperature rises permitted by *NEMA Standard ICS 1*. (3) The service limit current ratings represent the maximum rms current, in amperes, which the controller shall be permitted to carry for protracted periods in normal service. At service-limit current, temperature rises may exceed those obtained by testing the controller at its continuous current rating.

5.3.0 IEC Magnetic Contactors/ Motor Starters

This section covers IEC magnetic contactors and motor starters. Both rating and selection criteria are covered here.

5.3.1 Ratings

IEC Standard 60947 does not define any standard contactor or motor-starter sizes like NEMA. An IEC-rated contactor or motor starter (*Figure 30*) indicates that the contactor or motor starter has been evaluated by the manufacturer or a laboratory to meet the requirements of a number of defined applications called utilization categories. There are a number of categories (AC-1 to -8, -11 to -15, and -20 to -23) covering 600V or less AC current switchgear and control gear. The same is true for similar DC equipment (DC-1, -3, -5, -6, -11 to -14, and -20 to -23). Of these, the categories used for motor control or lighting control purposes are summarized in *Table 7*. *Table 8* lists and expands the definition of the most common categories used for motor control. Category A1 is included because most Category AC-3 and -4 devices have Category A-1 included on their nameplate. Category AC-2 is omitted because these motors are uncommon.

5.3.2 Selection

There may be several ratings for any particular contactor or motor starter for different categories and voltages. A designer can choose a preferred device for an application based on the device's ability to meet or exceed the required horsepower voltage rating and other factors, including performance. This technical data is available in the manufacturer's specifications. IEC contactors used in the U.S. are usually marked with voltage and horsepower ratings for maximum AC-3 rated operational current.

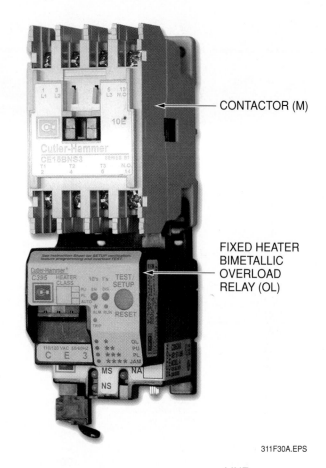

CONTACTOR (M)

FIXED HEATER BIMETALLIC OVERLOAD RELAY (OL)

311F30A.EPS

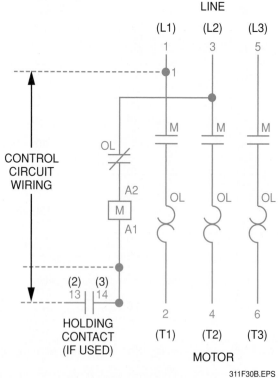

311F30B.EPS

Figure 30 ♦ Typical IEC-rated, nonreversing motor starter and IEC symbol representation.

Table 7 IEC Categories for Lighting and Motor Control

Category	Typical Application
AC-1	Noninductive or slightly inductive loads
AC-2	Slip-ring motor starting and switching off
AC-3	Squirrel-cage motor starting and switching off
AC-4	Squirrel-cage motor starting, plugging, and jogging
AC-5a	Switching electric discharge lighting loads
AC-5b	Switching large incandescent lighting loads
AC-6a	Switching transformer loads
AC-6b	Switching capacitor loads
AC-11	Control of AC electromagnetic circuits (auxiliary contacts)

Table 8 Common IEC Categories for Motor Control Devices

Category	Application
AC-1	These devices are used with noninductive or slightly inductive loads, such as resistive furnaces, fluorescent lights, and incandescent lights.
AC-3	These devices are used in squirrel-cage motors for starting and switching off while running at rated speed. They include contact-make capability for locked-rotor current and break at full-load current. They are occasionally used for jogging and plugging for limited times such as machine setup. During such periods, the number of operations should not exceed five per minute nor more than ten in a ten-minute period.
AC-4	These devices are used in squirrel-cage motors for starting and switching off while running at less than rated speed. They provide jogging, plugging stop, and plugging reverse, and include contact make-and-break capability for locked-rotor current. Very few applications in the industry are totally Category AC-4.

Most U.S. manufacturers provide tables that classify IEC contactors/motor starters in some sort of size ratings for horsepower, similar to NEMA, based on the maximum rated operational current in a range of standard line voltages. IEC contactor/motor starter selection is based on the percent that jogging and plugging (AC-4) is of the nonjogging and plugging (AC-3) condition in the duty cycle and the desired contact electrical life. In general, any time the duty cycle includes significant jogging or plugging, a larger size IEC contactor/motor starter is selected than would be needed for pure AC-3 applications, or an appropriate AC-4 rated device is selected. Like NEMA motor starters, the same motor nameplate data is required to select an IEC motor starter with an appropriate overload relay, as well as to set the adjustments on the overload relay.

5.4.0 Manual Motor Starters

Manual motor starters (*Figure 31*) are primarily used on small machine tools, fans, blowers, pumps, compressors, and conveyors. They are available as single-, double-, or three-pole devices. Instead of magnetically closed contacts, NEMA-rated and/or UL-approved motor starters are operated by a mechanically linked and latched toggle handle or pushbutton. The operating mechanisms are quick make-and-break toggle type switches that cannot be teased into a partially open or closed position. They generally come equipped with melting-alloy or bimetallic overload relays.

311F31.EPS

Figure 31 ◆ Typical NEMA manual motor starter.

AC-3 Rated Contactors

U.S. manufacturers usually supply AC-3 derating tables for plugging and jogging duty.

Manual Starters with Melting-Alloy Overloads

In versions of manual starters that use melting-alloy overloads, the latched contacts are tripped mechanically by the overload relay and must be manually reset with a separate button or by setting the starter to off (stop) and then back to on (start). In bimetallic overload versions, the reset function may be selectable as automatic or manual, depending on the manufacturer.

5.5.0 Contactor/Motor Starter Accessories

Manufacturers offer a variety of accessories that can be added to various contactor/motor starters to customize them for a particular application. *Figure 32* shows a number of these accessories.

- *Power-pole adder kits* – These kits can be used to increase the number of high-current power contacts actuated by the contactor/motor starter. Depending on the manufacturer, some power-pole adder kits are mounted on the sides or the top of the contactor/motor starter. In some cases, the internal springs or coil must be changed to accommodate the mechanical loads imposed by the extra poles.
- *Timer attachment (not shown)* – Mechanical or solid-state control circuit timer devices with adjustable on-delay or off-delay are available for side mounting on the contactor/motor starter. Some devices have field-selectable on- or off-delay. They are equipped with one or more single- or double-throw snap-switch control-circuit contacts.
- *Fuse kit* – When control circuit power tapped from the line inputs of a contactor/motor starter must be fused, side-mounted single- or double-fuse holder kits are available. Rated at 600V, they will usually accept $^{13}\!/_{32} \times 1\frac{1}{2}$ fuses up to 6A.
- *Transient suppression module* – Side-mounted transient suppression modules are available to reduce transient voltages and contact arcing in coil control circuits when the circuits are opened. This eliminates noise that interferes with operation of nearby electronic circuits. Most modules consist of an RC circuit that is designed to suppress coil voltage transients to approximately 200% of peak coil supply voltage.
- *Internal auxiliary contacts* – Some contactors/motor starters can have one or more internal N.O. or N.C. auxiliary contacts added for additional low-current circuits, including status feedback, or for electrical interlocking purposes. The devices are toggle snap switches with SB contacts.
- *External auxiliary contacts* – Side-mounted auxiliary contacts are available as single or double N.O. or N.C. contacts that are field convertible from N.O. to N.C. or vice versa. Some are available as DT contacts. The devices are toggle snap switches with SB contacts.

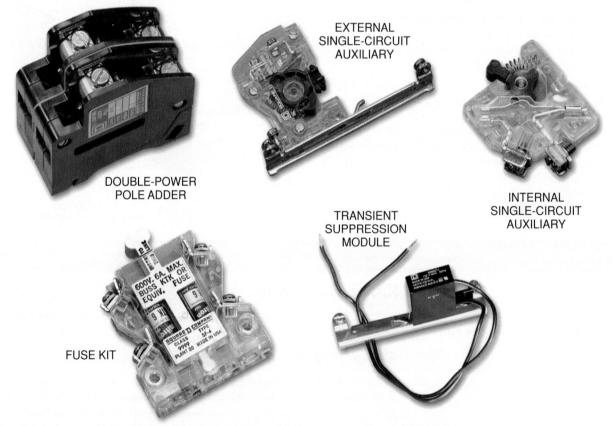

EXTERNAL
SINGLE-CIRCUIT
AUXILIARY

DOUBLE-POWER
POLE ADDER

INTERNAL
SINGLE-CIRCUIT
AUXILIARY

TRANSIENT
SUPPRESSION
MODULE

FUSE KIT

311F32.EPS

Figure 32 ◆ Typical contactor/motor starter accessories.

6.0.0 ◆ CONTROL TRANSFORMERS AND PILOT DEVICES

This section covers control transformers and several types of pilot devices widely used in motor control circuits. Pilot devices contain switch contacts. The opening, closing, or transfer of these contacts govern the operation of related relays or similarly controlled devices. Pilot devices are used to provide sequencing and automatic operation within certain parameters. Some commonly used types of pilot devices covered here include the following:

- Pushbutton switches
- Selector switches
- Pilot lights
- Temperature switches
- Pressure switches
- Limit switches
- Flow switches
- Float switches
- Foot switches
- Proximity switches/sensors
- Photoelectric switches/sensors

6.1.0 Control Transformers

Control transformers like the one shown in *Figure 33* provide the operating voltage for the motor control circuits and their components. Control

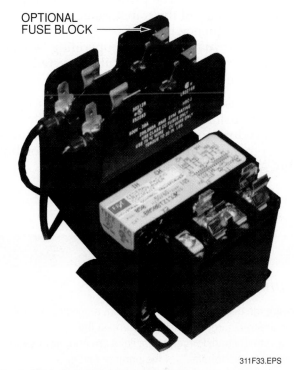

OPTIONAL FUSE BLOCK

311F33.EPS

Figure 33 ◆ Industrial control transformer.

transformers are step-down transformers that reduce the line voltage applied to the equipment to the operating voltage needed for the equipment's control circuits. They are designed to provide good transformer regulation where high inrush currents caused by contactors and relays are drawn. Secondary fuse protection kits are often used with control transformers for fusing the secondary winding to provide for control circuit protection. These kits consist of fuse blocks and either cartridge or glass fuses of the proper size. Stepping down a higher line voltage to a lower voltage for use in the control circuits has three advantages:

- Operator safety is increased at the control stations and other pilot devices.
- The reduced voltage decreases the chance of a fault occurring between lines of the control circuit wiring and ground.
- Use of components designed to operate at lower voltages lowers the cost for manufacturing the equipment.

Control transformers are made to accommodate many different primary-voltage to secondary-voltage combinations. Control transformers used in most residential HVAC equipment and other equipment normally operate to step down an applied primary voltage of 120V or 240VAC to 24VAC. Control transformers used for industrial applications typically operate to produce a 24V or 120VAC secondary voltage. A typical industrial control transformer contains two primary windings and one or two secondary windings. *Figure 34* shows a control transformer that has two primary windings and one secondary winding, with each primary winding having a voltage rating of 240V and the secondary winding a rating of 24V. For the purpose of discussion, assume that there is a turns ratio of 10:1 between each of the primary windings and the secondary winding. Terminals of most control transformers are identified using an industry-standard method. As shown, the terminals of one of the primary windings are marked H1 and H2 and for the other primary winding, they are marked H3 and H4. The terminals for the secondary winding are marked X1 and X2. The primary windings of most control transformers have the H2 and H3 terminals crossed as shown in *Figure 34*. This is done to make the physical connection between the terminals of the primaries easier.

If this control transformer must be connected to step down a primary voltage of 240V to produce a secondary voltage of 24V, the two primary windings are connected in parallel, as shown in

Figure 34(A). Since the two primary windings are connected in parallel, each will receive the same voltage. This will produce a turns ratio of 10:1 between the primary and secondary windings. When 240V is connected to the primary of a 10:1 ratio transformer, the secondary voltage produced is 24V.

If this same transformer is connected to step down a primary voltage of 480V to produce the secondary voltage of 24V, the primary windings are connected in series as shown in *Figure 34(B)*. In this connection, terminal H2 of one primary winding is connected to H3 of the other primary winding. This series connection of the two primary windings produces a turns ratio of 20:1. When 480V is connected to the primary windings, 24V is produced by the secondary winding.

Selecting a control transformer for a specific application requires that you know the following factors about the transformer and its related control circuit:

- *Inrush VA* – The product of the load voltage (V) times the current (A) that is required during circuit startup. It is determined by adding the inrush VA requirements for all the circuit load devices such as contactors, timers, relays, or pilot lights that will be energized at the same time.
- *Sealed (steady-state) VA* – The product of the load voltage (V) times the current (A) that is required to operate the circuit after the initial startup or under normal operating conditions. It is determined by adding the sealed VA requirements of all electrical components of the circuit that will be energized at any given time. The sealed VA requirements for each component can be obtained from the component manufacturer's data sheets.
- *Primary voltage* – The primary voltage applied to the primary of the transformer and its operational frequency.
- *Secondary voltage* – The operating voltage required for the control circuit connected to the transformer secondary.

Once these factors have been determined, the proper transformer can be selected by following the procedure normally given in the control transformer manufacturer's catalog or application bulletin. A typical procedure is described here.

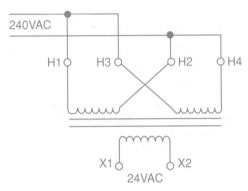

**(A) CONNECTED FOR 240-VOLT
TO 24-VOLT OPERATION**

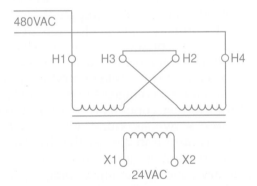

**(B) CONNECTED FOR 480-VOLT
TO 24-VOLT OPERATION**

311F34.EPS

Figure 34 ◆ Control transformer schematic shown connected for 240V/480V primary and 24V secondary operation.

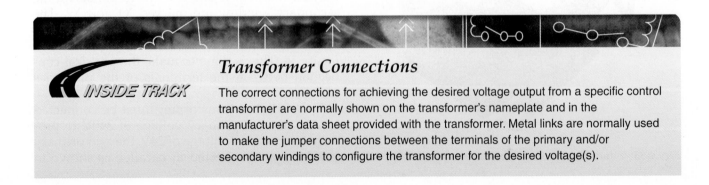

Transformer Connections

INSIDE TRACK

The correct connections for achieving the desired voltage output from a specific control transformer are normally shown on the transformer's nameplate and in the manufacturer's data sheet provided with the transformer. Metal links are normally used to make the jumper connections between the terminals of the primary and/or secondary windings to configure the transformer for the desired voltage(s).

Step 1 Determine the application inrush VA for the control transformer using the formula:

$$\text{Application inrush VA} = \frac{}{\sqrt{(\text{INRUSH VA})^2 + (\text{SEALED VA})^2}}$$

For example, assume that we have an inrush VA of 1,000VA and a sealed VA of 100VA. Using the formula, the application inrush VA is 1,005VA, calculated as shown:

$$\text{Application inrush VA} = \frac{}{\sqrt{(\text{INRUSH VA})^2 + (\text{SEALED VA})^2}}$$

$$\text{Application inrush VA} = \frac{}{\sqrt{(1,000)^2 + (100)^2}}$$

$$\text{Application inrush VA} = \frac{}{\sqrt{1,010,000}}$$

$$\text{Application inrush VA} = 1,004.987\text{VA} =$$

$$1,005\text{VA (rounded off)}$$

Step 2 Using *Table 9*, select a proper secondary voltage column to use. If the primary voltage is stable and does not vary more than 5% from nominal, use the 90% secondary voltage column of the table. If the primary voltage varies between 5% and 10% of nominal, use the 95% secondary voltage column. Note that to comply with NEMA standards, which require all magnetic devices to operate properly at 85% of rated voltage, the 90% secondary column is most often used in selecting a transformer.

Step 3 After determining the proper secondary voltage column to use, read down the column in *Table 9* until you find a value equal to or greater than the application inrush VA calculated in Step 1. For our example, an application inrush VA of 1,150 is closest to the 1,005VA. In no case should a value less than the application inrush VA be used. Then read left to the Transformer VA Rating column to find the proper transformer VA for your application. For this example, your transformer should have a VA rating equal to or greater than 200VA. As a final check, make sure that the transformer VA rating is equal to or greater than the total sealed requirements. If not, select a transformer with a VA rating equal to or greater than the total sealed VA.

Step 4 Using the transformer VA rating found in *Table 9*, refer to the applicable transformer manufacturer's product data catalog or bulletin to identify the model and part number for the transformer. Your selection is based on the required transformer VA rating and the primary and secondary voltage requirements.

6.2.0 Pushbutton Switches, Selector Switches, and Pilot Lights

Manually operated pushbutton switches and selector switches (*Figure 35*) are two widely used types of pilot devices. They are made in standard-duty and heavy-duty versions. Heavy-duty switches are able to carry higher continuous and make-break currents. Different types of pushbutton and lever switches are needed to serve the wide variety of industrial motor applications. For this reason, switch manufacturers designed many of their industrial pushbutton and lever switches

Table 9 Regulation Data Chart *(Reprinted by permission of National Electrical Manufacturers Association.)*

Transformer VA Rating	Application Inrush VA at 20% Power Factor		
	95% Secondary Voltage	90% Secondary Voltage	85% Secondary Voltage
25	100	130	150
50	170	200	240
75	310	410	540
100	370	540	730
150	780	930	1,150
200	810	1,150	1,450
250	1,400	1,900	2,300
300	1,900	2,700	3,850
350	3,100	3,650	4,800
500	4,000	5,300	7,000
750	8,300	11,000	14,000

so that they can be assembled using different parts that can be interchanged according to the customer's specifications to meet the requirements of a specific application. This modular approach to building a switch also lowers the manufacturer's costs by reducing the number and type of switches and switch components that must be manufactured and maintained in inventory.

6.2.1 Pushbutton Switches

The parts used to assemble a typical pushbutton switch include the operator (*Figure 36*), legend

SELECTOR
SWITCH

PUSHBUTTON
SWITCH

311F35.EPS

Figure 35 ◆ Typical pushbutton and selector switches.

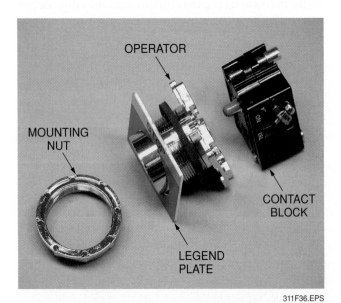

OPERATOR

MOUNTING
NUT

CONTACT
BLOCK

LEGEND
PLATE

311F36.EPS

Figure 36 ◆ Typical parts of a pushbutton switch.

plate, one or more contact blocks, and related mounting adapters and hardware. The operator is the part of a switch that is pressed or pulled by the person operating the switch in order to activate the switch contacts. Operators are made in many different shapes, sizes, and colors with red used to indicate a stop or off function. Operators can also be nonilluminated or illuminated to indicate when active. The legend is a plate that is placarded (marked) to show the function of the pushbutton, such as ON, OFF, and INCH. The contact block is the part of the switch containing the contacts that are activated when the operator is pressed.

As the name implies, the operator of a pushbutton switch is pushed to activate the electrical contacts. The switch contact block can have normally open (N.O.), normally closed (N.C.), or both N.O. and N.C. sets of contacts. Multiple contact blocks are often assembled (stacked) together to form a switch that has three, four, or more N.O. or N.C. sets of contacts. Normally open (N.O.) contacts usually are used to close the circuit when the operator is pressed in order to initiate a START or ON circuit function. Normally closed (N.C.) contacts are usually used to open the circuit when the operator is pressed in order to initiate STOP or OFF circuit functions. There are two types of pushbuttons: momentary and maintained. A normally open momentary pushbutton closes as long as the button is held down. A normally closed momentary pushbutton opens as long as the button is held down. A maintained pushbutton latches in place when the pushbutton is pressed. *Figure 37* shows the schematic symbols for pushbutton switches. *Figure 38* shows a simple line diagram of start and stop pushbutton switches used in a basic contactor control circuit.

6.2.2 Push-Pull Pushbutton Switches

Pushbutton switches can also be of the push-pull type. A push-pull switch is typically used to replace two separate pushbuttons, such as START and STOP pushbuttons. There are three types of push-pull switches:

- *Maintained* – This is a two-position switch that remains in the pulled or pushed position until manually actuated to the opposite position.
- *Momentary* – This is a three-position switch. A spring returns the switch to an intermediate position when pulled or pushed and released.
- *Momentary pull, maintained push* – This is a three-position switch. A spring returns the switch to an intermediate position when pulled. In the push position, it maintains its position until manually returned to the intermediate position.

MOMENTARY CONTACT

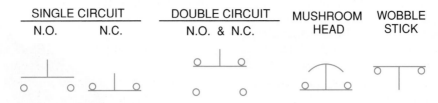

SINGLE CIRCUIT	DOUBLE CIRCUIT	MUSHROOM	WOBBLE
N.O. N.C.	N.O. & N.C.	HEAD	STICK

MAINTAINED CONTACT

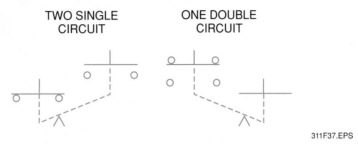

TWO SINGLE	ONE DOUBLE
CIRCUIT	CIRCUIT

311F37.EPS

Figure 37 ◆ Schematic symbols for pushbutton switches.

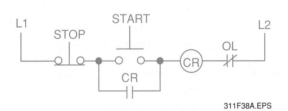

311F38A.EPS

311F38B.EPS

Figure 38 ◆ Stop and start pushbutton switches used in a basic control circuit.

6.2.3 Selector Switches

Selector switches are used to select one of several different positions. Standard-duty selector switches have either two or three positions; heavy-duty selector switches have two or more positions. Some have as many as 12 positions. The operator of a selector switch is rotated, instead of pushed, to activate the contacts for each switch position. Selector switches are assembled the same way as pushbutton switches, with the customer selecting the required operator, legend, and contact block assemblies suitable for the application. *Figure 39* shows the schematic symbols for a two-position and a three-position selector switch. Schematic diagrams and switch manufacturers' product catalogs commonly show the contact positions for each position of a selector switch using a truth table placed near the switch. As shown, an X is placed in the truth table if a contact is closed in any position.

Figure 39 also shows a simple line diagram of a three-position selector switch being used to control a contactor. In the OFF position, all the contacts of the selector switch are open, preventing the contactor coil from being energized. This is a safety position. In the HAND position, the upper contacts of the three-way switch are closed, enabling the contactor coil to be energized or de-energized by pressing the START or STOP pushbuttons, respectively. In the AUTO position, the lower contacts are closed, enabling the contactor coil to be energized whenever the contacts of the

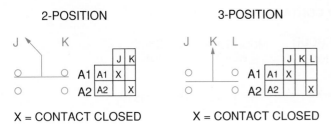

2-POSITION

3-POSITION

X = CONTACT CLOSED

X = CONTACT CLOSED

SCHEMATIC SYMBOL AND TRUTH TABLE FOR
2-POSITION AND 3-POSITION SELECTOR SWITCHES

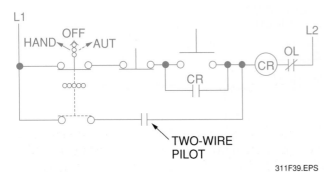

311F39.EPS

Figure 39 ◆ Selector switch schematic symbols and typical control circuit.

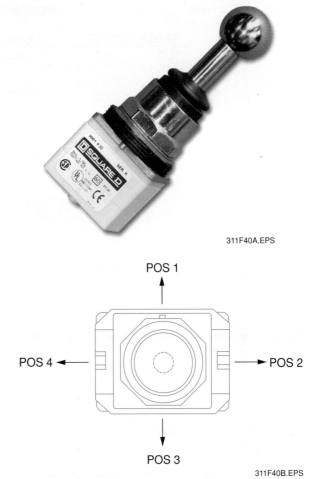

311F40A.EPS

311F40B.EPS

Figure 40 ◆ Joystick operated selector switch.

two-wire pilot device are closed. This device can be a temperature switch, pressure switch, limit switch, or similar device.

Some types of selector switches have a joystick operator (*Figure 40*). Joystick selector switches can have two to eight positions. This type of switch is normally used in applications where only one circuit is to be energized at one time, such as when operating a hoist or crane. In this application, movement of the operator for a five-position switch (four positions and center off) closes one circuit each in the up, down, left, and right positions, with all circuits open when the joystick is in the center position. Depending on design, the joystick operator may be a momentary contact with spring return of the handle to the center position, or it can be maintained in one of the four positions.

6.2.4 Pilot Lights

A pilot light is a small electric light used to visually show a specific condition of a circuit. Pilot lights can be separate assemblies, or they can be built into a pushbutton switch assembly. Pilot light assemblies, like the ones shown in *Figure 41*, typically have a polycarbonate or glass lens (cap). Colored lenses are made in red, green, amber, blue, clear, white, or yellow. Typically, pilot lights use a bayonet base incandescent lamp that operates at full voltage or at a reduced voltage via a

transformer or resistor. Depending on the control circuit arrangement, pilot lamps operate at a voltage ranging from 6VAC or DC to 120VAC or DC. Some pilot lights use a light-emitting diode (LED) instead of an incandescent lamp. Some are available with a local push-to-test capability, others with a remote test capability used to test the operation of the indicator lamp.

6.2.5 Pushbutton Stations

Pushbutton stations (*Figure 42*) are enclosures used to house one or more pushbuttons, selector switches, and/or pilot lights to protect them from dust, dirt, water, and corrosive fluids. They can be bought unassembled or with the pushbutton and/or lever switches installed. Cast metal, polyester, and stainless steel pushbutton stations are available in various NEMA enclosures and in various sizes. Always use the correct switch components and enclosure for the environment where they will be used.

Selecting the components for a typical pushbutton or lever switch and related switch station

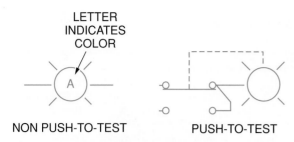

LETTER
INDICATES
COLOR

NON PUSH-TO-TEST   PUSH-TO-TEST

SCHEMATIC SYMBOLS

311F41A.EPS

INSTRUMENT TYPE   PUSH-TO-TEST TYPE
                         (LENS REMOVED)

311F41B.EPS

Figure 41 ◆ Typical pilot light assemblies.

involves using the switch manufacturer's catalog or data sheets to identify the proper components needed for each switch and switch station. You must make sure that the electrical characteristics of the selected switching device and its contacts are compatible with the requirements for your application. Switch manufacturers describe the

311F42.EPS

Figure 42 ◆ Example pushbutton station.

electrical characteristics of their pushbutton switches, lever switches, and contacts using codes defined in *NEMA Standard ICS 5* for categories of utilization and AC and DC contact switching ratings. For example, category entries such as AC-14 or DC-13 define typical applications of the device. Contact rating designations such as A600 or B600 define the continuous current-carrying thermal capability of the device or contact. The codes and related operating characteristics are given in *Tables 10* and *11*.

A general procedure for selecting a pushbutton/lever switch is given here.

Step 1 Select the style of switch operator:
  • Standard-duty or heavy-duty
  • Chrome, plastic, or other material
  • Oil, water, dust resistant, corrosion resistant

Installation of Pushbutton Switches in a Pushbutton Station

When installing pushbutton switches in a pushbutton station, follow these guidelines:

• *STOP button location* – On all single row pushbutton stations, the STOP button shall be located below or to the right of all other associated buttons, including lights and selector switches.

• *Pushbuttons used with multi-speed motors (mounted vertically)* – The lowest button shall be the STOP button. The lowest-speed button shall be above the STOP button, followed by those for consecutively higher speeds.

• *Pushbuttons used with multi-speed motors (mounted horizontally)* – The right-hand button shall be the STOP button. The lowest-speed button shall be to the left of the STOP button, followed by those for consecutively higher speeds.

Table 10 Rating Codes for AC and DC Control Circuit Contacts *(Reprinted by permission of National Electrical Manufacturers Association.)*

Rating Codes for AC Control Circuit Contacts at 50 and 60 Hertz(†)

Contact Rating Code Designation*	Thermal Continuous Test Current (Amperes)	120 Volt		240 Volt		480 Volt		600 Volt		Maximum Volt-Amperes	
		Make	Break	Make	Break	Make	Break	Make	Break	Make	Break
A150	10	60	6.0	—	—	—	—	—	—	7200	720
A300	10	60	6.0	30	3.00	—	—	—	—	7200	720
A600	10	60	6.0	30	3.00	15	1.50	12	1.20	7200	720
B150	5	30	3.00	—	—	—	—	—	—	3600	360
B300	5	30	3.00	15	1.50	—	—	—	—	3600	360
B600	5	30	3.00	15	1.50	7.50	0.75	6	0.60	3600	360
C150	2.5	15	1.5	—	—	—	—	—	—	1800	180
C300	2.5	15	1.5	7.5	0.75	—	—	—	—	1800	180
C600	2.5	15	1.5	7.5	0.75	3.75	0.375	3.00	0.30	1800	180
D150	1.0	3.60	0.60	—	—	—	—	—	—	432	72
D300	1.0	3.60	0.60	1.80	0.30	—	—	—	—	432	72
E150	0.5	1.80	0.30	—	—	—	—	—	—	216	36

(*)The numerical suffix designates the maximum voltage design values, which are to be 600, 300, and 150 volts for suffixes 600, 300, and 150, respectively. The test voltage is to be 600, 240, or 120 volts. (†) For maximum ratings at voltages between the maximum design value and 120 volts, the maximum make and break ratings are to be obtained by dividing the volt-amperes rating by the application voltage. For voltages below 120 volts, the maximum make current is to be the same as for 120 volts, and the maximum break current is to be obtained by dividing the break volt-amperes by the application voltage, but these currents are not to exceed the thermal continuous test current.

Rating Codes for DC Control Circuit Contacts

Contact Rating Code Designation*	Thermal Continuous Test Current (Amperes)	Maximum Make or Break Current, Amperes			Maximum Make or Break Volt-Amperes At 300 Volts or Less
		125 Volt	250 Volt	301 to 600 Volt	
N150	10.0	2.2	—	—	275
N300	10.0	2.2	1.1	—	275
N600	10.0	2.2	1.1	0.40	275
P150	5.0	1.1	—	—	138
P300	5.0	1.1	0.55	—	138
P600	5.0	1.1	0.55	0.20	138
Q150	2.5	0.55	—	—	69
Q300	2.5	0.55	0.27	—	69
Q600	2.5	0.55	0.27	0.10	69
R150	1.0	0.22	—	—	28
R300	1.0	0.22	0.11	—	28

(*)The numerical suffix designates the maximum voltage design values, which are to be 600, 300, and 150 volts for suffixes 600, 300, and 150, respectively. Test voltage shall be 600, 240, or 120 volts. (†) For maximum ratings at 300 volts or less, the maximum break ratings are to be obtained by dividing the volt-ampere rating by the application voltage, but the current values are not to exceed the thermal continuous test current.

Table 11 Utilization Categories for Control Circuit Switching Elements

Utilization Categories for Switching Elements		
Kind of Current	**Category**	**Typical Applications**
Alternating Current	AC-12	Control of resistive loads and solid-state loads with optical isolation
	AC-13	Control of solid-state loads with transformer isolation
	AC-14	Control of small electromagnetic loads (max. 72VA closed)
	AC-15	Control of electromagnetic loads (greater than 72VA closed)
Direct Current	DC-12	Control of resistive loads and solid-state loads with optical isolation
	DC-13	Control of electromagnets
	DC-14	Control of electromagnet loads having economy resistor in circuit

Step 2 Select the type of switch operator:
- Pushbutton or lever
- Nonilluminated or illuminated
- Color of pushbutton

Step 3 Select the contact block(s) needed:
- Normally open or normally closed
- Standard or hazardous location
- Standard-duty or heavy-duty

Step 4 Select appropriate operator identification or legend nameplate and color.

Step 5 Select the proper station enclosure for the type and number of operators involved.

6.2.6 Using Multiple Pushbutton Stations

If motors are required to be started from more than one location, additional pushbutton stations must be installed in the circuit. In doing so, the START buttons in these stations must be connected in parallel with the original START button, and the STOP buttons must be connected in series with the original STOP button as shown in *Figure 43*. The auxiliary contactor must also be connected in parallel with the START buttons. For three or more control stations, all the START buttons must be connected in parallel with the auxiliary contactor and all the STOP buttons connected in series.

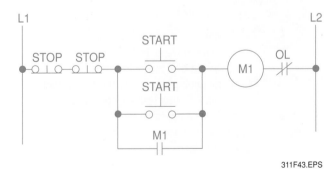

311F43.EPS

Figure 43 ◆ Multiple STOP and START pushbuttons.

6.3.0 Temperature Switches

Temperature switches (*Figure 44*) are used to monitor temperatures in the control circuits for applications such as heating/cooling systems, damper systems, fire alarm systems, and process control systems. Temperature switch contacts open or close in response to a rise or fall in temperature detected by their sensing element. The temperature where the opening or closing of the contacts occurs is normally determined by operator-adjustable **setpoints**. The temperature-sensing elements used in mechanical temperature switches

OPENS ON	CLOSES ON
TEMPERATURE RISE	TEMPERATURE RISE
(HIGH-TEMPERATURE	(MINIMUM TEMPERATURE
SWITCH OR ALARM)	IS SATISFIED)

SCHEMATIC SYMBOLS

311F44A.EPS

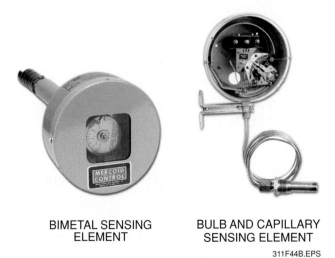

BIMETAL SENSING	BULB AND CAPILLARY
ELEMENT	SENSING ELEMENT

311F44B.EPS

Figure 44 ◆ Typical temperature switches.

are usually either a bimetallic element or a bulb and capillary element. There are also electronic temperature-sensing elements, but these tend to be used more with temperature controller units rather than switches.

A bimetallic element is made of two different metals bonded together; one is usually copper or brass (*Figure 45*). The other, a special metal called *Invar* steel, contains 36% nickel. When heated, the copper or brass has a more rapid expansion rate than the Invar steel and changes the shape of the element. The movement that occurs when the bimetal changes shape is used to open or close the switch contacts. Although bimetal elements are constructed in various shapes, the coil-wound element is the most commonly used. In one type of widely used bimetal-operated switch, the coiled bimetal strip is attached to a sealed glass tube containing a small pool of mercury (a conductor). When the bimetal strip causes the tube to be tipped in one direction, the mercury moves across the switch contacts, closing the contacts. When the tube is tipped in the opposite direction, the mercury moves to the other end of the tube, causing the switch contacts to be open. In some other

types of bimetal-operated switches, the bimetal strip is connected directly to one of the switch contacts. This moving contact makes or breaks contact with a stationary switch contact in response to the expansion and contraction of the bimetal strip.

Temperature switches with bulb and capillary sensing elements consist of a tube (capillary tube) that connects the main body of the switch to a remotely located bulb that is partially filled with a liquid, gas, or vapor. This sensing bulb can be clamped to a pipe or duct, inserted in a cooling/ heating coil, or placed in a tank well for sensing product temperature. The switch operates on the principle that when the temperature increases at the bulb, expansion of the bulb media (liquid, gas, or vapor) takes place. This causes a force to be transmitted through the capillary tube and exerted on the mechanism that operates the switch contacts. Two types of switching mechanisms are commonly used: Bourdon tube and bellows. *Figure 46(A)* shows a C-type Bourdon tube. It is a flattened metal tube that is open at one end and closed at the other. The tube straightens out when a rise in pressure is applied to the open end of the tube and curls inward with a decrease in pressure. This movement is transmitted by a mechanical linkage to the switch contacts, causing them to open or close. Spiral and helical Bourdon tubes are also used that work on the same principle as the C-type. However, both of these tubes produce more tip travel and higher torque capacity at the free end than the C-type.

In a bellows temperature switch, the pressure change in the bulb media is transmitted through the capillary tube for application to a bellows inside the temperature switch, shown in *Figure 46(B)*. One end of the bellows is closed, and the other end is connected to the bulb pressure source via the capillary tube. The bellows is a round

BIMETAL

CONTACTS
OPEN

MERCURY

NICKEL (INVAR)
ALLOY

COPPER OR
BRASS ALLOY

CONTACTS
CLOSED

MERCURY

311F45.EPS

Figure 45 ◆ Bimetal strips used in temperature switches.

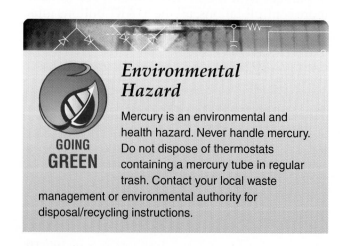

Environmental Hazard

GOING GREEN

Mercury is an environmental and health hazard. Never handle mercury. Do not dispose of thermostats containing a mercury tube in regular trash. Contact your local waste management or environmental authority for disposal/recycling instructions.

device with several folds, much like an accordion, that expand or contract with the applied pressure. The expanding bellows moves against a calibrated spring switch mechanism that opens or closes the switch contacts. Usually, the spring tension is adjustable so that the switch can be adjusted for different temperature set points.

To properly select temperature switches, you need to know and understand several terms used in switch manufacturers' catalogs:

- *Allowable temperature limits* – The maximum and minimum temperatures to which a temperature switch may be exposed without altering its performance characteristics. It is important that a temperature switch not be exposed to or used at temperatures beyond the manufacturer's allowable limits.
- *Cut-in temperature* – The temperature of the sensed medium at which a temperature switch is actuated to energize the load. The cut-in temperature may be the trip or reset point.

- *Cut-out temperature* – The temperature of the sensed medium at which a temperature switch is actuated to de-energize the load. The cut-out temperature may be the trip or reset point.
- *Operating temperature differential* – The difference between the cut-in and cut-out temperatures. The operating temperature differential is also called the deadband.
- *Operating temperature range* – The range between the maximum and minimum temperature settings at which a temperature switch will continue to operate within the manufacturer's specifications.
- *Remote temperature sensing* – The construction and installation of a temperature switch in which the sensing unit and the switch mechanism are thermally isolated so that the temperature of the sensed medium does not affect the performance of the switch mechanism.

6.4.0 Pressure Switches

Pressure switches (*Figure 47*) use mechanical motion in response to pressure changes to open or close contacts when a predetermined pressure level is reached. Depending on the switch design

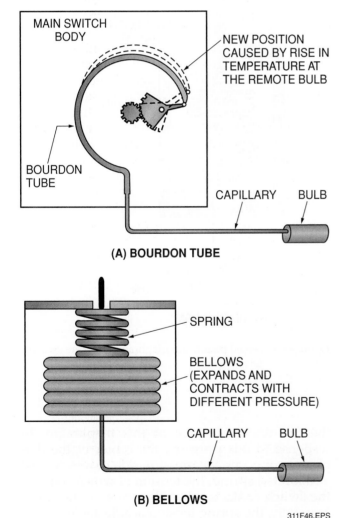

Figure 46 ◆ Bourdon tube and bellows temperature switch mechanisms.

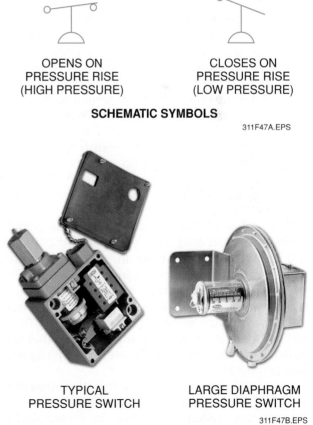

Figure 47 ◆ Typical pressure switches.

Bourdon Tube Pressure Switches

Care must be taken not to apply pressures beyond the rating of a Bourdon pressure switch. Exposure to excessive pressures can bend the Bourdon tube beyond its ability to return to its original shape.

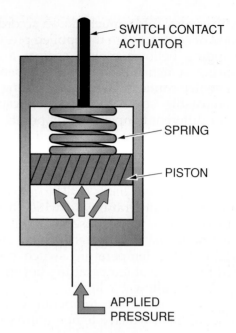

(A) SEALED PISTON

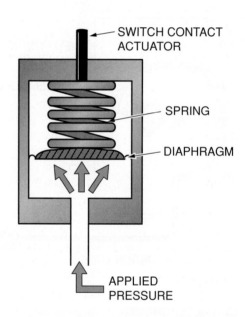

(B) DIAPHRAGM

311F48.EPS

Figure 48 ◆ Sealed piston and diaphragm pressure switch mechanisms.

and application, N.O. and N.C. contacts may be activated on positive, negative (vacuum), or differential pressures. Single-stage switches are used in a wide variety of machine and process applications to protect the equipment from low or high pressures or to monitor the system pressure. Dual-stage pressure switches are made that contain two pressure switches. They are used in applications where it is practical to monitor two separate pressures using a single device. For example, in a refrigeration system, a dual pressure switch is commonly used to monitor the compressor suction and head pressures for a low-pressure and high-pressure condition, respectively. Differential pressure switches can signal that a predetermined pressure difference has been reached by an increasing difference between two pressures or a decreasing difference between two pressures.

Both adjustable and nonadjustable types of pressure switches are made. Use of an adjustable type allows a particular model switch to be used in more than one application. Different kinds of pressure-responsive elements are used in pressure switches. They can be a Bourdon tube, bellows, sealed piston, or diaphragm. Bourdon tube and bellows switch sensors operate in the same way as described earlier for use in temperature switches.

In a sealed piston pressure switch, the switch contacts are actuated by a piston that moves up and down in its cylinder in response to the applied pressure, as shown in *Figure 48(A)*. The top of the piston moves against a calibrated spring, the tension of which actuates the switch contacts when the source pressure increases. Normally, the spring tension is adjustable so that the switch can be adjusted to operate at different pressure setpoints.

In a diaphragm pressure switch, the edge of a disk-shaped diaphragm is firmly attached to the case of the switch, as shown in *Figure 48(B)*. The pressure being monitored is applied against

the full underside area of this diaphragm. In response to this pressure, the center of the diaphragm opposite the pressure side moves against a calibrated spring, the tension of which actuates the switch contacts when the pressure increases. Normally, the spring tension is adjustable so that the switch can be adjusted to operate at different pressure setpoints.

Pressure Switches

To properly select pressure switches, you need to understand several terms used in switch manufacturers' catalogs:

- *Allowable pressure limits* – The maximum and minimum pressures, stated by the manufacturer, to which a pressure switch may be exposed for brief or extended periods without altering the performance characteristics of the switch. Maximum allowable pressure is also referred to as *overrange* and includes surge pressures.
- *Cut-in pressure* – The pressure at which the switch is actuated to energize the load. The cut-in pressure may be the trip or reset point.
- *Cut-out pressure* – The pressure at which the switch is actuated to de-energize the load. The cut-out pressure may be the trip or reset point.
- *Deadband* – The pressure difference between the setpoint at which a switch activates when the pressure increases and the point at which the switch resets when the pressure drops.
- *Drift* – An inherent change in operating value for a given setting over a specified number of operations and specified environmental conditions.
- *Maximum static pressure* – The continuous pressure that the pressure-containing envelope of a pressure switch sustains without rupture. The maximum static pressure is sometimes called the rated static pressure.
- *Operating differential pressure* – The difference between cut-in pressure and cut-out pressure. The operating differential pressure is also called the deadband.
- *Operating pressure range* – The range between the maximum and minimum pressure settings at which a pressure switch will operate and continue to operate within the manufacturer's specifications.
- *Proof pressure* – The nondestructive static test pressure, in excess of the maximum allowable pressure, which causes no permanent deformation or malfunction.
- *Pulsation snubber* – A device used with pressure switches to reduce the effect of pressure surges within the pressure system on the pressure responsive element of the switch.

6.5.0 Mechanical Limit Switches

Mechanical limit switches detect the position of an object by having the object make direct physical contact with the switch actuator. They commonly are used in applications in which it is desired to limit the travel of machine tools, detect moving items on a conveyor belt, monitor an object's position, or detect that machinery safety guards are properly in place. Limit switches come in many designs and sizes with single-pole, double-throw (SPDT) and double-pole, double-throw (DPDT) versions being the most common. There are two types of limit switches: rotary lever-actuated (*Figure 49*) and plunger-actuated (*Figure 50*). Both consist of an actuator, switch body, and terminals. The switch actuator is the part of the switch that, when moved by physical contact with an object, operates the switch contacts. The switch body houses the electrical contacts, and the terminals are the point of switch connection for the circuit wiring.

A rotary lever-actuated limit switch works on the principle that a cam or plate hits the end of the lever arm, which rotates a shaft that operates the switch contacts. In some rotary lever-actuated limit switches, the actuator attached to the shaft can be interchanged with a variety of different kinds of actuators (*Figure 51*). This allows the same switch to be used in many different applications. A plunger-activated limit switch works on the principle that a cam or plate hits the end of the plunger, which is pressed in to operate the contacts of the switch.

Limit switch operating heads can be of two types: momentary contact switch or maintained contact switch. Momentary contacts, sometimes called spring return contacts, return from the actuated position to their normal (nonactuated) position when the actuating force is removed. Maintained contacts remain in the actuated position until actuated to another position even after the actuator is released. They are reset only by further mechanical action of the operating head. For example, the contacts may be reset by shaft rotation in the opposite direction.

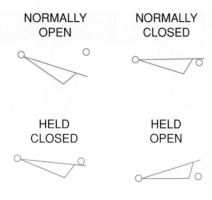

NORMALLY OPEN　　NORMALLY CLOSED

HELD CLOSED　　HELD OPEN

SCHEMATIC SYMBOLS

311F49A.EPS

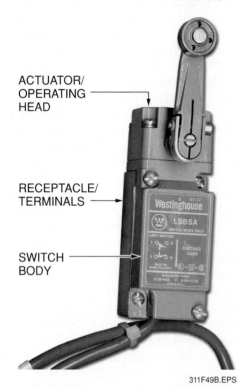

ACTUATOR/ OPERATING HEAD

RECEPTACLE/ TERMINALS

SWITCH BODY

Westinghouse

LSBSA
SWITCH BODY ONLY
LIMIT SWITCH

RATING A600

311F49B.EPS

Figure 49 ◆ Rotary lever-actuated limit switch.

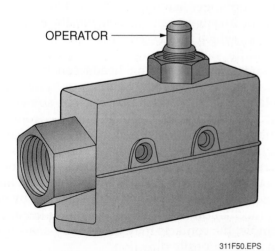

OPERATOR

311F50.EPS

Figure 50 ◆ Plunger-actuated limit switch.

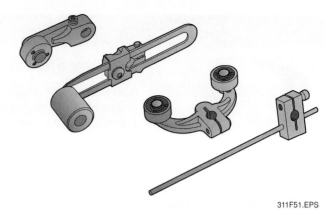

311F51.EPS

Figure 51 ◆ Rotary lever-actuated limit switch actuators.

Manufacturers of limit switches describe the position or motion of the switch actuator in specific terms in their product literature. These terms are shown in *Figure 52* and are described in the following list:

- *Movement differential* – The distance or angle from the operating position to the releasing position.
- *Free (initial) position* – The switch actuator is positioned such that the limit switch contacts are in their normal (untriggered) position.
- *Operating position or point* – The position of the actuator at which the contacts change from their untriggered position to the operated (triggered) position.
- *Operating torque* – The minimum force that must be applied to the actuator to cause the contacts to change from the untriggered to the triggered state.
- *Overtravel* – In the case of the rotary lever-actuated limit switch, it is the distance or angle through which the actuator moves when traveling past the triggered position. In a plunger-actuated limit switch, the overtravel distance is the safety margin for the sensor to avoid breakage.
- *Pre-travel* – The distance or angle through which the actuator moves from the free position to the position just before the contacts change state. The contacts are still at their normal (untriggered) position.
- *Release position or point* – The position of the actuator at which the contacts change from the triggered position and return to the normal untriggered position.
- *Release torque* – The value to which the torque on the actuator must be reduced to allow the contacts to change from the triggered position to the normal untriggered position.

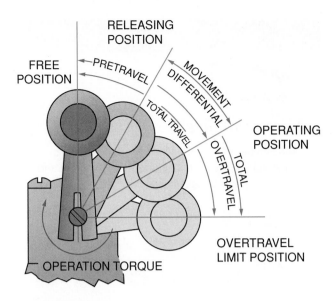

ROTARY LEVER-ACTIVATED LIMIT SWITCH

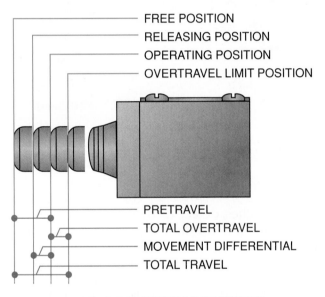

PLUNGER-ACTIVATED LIMIT SWITCH

311F52.EPS

Figure 52 ◆ Positions of limit switch actuators.

Depending on the limit switch design, the normally open and normally closed contacts may or may not be conducting current simultaneously during operation. SPDT limit switches can have make-before-break, break-before-make, and simultaneous make-and-break contacts. With make-before-break contacts, the normally open contact closes before the normally closed contact opens. With break-before-make contacts, the normally closed contact opens before the normally open contact closes. With simultaneous make-and-break contacts, the normally closed contact

opens at the same time the normally open contact closes. Manufacturers of limit switches often use contact function diagrams like the one shown in *Figure 53* to define the operation of their switches and show any overlap or nonoverlap of contact operation.

When selecting and installing limit switches, the following guidelines should be taken into consideration:

- A rotary lever-actuated limit switch is the best choice for most applications. It can be used in any application in which the cam moves perpendicular to the lever's rotational shaft.
- A plunger-actuated limit switch is the best choice to monitor short, controlled machine movements, or where space or mounting restrictions will not permit the use of a lever-actuated switch.
- The switch contacts must be selected according to the proper voltage and current size for the load. If the load current exceeds the switch contact rating, a relay, contactor, or motor starter must be used to interface the limit switch with the load.
- The switch contacts must be connected to the same polarity. Otherwise, arcing or welding of the contacts may occur. Do not connect voltages from different sources to the contacts of the same switch unless the switch is designed for such service.
- Mount the switch so that there is no possibility of unintentional actuation by the movement of an operator or moving parts of the machine or equipment. Mount it firmly in an easily accessible location with suitable clearances to allow for service and replacement. Do not place the switch in a location where machining chips or other materials can accumulate under normal operating conditions or where the temperature and atmospheric conditions are beyond those for which the switch has been designed.
- Do not expose the switch to oils, coolants, or other liquids unless designed for such service. If liquid entry is possible, the switch should be mounted face down to prevent seepage through the seals on the operating head. All conduit connections must also be tightly sealed.
- For rotary lever-actuated limit switches, it is important that the correct lever actuator be selected. The manufacturer's recommendations in selecting and applying limit switches and actuators should always be followed. It is important that the selected actuator be securely fastened to the switch body shaft to prevent it from slipping.

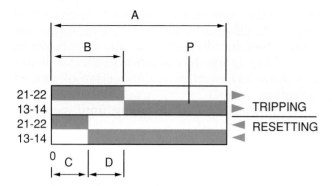

A = Maximum travel of the operator in mm or degrees
B = Tripping travel of the contact
C = Reset travel
D = B = C = Differential travel
P = Point from which positive opening is assured

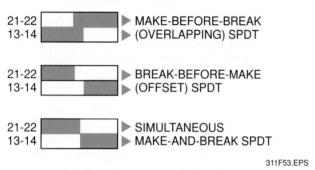

311F53.EPS

Figure 53 ◆ Example of a limit switch contact function diagram.

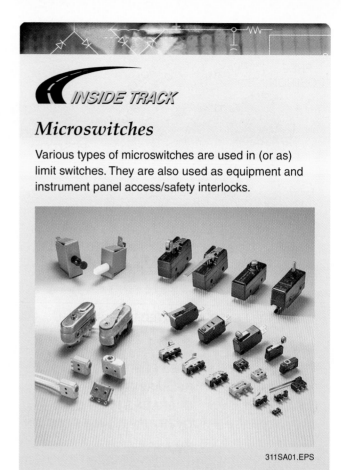

INSIDE TRACK

Microswitches

Various types of microswitches are used in (or as) limit switches. They are also used as equipment and instrument panel access/safety interlocks.

311SA01.EPS

• Where relatively fast motions are involved with a rotary lever-actuated limit switch, the cam arrangement should be such that the actuator does not receive a severe or sharp impact. The cam should be tapered to extend the time it takes to engage the electrical contacts. This prevents wear on the switch and allows the contacts a longer closing time, giving relays, valves, and other related devices enough time to operate.

6.6.0 Flow Switches

Flow switches detect (prove) the presence of a liquid flowing through a pipe or airflow in a duct. Mechanical flow switches (*Figure 54*) have a vane that extends into the pipe or duct. When the force of the liquid flowing in the pipe or air flowing in the duct is sufficient to overcome the spring tension of the vane, the vane moves and actuates the switch contacts. When there is a loss or reduction in flow below the predetermined vane actuation setpoint, the vane returns to the

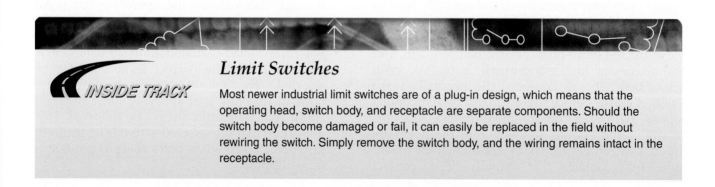

INSIDE TRACK

Limit Switches

Most newer industrial limit switches are of a plug-in design, which means that the operating head, switch body, and receptacle are separate components. Should the switch body become damaged or fail, it can easily be replaced in the field without rewiring the switch. Simply remove the switch body, and the wiring remains intact in the receptacle.

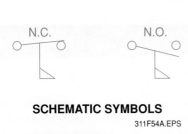

SCHEMATIC SYMBOLS

311F54A.EPS

SWITCH
BODY

VANE

311F54B.EPS

Figure 54 ◆ Vane liquid flow switch.

SCHEMATIC SYMBOLS

311F55A.EPS

SWITCH
BODY

FLOAT

311F55B.EPS

Figure 55 ◆ Liquid level float switch.

central position. For some flow switches, the spring tension on the vane is adjustable. This allows the switch vane to be adjusted so that it actuates the switch contacts at different flow rates. The N.O. or N.C. switch contacts may be used to start or stop pumps, open or close valves, actuate a warning lamp, or sound an alarm.

6.7.0 Float Switches

Float switches are used to maintain or monitor the level of a liquid in a storage tank or other vessel. For example, when the level of a liquid reaches a predetermined high (or low) point in a tank, the float switch operates to shut off or turn on a pump. Another example is that of a float switch used to activate an alarm, shut down a pump, or both as it monitors a safe level or a not-to-exceed level of a liquid in a tank. There is a wide variety of float switches including mechanical, electronic, ultrasonic, and optical designs. A basic mechanical type of float switch is shown in *Figure 55*. It

consists of a float and hermetically sealed, magnetically operated snap switch. As the float moves up and down with the level of the liquid in a tank, the N.O. or N.C. switch contacts are activated when the liquid level reaches a predetermined height. Two switches can be used to monitor and maintain minimum and maximum liquid levels.

6.8.0 Foot Switches

A foot switch (*Figure 56*) is a control device operated by a foot pedal. It is used where the process or machine requires that the operator have both hands free. Foot switches usually have momentary contacts but some are available with latches that enable them to be used as maintained-contact devices. Most foot switches typically have two positions, a toe-operated position and a spring-loaded off position. Some foot switches are available with three positions, two positions that allow for toe and heel control and a spring-loaded off position.

Use of Foot Switches

INSIDE TRACK

When using a foot switch in applications such as power presses, additional operator protection, such as point-of-operation guarding, must be provided. This is necessary since the operator's hands and other parts of the body are free to enter the pinch-point area and serious injury can occur.

SCHEMATIC SYMBOLS

311F56.EPS

Figure 56 ◆ Foot-operated switch.

In applications in which more than one foot switch is required, as when two or more persons are operating the machine, it is required that the foot switches be wired in series, making it necessary that each operator's foot switch be actuated before the machine can operate.

The use and selection of a foot switch for an application is based on the customer's knowledge of the conditions and factors present during the setup, operation, and maintenance of the particular machine or process with which the switch is to be used. When selecting a foot switch, always refer to the applicable ANSI standards and OSHA regulations. Additional information can be found in the National Safety Council's *Accident Prevention Manual*.

6.9.0 Jogging and Plugging Switches

Many motor control applications require the use of switches in the control circuit to initiate the jogging (inching) or plugging of a motor. Jogging a motor involves the frequent closure of a switch to stop and start the motor for short periods of time, which is required when it is necessary to precisely

position a crane or machine tool. There are many ways that a pushbutton switch can be connected in a control circuit to facilitate jogging. *Figure 57* shows a line diagram of one basic control circuit where a pushbutton switch is used to jog a motor. In this circuit, the normally closed contacts of the JOG pushbutton switch are connected in series with the holding circuit contact (M) of the magnetic starter. When the JOG pushbutton is pressed, the normally open contacts energize the starter magnet. At the same time, the normally closed contacts disconnect the holding circuit. Therefore, when the JOG pushbutton is released, the starter immediately opens to disconnect the motor from the line.

Plugging is an operation in which a motor is brought to a rapid stop (braked) by reversing the phase sequence of the power applied to the motor. Plugging typically is done in machine tool applications when the tool must be stopped rapidly at some point in its cycle of operation in order to prevent inaccuracies in the work or damage to the machine. Plugging can only be done if the driven machine and its load will not be damaged by the reversal of the motor.

A plugging switch (*Figure 58*) is connected mechanically to the shaft of the motor or driven machinery. The rotating motion of the motor is transmitted to the plugging switch contacts either by a centrifugal mechanism or by magnetic induction. The contacts of a plugging switch are designed to open and close as the shaft speed of the plugging switch varies.

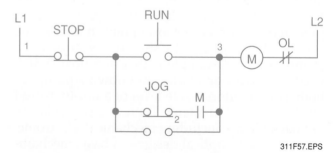

311F57.EPS

Figure 57 ◆ Basic motor jogging control circuit.

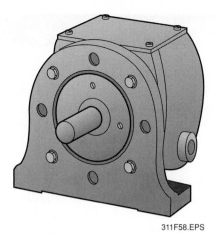

Figure 58 ◆ Plugging switch.

311F58.EPS

There are several ways that a plugging (speed) switch can be connected in a control circuit to facilitate plugging. *Figure 59* shows a line diagram of a basic control circuit that uses a plugging switch to plug a motor to a stop from one direction only. In this circuit, the forward rotation of the motor closes the normally open plugging switch contact. When the stop button is pushed, the forward starter de-energizes. At the same time, the reverse starter is energized through the plugging switch and the normally closed forward interlock. This reverses the motor connections, and the motor is braked to a stop. When the motor is stopped, the plugging switch opens to disconnect the reverse contactor.

Some plugging switches have an anti-plugging protection capability. This means the switch design is such that it prevents the application of a counter torque until the motor speed is reduced to an acceptable value. With this switch, a contact on the switch opens the control circuit of the contactor used to reverse the rotation of the motor and is prevented from closing until the motor speed is reduced. Then the other contactors can be energized.

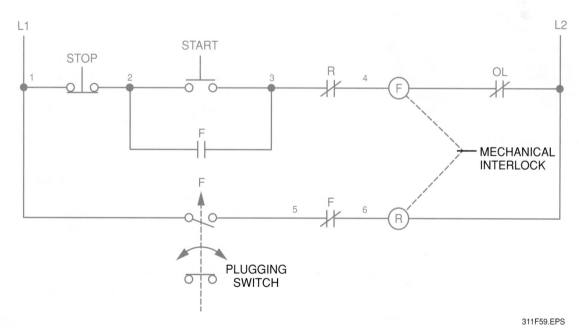

Figure 59 ◆ Basic motor plugging control circuit.

311F59.EPS

Plugging

Capacitor-start motors cannot be plugged to a stop because they cannot be reversed at full speed.

6.10.0 Proximity Switches/Sensors

Proximity switches, commonly called **proximity sensors**, are encapsulated solid-state devices used where it is required or desired to detect the presence of objects (targets) within a short sensing range without having direct physical contact with the target(s). They can detect all sizes of objects, including very small ones, in applications such as presence, passage, flow of parts, end of travel, rotation, and counting. Proximity switches are commonly used in place of mechanical limit switches, float switches, or level switches for many applications. This is because they have faster switching speeds and no moving parts, making them more reliable and accurate. Being solid-state, they are compatible with electronic controllers and automated systems. Proximity switches are intended for use in circuits with rated voltages not exceeding 250 or 300VDC. Two widely used proximity switches are the inductive proximity sensor and the capacitive proximity sensor. Physically,

both types of sensors look the same. *Figure 60* shows two circularly shaped proximity sensors. Block-shaped sensors are also available.

6.10.1 Inductive Proximity Sensors

An inductive proximity sensor is used to detect only metal targets within its sensing field, which is typically 0.5mm to 40mm (0.020" to 1.575"). Its sensitivity depends on the size of the target and the metallic material of which the target is made. The internal circuitry of a solid-state inductive proximity sensor consists of an oscillator circuit, output driver, and output switch (*Figure 61*). The oscillator circuit produces radio frequency (RF) oscillations. These oscillations are present at a coil (part of the oscillator circuit) that is located at the sensing face of the proximity sensor. This coil has two functions: it determines the tuning of the oscillator circuit, and it radiates an electromagnetic field created by the RF oscillations into the sensing field in front of the switch face.

When a target enters the sensing field, the electromagnetic field emitted by the proximity sensor causes **eddy currents** to be induced in the target. These eddy currents disrupt the electromagnetic field at the sensor face coil, causing a change in the tuning of the oscillator circuit. As a result, the oscillations cease. The output driver senses the loss of the oscillations and commands the output switch to operate, producing an ON or OFF output signal depending on the switch design. After the target passes out of the sensing field, the oscillator signal is again generated, causing the switch output signal to be returned to its normal inactive state.

311F60.EPS

Figure 60 ◆ Examples of typical proximity sensors.

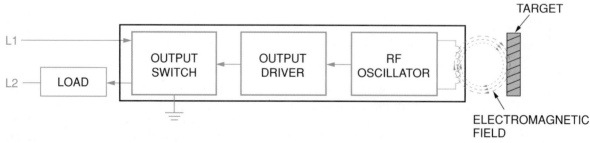

311F61.EPS

Figure 61 ◆ Block diagram of an inductive proximity sensor.

6.10.2 Capacitive Proximity Sensors

The capacitive proximity sensor can be used to detect the absence or presence of both metallic and nonmetallic targets within its sensing field, which is typically 0.3mm to 15mm (0.012" to 0.59"). Target detection with this type of sensor is based on the principle that when a target moves into the sensing field, it changes the coupling capacitance of the sensor's sensing capacitor. Remember that a capacitor consists of two plates separated by an insulator (dielectric) and that any change in a capacitor's **dielectric constant** causes a change in the capacitor's capacitance value. The internal circuitry of a solid-state capacitive proximity sensor consists of an oscillator, output driver, and output switch circuit (*Figure 62*). The sensing capacitor is part of the oscillator's tuning circuit. It is formed by two small plates separated by a dielectric (air) and is located behind the front of the sensing face of the sensor. In the absence of a target, the tuning of the oscillator is such that the oscillator is inoperative. Note that this is just the opposite of the inductive proximity sensor.

Capacitive proximity sensors can detect any target that has a dielectric constant slightly greater than that of air. Air has a dielectric constant of 1. When a target with a dielectric constant greater than 1 moves into the sensing field, its presence modifies the coupling capacitance. This change in capacitance tunes the oscillator circuit so that it begins to generate oscillations. The output driver senses these oscillations and commands the output switch to operate, producing an ON or OFF output signal depending on the switch design. After the target passes out of the sensing field, the oscillator signal again ceases, causing the switch output signal to return to its normal inactive state.

6.10.3 Load-Powered Proximity Sensors

There are two categories of proximity sensors: load-powered sensors and line-powered sensors. Load-powered sensors are two-wire sensors (excluding ground) that are connected in series with the controlled load (*Figure 63*). They draw their operating current, also commonly called leakage or residual current, through the load. When the sensor is in the open state (absence of a target), it must draw a minimum operating current, sometimes called the off-state leakage current, through the de-energized load device in order to power the proximity sensor's electronics. However, this operating current must be low enough so that it will not inadvertently energize the load.

When the sensor is in the closed state (target present), the current flowing in the control circuit is a combination of the proximity sensor operating current and the current drawn by the energized load. This means that the proximity sensor must have a current rating high enough to carry the

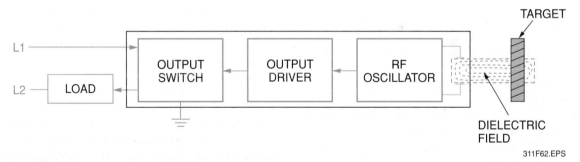

Figure 62 ◆ Block diagram of a capacitive proximity sensor.

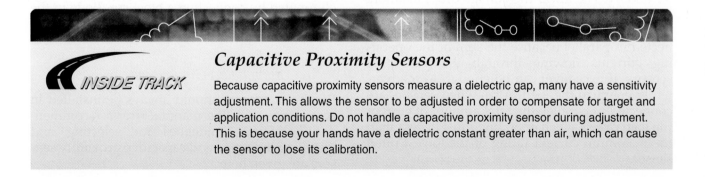

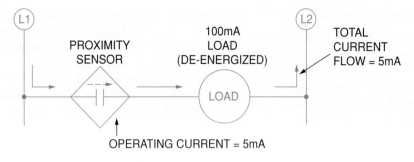

(A) PROXIMITY SENSOR OFF (NO TARGET DETECTED)

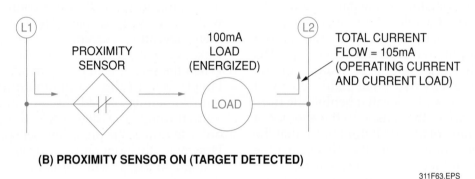

(B) PROXIMITY SENSOR ON (TARGET DETECTED)

311F63.EPS

Figure 63 ◆ Load-powered proximity sensor circuit.

load current. Another factor to consider is that the current draw of the load must be high enough to ensure that the sensor will operate when the load is energized. This current level is called the minimum holding current. If the current drawn by the load falls below the required minimum holding current, the sensor will be inoperative.

Two-wire proximity sensors can be connected in series or in parallel like other pilot devices. However, it is recommended that no more than three devices be connected in series or in parallel because they can affect the operation of the load. No more than three sensors should be connected in series because the sum of the individual voltage drops across each of the sensors can lower the voltage available to the load sufficiently enough to render it inoperative. Sensor voltage drops vary for different sensors. Typically, two-wire sensors have a voltage drop ranging anywhere between 3V and 8V. The lower the voltage drop, the more sensors that can be connected in series.

No more than three sensors should be connected in parallel because the sum of the operating currents flowing through the individual sensors, if excessive, can be enough to inadvertently energize the load. Also, when sensors are connected in parallel, the total amount of current the load draws when operating must be less than the maximum current rating of the lowest rated sensor.

6.10.4 Line-Powered Proximity Sensors

Line-powered proximity sensors are three-wire sensors (excluding ground) that derive their power from the power lines, not the load (*Figure 64*). They have two wires for the power supply and one output wire. Some four-wire models are also available. These produce complementary N.O. and N.C. outputs. The operating current for a line-powered sensor drawn from the power lines is called the burden current.

Connecting line-powered sensors in series and in parallel can affect the operation of the load. Connecting these sensors in series can affect the load because the first sensor in the line must carry the load current plus the operating currents for each of the following (downstream) sensors. When connected in parallel, there is a danger that a nonconducting sensor may be damaged by reverse polarity. Note that this problem is usually eliminated by connecting a blocking diode in the output line of each sensor.

6.10.5 Proximity Sensor Installation Guidelines

Proximity sensors should always be installed in accordance with the manufacturer's recommendations. Some guidelines for selecting and installing inductive and capacitive proximity sensors are given here:

- The output switch in inductive and capacitive proximity sensors can use either a triac, SCR, or transistor as the switching device. Proximity sensors with triacs are used to switch AC loads. Those with SCRs and transistors are used to switch high-power DC loads and low-power DC loads, respectively. Many proximity sensors are designed for use in either AC or DC circuits. For DC and AC/DC versions, the switch can be connected to either positive (PNP) or negative (NPN) logic inputs.
- A shielded (flush mount) inductive sensor can be fully embedded in a metal mounting block without affecting the range. A nonshielded sensor needs a clearance around it called a metal-free zone that is determined by the sensing range. Otherwise, the sensor will sense the metal mounting and operate continuously.

- When mounting two or more sensors near each other, make sure to mount them in accordance with the manufacturer's instructions so that their radiated detection fields do not interfere with each other. A rule of thumb for flush-mounted sensors is to use a distance equal to or greater than twice the diameter of the sensors. Note that when sensors of different diameters are used, you should base the separation distance on the largest diameter. For nonflush mounted sensors, use a distance equal to or greater than three times the diameter of the sensors.
- When inductive and capacitive sensors are installed next to each other, their sensing fields can cause false readings on each other. For this reason, leave a space three times the diameter of the largest sensor between them. When installed opposite to each other, leave a distance of six times the rated sensing distance.
- Where metal chips or similar debris are created, the sensor should be mounted to prevent the chips from building up on the sensor face.
- Avoid installing inductive sensors in locations where the magnetic field from nearby electrical wiring can affect sensor operation.
- Avoid installing sensors in locations where electrical interference generated by nearby motors, solenoids, or relays can have an effect on sensor operation.

6.11.0 Photoelectric Switches/Sensors

Photoelectric sensors (*Figure 65*), like proximity sensors, are solid-state devices used to detect the absence or presence of objects (targets) within their sensing range. Generally, they can detect objects more quickly and at greater distances than proximity switches. A photoelectric sensor sends

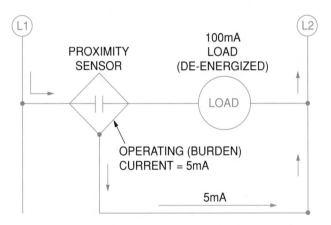

(A) PROXIMITY SENSOR OFF (NO TARGET DETECTED)

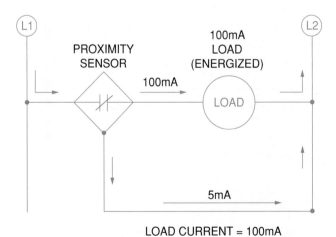

(B) PROXIMITY SENSOR ON (TARGET DETECTED)

311F64.EPS

Figure 64 ◆ Line-powered proximity sensor circuit.

311F65.EPS

Figure 65 ◆ Photoelectric sensor.

out a beam of light, which is picked up by a photodetector. When an object (target) moves into the path of the light beam, the beam is interrupted, indicating that a target is detected.

Figure 66 shows a block diagram for a basic photoelectric switch. As shown, a photoelectric sensor consists of transmitter and receiver circuits and an output switch that is turned on or off by the receiver output. Depending on the application, these circuits can be housed together in the same encapsulated enclosure or be in separate enclosures. Basically, the transmitter emits a beam of light, which is picked up by the receiver. When an object (target) moves into the path of the light beam, the beam is interrupted, indicating that a target has been detected.

The transmitter circuit operates to generate **infrared (IR)** light that is emitted from an LED through a lens placed just behind the face of the sensor. By sensor design, the emitted infrared light beam is either an unmodulated beam or modulated beam. An unmodulated beam is one that is emitted continuously; a modulated beam is one that is turned on and off at a very high frequency. Modulated-beam sensors tend to be more popular since they do not respond to ambient light or other forms of light noise. Unmodulated-beam sensors are typically used where the scanning range is very short and where dirt, dust, and bright ambient light are not a problem.

A photodetector (photodiode or phototransistor) placed just behind a lens at the face of the sensor is the front end of the sensor's receiver circuit. Being light sensitive, the photodetector detects any received infrared light and, in response, produces an output signal. This signal is further processed in the receiver to amplify it to a useable level. In the receiver of a modulated-beam sensor, the amplified received signal is also applied to a demodulator circuit. There, the characteristics of the received signal are compared against those of the original modulated transmitted signal. This comparison helps to select the desired received signals and reject unwanted noise signals. The receiver output is sent to the output switch commanding it on or off, depending on the application.

Photoelectric sensors generate an output any time an object is detected. If this occurs when the photodetector sees light, the sensor is classified as working in the light operate mode. This means that the sensor's output switch is energized when the target is missing. If the sensor generates an output when the photodetector does not see light, the sensor is classified as working in the dark operate mode. This means the sensor's output switch is energized when a target is present (breaks the beam).

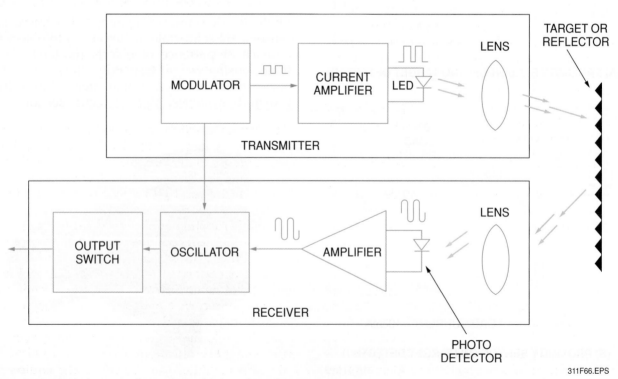

Figure 66 ◆ Block diagram of a modulated-beam photoelectric sensor.

311F66.EPS

Photoelectric sensors can be used in several different scanning arrangements to meet the requirements of different applications. These include:

- *Direct scan (thru-beam)* – The transmitter and receiver are separate assemblies located opposite each other. The light beam from the transmitter is aimed directly at the receiver. The target to be detected passes between them, interrupting the light beam. This is the most basic approach.
- *Retroreflective scan* – The transmitter and receiver are housed in one assembly. The light beam transmitted from the transmitter is reflected to the receiver by means of a reflector located opposite the sensor. The target to be detected passes between the sensor and reflector, interrupting the light beam.
- *Polarized scan* – The transmitter and receiver are housed in one assembly. The light beam transmitted from the sensor is filtered by a special lens so that it projects in one plane only. The receiver responds only to depolarized light reflected from a corner cube reflector or polarized sensitive tape.
- *Specular scan* – The transmitter and receiver are separate assemblies placed at equal angles from a highly reflective surface, requiring the receiver to be positioned precisely to receive reflected light. The transmitter sends the signal to the receiver by reflecting the signal off the reflective surface to be detected.
- *Diffused scan* – The transmitter and receiver are housed in one assembly. The light beam transmitted from the transmitter is reflected back to the receiver by the target to be detected.
- *Convergent scan* – The transmitter and receiver are housed in one assembly. The light beam transmitted from the transmitter is simultaneously focused and converged to a fixed point in front of the sensor. The detection point is fixed such that targets before or beyond the focal point are not detected.

Photoelectric sensors should be selected and installed in accordance with the manufacturer's recommendations. General guidelines for connecting and installing load-powered and line-powered photoelectric sensors are the same as those described for proximity sensors.

7.0.0 ◆ DRUM SWITCHES

Drum switches are totally enclosed, multi-pole switches used in machine operation applications to start, stop, and reverse the direction of rotation of single-phase, three-phase, or DC motors. As shown in *Figure 67*, the switching elements are contained in a cylindrical housing that resembles

311F67.EPS

Figure 67 ◆ Drum switch.

a drum, from which it gets its name. Drum switches are rated by maximum horsepower and are made in several sizes and types of enclosures. They can have either maintained or momentary (spring return to center) contact operation. Drum switches do not contain protective overloads; therefore, circuit overload protection must be provided by installing a manual or magnetic starter in line before the drum switch.

When the motor being controlled by a drum switch is not running in forward or reverse, the handle is in the OFF (center) position. To reverse the direction of a running motor, the handle must first be moved to the OFF position until the motor stops, then moved to the reverse position. Motor reversing occurs in single-phase motors by the drum switch contacts causing the starting winding connections to be reversed. For a three-phase motor, it occurs because the switch contacts reverse two of the three motor leads. For a DC motor, it is done by the switch contacts reversing the direction of current through the motor fields without changing the direction of the current through the armature or vice versa.

8.0.0 ◆ ENCLOSURES

The correct selection and installation of an enclosure for a particular application can contribute considerably to the length of life and trouble-free operation. To shield electrically live contactors/motor starters and pilot devices from accidental contact, some form of enclosure is always necessary. This function is usually filled by a general-purpose sheet-steel cabinet. However, as specified by the *NEC®*, dust, moisture, or explosive gases make it necessary to employ a special enclosure to protect contactors/motor starters and pilot devices from corrosion or the surrounding equipment from explosion.

When selecting and installing any electrical equipment, it is always necessary to carefully consider the conditions under which the equipment must operate so that compliance with the *NEC®* or local codes can be accomplished. A general-purpose enclosure does not afford the required protection in many applications.

Underwriters Laboratories, Inc. has defined the requirements for protective enclosures according to various hazardous conditions, and NEMA has standardized enclosures from these requirements. NEMA enclosure types (*Figure 68*) and the IEC IP enclosure types that they conform to include the following:

- *General-purpose (NEMA Type 1–IP40)* – A general-purpose enclosure is intended primarily to prevent accidental contact with the enclosed apparatus. It is suitable for general-purpose applications indoors where it is not exposed to unusual service conditions. A NEMA Type 1 enclosure serves as protection against dust, light, and indirect splashing but is not dust-tight.

- *Dust-tight, rain-tight (NEMA Type 3–IP52)* – This enclosure is intended to provide suitable protection against rain and dust. It is suitable for application outdoors, such as for construction work.

- *Rainproof, sleet-resistant (NEMA Type 3R–IP52)* – This enclosure protects against interference in operation of the contained equipment due to rain and resists damage from exposure to sleet. It is designed with conduit hubs and external mounting, as well as drainage provisions.

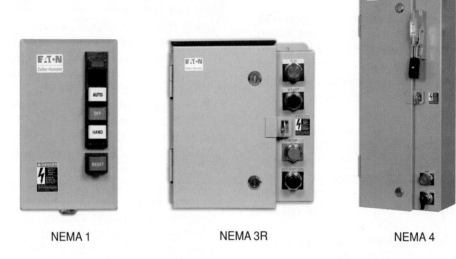

NEMA 1   NEMA 3R   NEMA 4

NEMA 7 AND 9 BOLTED

NEMA 12

311F68.EPS

Figure 68 ◆ Typical NEMA enclosures.

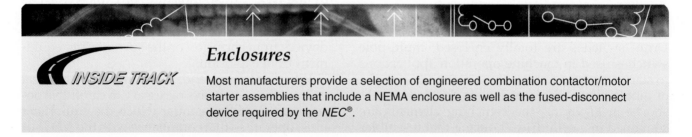

Enclosures

INSIDE TRACK

Most manufacturers provide a selection of engineered combination contactor/motor starter assemblies that include a NEMA enclosure as well as the fused-disconnect device required by the *NEC®*.

- *Watertight (NEMA Type 4–IP65)* – This enclosure is designed to withstand high-pressure hose-directed water and windblown dust or rain. It is constructed of steel with a gasketed cover. Pilot device holes are located on the enclosure cover and are provided with hole plugs.
- *Watertight, corrosion-resistant (NEMA Type 4X–IP65)* – These enclosures are similar to NEMA Type 4 enclosures except that they are made of a material that is highly resistant to corrosion. They are used in applications such as meatpacking and chemical plants, where contaminants would ordinarily destroy a steel enclosure over a period of time.
- *Hazardous locations, Class I (NEMA Type 7C or D)* – These enclosures are designed to meet the application requirements of *NEC®* Class I hazardous locations in which flammable gases or vapors are or may be present in the air in quantities sufficient to produce explosive or ignitable mixtures. They must withstand the pressure generated by explosion of internally trapped gases and be able to contain the explosion so that gases in the surrounding atmosphere are not ignited. The letter(s) following the type number indicate(s) the particular group(s) of hazardous locations (as defined in the *NEC®*) for which the enclosure is designed. The designation is incomplete without the suffix letter(s).
- *Hazardous locations, Class II (NEMA 7 Type 9E, F, or G)* – These enclosures are designed to meet the application requirements of *NEC®* Class II locations for operation in the presence of combustible dust. The letter(s) following the type number indicate(s) the particular group(s) of hazardous locations (as defined in the *NEC®*) for which the enclosure is designed. The designation is incomplete without the suffix letter(s).
- *Industrial use (NEMA Type 12–IP62)* – This type of enclosure is designed to exclude dust, lint, fibers and other flying materials, and oil or coolant seepage. There are no conduit openings or knockouts in the enclosure, and mounting is by means of flanges or mounting feet. Pilot device holes are located on the enclosure cover and are provided with hole plugs.
- *Oil-tight, dust-tight (NEMA Type 13–IP62)* – These enclosures are generally made of cast iron and are used in the same areas as NEMA Type 12 enclosures. The main difference is that, due to its cast housing, a conduit entry is provided as an integral part of the NEMA Type 13 enclosure, and mounting is by means of blind holes rather than mounting brackets. Pilot device holes are located on the enclosure cover and are provided with hole plugs.

IEC-compliant enclosures are identified by a two-digit IP protection rating system that is defined in *Table 12*. Additional information for selecting enclosures for use in specific locations other than hazardous locations can be found in *NEC Section 430.91* and *NEC Table 430.91*.

Table 12 IEC Enclosure IP Two-Digit Protection Code Definitions

First Digit	Description	Second Digit	Description
0	No protection	0	No protection
1	Protection against solid objects greater than 50mm	1	Protection against vertically falling drops of water
2	Protection against solid objects greater than 12mm	2	Protection against dripping water when tilted up to 15°
3	Protection against solid objects greater than 2.5mm	3	Protection against spraying water
4	Protection against solid objects greater than 1mm	4	Protection against splashing water
5	Total protection against dust; limited ingress (dust protected)	5	Protection against water jets
6	Total protection against dust (dust tight)	6	Protection against heavy seas
		7	Protection against the effects of immersion
		8	Protection against submersion

9.0.0 ◆ DIAGRAMS

The wiring scheme for motor control circuits and their components can be shown in several forms. These include wiring diagrams, circuit schedules, control ladder diagrams, and logic diagrams.

9.1.0 Wiring Diagrams

Wiring diagrams, commonly called connection diagrams, show the actual point-to-point wiring connections for a motor-driven device and its control circuits. This type of diagram is normally found in the service literature for a specific piece of factory-wired equipment. Connection diagrams are useful when it is necessary to trace and locate a specific wire connected between two components in a circuit, such as might be required when it is necessary to find and replace an open or grounded wire.

Figure 69 shows an example of a typical connection diagram. The format of connection diagrams varies depending on the type of equipment and manufacturer. Some connection diagrams show power-circuit wiring as heavy-weight lines and control circuit wiring as lighter-weight lines. Some also show the actual color of each wire in the point-to-point wiring scheme. Where applicable, the component terminal identification number or letter to which a specific wire is to be connected is shown. To help the technician interpret the connection diagram, it normally has a legend that describes any nonstandard symbols and abbreviations that are used on the diagram.

9.2.0 Circuit Schedules

Circuit schedules, sometimes called wire lists, show actual wire or cable point-to-point connections for a motor and its control circuits. They are used mainly when installing a new system or modifying an existing one. Connection information in a circuit schedule is typically presented in tabular form; each individual wire or cable is given a sequence or circuit identification number. The information given normally includes the connection points for the wire and the type and size of the wire. Some circuit schedules also include the color of the wire and type(s) of termination used at each end of the wire. *Table 13* shows an example of a typical circuit schedule.

9.3.0 Control Ladder Diagrams

Control circuits that use various discrete components such as pilot devices and relay coils/contacts and/or those that use a microprocessor to energize contactor/motor starter coils are generally presented as control ladder (line) diagrams (*Figure 70*). Without identifying them as such, ladder diagrams have been used in many illustrations of control circuits presented earlier in this module. A ladder diagram shows the operational sequence of an electrical circuit using single lines and symbols.

9.3.1 Sections of a Ladder Diagram

All ladder diagrams are usually presented in three sections between the two wires from the source of the control voltage, as shown in *Figure 71*. The three sections consist of a signal source section, a decision section (or combined signal/decision section), and an action section.

- *Signal section* – Pilot devices are the source of the control signal. These devices can be manual (pushbuttons or foot switches), mechanical (limit switch), or automatic reset devices (flow, pressure, or temperature switches).

Table 13 Example of a Circuit Schedule

Circuit/Cable ID	Cable Type	Length	From	To
P108	3/c #10 w/grd	140'	MCC 10 Section 4A	M108
C108	9/c #14	120'	MCC 10 Section 4A	PLC10-TB1
C108A	12/c #14	160'	MCC 10 Section 4A	JB108
C108B	5/c #14	40'	JB108	HS108
C108C	3/c #14	30'	JB108	HSS108A
C108D	3/c #14	10'	HSS108A	HSS108B
C108E	3/c #14	30'	JB108	AA108A
C108F	3/c #14	80'	JB108	AA108B
C108G	5/c #14	70'	JB108	SSL108
C108H	3/c #14	10'	JB108	ZS108A
C108J	3/c #14	10'	ZS108A	ZS108B
C108K	3/c #14	60'	JB108	ZS108C
C108L	3/c #14	10'	ZS108C	ZS108D

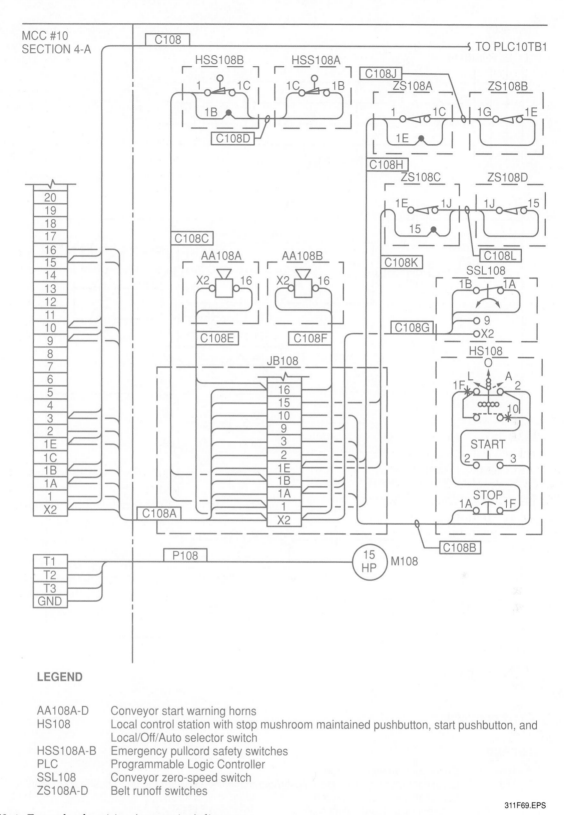

Figure 69 ◆ Example of a wiring (connection) diagram.

LEGEND

AA108A-D	Conveyor start warning horns
HS108	Local control station with stop mushroom maintained pushbutton, start pushbutton, and Local/Off/Auto selector switch
HSS108A-B	Emergency pullcord safety switches
PLC	Programmable Logic Controller
SSL108	Conveyor zero-speed switch
ZS108A-D	Belt runoff switches

311F69.EPS

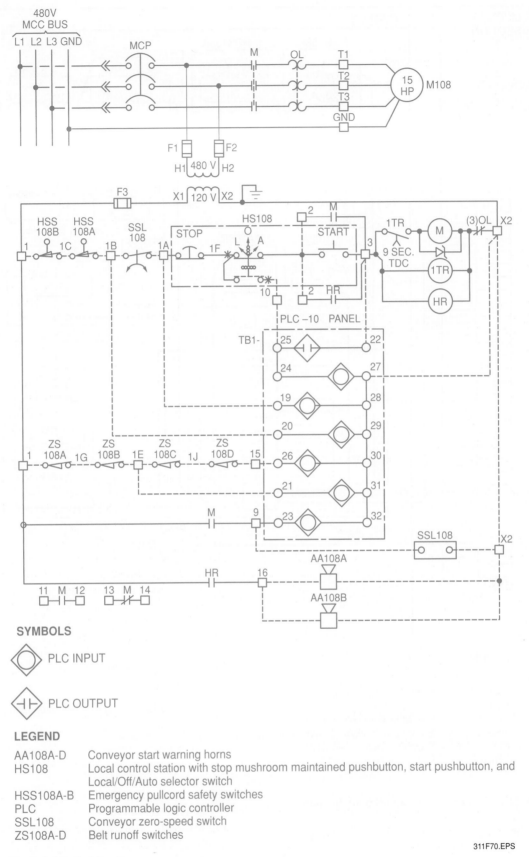

SYMBOLS

⬦ PLC INPUT

⬦ PLC OUTPUT

LEGEND

AA108A-D	Conveyor start warning horns
HS108	Local control station with stop mushroom maintained pushbutton, start pushbutton, and Local/Off/Auto selector switch
HSS108A-B	Emergency pullcord safety switches
PLC	Programmable logic controller
SSL108	Conveyor zero-speed switch
ZS108A-D	Belt runoff switches

311F70.EPS

Figure 70 ◆ Example of a ladder diagram.

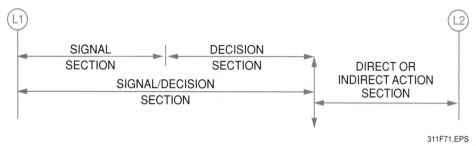

Figure 71 ◆ Typical ladder diagram sections.

- *Decision section* – The decision section of the circuit adds, sorts, selects, and redirects the signals from the pilot devices to the load. The function of this area must be logical. In many cases, the pilot devices are part of the decision section. The way pilot device contacts are connected in the circuit provides the circuit logic. There are three basic logic functions: AND, OR, and NOT. The electrical relay logic symbols for these logic functions, their related truth tables, and logic symbols are shown in *Figure 72* where contacts A and B represent pilot device contacts. The status of each set of contacts (open or closed) is represented in the truth table by a zero (0) when the contacts are open and by a one (1) when they are closed. The status of the output, on (1) or off (0), called *Y*, is shown for all the possible combinations of contact A and B positions.
- The logic symbols used in *Figure 72* for the AND, OR, NOT, NAND, and NOR logic functions are a version of NEMA logic symbols that are typical of those used for a motor control circuit. It's important to point out that the particular logic symbols used on a diagram can be different depending on whether the diagram is related to electrical, electronic, hydraulic, pneumatic, or some other type of equipment. Even for similar types of equipment, the symbols used by different manufacturers may vary.
- As shown in *Figure 72*, the electrical equivalent of an AND function is two normally open sets of contacts (or switches) connected in series. As indicated by its truth table, the AND function has an output that is on (1) only when both contacts A and B are closed (1). This is because the related coil CR is energized, causing its normally open contacts to close. For all the other possible contact position combinations, the Y output remains off (0). This is because coil CR is de-energized, resulting in its normally open contacts remaining open.
- The electrical equivalent of an OR function is two normally open sets of contacts connected in parallel. An OR function is so called because it will produce a 1 output (on) whenever contact

A *or* contact B is closed (1). Output Y will be a 0 only when both contacts A and B are open (0).
- The third basic logic function is the NOT function, commonly called an inverter. Note that it has only one normally open contact. The function of the NOT circuit is simple. It operates to produce an output that is always opposite the contact position. Thus, when contact A is closed (1), output Y is 0 (off): when contact A is open (0), output Y is 1 (on). For the circuit shown, the inversion is achieved by the open (normally closed) contacts of coil CR.
- A NOR (NOT-OR) function is the same as an inverted OR function. It consists of an OR function combined with an inverter (NOT) function. Comparison of the NOR truth table with the OR truth table shows that for the same contact A and B positions, the output Y for the NOR is always opposite (inverted) from that of the basic OR function.
- A NAND (NOT-AND) function is the same as an inverted AND function. It consists of an AND function combined with an inverter (NOT) function. Comparison of the NAND truth table with the AND truth table shows that for the same contact A and B positions, the output Y for the NAND is always opposite from that of the basic AND function.
- Except for the NOT function, any of the functions just described can have more than two contacts added in series or in parallel with those shown in the figure, and the truth table can be expanded to include them. For example, if you have three series-connected contacts in an AND function, all three contacts must be closed to get a Y output of 1. Otherwise, the output is 0. Similarly, three parallel-connected contacts in an OR function would mean that the Y output is 1 if any one of the three contacts is closed.
- Up to this point, relay coils and contacts have been used to describe the operation of the basic AND, OR, NOT, NOR, and NAND logic functions. These logic functions are commonly incorporated electronically into integrated chips, PLCs, and similar devices used in modern solid-state motor control equipment.

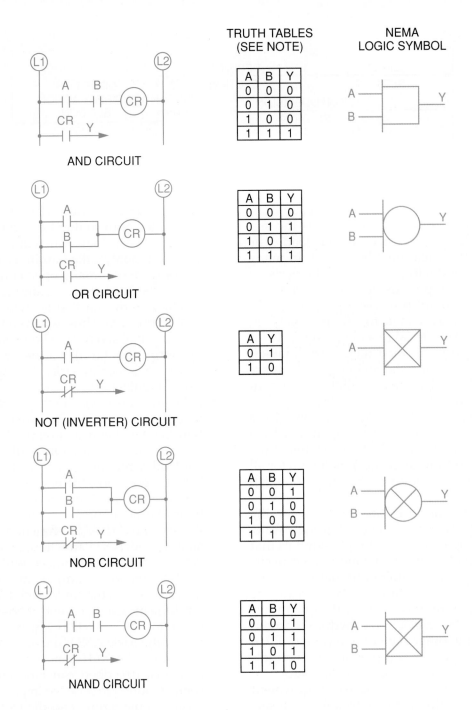

AND CIRCUIT

A	B	Y
0	0	0
0	1	0
1	0	0
1	1	1

OR CIRCUIT

A	B	Y
0	0	0
0	1	1
1	0	1
1	1	1

NOT (INVERTER) CIRCUIT

A	Y
0	1
1	0

NOR CIRCUIT

A	B	Y
0	0	1
0	1	0
1	0	0
1	1	0

NAND CIRCUIT

A	B	Y
0	0	1
0	1	1
1	0	1
1	1	0

NOTE:
When contacts A and/or B are shown as zero (0) in the truth table, it means that the contacts are open. If shown as one (1), it means that the contacts are closed.

311F72.EPS

Figure 72 ◆ Basic logic functions.

- *Action section* – The action section is usually the energization of a relay or contactor coil. As shown in the truth tables, when Y is 1, power is applied to the coil, and when it is 0, the coil is de-energized.

As a simple example, examination of *Figure 73* reveals an AND function overlapped by an OR function. The AND function is the STOP/START switch and MOL contact. Both the STOP contacts and the START contacts and the MOL contacts must be closed to energize the contactor coil M. Once energized, an OR function consisting of the START switch or the M holding contacts keeps the coil energized.

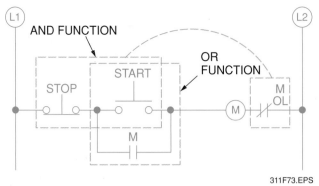

Figure 73 ◆ Example AND function and OR function in a motor-starter control circuit.

9.3.2 Conventions Used on Ladder Diagrams

Any ladder diagram normally references the type of machine or process to which the diagram applies. Careful study of the applicable ladder diagram can help the technician determine or confirm the correct sequence of operation for a particular piece of motor control equipment. As previously described, ladder diagrams are divided into sections horizontally and read as follows:

- *Left to right* – Control devices are connected between L1 and the operating coil (or load) from left to right in accordance with the following rules:
  - All control functions shall occur on the left side (L1 side) of the coil.
  - Only overload contacts are connected on the right side (L2 side) of the coil.
  - Stop and shutdown circuits should occur before start and run circuits (closer to L1).
  - Where practical, external connections such as buttons and switches should occur before internal connections such as relay connections.
- *Top to bottom* – Ladder diagrams are usually reference-line numbered from top to bottom with the first function starting on line one and subsequent or dependent functions proceeding down the ladder until the last function occurs.

Figure 74 shows the conventions that should be used on a typical ladder diagram. However, diagrams furnished by manufacturers or designers vary widely and not all the conventions pointed out on the figure may appear in all diagrams for similar functions. The conventions that should be used on a diagram are as follows:

- *Component identification* – The diagram should identify by reference designator (PB1, S1, CR1) and/or by function (low-pressure switch, on-delay timer) the various control circuit devices and loads. If a control device is remotely located from the main equipment enclosure, the diagram should also identify the location of a device.
- *Line fuse* – Many codes require that line-voltage control circuits must be fused if they exist outside the enclosure for the applicable relays or contactors, or if a step-down transformer is not power limited or is located outside the enclosure.
- *Step-down transformer and fuse* – In many cases, a low-voltage step-down transformer along with a fuse may be required for control circuits. If not, these devices are omitted and line voltage is used as the control voltage.
- *Component terminal numbers* – If a component has terminals that are marked with numbers or letters, they should be shown on the diagram.
- *Ladder line numbers* – The horizontal lines on the ladder should be numbered sequentially from top to bottom for contact cross-reference purposes. This aids in locating all contacts of a relay, contactor, or other device.
- *Mechanical linkage symbols* – If the items are located on adjacent lines, a dashed line may be used to indicate the linkage. An alternate approach is the arrowed box containing the ladder line reference to the location of the other linked device. Appropriately numbered arrowed boxes are used at both locations of the linked item.
- *Wire numbers* – Unique wire numbers are assigned to each wire or group of wires that are electrically common. Labeling the wires in the actual equipment with these numbers eases the installation and future troubleshooting.
- *Line references for contacts* – Relays and most contactors/motor starters have one or more sets of control circuit contacts. At the right-hand side of the line containing the coil for the device, the location of any control contacts not shown on the same line are referenced by ladder line numbers, separated by commas. Underlined line numbers are for any N.C. contacts, and nonunderlined numbers are for any N.O. contacts.

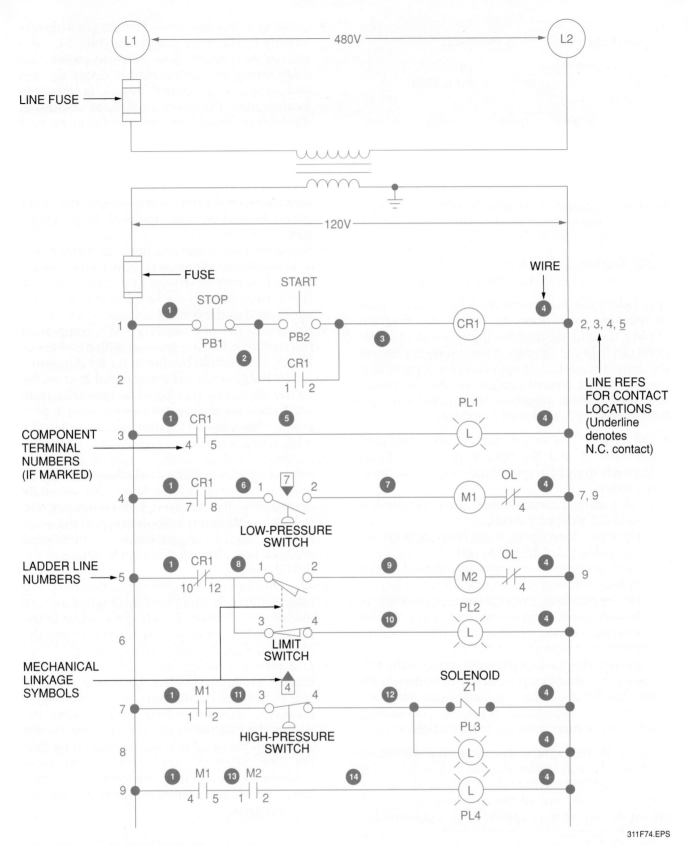

Figure 74 ◆ Typical ladder diagram conventions.

311F74.EPS

Fail-Safe Designs

Motor controls must be fail-safe. This means that in the event of a power loss, the equipment should be designed so that it cannot be re-energized upon the restoration of power without operator intervention. This is necessary in order to protect personnel from injury and/or the equipment from damage should the power be inadvertently reapplied, such as often occurs during power failures associated with electrical storms. Relays, contactors, and similar control devices should be wired so that if they fail, the operation of the equipment will cease, or an appropriate alarm will be generated to alert personnel to the failure. Otherwise, the failure of a device that provides for personnel and/or equipment protection can go undetected.

9.4.0 Logic Diagrams

Logic diagrams show a pictorial representation of circuit elements in the form of logic symbols in order to convey the logic behind the operation for a particular system, machine, or network. Typically they are used to communicate logic requirements to a control systems supplier or programmer. They are also used collectively by engineers, designers, technicians, and operators to agree on the rules of system operation before programming and circuit design of a system begin. Logic diagrams are useful for technicians maintaining a piece of equipment because they show the decision-making processes that are made by the equipment with regard to system operation and sequencing. These decision processes often are not obvious by the examination of a related ladder or schematic diagram because many are accomplished solely by the system software.

The logic symbols used on logic diagrams do not represent the actual types of electronic components used, but represent only their logical functions (AND, OR, NOR, NAND, NOR, INVERTER). As mentioned earlier, there are many styles of logic symbols used on logic diagrams. The specific symbols used are different, depending on whether the diagram is related to electrical, electronic, hydraulic, pneumatic, or some other type of equipment. Even for similar types of equipment, the symbols used by different manufacturers may vary. *Figure 75* shows a typical logic diagram of a motor control circuit. The symbols used on this diagram are the same NEMA logic symbols described earlier in this module.

The presentation of information on a logic diagram typically flows from the top of the page to the bottom. Logic functions associated with permissive signals normally are shown first; logic functions associated with the generation of running signals are shown second; and logic functions associated with the generation of miscellaneous signals, such as alarm signals, are shown last. Permissive signals typically are ones generated by safety switches, stop switches, mode switches, or related software that function to enable or inhibit machine or system operation, depending on their status. Running signals are those signals that cause the machine to operate and control its operation.

The logic flow path shown on the diagram is from left to right. Sometimes the output line from the last logic element on the right-hand side of the drawing may loop back to the left side of the drawing if this is necessary to continue showing the remainder of a logic function. Labels shown on the left side of the diagram normally identify a device name or signal name input to the logic element; those on the right side show the output or action from a logic element. Note that the inputs and outputs shown on a logic diagram can be applied from or sent to hardware devices, while others may be applied from or sent to the system-controlling software. Understanding the information shown on a logic diagram will help the electrician to:

- Understand how a programmable logic controller (PLC) controls the operation of a system or device without the need for sorting through the specific PLC programming details.
- Understand the details of system operation that are not obvious by using a ladder diagram.
- Generate a ladder diagram.
- Determine the status of field devices shown on the drawings.
- Determine how and where to implement wiring modifications because of changes in operating requirements.
- Explain to others how the equipment works.

9.5.0 Relating Diagrams to Equipment Wiring and Operation

Up to this point, you have been introduced to the different types of diagrams that can be used to show the wiring and operation for a system or piece of equipment. The diagrams and circuit

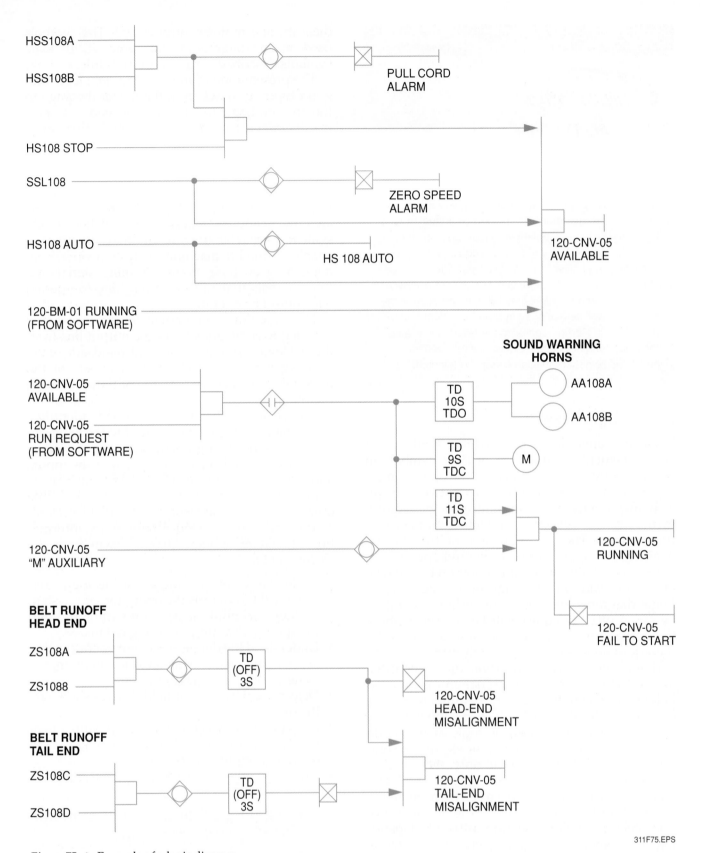

Figure 75 ◆ Example of a logic diagram.

311F75.EPS

schedules shown earlier in this module all are related to a mill conveyor system, designated as 120-CNV-05. It has a 15hp head-mounted drive motor that serves to transfer screened oversize materials to a ball mill, designated as 120-BM-01. Operation of this conveyor system is described in the following section. Based on this description of system operation, you should examine the different diagrams and relate what they show to the actual equipment and its operation. The connection diagram and circuit schedule (*Figure 69* and *Table 13*) show the wiring connections between the different components of the conveyor system. The ladder and logic diagrams (*Figures 70* and *75*) show the control sequence and design logic, respectively, pertaining to the equipment operation.

9.5.1 Relating Ladder Diagrams to System Operation

Refer to the ladder diagram shown in *Figure 70* while following the description of conveyor system operation given here. Placing the LOCAL/OFF/AUTO switch (part of HS108) in the AUTO position selects the normal operation mode for the conveyor. In this mode, the conveyor is controlled by the programmable logic controller (PLC10) that functions to provide sequential interlocking for the 120-BM-01 ball mill.

Emergency pull-cord safety switches (HSS108A and HSS108B) prevent the conveyor motor from starting if one or both are activated (contacts open). If activated when the conveyor motor is running, interruption of control power will cause the conveyor motor to stop immediately. It will also cause the PLC to generate an alarm.

A conveyor zero-speed switch (SSL108) is mounted on the tail end of the belt pulley. It provides protection against belt freeze-up or slippage. This switch is energized when the conveyor motor is started. It contains an internal time-on delay circuit whose contacts allow sufficient time (typically one to five seconds depending on belt length) for the conveyor belt to accelerate to normal speed after the conveyor is turned on. If the conveyor belt is rotating at the correct speed at the end of the delay period, the contacts remain closed, allowing the conveyor motor to continue operation. Should belt rotation slow down below the correct speed at any time or if belt rotation does not begin within the delay period, the switch contacts open, opening the control circuit and causing the motor to turn off and an alarm to be generated via the PLC.

Belt runoff switches (ZS108A through ZS108D) serve to detect belt misalignment (side travel) at the head end (ZS108A and ZS108B) and tail end (ZS108C and ZS108D) of the conveyor. Should a misalignment be detected by one or more of the switches, the switch signals cause the PLC to generate an alarm. They do not cause the conveyor motor to be turned off.

Conveyor start warning horns (AA108A and AA108B) provide a local area alarm that sounds for ten seconds as controlled by horn relay HR when a motor start command is initiated. This warns people that the conveyor is about to be started. Starting of the conveyor motor is delayed for nine seconds by time delay relay 1TR after a start command is issued.

Placing the LOCAL/OFF/AUTO switch (part of HS108) in the LOCAL position allows the conveyor to be operated for maintenance purposes independent of PLC control and its sequential interlocking. The safety features provided by the emergency pull-cord switches, zero-speed switch, and conveyor start warning horns remain active when in the local mode. The functions of belt runoff switches are inactive during the local mode of operation.

9.5.2 Relating Logic Diagrams to System Operation

Remember that the purpose of a logic diagram is to convey the design logic behind the operation for a particular system, machine, or network. It is important to point out that many of the functions shown on a logic diagram may be accomplished by the system operating software, rather than by hardware. For this reason, there is not always a direct correlation between what is shown on the logic diagram and what is shown on the related ladder diagram. In order to get the entire picture of how a particular system or machine operates, you should refer to both types of diagrams.

Refer to the logic diagram shown in *Figure 75* for the conveyor system. The top part of the logic diagram shows the logic elements that develop the inputs to a four-input AND. As shown, the field inputs to this part of the logic diagram are applied from the system emergency pull cord safety switches (HSS108A and HSS108B), the STOP button (part of HS108), the conveyor zero-speed switch (SSL108), and the AUTO switch (part of HS108). Also applied from the system software is an input (interlock input) signaling that the 120-BM-01 ball mill is running. Each of these permissive signal inputs is applied from a contact that must be closed (supplying a logical one) in order for the conveyor to run.

Tracing the emergency pull cord safety switch and the STOP button inputs through their respective two-input AND logic gates shows that when all three input signals are a logical one, the related input to the four-input AND is a logical one. The remaining three inputs applied to the four-input AND from the conveyor zero-speed switch, AUTO switch, and the system software are all a logical one. This causes the output signal from the four-input AND to be a logical one. This signal (*120-CVN-05 available*) is applied as an input to the next group of decision logic. Should any one of the input signals be a logical zero, the output from the four-input AND will be a logical zero, indicating that the conveyor system is not available for use.

The middle of the diagram shows the development of the system running control signals. The first logic decision made here is whether the conveyor system is ready to run and whether the software or operator wants it to run. As shown, this decision is made by a two-input AND gate. The decision depends on the status of its input signals, *120-CVN-05 available* and *120-CVN-05 run request* applied from the software. If both inputs are a logical one, the output from the AND is a logical one. It causes a PLC contact closure that begins the conveyor starting sequence by enabling a ten-second, timed open, timing function that causes the conveyor start warning horns to sound for ten seconds. A look at the ladder diagram shows that this function is accomplished in the conveyor system hardware by the ten-second horn relay (HR) and the conveyor start warning horns (AA108A and AA108B). Simultaneously, a nine-second, timed closed, timing function is also enabled that causes the conveyor motor contactor (M) to energize after a nine-second delay, causing the conveyor motor to turn on. In the hardware, this function is performed by nine-second timing relay 1TR and contactor M.

The remainder of the decision logic shown in this section consists of an 11-second, time closed, time delay (TD) function; a PLC input; a two-input AND gate; and an inverter. The purpose of this logic is to verify that when the conveyor is commanded to run, it is indeed running. If the input to the 11-second TD function is a logical one for 11 seconds after conveyor turn on, the TD function outputs a logical one to the two-input AND. The purpose for the 11-second delay is to allow time for contactor M to pull in and the internal time delay of the conveyor zero-speed switch (SSL108) to expire. The second input to the two-input AND (*input 120-CNV-05 M aux*) via the PLC is the status of the contactor M auxiliary contact signal. With both inputs to the two-input AND at a logical one, it outputs a logical one signal (*120-CNV-05 running*) to the software that verifies the conveyor is running. If either the TD function output or the *120-CNV-05 M aux* signal input is a logical 0, the AND gate output (*120-CNV-05 running*) will be a logical zero, indicating that the conveyor is not running. The logical zero output from the AND gate is applied through an inverter to generate a logical one *120-CVN-05 fail to start signal* for application to the software.

The bottom of the diagram shows the decision logic used to determine if a belt runoff misalignment condition exists, and if it does, whether the problem is caused by a head-end misalignment or a tail-end misalignment. The individual status of belt runoff switches ZS108A and B (head end) and Z108C and D (tail end) are applied as inputs to two-input AND gates. The outputs from the head-end and tail-end AND gates are applied via PLC inputs to three-second, timed off, time-delay (TD) functions. The purpose for using the three-second, timed off TD functions is to eliminate the generation of nuisance alarms.

When no conveyor belt misalignment exists, the inputs applied to both three-second TD functions are a logical one, causing the outputs from both three-second TD functions to be a logical one. Should a head-end belt misalignment exist (or both a head-end and tail-end belt misalignment exist) for more than three seconds, the outputs from both three-second TD functions are a logical zero. Both are a logical zero in this case because the tail-end run-off switches (Z108C and D) are in series with the head-end run-off switches (ZS108A and B), as shown on the ladder diagram. Therefore, if the head-end input is open, the tail-end input will also be open. When only a tail-end belt misalignment occurs, the output of the head-end three-second TD function will be a logical one, and the output of the tail-end three-second TD function will be a logical zero. The statuses of the three-second TD function outputs for the various belt conditions are summarized in *Table 14*.

As shown, the output from the head-end three-second TD function is applied as an input to an inverter and to one input of a two-input AND gate. The output from the tail-end three-second TD function is applied through an inverter as the second input to the two-input AND gate. The resultant logic status of the inverter output signal (*120-CNV-05 head-end misalignment*) and the AND gate output signal (*120-CNV-05 tail-end misalignment*) for the different belt conditions is shown in the truth table given above. These signals are used by the PLC software to determine the condition of belt alignment.

Table 14 Three-Second TD Function Outputs

Belt Alignment Condition	Logical Signal Output from Head-End Three-Second Timed Off Function	Logical Signal Output from Tail-End Three-Second Timed Off Function	120-CNV-05 Head-End Misalignment Logical Signal Output	120-CNV-05 Tail-End Misalignment Logical Signal Output
No belt misalignment	1	1	0 (no alarm)	0 (no alarm)
Head-end belt misalignment or both head-end and tail-end belt misalignment	0	0	1 (alarm)	0 (no alarm)
Tail-end belt misalignment	1	0	0	1 (alarm)

10.0.0 ◆ *NEC*® REGULATIONS FOR THE INSTALLATION OF MOTOR CONTROL CIRCUITS

NEC Article 430 governs the installation of motors, motor circuits, and controllers. Where equipment incorporates a hermetic refrigerant compressor, such as in refrigeration or air conditioning equipment, *NEC Article 440* also applies. Other articles that pertain to motor circuits involving special equipment and hazardous locations are referenced in *NEC Table 430.5*. You have studied the *NEC*® requirements for motor circuits in your earlier training on motors. *Figure 76* shows a summary of the *NEC Article 430* requirements that apply to the installation of motors, motor circuits, and their controllers. Detailed information can be found in the *NEC*® under the articles or sections referenced in the diagram.

Some *NEC Article 430* requirements pertaining to motor controllers and related control circuit devices are overviewed here:

- A magnetic starter cannot serve as a disconnecting means.
- Controllers must be able to start and stop the motors they control and be able to interrupt the locked rotor current of the motor.
- A controller must have a horsepower rating not lower than the rating of the motor it controls. Controllers used with Design B energy-efficient motors rated up to 100hp shall be marked by the manufacturer as rated for use with Design B energy-efficient motors or have a horsepower rating not less than 1.4 times the rating of the motor. For Design B energy-efficient motors

over 100hp, the rating shall be 1.3 times the motor rating.
- Branch circuit protective devices (circuit breaker, fuse) can be used as the controller for small stationary motors (⅛hp or less) and portable motors (⅓hp or less). A general-use switch rated at not less than twice the full loaded motor current may be used with stationary motors up to 2hp, rated at 300V or less.
- Where a controller does not also serve as both the controller and disconnect switch, and a fused switch provides the required disconnect as well as the circuit short circuit and overload protection, the controller need not open all conductors to the motor. It needs to open only enough conductors to stop and start the motor.
- The rating of time-delay fuses used in a combination fuseholder and switch as the motor-running protective devices must not exceed 125% or, in some cases, 115% of the full-load motor current.
- *NEC Table 110.20* provides the selection criteria for selecting the type of enclosure to use for controllers in nonhazardous (classified) locations.
- All conductors of a remote motor control circuit outside of the controller, such as to remote pushbutton stations, must be installed in a raceway or otherwise protected.
- Where one side of the motor control circuit is grounded, the circuit must be wired so that an accidental ground in the control device or related wiring will not start the motor or bypass manually operated shutdown devices or automatic safety shutdown devices.

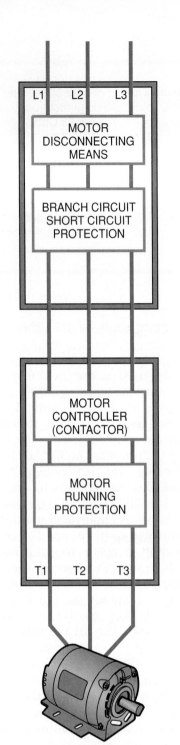

NEC Article 430 Part IX
Sections 430.101 through 430.113

Disconnects motor and controllers from circuit.
1. Continuous rating of 115% or more of motor FLC. Also see *NEC Article 430 Part II*.
2. Disconnecting means shall be as listed in *NEC Section 430.109*.
3. Must be located in sight of motor location and driven machinery. The controller disconnecting means can serve as the disconnecting means if the controller disconnect is located in sight of the motor location and driven machinery.

NEC Article 430 Part IV
Sections 430.51 through 430.58

Protects branch circuit from short circuits or grounds.
1. Must carry starting current of motor.
2. Rating must not exceed values in *NEC Table 430.52* unless not sufficient to carry starting current of motor.
3. Values of branch circuit protective devices shall in no case exceed exceptions listed in *NEC Section 430.52*.

NEC Article 430 Part VII
Sections 430.81 through 430.90

Used to start and stop motors.
1. Must have current rating of 100% or more of motor FLC.
2. Must be able to interrupt LRC.
3. Must be rated as specified in *NEC Section 430.83*.

NEC Article 430 Part III
Sections 430.31 through 430.44

Protects motor and controller against excessive heat due to motor overload.
1. Must trip at following percent or less of motor FLC for continuous motors rated more than one horsepower.
   (a) 125% FLC for motors with a marked service factor of not less than 1.15 or a marked temperature rise of not over 40°C.
   (b) 115% FLC for all others. (See *NEC* for other types of protection.)
2. Three thermal units required for any three-phase AC motor.
3. Must allow motor to start.
4. Select size from FLC on motor nameplate.

NEC Article 430 Part II
Sections 430.21 through 430.29

Specifies the sizes of conductors capable of carrying the motor current without overheating.
1. To determine the ampacity of conductors, switches, branch circuit overcurrent devices, etc., the full-load current values given in *NEC Tables 430.247 thru 430.250* shall be used instead of the actual current rating marked on the motor nameplate. *(See NEC Section 430.6.)*
2. According to *NEC Section 430.22*, branch circuit conductors supplying a single motor shall have an ampacity of not less than 125% of motor FLC, as determined by *NEC Section 430.6(A)(1)*.

311F76.EPS

Figure 76 ◆ Summary of requirements for motors, motor circuits, and controllers.

11.0.0 ◆ CONNECTING MOTOR CONTROLLERS FOR SPECIFIC APPLICATIONS

Motor controllers and their related control circuits can be connected in many ways to satisfy the requirements needed for specific applications. This section gives a few examples of basic control circuit wiring of controllers for some common applications including:

- Controlling an air compressor motor
- Controlling a pump motor
- Controlling two pump motors
- Controlling a motor from multiple locations
- Controlling the reversing of a three-phase motor
- Controlling conveyor system motors

11.1.0 Controlling an Air Compressor Motor

Controlling an air compressor motor is done in a similar way as controlling an AC motor with a contactor. However, an N.C. pressure switch (high-pressure switch) is added in the control circuit to cause the compressor motor to turn on and off, depending on the amount of air pressure in the system. *Figure 77* shows a line diagram for this circuit. In this circuit, setting the ON/OFF switch to ON allows current to pass from L1 through the N.C. pressure switch contacts, the contactor coil C, N.C. overloads, and on to L2. This energizes the contactor coil, closing its N.O. power contacts (not shown). This turns on the air compressor motor.

The air compressor motor will continue to run until the air pressure in the system builds up to and exceeds the set point of the pressure switch. When this occurs, the N.C. contacts of the pressure switch open, causing the contactor coil to de-energize and its related N.O. power contacts to open. This turns off the air compressor motor. It will remain turned off until the pressure in the system falls below the pressure switch reset point, causing the N.C. contacts of the pressure switch to close again. This energizes the contactor coil again

and turns on the air compressor motor. This automatic sequence of turning on and off the air compressor motor in response to the system pressure level continues until the ON/OFF switch is set to OFF.

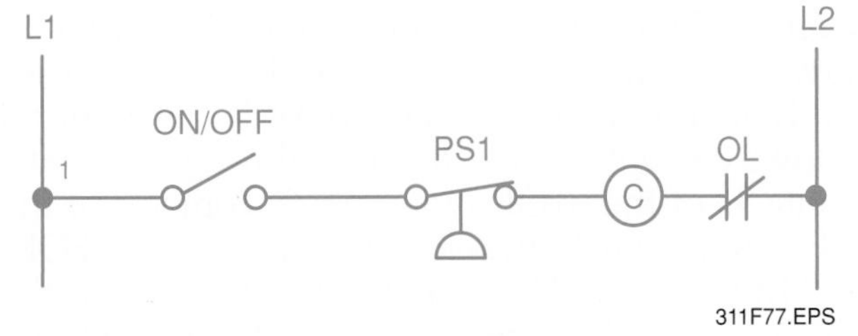

11.2.0 Controlling One-Pump Motors

A control circuit used for controlling a pump that maintains the level of water (or any liquid) in a tank or other vessel can use one or more float switches. *Figure 78* shows the line diagram for a liquid pump being controlled by a HAND/OFF/AUTO switch and a single float switch. When the HAND/OFF/AUTO switch in this circuit is set to OFF, no current can flow through the contactor coil to L2; therefore, the pump is turned off. When the switch is placed in the HAND position, current from L1 can pass through the N.O. switch contacts, the contactor coil C, N.C. overloads, and on to L2. This energizes the contactor coil, closing its N.O. power contacts (not shown), thus turning on the pump motor. The pump motor will continue running until the HAND/OFF/AUTO switch is set to OFF. This provides a manual mode of operation.

When the HAND/OFF/AUTO switch is set to AUTO, current can flow from L1 through the closed switch contacts, the N.C. float switch contacts, the contactor coil, N.C. overloads, and on to L2. This energizes the contactor coil, closing its N.O. power contacts, causing the pump motor to turn on. The pump will continue running until either the HAND/OFF/AUTO switch is set to OFF or the water in the tank reaches a level high enough to actuate the float switch, causing its N.C. contacts to open. When this occurs, the contactor coil is de-energized and its N.O. power contacts open. This turns off the pump motor. The pump motor will remain turned off until the water level in the tank falls below the float switch reset point, causing the N.C. contacts of the float switch to close again. In turn, the contactor coil is energized again and turns on the pump motor.

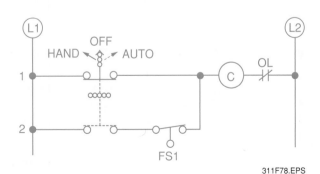

311F77.EPS

Figure 77 ◆ Control circuit for an air compressor.

311F78.EPS

Figure 78 ◆ Control circuit for a pump motor.

This automatic sequence of turning on and off the pump motor in response to the water level in the tank continues until the HAND/OFF/AUTO switch is set to OFF.

11.3.0 Controlling Two-Pump Motors

For the purpose of an example, assume that there is a requirement that involves controlling two-pump motors so that they alternately operate to automatically fill and drain a tank, based on the level of the liquid in the tank. *Figure 79* shows one way of connecting such a circuit. Pressing the START pushbutton in this circuit allows current to flow from L1 through the STOP switch N.C. contacts, the closed START switch contacts, the control relay coil CR, and on to L2. This energizes the control relay, causing its three sets of N.O. contacts to close. Assuming the tank is empty or only partially full, coil M2 is de-energized. This allows current to flow from L1 through the closed control relay contacts, closed M2 auxiliary contacts, the N.C. float switch contacts, coil M1, N.C. overloads, and on to L2. This energizes coil M1, closing its N.O. power contacts (not shown) to turn on the fill pump motor. It also causes its N.C. auxiliary contacts to open. This provides an electrical interlock in the control circuit for coil M2 that prevents coil M2 from being energized at the same time as coil M1. The fill pump motor will continue to operate until the water in the tank reaches a level high enough to actuate the float switch, causing its N.C. contacts to open and its N.O. contacts to close. This causes coil M1 to de-energize and its related N.O. power contacts to open, turning off the fill pump motor.

Activation of the float switch also allows current to flow from L1 through the closed control relay contacts, closed auxiliary M1 contacts, the closed N.O. float switch contacts, coil M2, N.C. overloads, and on to L2. This energizes coil M2, closing its N.O. power contacts (not shown) to turn on the drain pump motor. It also causes its N.C. auxiliary contacts to open. This provides an electrical interlock in the control circuit for coil M1 that prevents coil M1 from being energized at the same time as coil M2. The drain pump motor will continue to operate until the water in the tank is drained to a level that is low enough to actuate the float switch, causing its N.C. contacts and N.O. contacts to close and open again, respectively. This causes coil M2 to de-energize and its related N.O. power contacts to open, turning off the drain pump motor. It also allows the fill pump motor to be started again, as described earlier. This cycle of turning on and off the fill and drain pumps will continue until the STOP pushbutton is pressed. Pressing the STOP pushbutton opens its N.C. contacts, removing power from the control relay coil. This causes all its N.O. contacts to open, preventing both coils M1 and M2 from being energized.

11.4.0 Controlling a Motor from Multiple Locations

It is often required to start and stop a motor from more than one location. *Figure 80* shows a line diagram for a control circuit where a magnetic starter may be energized to start and stop a motor from any one of three locations. In this circuit, pressing any one of the START pushbuttons causes coil M1 to energize. This causes the auxiliary contacts M1 to close, adding memory to the circuit until coil M1 is de-energized by pressing any one of the stop buttons. The important thing to remember is that no matter how many stations, the START pushbuttons must all be connected in parallel and the STOP pushbuttons in series.

11.5.0 Controlling the Reversal of a Three-Phase Motor

Reversing of three-phase motors is a common requirement. It is usually done by reversing the L1 and L3 power leads to the motor via the operation of a magnetic reversing starter and related control circuit (*Figure 81*). Circuit operation is started by pressing either the FORWARD or REVERSE pushbutton. Pressing the FORWARD pushbutton allows current to flow from L1 through the N.C. contacts of the STOP switch, N.C. contacts of the REVERSE switch, the closed contacts of the FORWARD switch, N.C. auxiliary contacts of the reverse coil, the forward coil F, the N.C. closed

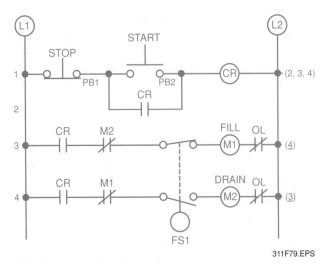

Figure 79 ◆ Control circuit for two pumps.

311F79.EPS

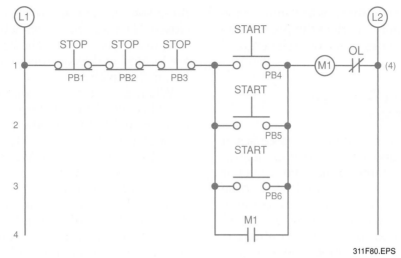

Figure 80 ◆ Control circuit for starting and stopping a single motor from three locations.

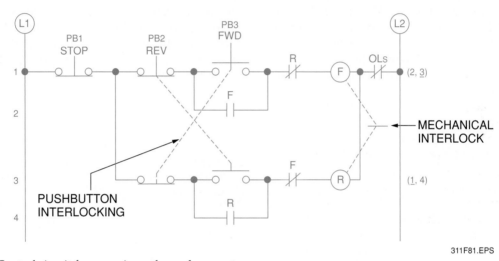

Figure 81 ◆ Control circuit for reversing a three-phase motor.

overload contacts, and on to L2. This energizes coil F, closing its N.O. forward-power contacts (not shown), causing the motor to turn on and run in the forward direction. It also activates the N.O. and N.C. auxiliary contacts of coil F. The N.O. auxiliary contacts close, providing memory for the FORWARD switch. The N.C. auxiliary contacts (located in the coil R control circuit) open, providing an electrical interlock that prevents coil R from being energized simultaneously with coil F.

Pressing the REVERSE pushbutton allows current to flow from L1 through the N.C. contacts of the STOP switch, N.C. contacts of the FORWARD switch, the closed contacts of the REVERSE switch, N.C. auxiliary contacts of the forward coil, the reverse coil R, the N.C. closed overload contacts, and on to L2. This energizes coil R, closing its N.O. reverse-power contacts (not shown), causing the motor to turn on and run in the reverse direction. It also activates the N.O. and

N.C. auxiliary contacts of coil R. The N.O. auxiliary contacts close, providing memory for the REVERSE switch. The N.C. auxiliary contacts (located in the coil F control circuit) open, providing an electrical interlock that prevents coil F from being energized simultaneously with coil R. Pressing the STOP pushbutton de-energizes, as appropriate, coil F or coil R, stopping motor operation.

This circuit uses three methods of interlocking. One method is a built-in mechanical interlock protection incorporated in the magnetic reversing starter. Another method is the pushbutton interlocking of the FORWARD and REVERSE pushbuttons. Pushbutton interlocking uses both N.O. and N.C. contacts mechanically connected on each pushbutton. This provides a safety backup to ensure that both the forward and reverse coils cannot be energized simultaneously. As has already been pointed out, the third method involves the

N.C. auxiliary contacts for the forward and reverse coils being used as safety interlocks to prevent the forward and reverse coils from being energized simultaneously.

11.6.0 Controlling Conveyor System Motors

In manufacturing or other controlled processes, it is often required to sequence the operation of multiple conveyor belt systems so that one conveyor system is turned on before a second one can be started, a second one must be turned on before a third one can be started, and so on. *Figure 82* shows one basic method for wiring a control circuit to sequentially turn on three conveyor systems. Pressing the START pushbutton energizes the conveyor 1 starter coil M1 via current flow through the closed contacts of the STOP and START switches and N.C. overload contacts. When coil M1 energizes, it closes its N.O. power contacts (not shown), causing the conveyor 1 motor to turn on. Energizing coil M1 also closes its

three sets of N.O. auxiliary contacts. One set of closed N.O. auxiliary contacts provides memory for the START switch. The second and third sets are located in the control circuits for coils M2 and M3, respectively.

When the set of N.O. auxiliary contacts in the coil M2 control circuit closes, it provides a current path from L1 through the closed N.O. contacts, through coil M2, and the N.C. overload contacts to L2. This energizes coil M2. Energizing coil M2 closes its N.O. power contacts, causing the conveyor 2 motor to turn on. Energizing coil M2 also closes its N.O. set of auxiliary contacts located in the coil M3 control circuit. With both the M1 and M2 sets of N.O. auxiliary contacts closed in the coil M3 control circuit, coil M3 energizes. This closes the coil 3 N.O. power contacts, causing conveyor motor 3 to turn on. Since there are no additional conveyors to energize, the N.O. auxiliary contacts for coil M3 are not used. Pressing the STOP pushbutton de-energizes, as applicable, coils M1, M2, and M3, stopping all conveyor motor operation.

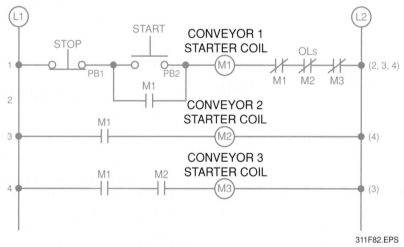

311F82.EPS

Figure 82 ◆ Control circuit for turning on three conveyor systems in sequence.

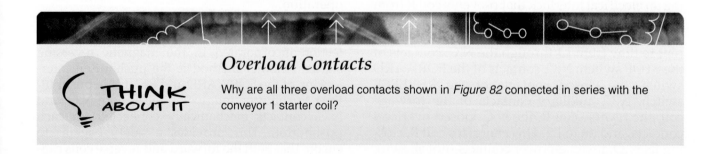

Overload Contacts

Why are all three overload contacts shown in *Figure 82* connected in series with the conveyor 1 starter coil?

1. Contactors without overload protection may be used to control _____.
   a. fractional horsepower single-phase motors
   b. single-phase motors
   c. three-phase motors
   d. noninductive loads

2. Relay contacts that are defined as being normally open (N.O.) have contacts that are _____.
   a. open when the relay coil is energized
   b. not being used in a particular application
   c. open when the relay coil is de-energized
   d. closed when the relay coil is de-energized

3. The voltage level at which the armature spring of a power relay overcomes the magnetic field of the coil causing the contacts to change state is called the _____ voltage.
   a. dropout
   b. seal-in
   c. pickup
   d. transition

4. IEC-rated contactors and motor starters perform the same functions as NEMA-rated contactors and motor starters but are usually _____ than the NEMA devices for the same horsepower or current ratings.
   a. physically larger
   b. more application sensitive
   c. more expensive
   d. better designed

5. When replacing the contacts of a magnetic contactor, all moving and stationary contacts for all poles must be replaced.
   a. True
   b. False

6. Overload protection for motors is provided by _____.
   a. circuit breakers
   b. general-purpose, current-limiting fuses
   c. an overload relay
   d. backup current-limiting fuses

7. All of the following are NEMA-rated classes for the trip time of an overload relay *except* _____.
   a. Class 10
   b. Class 20
   c. Class 30
   d. Class 40

8. The thermal units for overload relays should always be selected based on the motor _____.
   a. locked rotor amps
   b. full-load currents from *NEC Tables 430.247 through 430.250*
   c. nameplate full-load current
   d. horsepower

9. The part of a melting-alloy overload that responds to an overload current is known as the _____.
   a. bimetal element
   b. eutectic alloy
   c. ratchet wheel
   d. solder pot

10. A NEMA-rated motor starter consists of _____ assembled and wired in the same enclosure.
    a. a disconnect switch and magnetic contactor
    b. a magnetic contactor and stop and start pushbuttons
    c. one or more magnetic contactors with associated overload relays
    d. a circuit breaker and magnetic contactor

11. All of the following are considerations when selecting overload relays *except* the _____.
    a. type of motor
    b. full-load current rating of the motor
    c. maximum horsepower rating for non-plugging and nonjogging duty
    d. manufacturer of the motor

12. The motor starter that must be used with a 230V, single-phase, 60Hz, 10hp motor not used for plugging or jogging applications is the _____.
    a. NEMA Size 2
    b. NEMA Size 3
    c. NEMA Size 4
    d. NEMA Size 5

13. When selecting a control transformer for a specific application, you have calculated its application inrush to be 1,850VA. If the primary input voltage is expected to vary between 5% and 10% from nominal, a _____ transformer should be used.
    a. 150VA
    b. 250VA
    c. 300VA
    d. 350VA

14. Pushbutton switches with normally open (N.O.) contacts are usually used to initiate a STOP or OFF circuit function.
    a. True
    b. False

15. A manufacturer states that the contacts of a particular switch are rated as A600. In a 120VAC application, the maximum break current for these contacts is _____.
    a. 1A
    b. 1.5A
    c. 3A
    d. 6A

16. A manufacturer lists the utilization category for a particular switching device as AC-15. The intended application for this device is control of _____.
    a. electromagnetic loads greater than 72VA closed
    b. solid-state loads with transformer isolation
    c. small electromagnetic loads up to 72VA closed
    d. resistive loads and solid-state loads with optical isolation

17. A line-powered proximity sensor draws its operating current through the load.
    a. True
    b. False

18. The scanning arrangement for a photoelectric sensor where the light beam transmitted from the transmitter is reflected back to the receiver by the target is called _____.
    a. direct scan
    b. retroreflective scan
    c. diffused scan
    d. convergent scan

19. Ladder diagrams include all of the following *except* a(n) _____.
    a. signal section
    b. decision section
    c. action section
    d. termination section

20. *NEC Article 430* _____ covers motor and controller circuit overload protection.
    a. *Part III*
    b. *Part IV*
    c. *Part VII*
    d. *Part X*

Summary

Magnetic contactors and motor starters use a small control current to energize or de-energize the load connected to it. Motor starters are contactors that contain overload protection. Various auxiliary devices can be added to contactors and motor starters to expand their capability.

Many types of pilot (control) devices, ranging from simple pushbutton switches to proximity sensors, are used in motor control circuits. The device(s) selected for use depends on the specific application. Manufacturers' specifications sheets describe required amperage, voltage, and sizing information for control devices. For safety and proper operation in the intended environment, their installation requires that they be installed in the proper position and location.

Knowledge of the definitions, symbols, and types of wiring diagrams that have been presented throughout this module must be understood. This is necessary to properly select, size, and install electromagnetic relays, contactors, motor starters, and the pilot devices in their related control circuits.

Notes

Trade Terms
Introduced in This Module

Actuator: The mechanism of a switch or switch enclosure that operates the contacts.

Auxiliary contact: A contact device that is mechanically connected to a contactor. It can be used to give an electrical signal related to the contactor's position.

Cam: A machine part or component that applies a force to the actuator of a limit switch, causing the actuator to move as intended.

Dashpot: A device that provides a time delay by controlling how fast air or liquid is allowed to pass into or out of a container through an orifice (opening) that is either fixed or variable in diameter.

Deadband: The temperature or pressure difference between the setpoint at which a switch activates when the temperature or pressure increases and the point at which the switch resets when the temperature or pressure decreases.

Dielectric constant: The ratio of the capacitance of a capacitor with a given dielectric to the capacitance of an identical capacitor having air for its dielectric.

Dropout voltage: The coil voltage at which the armature return spring of an electromechanical device overpowers the magnetic field of the coil and the contacts of the device change state.

Eddy current: Current induced in the body of a metallic material by an oscillating magnetic field.

Electromechanical relay (EMR): A switching device with a set of contacts that are closed by a magnetic effect.

Infrared (IR): A portion of the light spectrum that is not visible (wavelengths beyond 690 nanometers). IR radiation is emitted by certain light-emitting diodes (LEDs).

International Electrotechnical Commission (IEC): An international organization that produces recommended performance and safety standards for electrical products.

Jogging (inching): Repeated closure of the circuit used to start and stop a motor. Inching is used to accomplish small movements in the driven machine.

Line-powered sensor: A three-wire sensor that draws its operating current (burden current) directly from the line. Its operating current does not flow through the load.

Load-powered sensor: A two-wire sensor in series with the load that draws its operating current (residual current) through the load.

Maintained contact switch: A switch with contacts that remain in the operating position when the actuating force is removed. The switch has provisions for resetting.

Momentary contact switch: A switch with contacts that return from the operating position to the release position when the actuating force is removed.

Motor control circuit: A circuit that carries the electric signals that direct the performance of a controller or other device, but does not carry the main power current.

Operator: A switch component that is pressed, pulled, or rotated by the person operating the circuit in order to activate the switch contacts.

Pickup voltage: The minimum allowable coil control voltage that will cause an electromechanical device to energize.

Pilot devices: Devices such as electrical pushbuttons and switches that contain contacts which govern the operation of related relays or similarly controlled devices.

Plugging: Braking a motor by reversing the phase sequence of the power to the motor. It is used to achieve both rapid stop and quick reversal.

Proximity sensors (switches): Sensing switches that detect the presence or absence of an object using an electronic noncontact sensor.

Seal-in voltage: The minimum allowable coil control voltage that will keep an electromechanical device energized. It is usually less than the pickup voltage.

Setpoint: The desired value to which a switching device is set or adjusted.

This module is intended to present thorough resources for task training. The following reference works are suggested for further study. These are optional materials for continued education rather than for task training.

Electrical Motor Controls. Gary Rockis and Glen Mazur. Homewood, IL: American Technical Publishers, Inc., 1997.

National Electrical Code® Handbook, Latest Edition. Quincy, MA: National Fire Protection Association.

NFPA 70E Recommended Practice for Electrical Equipment Maintenance. Quincy, MA: National Fire Protection Association, 2004.

CONTREN® LEARNING SERIES – USER UPDATE

NCCER makes every effort to keep these textbooks up-to-date and free of technical errors. We appreciate your help in this process. If you have an idea for improving this textbook, or if you find an error, a typographical mistake, or an inaccuracy in NCCER's Contren® textbooks, please write us, using this form or a photocopy. Be sure to include the exact module number, page number, a detailed description, and the correction, if applicable. Your input will be brought to the attention of the Technical Review Committee. Thank you for your assistance.

Instructors – If you found that additional materials were necessary in order to teach this module effectively, please let us know so that we may include them in the Equipment/Materials list in the Annotated Instructor's Guide.

Write: Product Development and Revision
National Center for Construction Education and Research
3600 NW 43rd St., Bldg. G, Gainesville, FL 32606

Fax: 352-334-0932

E-mail: curriculum@nccer.org

Craft ______________________ Module Name ______________________

Copyright Date ____________ Module Number ____________ Page Number(s) ____________

Description

____

____

____

____

(Optional) Correction

____

____

____

(Optional) Your Name and Address

____

____

____

Glossary

Actuator: The mechanism of a switch or switch enclosure that operates the contacts.

American Wire Gauge (AWG): The United States standard for measuring wires.

Ampacity: The current in amperes that a conductor can carry continuously under the conditions of use without exceeding its temperature rating.

Ampere rating: The current-carrying capacity of an overcurrent protective device. The fuse or circuit breaker is subjected to a current above its ampere rating; it will open the circuit after a predetermined period of time.

Ampere squared seconds (I^2t): The measure of heat energy developed within a circuit during the fuse's clearing. It can be expressed as melting I^2t, arcing I^2t, or the sum of them as clearing I^2t. I stands for effective let-through current (rms), which is squared, and t stands for time of opening in seconds.

Ampere turn: The product of amperes times the number of turns in a coil.

Amperes interrupting capacity (AIC): The maximum short circuit current that a circuit breaker or fuse can safely interrupt.

Appliance: Utilization equipment, generally other than industrial, normally built in standardized sizes or types, that is installed or connected as a unit to perform one or more functions such as clothes washing, air conditioning, food mixing, deep frying, etc.

Appliance branch circuit: A branch circuit supplying energy to one or more outlets to which appliances are to be connected. Such circuits are to have no permanently connected lighting fixtures that are not part of an appliance.

Approved: Acceptable to the authority having jurisdiction.

Arcing time: The amount of time from the instant the fuse link has melted until the overcurrent is interrupted or cleared.

Attenuation: The reduction of a signal due to system losses.

Autotransformer: Any transformer in which primary and secondary connections are made to a single winding. The application of an autotransformer is a good choice where a 480Y/277V or 208Y/120V, three-phase, four-wire distribution system is used.

Auxiliary contact: A contact device that is mechanically connected to a contactor. It can be used to give an electrical signal related to the contactor's position.

Backbone: A high-capacity network link between two sections of a network.

Bandwidth: The range of frequencies that will pass through a particular portion of a system.

Branch circuit: The circuit conductors between the final overcurrent device protecting the circuit and the outlet(s).

Cable: An assembly of two or more insulated or bare wires.

Cam: A machine part or component that applies a force to the actuator of a limit switch, causing the actuator to move as intended.

Candela (cd): The International System (SI) unit of luminous intensity. It is roughly equal to 12.57 lumens. The name has been retained from the early days of lighting when a standard candle of a fixed size and composition was used as the basis for evaluating the intensity of other light sources. See candlepower.

Candlepower (cp): The measure of light intensity measured in candelas (cd). It is used to indicate the intensity of a light source in a given direction. The higher the intensity of light, the more candelas it represents and the higher the candlepower.

Capacitance: The storage of electricity in a capacitor or the opposition to voltage change. Capacitance is measured in farads (F) or microfarads (μF).

Circuit interrupter: A non-automatic, manually operated device designed to open a current-carrying circuit without injury to itself.

Clearing time: The total time between the beginning of the overcurrent and the final opening of the circuit at rated voltage by an overcurrent protective device. Clearing time is the total of the melting time and the arcing time.

Coaxial cable: A type of cable that has a center conductor separated from the outer conductor by a dielectric. Coaxial cable is used to transmit very small currents or high-frequency signals.

Cold sequence metering: A service installation where the service disconnect is installed ahead of the metering in order for power to be disconnected to allow for safe maintenance of the metering.

Conduit: A tubular raceway such as electrical metallic tubing (EMT); rigid metal conduit, rigid nonmetallic conduit, etc.

Conduit body: A separate portion of a conduit or tubing system that provides access through removable covers to the interior of the system at a junction of two or more sections of the system or at a terminal point of the system.

Continuous load: A load in which the maximum current is expected to continue for three hours or more.

Cross connect: A facility for connecting cable runs, subsystems, and equipment using patch cords or jumper cable.

Current-limiting device: A device that will clear a short circuit in less than one half cycle. Also, it will limit the instantaneous peak let-through current to a value substantially less than that obtainable in the same circuit if that device were replaced with a solid conductor of equal impedance.

Dashpot: A device that provides a time delay by controlling how fast air or liquid is allowed to pass into or out of a container through an orifice (opening) that is either fixed or variable in diameter.

Deadband: The temperature or pressure difference between the setpoint at which a switch activates when the temperature or pressure increases and the point at which the switch resets when the temperature or pressure decreases.

Delay skew: A change of timing or phase in a transmission signal.

Delta-connected: A three-phase transformer connection in which the terminals are connected in a triangular shape like the Greek letter delta (Δ).

Demand factors: The ratio of the maximum demands of a system, or part of a system, to the total connected load of a system or the part of the system under consideration.

Device: A unit of an electrical system that is intended to carry but not utilize electric energy.

Dielectric constant: The ratio of the capacitance of a capacitor with a given dielectric to the capacitance of an identical capacitor having air for its dielectric.

Diffused lighting: Lighting on a work surface or object that is not predominantly incident from any particular direction.

Diffuser: A part of a lighting fixture that redirects or scatters the light produced by the fixture lamp to provide diffused lighting.

Dropout voltage: The coil voltage at which the armature return spring of an electromechanical device overpowers the magnetic field of the coil and the contacts of the device change state.

Eddy current: Current induced in the body of a metallic material by an oscillating magnetic field.

Electromechanical relay (EMR): A switching device with a set of contacts that are closed by a magnetic effect.

ELFEXT: Equal level far end crosstalk; a value derived by subtracting the attenuation of the interfering pair from the far end crosstalk (FEXT) that it has caused in the interfered pair.

Equipment: A general term including material, fittings, devices, appliances, fixtures, apparatus, and the like used as a part of (or in connection with) an electrical installation.

Ethernet: A local area network (LAN) developed by Xerox Corporation that uses a bus or star topology. Ethernet is currently the most popular type of LAN.

Explosionproof: Designed and constructed to withstand an internal explosion without creating an external explosion or fire.

Explosionproof apparatus: Apparatus enclosed in a case that is capable of withstanding an explosion of a specified gas or vapor that may occur within it; also capable of preventing the ignition of a specified gas or vapor surrounding the enclosure by sparks, flashes, or explosion of the gas or vapor within; which operates at such an external temperature that a surrounding flammable atmosphere will not be ignited thereby.

Fast-acting fuse: A fuse that opens on overloads and short circuits very quickly. This type of fuse is not designed to withstand temporary overload currents associated with some electrical loads (inductive loads).

Firestopping: A material or mechanical device used to block openings in walls, ceilings, or floors to prevent the passage of fire or smoke.

Flux: The rate of energy flow across or through a surface. Also a substance used to promote or facilitate soldering or welding by removing surface oxides.

Footcandle (fc): The unit used to measure how much total light is reaching a surface, such as a wall or table. One lumen falling on one square foot of surface produces an illumination level of one footcandle. One footcandle also equals 10.76 lux. Lux is the standard international unit of illuminance and is equal to one lumen per square meter.

General-purpose branch circuit: A branch circuit that supplies a number of outlets for lighting and appliances.

Gigabit: An extremely fast rate of data transfer equal to 1,000,000,000 bits per second.

Hazardous (classified) location: A location in which ignitable vapors, dust, or fibers may cause a fire or explosion.

Individual branch circuit: A branch circuit that supplies only one piece of utilization equipment.

Induction: The production of magnetization or electrification in a body by the mere proximity of magnetized or electrified bodies, or the production of an electric current in a conductor by the variation of the magnetic field in its vicinity.

Inductive load: An electrical load that pulls a large amount of current—an inrush current—when first energized. After a few cycles or seconds, the current declines to the load current.

Infrared (IR): A portion of the light spectrum that is not visible (wavelengths beyond 690 nanometers). IR radiation is emitted by certain light-emitting diodes (LEDs).

Innerduct: A type of high-density polyethylene (HDPE) conduit that is available in both corrugated and smooth types, and is often used to run communications cable.

International Electrotechnical Commission (IEC): An international organization that produces recommended performance and safety standards for electrical products.

Jogging (inching): Repeated closure of the circuit used to start and stop a motor. Inching is used to accomplish small movements in the driven machine.

Kilovolt-amperes (kVA): 1,000 volt-amperes (VA).

Line-powered sensor: A three-wire sensor that draws its operating current (burden current) directly from the line. Its operating current does not flow through the load.

Load-powered sensor: A two-wire sensor in series with the load that draws its operating current (residual current) through the load.

Loss: The power expended without doing useful work.

Magnetic field: The area around a magnet in which the effect of the magnet can be felt.

Magnetic induction: The number of magnetic lines or the magnetic flux per unit of cross-sectional area perpendicular to the direction of the flux.

Maintained contact switch: A switch with contacts that remain in the operating position when the actuating force is removed. The switch has provisions for resetting.

Megabit: A rate of data transfer equal to 1,000,000 bits per second.

Melting time: The amount of time required to melt a fuse link during a specified overcurrent.

Momentary contact switch: A switch with contacts that return from the operating position to the release position when the actuating force is removed.

Motor control circuit: A circuit that carries the electric signals that direct the performance of a controller or other device, but does not carry the main power current.

Multi-outlet assembly: A type of surface or flush raceway designed to hold conductors and receptacles, assembled in the field or at the factory.

Mutual induction: The condition of voltage in a second conductor because of current in another conductor.

NEC® dimensions: These are dimensions once referenced in the *National Electrical Code®*. They are common to Class H and K fuses and provide interchangeability between manufacturers for fuses and fusible equipment of given ampere and voltage ratings.

NEXT: Near end crosstalk; a measurement of the interference (crosstalk) between two wire pairs.

Operator: A switch component that is pressed, pulled, or rotated by the person operating the circuit in order to activate the switch contacts.

Outlet: A point on the wiring system at which current is taken to supply utilization equipment.

Overcurrent: Any current in excess of the rated current of equipment or the ampacity of a conductor. It may result from overload, short circuit, or ground fault.

Overload: Can be classified as an overcurrent that exceeds the normal full-load current of a circuit.

Patch cord: A connecting cable used with cross-connect equipment and work area cables. It has three conductive portions separated by insulators.

Peak let-through (Ip): The instantaneous value of peak current let-through by a current-limiting fuse when it operates in its current-limiting range.

Pickup voltage: The minimum allowable coil control voltage that will cause an electromechanical device to energize.

Pilot devices: Devices such as electrical pushbuttons and switches that contain contacts which govern the operation of related relays or similarly controlled devices.

Plugging: Braking a motor by reversing the phase sequence of the power to the motor. It is used to achieve both rapid stop and quick reversal.

Power transformer: A transformer that is designed to transfer electrical power from the primary circuit to the secondary circuit(s) to step up the secondary voltage at less current or step down the secondary voltage at more current, with the voltage-current product being constant for either the primary or secondary.

Proximity sensors (switches): Sensing switches that detect the presence or absence of an object using an electronic noncontact sensor.

Rating: A designated limit of operating characteristics based on definite conditions. Such operating characteristics as load, voltage, frequency, etc., may be given in the rating.

Reactance: The opposition to AC due to capacitance and/or inductance.

Receptacle: A contact device installed at an outlet for connection as a single contact device. A single receptacle is a single contact device with no other contact device on the same yoke. A multiple receptacle is a single device containing two (duplex) or more receptacles.

Receptacle outlet: An outlet where one or more receptacles are installed.

Rectifiers: Devices used to change alternating current to direct current.

Root-mean-square (rms): The effective value of an AC sine wave, which is calculated as the square root of the average of the squares of all the instantaneous values of the current throughout one cycle. Alternating current rms is that value of an alternating current that produces the same heating effect as a given DC value.

Seal-in voltage: The minimum allowable coil control voltage that will keep an electromechanical device energized. It is usually less than the pickup voltage.

Sealing compound: The material poured into an electrical fitting to seal and minimize the passage of vapors.

Sealoff fittings: Fittings required in conduit systems to prevent the passage of gases, vapors, or flames from one portion of the electrical installation to another through the conduit. Also referred to as sealing fittings or seals.

Semiconductor fuse: Fuse used to protect solid-state devices.

Service: The electric power delivered to the premises.

Service conductors: The conductors between the point of termination of the overhead service drop or underground service lateral and the main disconnecting device in the building or on the premises.

Service drop: The overhead conductors through which electrical service is supplied between the last power company pole and the point of their connection to the service-entrance conductors located at the building or other support used for the purpose.

Service entrance: All components between the point of termination of the overhead service drop or underground service lateral and the building's main disconnecting device, except for metering equipment.

Service equipment: The necessary equipment, usually consisting of a circuit breaker or switch and fuses and their accessories, located near the point of entrance of supply conductors to a building and intended to constitute the main control and cutoff means for the electric supply to the building.

Service factor: The number by which the horsepower rating is multiplied to determine the maximum safe load that a motor may be expected to carry continuously at its rated voltage and frequency.

Service lateral: The underground service conductors between the street main, including any risers at a pole or other structure or from transformers, and the first point of connection to the service-entrance conductors in a terminal box, meter, or other enclosure with adequate space, inside or outside the building wall.

Setpoint: The desired value to which a switching device is set or adjusted.

Short circuit current: Can be classified as an overcurrent that exceeds the normal full-load current of a circuit by a factor many times greater than normal. Also characteristic of this type of overcurrent is that it leaves the normal current-carrying path of the circuit—it takes a shortcut around the load and back to the source.

Single phasing: The condition that occurs when one phase of a three-phase system opens, either in a low-voltage or high-voltage distribution system. Primary or secondary single phasing can be caused by any number of events. This condition results in unbalanced loads in polyphase motors and unless protective measures are taken, it will cause overheating and failure.

Terminal: A point at which an electrical component may be connected to another electrical component.

Torque: A force that produces or tends to produce rotation. Common units of measurement of torque are foot-pounds and inch-pounds.

Transformer: A static device consisting of one or more windings with a magnetic core. Transformers are used for introducing mutual coupling by induction between circuits.

Turn: The basic coil element that forms a single conducting loop comprised of one insulated conductor.

Turns ratio: The ratio between the number of turns between windings in a transformer; normally the primary to the secondary, except for current transformers, in which it is the ratio of the secondary to the primary.

UL classes: Underwriters Laboratories has developed basic physical specifications and electrical performance requirements for fuses with voltage ratings of 600V or less. These are known as UL standards. If a type of fuse meets with the requirements of a standard, it can fall into that UL class. Typical UL classes are R, K, G, L, H, T, CC, and J.

Utilization equipment: Equipment that utilizes electric energy for electronic, chemical, heating, lighting, electromechanical, or similar purposes.

Voltage rating: The maximum value of system voltage in which a fuse can be used, yet safely interrupt an overcurrent. Exceeding the voltage rating of a fuse impairs its ability to safely clear an overload or short circuit.

Work plane: The distance above the floor where a task is to be performed. In offices and schools, this is usually 30". In kindergartens, it would be lower, and in some manufacturing areas, it would be much higher.

Module 26301-08

John Traister, 301F01
Topaz Publications, Inc., 301SA01–301SA04, 301SA06
Mike Powers, 301SA05, 301SA07
Miller Electric Manufacturing Company, 301F02A, 301F02B

Module 26302-08

Topaz Publications, Inc., 302SA01–302SA05, 302SA07
National Fire Protection Association, 302T01
Reprinted with permission from NFPA 70®-2008, the *National Electrical Code*® Copyright © 2007, National Fire Protection Association, Quincy, MA 02269. This reprinted material is not the complete and official position of the National Fire Protection Association on the referenced subject, which is represented only by the standard in its entirety.
Mike Powers, 302SA06
U.S. Department of Commerce, National Oceanic and Atmospheric Administration, 302A01

Module 26303-08

Associated Builders and Contractors, Module divider
Hubbell Incorporated, TOC, 303F05–303F07, 303F09, 303F11, 303F12, 303F13A–C, 303F14–303F18, 303SA06, 303SA08, 303F29, 303F36–303F39
Crescent–Stonco, 303F13D (left), 303SA02
TCP Incorporated, 303F13D (right)
Topaz Publications, Inc., 303SA01, 303F22B
Philips Lighting Company, 303F19C
OSRAM SYLVANIA, 303F20B
Used with permission, Acuity Brands Lighting, 303F21, 303F22C
Images provided by LEDtronics, www. LEDtronics.com, 303SA03A, 303F25

Dan Lamphear, 303SA03B
Wiedamark, 303SA04, 303F23
Mitsubishi Electric Power Products, Inc., 303F22A
Lighting Science Group Corporation, 303F22D, 303F24, 303F27, 303SA05
The Sky Factory, 303F26
Beta Lighting, 303F28
Greenlee Textron, Inc., a subsidiary of Textron Inc., 303SA07
Tim Dean, 303SA10

Module 26304-08

John Traister, 304F01–304F04, 304F08 (line art), 304F11A–B, 304F12–304F16
Tim Dean, 304SA01
Topaz Publications, Inc., 304SA02, 304SA03, 304F05 (bottom), 304F06, 304F07, 304F08 (photo), 304F09, 304F10, 304F11C
Dan Lamphear, 304SA04
Ridge Tool Co. / RIDGID®, 304F05 (top)

Module 26305-08

John Traister, 305F04–305F06, 305SA01, 305F08, 305F16
Ed Cockrell, 305F07
Topaz Publications, Inc., 305T02, 305T03
Square D/Schneider Electric, 305SA02

Module 26306-08

Topaz Publications, Inc., 306SA01, 306F02, 306SA04, 306F08, 306F10, 306F26, 306F32
Square D/Schneider Electric, 306F03, 306F04
Jim Mitchem, 306SA02, 306SA06, 306F12, 306F18, Appendix B
Mike Powers, 306SA03, 306SA05
Fluke Corporation, Reproduced with permission, 306F05
Boltswitch, Inc., 306F09
Federal Pacific Transformer Co., 306F11
Courtesy of General Electric Company, 306F31

Module 26307-08

John Traister, 307F02–307F08, 307F10–307F14, 307F19, 307F22, 307F23, 307F26, 307F27
Dan Lamphear, 307SA01
Jim Mitchem, 307SA02, 307SA04
Topaz Publications, Inc., 307SA03
Phenix Technologies, 307SA05

Module 26308-08

Al Hamilton, 308F01, 308F06, 308F07–308F18
Topaz Publications, Inc., 308SA01–308SA04
Tim Dean, 308F03

Module 26309-08

Topaz Publications, Inc., 309F01, Appendix
John Traister, 309F03–309F19, 309F21–309F23, 309T01
Baldor Electric Company, 309SA01

Module 26310-08

Mike Powers, 310F01, 310F02–310F07, 310F19
Carlon, 310SA01, 310SA03
Topaz Publications, Inc., 310SA02, 310F13, 310F15, 310F16, 310F18, 310SA04, 310F20, 310F22
Tim Dean, 310F08–310F10

Image courtesy of Siemon®, 310F11, 310F21, 310SA05
LANshack, 310F14
Hometech Solutions, 310F17

Module 26311-08

Topaz Publications, Inc., 311F01, 311F02, 311F06, 311F09–311F15, 311F19, 311F21, 311F23, 311F24, 311F27 (photos), 311F28, 311F30 (photo), 311F31, 311F32, 311F35, 311F36, 311F38 (photo), 311F40 (photo), 311F41 (photo), 311F42, 311F47 (bottom left), 311F49 (photo), 311F60, 311F67
John Traister, 311F22
Square D/Schneider Electric, 311F25, 311F65
Federal Pacific Transformer Co., 311F33
Reprinted from NEMA ICS 5-2000 *Industrial Control and Systems: Control Circuit and Pilot Devices*, by permission of the National Electrical Manufacturers Association, 311T09, 311T10
© Dwyer Instruments, Inc., 311F44 (photo), 311F47 (bottom right), 311F54 (photo), 311F55 (photo)
Courtesy of Honeywell International, Inc., 311SA01
Eaton's Electrical Group, manufacturer of Cutler-Hammer electrical control and distribution products, 311F68

Index

Figures are indicated with *f*.

A

Abbreviations, electrical drawings, 6.32–6.34, 6.34*f*
AC adjustable-voltage motors, 9.11
Accent (directional) lighting, 3.3, 3.3*f*
AC-3 rated contactors, 11.28, 11.29
Action sections, control ladder diagrams, 11.61*f*, 11.62
Actuators,11.78
   mechanical limit switches and, 11.43–11.45, 11.44*f*, 11.45*f*
Adjustable magnetic trips, panelboards, 6.19, 6.44, 6.44*f*
Adjustable pressure switches, 11.42
Administrative subsystem conduit systems, 10.5–10.6
Aerial distribution systems, campus backbone systems, 10.3
Aesthetic lighting, 3.35
Agilent cable certification testers, 10.25*f*
Air circuit breakers (ACBs), 6.31, 6.49
Air compressor motors, control circuits for, 11.71, 11.71*f*
Air conditioning loads, 1.17, 1.18
Air conditioning nameplates, 1.17, 1.18
Air gaps, switchboard spacing requirements, 6.4
Airport hangars, as hazardous locations, 4.22, 4.23*f*
Allowable pressure limits, pressure switches, 11.43
Allowable temperature limits, temperature switches, 11.41
Aluminum conductors, 2.2, 2.3*f*
Ambient temperature compensation, thermal overload relays, 11.20, 11.21
American Standard C37.2-1970, 6.50
American Wire Gauge (AWG), 2.23
   B & S gauge, 2.13
   wire sizes, 2.2, 2.13–2.14
Ampacity, 1.25
   branch circuits and, 1.2, 1.8–1.9, 2.6–2.7
   cable, 6.16
   derating branch circuit conductors and, 1.4–1.5
   derating conductors in conduit and, 1.4–1.5
   sizing conductors and, 2.6–2.7

Ampere ratings, 5.38
   circuit protection and, 5.3, 5.20
   fuses, 5.3, 5.5, 9.12
   horsepower and, 9.9–9.10
   motor disconnect switches and, 9.13
Amperes interrupting capacity (AIC), 5.2, 5.38
Ampere squared seconds (I^2t), 5.38
   UL fuse classes and, 5.10
Ampere turns, 7.41
   transformers and, 7.7
AND functions, 11.61–11.62, 11.62*f*, 11.63*f*
ANSI/IEEE classifications for medium-voltage fuses, 5.11, 7.22–7.23, 7.24
Anti-plugging protection, 11.49
Anti-seize compounds, 4.15
Appliance branch circuits, 1.25
   circuit protection and, 5.20–5.21
   small appliance loads and, 1.13
Appliances, 1.25
   cooking appliances, 1.14–1.15
   locating receptacles for, 2.6–2.7
   multi-outlet assemblies and, 1.10–1.11
   receptacle loads and, 2.7
   small appliance loads, 1.13
Approved, 4.26
   explosionproof equipment and, 4.10
Arc fault circuit interrupters (AFCIs), 5.19
Arc faults, 5.4
Arc flash
   safety and, 5.19
   service point and, 8.21
Architectural LED lighting, 3.22–3.23, 3.23*f*
Arcing
   electromechanical relays and, 11.7–11.8
   ground faults and, 6.19
   high-voltage limiting switches and, 11.45
Arcing effects, installing electrical services, 8.3
Arcing times, 5.5, 5.38
Arc protection circuits, electromechanical relays, 11.8–11.9, 11.9*f*
Arc voltages, fuses, 5.11, 5.12
Armature overtravel, 11.5, 11.6, 11.44, 11.45*f*

Askarel-insulated transformers, *NEC* requirements for, 7.19
ASTM International Standard F855-04, Temporary Protective Grounds to be Used on De-energized Electric Power Lines and Equipment, 6.9
Atmospheric hazards, 4.2–4.3, 4.3*f*, 4.5
Attenuation, 10.20, 10.28
AT&T Standard 258A, 10.14
Authority Having Jurisdiction (AHJ), 8.2
Automatic reset, bimetallic overload relays, 11.21
Autotransformers, 7.15, 7.15*f*, 7.41
Auxiliary contacts, 11.16, 11.78
Average ambient temperatures, 2.24

B

Backbones, 10.28
   subsystems as, 10.2
Backup current-limiting fuses, 5.11
Bags, as support structures, 10.12
Ballasts
   fluorescent dimming, 3.40–3.42, 3.41*f*, 3.42*f*
   fluorescent lighting fixture ballasts, 1.9–1.10
   high-intensity discharge (HID) lighting fixture ballasts, 3.13, 3.42, 3.43*f*
   replacing, 3.12
Bandwidths, 10.20, 10.28
Bare conductors, 2.2
Basic impulse insulation level (BIL), 6.49
   switch ratings and, 6.20
   transformer nameplate data and, 6.27
Bee-lining, 10.7
Bellows pressure switches, 11.42
Bellows temperature switches, 11.40–11.41, 11.41*f*
Bell System PBX installations, 10.2
Bimetallic elements, temperature switches, 11.39–11.40, 11.39*f*, 11.40*f*
Bimetallic overload relays, 11.19, 11.20–11.21, 11.21*f*
Blown main fuse detector, bolted pressure switches, 6.23
Boilers
   horizontal trunks and, 10.12
   riser pathways and, 10.9